Macromolecular Physics

VOLUME 2

Crystal Nucleation, Growth, Annealing

Macromolecular Physics

Bernhard Wunderlich

Rensselaer Polytechnic Institute
Troy, New York

VOLUME 2

Crystal Nucleation, Growth, Annealing

ACADEMIC PRESS New York San Francisco London 1976

A Subsidiary of Harcourt Brace Jovanovich, Publishers

ACADEMIC PRESS, INC.
111 Fifth Avenue, New York, New York 10003

United Kingdom Edition published by
ACADEMIC PRESS, INC. (LONDON) LTD.
24/28 Oval Road, London NW1

Library of Congress Cataloging in Publication Data

Wunderlich, Bernhard, (date)
Macromolecular physics.

Includes bibliographies.
CONTENTS: v. 1. Crystal structure, morphology, defects.--v. 2. Crystal nucleation, growth, annealling.
1. Macromolecules. 2. Crystals. I. Title.
QD381.W86 548 72-82632
ISBN 0-12-765602-2 (v. 2)

PRINTED IN THE UNITED STATES OF AMERICA

Contents

List of Tables

Preface

This volume is the continuation of the discussion of crystals of linear macromolecules. It deals with the transformation of matter into the crystalline state. Volume 1 dealt with "Crystal Structure, Morphology, Defects," a subject area that was beginning to be understood by 1971, the closing date for literature treated in Volume 1. At the time of this writing, little additional work except for more examples could be added in this area, a sign of the maturity of our knowledge about crystalline macromolecules. When the material for the present volume was assembled, making use of the literature up to mid-1974, a different picture arose. "Crystal Nucleation, Growth, and Annealing" are subjects that are still, in many cases, in flux. The classical treatments of crystallization as found for example in Stuart's "Die Physik der Hochpolymeren" or Mandelkern's "Crystallization of Polymers" do present a firm base for the discussion, but, in the light of the research of the last ten years, are inadequate to describe the known phenomena.

This book is a first attempt at a unified treatment of the processes involved in crystallization of linear macromolecules. It represents an effort to collect sufficient material to show that crystallization of macromolecules cannot be treated as a simple extension of small-molecule crystallization. The long-chain nature leads to a special nucleation process, molecular nucleation, not found in small-molecule crystals. Another new aspect is the possibility of crystallization during polymerization, which leads to a series of previously unknown effects of nucleation, crystallization, and annealing. Both topics are treated for the first time in a comprehensive fashion (Sections 5.3, 5.1.2, 6.4, and 7.3.4).

The documentation is done, as in the first volume, by frequent references to the original literature and other sections of the two volumes. For the more rapidly developing areas an attempt was made to cover all present literature. Well-established areas were documented with a critically selected series of references. An attempt was made to trace the basic ideas to their origin. Whenever possible, tables or footnotes are used to present larger bodies of information or larger numbers of references in abbreviated form. In particular, crystallization of many macromolecular melts are collected in Table VI.8, and annealing data are presented in Tables VII.2–VII.4. Almost all drawings were newly made to present a uniform style and to permit, in many cases, the combination of data from several sources.

Since the subject of crystals of macromolecules cuts across many disciplines, several introductory sections were included in Volume 1 (Polymer Science, Chapter I; Crystallography, Section 2.2; Molecular Structure, Section 2.3; Crystal Morphology, Section 3.1; Crystal Defects, Section 4.1). The present book builds on the material of Volume 1. Each chapter consists of a general section describing history and theory of the subject, followed by specific examples.

Volumes 1 and 2 parallel the graduate lecture courses at Rensselaer, and have also been presented as audio courses.†

Volume 2 concludes the description of crystals and their formation. Because of time and space limitations, it was not possible to include (as promised in the preface of Volume 1) the subject of melting; this will be treated in detail in Volume 3, subtitled "Thermal Properties." Volume 3 will also deal with heat capacities, thermal conductivity, and related topics. An early extensive summary of this topic was presented in a review [*Adv. Polymer Sci.* **7**, 151–368 (1970)], and a third audio course‡ treats some basic material of this topic.

The author hopes that still more volumes will be forthcoming to allow for a truly comprehensive treatment of the subject.

† Crystals of Linear Macromolecules, Rensselaer Audio Course 72.668 and Crystallization of Linear Macromolecules, Rensselaer Audio Course 72.669; Division of Continuing Studies, Rensselaer Polytechnic Institute, Troy, New York 12181.

‡ Thermal Analysis of Linear Macromolecules, Rensselaer Audio Course 72.666; Division of Continuing Studies, Rensselaer Polytechnic Institute, Troy, New York 12181.

Acknowledgements

As in the first volume, I should like to thank colleagues and students who have supported this work by contributing discussions, preprints, and photographs. The basis of Volume 2 was a lecture course of the same title given to students at Rensselaer in 1972. The bulk of the material was assembled during a sabbatic leave in 1973–1974. Initial testing was possible in a second lecture course in 1974. Many of the illustrations in this book were possible only by the contribution of original prints by the authors listed in the legends. In addition, permission for reproduction was generously granted by the following copyright holders:

American Chemical Society, Washington (Figs. VI.26 and VI.73c).
American Institute of Physics, New York (Figs. VI.29, VI.30, VI.86, and VI.87A,B).
Marcel Dekker, Inc., New York (Figs. V.26, V.34, V.38, VI.24a,b, VI.25, VI.34, VI.35, VI.36, VI.37, VI.83, and VI.84).
Hüthig and Wepf Verlag, Heidelberg (Figs. VI.62, VII.12, and VII.22).
North-Holland Publishing Co., Amsterdam (Figs. V.19 and VI.2).
Springer Verlag, Heidelberg (Fig. VI.38).
Dr. Dietrich Steinkopff Verlag, Darmstadt (Figs. VI.27, VI.46, VI.80 A and B).
Wiley and Sons, Inc., New York (Figs. V.18, V.23, V.25, V.48, V.49, VI.70, VI.71, VI.72, VI.73a,b, VI.74a–c, VI.75, VI.76, VI.79, VI.81, VI.82, VII.6, VII.7, VII.8, VII.13, VII.14, VII.21, VII.26, VII.27, and VII.28).

Contents of Volume 1

Macromolecular Physics

VOLUME 2

Crystal Nucleation, Growth, Annealing

CHAPTER V

The Nucleation Step

5.1 Primary Nucleation of Crystals

In Volume I of this book, which contains Chapters I–IV, macromolecular crystals were characterized on a molecular level by their degree of long-range order and the position of the atoms of the molecules within the crystal lattice (Sect. 2.4). The reasons for the existence of particular crystal structures were traced to the hierarchy of potential energy effects of covalent bonding, rotational isomerism, and closest packing (Sect. 2.3). The crystal morphology was found to be linked closely to the macroconformation of the molecules (Sect. 3.2); lamellar and fibrous crystals being the most common of the well-formed habits (Sects. 3.3 and 3.8, respectively). Defects were shown to be a major impediment to the full understanding of the crystalline state of linear macromolecules (Sects. 4.2 and 4.3).

This volume, which contains Chapters V–VII, deals with the processes leading to the crystalline state. A full description of crystallization would need a treatment of the initial state, the process of crystallization, and the final crystalline state. The final state is discussed in detail in Volume I; the actual process of crystallization is presented in this volume. Of the initial states of matter, the gaseous and crystalline solid states are well known; they are of interest, however, only for the formation of macromolecular crystals from the monomer (see Sect. 6.4.). Much less is known about the liquid state. In particular, the much wider studied case of crystallization of the already completed macromolecule starts usually from molecularly disordered and mobile solution or melt. In a sufficiently dilute solution, the macromolecules are well separated from each other and can be treated as isolated random coils as outlined in Sect. 1.4. Figure I.3 illustrates the macroconformation of such dissolved macromolecules. A more detailed discussion of polymer solution properties can be found, for example, in the treatise by Morawetz (1965).

As a solution becomes more concentrated, the description of the macromolecules is more complicated because of increasing need for interpenetration of the macromolecules, the most complex description being needed for the pure macromolecular melt. At present the structure of the melt is not fully understood, but under active investigation.† On one hand, it was shown by Flory (1953) that the partition function of a mixture of a randomly coiling macromolecule with solvents can be separated into one factor arising from the intramolecularly governed conformations of the chains and a second factor connected with the effects of interaction and mixing with the solvent. In the absence of an added solvent, each molecule is dispersed in other molecules of the same kind which of necessity makes the second factor one and should lead to a substantially unperturbed molecule. In this description the molecule has nothing to gain by expanding or contracting. On the other hand, Lindenmeyer (1973) argues that there exists a thermodynamic driving force which tends to limit interpenetration of molecules so that polymer molecules in the melt should be more nearly irregularly folded molecules than random coils.

A completely different approach to the problem of understanding the polymer melt structure was taken by Stuart (1943) who carried out model experiments with strings of beads. Due to the geometry of the model molecules, a predominantly parallel alignment of neighboring chains was observed. A similar structure arose on expanding the kink model of crystals (Sect. 4.2.3) to the description of the short-range order expected in the melt. Figure V.1 shows a two-dimensional representation of the number of kinks necessary to account for a typical liquid structure.‡ Over short distances the amorphous polymer chains are thus assumed to form bundles which are changing in their association and orientation with time. Somewhat differing models of the amorphous state based on bundle-type aggregation were given by Kargin *et al.* (1957) and Yeh (1972) (see also Sect. 6.1.7).

Experimental evidence has been used to support both the random coil and the bundle models. The bundle model seems to have support through the

† See, for example, *Symp. Phys. Structure Amorphous State. Polym. Preprints, Div. Polym. Chem. Inc. Amer. Chem. Soc.* **15** (2), 1–45, 324–343 (1974).

‡ For more detailed analysis of the kink model as applied to the amorphous structure, see the review by Pechhold *et al.* (1973) and earlier papers by Pechhold and Blasenbrey (1967, 1970), Pechhold (1968), and Scherr *et al.* (1970).

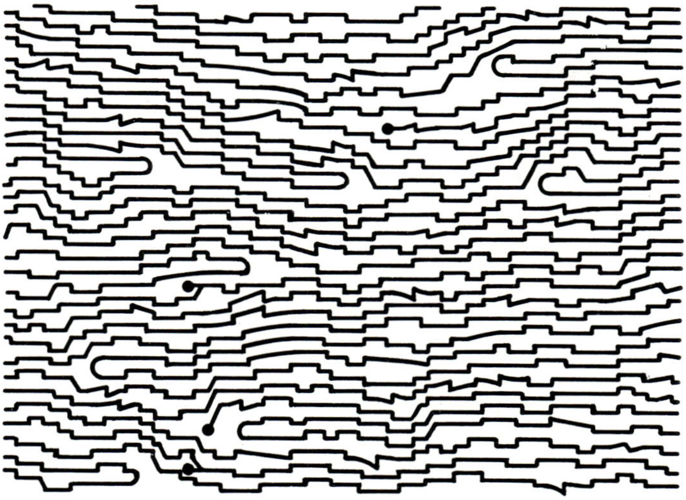

Fig. V.1 *Two-Dimensional Representation of a Small Section of Macromolecular Melt Based on the Kink Model.*
Drawn after Pechhold (1968).

observation of a 50–100 Å range order detected by electron diffraction (Ovchinnikov *et al.*, 1969; Petermann and Gleiter, 1973), electron microscopy (see Sect. 6.1.7 and Figs. VI. 35–VI.37), far infrared absorption (Amrhein and Frischkorn, 1973) and packing considerations (Robertson, 1965). Fischer *et al.* (1974) partially refuted the interpretation of these experiments and presented new data on light scattering, small angle X-ray scattering and small-angle neutron scattering which made a random coil macroconformation more likely. In particular neutron scattering on polymers dissolved in perdeuterated melts of otherwise identical molecules carried out on polystyrene (Benoit, 1974; Ballard *et al.*, 1973; Benoit *et al.*, 1973), polyethylene (Fischer *et al.*, 1974) and poly(methyl methacylate) (Kirste *et al.*, 1973) are in accord with a largely unchanged macroconformation on changing from solution in a solvent of small molecules to the pure melt. The radius of gyration of molecules in the melt was found to agree with that in a Θ solvent, and the scattering intensity as a function of scattering angle agrees well with that of a Gaussian random coil. For the present discussion of crystallization one can thus suggest that perhaps the only change from solution to melt crystallization is an increase in interpenetration of molecules. Little is known at present about the degree of interpenetration of the molecules, the mechanism of disentanglement of the molecules on crystallization, and the influence of short-range order on the overall melt structure and crystallization process. In this respect the initial states of the crystallization experiments are not well defined.

Turning now to the thermodynamic description of the crystallization process, one can write

$$\Delta G = G_{\text{crystal}} - G_{\text{melt}} \tag{1}$$

$$\Delta G = \Delta H - T\,\Delta S \tag{2}$$

where ΔG is the free enthalpy (Gibbs free energy) of crystallization and ΔH and ΔS are the enthalpy and entropy of crystallization. Equation (2) applies to the overall macroscopic transformation. Crystallization becomes possible as soon as ΔG assumes negative values. If the phase transition would strictly follow Eq. (2), crystals grown on cooling would have to be distributed uniformly over the volume which has reached the melting temperature T_m at which ΔG is zero. Polymer melts and solutions crystallize, in contrast, normally only at much lower temperature T_c, they supercool. In addition, crystals grow from a nucleus rather than generate uniformly throughout the random phase.

The reason for supercooling and the need for nucleation can be found by analyzing the crystallization process in more detail. Any crystal must have its beginning as a small crystal with a large specific surface area (measured in cm^2/g). The overall free enthalpy of a crystal is thus written more precisely as

$$G_{\text{crystal}} = G_{\text{bulk}} + \sum \gamma A \tag{3}$$

Where G_{bulk} is the free enthalpy of the crystal disregarding surface effects, γ represents the specific surface free energy, and A is the corresponding surface area.†

---

† Surface and surface free energy of solids need careful definition to avoid ambiguities in the later discussions of nucleation, crystal growth, and melting. An advantageous definition of the surface was given by Gibbs (1878); he chose . . . "some point in or very near to the physical surface of discontinuity, and imagine(d) a geometrical surface to pass through this point and all other points which are similarly situated with respect to the condition of the adjacent matter." The most helpful condition to locate the surface of a solid is to require the surface density of the principle constituent of the solid to vanish at the surface.

The thermodynamic functions of the surface have also been given a precise meaning by Gibbs (1878). He considered an imaginary volume to be generate around a surface element. For simplicity, this surface element may be chosen to be of unit size. Any extensive quantity of the imaginary volume such as energy E, entropy S, free energy F, or number of moles N_i can be uniquely subdivided into three parts:

$$\begin{aligned} E &= E_1 + E_2 + e_s \\ S &= S_1 + S_2 + s_s \\ F &= F_1 + F_2 + f_s \\ N_i &= N_{i1} + N_{i2} + n_{is} \end{aligned}$$

The subscripts 1 and 2 indicate contributions of phase 1 and 2, located on the different sides of the unit surface. By assigning the values of bulk energy, entropy, free energy and number of moles to the respective volumes on sides 1 and 2, the specific surface quantities (subscripts s) are unambiguously fixed. They represent excess quantities over what energy,

The free enthalpy of crystallization (Eq. 1) becomes

$$\Delta G = \Delta G_c + \sum A\gamma \tag{4}$$

with ΔG_c representing the bulk free enthalpy change. In both Eqs. (3) and (4) the sum has to be extended over all surfaces which in general have different specific surface free energies. The specific surface free energies are positive for all temperatures of interest to crystallization. For temperatures below T_m where ΔG_c is negative, ΔG exhibits a maximum value. This initial process leading from the amorphous state to a growing crystal is called primary nucleation. Figure V.2 represents the process schematically. Before a thermodynamically stable crystal of sufficient size can be grown, a primary nucleus must be formed via a path of positive ΔG. The maximum in ΔG corresponds to the critical size nucleus. Nuclei to the left are called subcritical nuclei or embryos, nuclei to the right are called supercritical as long as their ΔG is still positive. Nuclei with a negative ΔG are called stable nuclei or small crystals. This breakdown of the overall crystallization into a nucleation step followed by growth was already clearly stated by Tammann (1903). The free enthalpy barrier to crystallization can only be overcome by local random fluctuation of order. The larger the required size of the nucleus, the longer will be the time needed for the nucleation process.

Primary nucleation is called homogeneous nucleation if no preformed nuclei or foreign surfaces are present. A foreign surface reduces frequently the nucleus size needed for crystal growth since the creation of the interface

---

entropy, free energy, and number of moles would have been if both phases were uniform to the surface.

Locating the surface, as suggested above, at the point where the principle constituent of the solid vanishes ($n_{is} = 0$) leads to equality of surface tension γ and specific surface free energy f_s when there is no other component adsorbed in the surface. One must, however, always consider the anisotropy of γ; and, in addition, when one is not willing to neglect quantities of the order of the ratio of the thickness of the inhomogeneous surface layer to the linear dimension of radii of curvature of the surface, three more points must be considered [see Herring (1953) for further details]: (a) the change of γ with curvature; (b) the inhomogeneous volume stresses induced by the presence of the surface and the distinction between surface tension and surface stress; and (c) the precise definition of chemical potentials in the solid.

The surface stress measures the work required to deform a surface, while surface tension (specific surface free energy) measures the work required to create new surface (Gibbs, 1878). For liquids both quantities are identical since displacements imposed on atoms at a boundary of the surface will not affect the surface beyond the distance of local order. In solids the presence of surface stresses causes small particles to have compensating volume stresses which are in general inhomogeneously distributed. The energy associated with these volume stresses, however, is usually negligible compared to the surface free energy down to dimensions smaller than those to which macroscopic concepts like surface tension and elasticity can be applied (Herring, 1953).

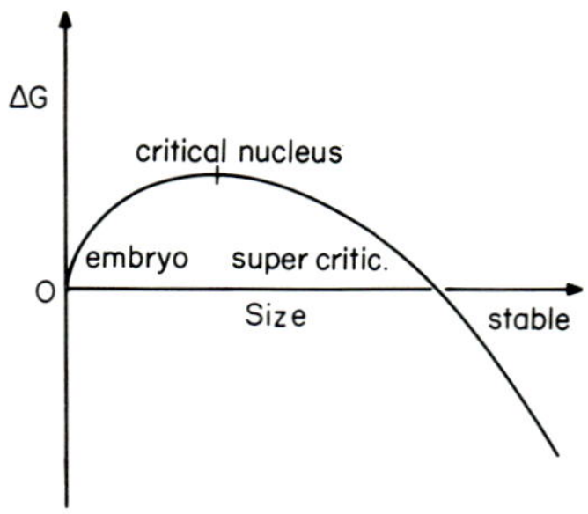

Fig. V.2 *Schematic Representation of the Change in Free Enthalpy ΔG as a Function of Size Illustrating the Nucleation Process.*
Nucleation theory leads frequently to expressions in the form $\Delta G/kT = Ai^{2/3} - Bi$, where i is the size in number of elements in the transformed region, A is proportional to the surface free energy, and B to the bulk free energy in the absence of surface (see text).

between polymer crystal and substrate may be less hindered than the creation of the corresponding free polymer crystal surface. The resulting enhanced nucleation process is called heterogeneous nucleation. A special type of primary nucleation is called self-nucleation. This term describes primary nucleation caused by polymeric crystals which are chemically identical to the crystallizing polymer, but have survived the prior dissolution or melting step. Sections 5.1.1–5.1.4 deal with the theory of primary nucleation and the experimental observations of homogeneous nucleation, heterogeneous nucleation, and self-nucleation.

It is obvious from Eq. (4) that any progress of crystallization which increases the crystal surface has a more positive free enthalpy of crystallization ΔG. Only when a motif can be added into an appropriately shaped corner is there no net change in surface area. On a molecularly smooth crystal surface, a new layer can only be grown after secondary nucleation, a process similar to primary nucleation, but with a somewhat lower free enthalpy barrier since the surface area which must be newly created is smaller. Analogously, tertiary nucleation can be defined as the initial step to start a new row of crystal growth at an edge. Figure V.3 illustrates schematically the three nucleation steps. Secondary and tertiary nucleation and their influence on polymer morphology are described in Sect. 5.2. Tertiary nucleation does present the smallest free enthalpy barrier to growth. A growing monomolecular layer on the surface of a crystal has usually a sufficient number of irregularities to provide new crystallization sites (see Sects. 5.1.1 and 6.1.2.2, Fig. V.9).

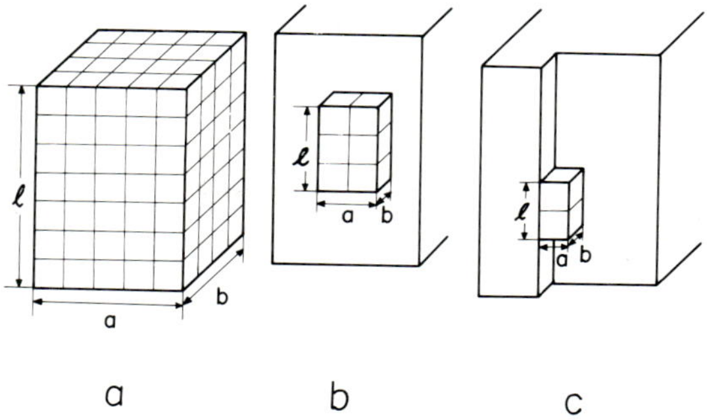

Fig. V.3 *Types of Crystal Nuclei.*
(a) Primary nucleus, (b) secondary nucleus, (c) tertiary nucleus.

An additional complication in polymer crystallization is caused by the size of macromolecules. While in crystallization of small molecules the elementary step in crystal growth can be identified with the addition of a single molecule to the growing crystal, a macromolecule must be crystallized in many thousands of cooperative steps. Crystallizing the first portion of a molecule on a crystal surface leads to an additional free enthalpy barrier to crystallization which is different in origin, but similar in effect to the crystal nucleation barriers. This free enthalpy barrier which retards the addition of each new molecule to the growing crystal is called molecular nucleation and will be discussed in Sect. 5.3. Crystal growth and crystal perfection will be treated in the subsequent chapters.

5.1.1 Nucleation Theory Concept

The classical concept of crystal nucleation† was developed by Gibbs (1878) [see, for example, Gibbs (1906) p. 252], Volmer (1939), and Kossel (1927). It is based on the assumption that fluctuations in the supercooled phase can overcome the nucleation barrier caused by the surface of the crystal. The probability of the presence of a nucleus of given size at constant volume and energy is, according to Boltzmann's law, a function of the change in entropy, proportional to $\exp(\Delta S/k)$. At constant pressure and temperature, the probability for the presence of a nucleus of given size is proportional to $\exp(-\Delta G/kT)$.

† For general surveys of nucleation see Zettlemoyer (1969) and Michaels (1966) for small molecules and Price (1969) for macromolecules.

The rate of nucleation I^* has been derived on these assumptions by Turnbull and Fisher (1949) using absolute reaction rate theory (see Glasstone *et al.*, 1941) to be

$$I^* = (NkT/h)\exp[-(\Delta G^* + \Delta G_\eta)/kT] \tag{5}$$

The rate of nucleation I^* is given in nuclei per second to at least one order of magnitude and refers to the number of uncrystallized elements N able to participate in nucleation by a single step (see below). ΔG^* stands for the free enthalpy of crystallization of a nucleus of critical size. Its variation with temperature for typical parameters is shown in Fig. V.4. A more detailed

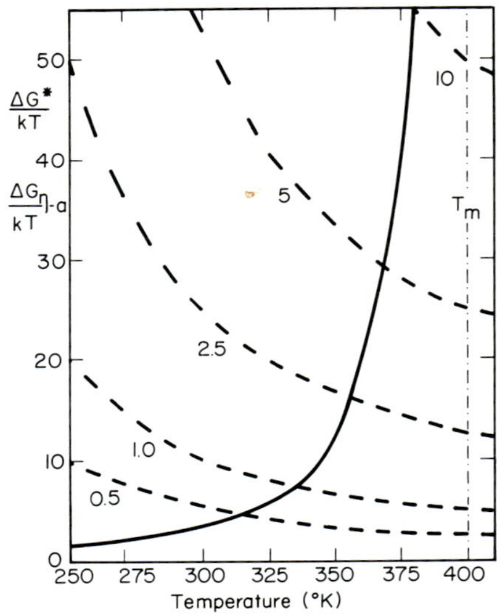

Fig. V.4 *Change of Exponents of Eq. (5) as a Function of Temperature.*
Solid line: $\Delta G^*/kT$: Free enthalpy of crystallization of a nucleus of critical size [see Eq. (21)]. Parameters used for its computation were $\gamma = 5$ erg/cm^2, $T_m = 400$°K, $\gamma_e = 50$ erg/cm^2, $\Delta h_f = 50$ cal/g, $\rho_c = 1.00$ g/cm^3. Broken lines: $(\Delta G_\eta/kT) - a$: Estimate of the temperature dependence of the free enthalpy of activation for diffusion across the phase boundary as given by Eq. (6). Different values for $b \times 10^{-3}$ °K are indicated in the graph. T_0 was assumed to be 200°K.

discussion of ΔG^* will be given below. The second term in the exponent ΔG_η represents the free enthalpy of activation which governs the short distance diffusion of the crystallizing element across the phase boundary. Its temperature dependence is expected to be similar to the temperature dependence of viscosity. At high temperature, ΔG_η is almost constant, at low temperature it increases rapidly as the glass transition temperature is approached. The simplest mathematical expression which takes this fact into account is

$$\Delta G_\eta/kT = a + [b/(T - T_0)] \tag{6}$$

where T_0 is a temperature where no further transport across the phase boundary can take place.† Figure V.4 shows curves of $(\Delta G_\eta/kT) - a$ for several values of b (dashed curves). The preexponential factor of Eq. (5) represents the approximate transmission coefficient for the crystallizing element. At room temperature its value should be about 6.2×10^{12} sec^{-1}. Its temperature dependence is negligible compared to the exponential terms and for data analysis it is usually combined with $\exp(-a)$ and treated as a constant. Figure V.5 illustrates the general shape of I^* as a function of temperature.

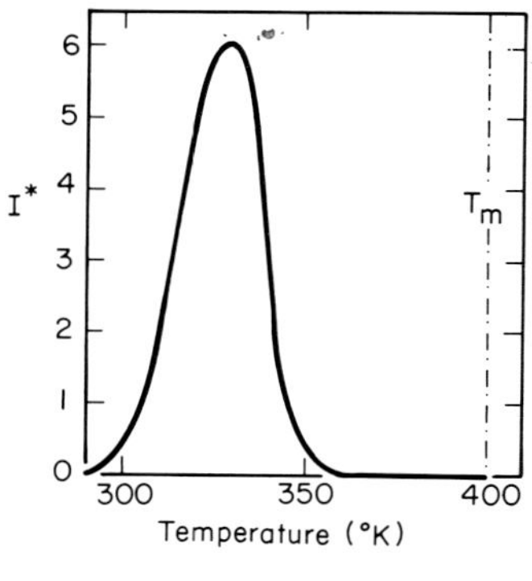

Fig. V.5 *Nucleation Rate as a Function of Temperature.* Data as in Fig. V.4. $b = 2500$, $a = 56.6$ (I^* in nuclei per second for one mole of uncrystallized elements).

The nucleation rate increases rapidly with decreasing temperature over a relatively narrow temperature interval due to the rapid decrease in ΔG^*. For polymers this region of fast primary nucleation begins typically at a supercooling of 30–100°C. At larger degrees of supercooling the nucleation rate

† The WLF equation which universally describes the temperature dependence of polymer melt viscosity has been similarly used (Williams *et al.*, 1955): $\Delta G_\eta/kT = 2.07 \times 10^3/(51.6 + T - T_g)$; T_g = glass transition temperature.

decreases again due to the increase in ΔG_η when the glass transition temperature is approached (see Fig. V.4). The maximum level of nucleation rate is critically dependent on the ease with which the crystallizing unit can cross the phase boundary. Many polymers can be quenched sufficiently rapidly by simple conduction cooling to preclude crystallization completely and yield a glassy solid.

A more detailed picture of the free enthalpy change during the addition of one crystallization element in the course of nucleation is given by the schematic diagram of Fig. V.6. To derive Eq. (5) one assumes steady state at every step.†

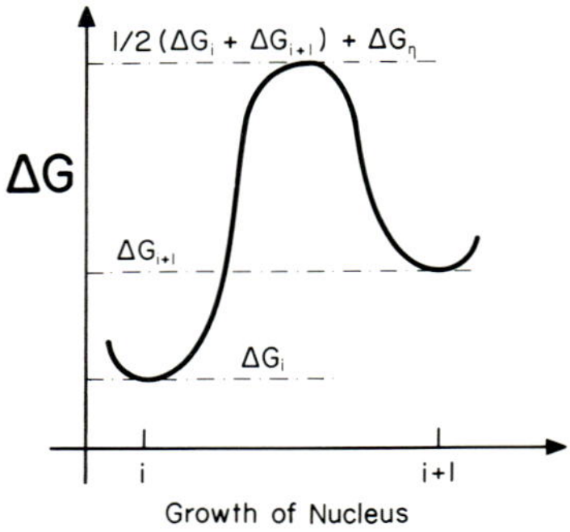

Fig. V.6 *Free Enthalpy Change during Nucleation.*
Increase in nucleus size by one crystallizable element on going from left to right [Eq. (7)].

$$\beta_i + \alpha \rightleftharpoons \beta_{i+1} \tag{7}$$

The uncrystallized molecule or segment of a molecule is represented by α, and β_i stands for a nucleus already containing i such molecules or segments of molecules (see Fig. V.3a). The nucleus of critical size (see Fig. V.2) is marked by an asterisk β^*. The free enthalpy of each step is determined relative to the state of α, so that ΔG_i and ΔG_{i+1} are the respective free enthalpies of crystallization. The free enthalpy difference between that of the activated complex (at the maximum free enthalpy between β_i and β_{i+1}) and the mean ΔG_i and ΔG_{i+1} is ΔG_η (see Fig. V.6). Its value is assumed constant for all steps of nucleation. The forward and reverse rates at each step are then given

† For additional discussions of nucleation and growth based on the absolute reaction rate theory see the derivation of Eq. (75) of Sect. 5.2.2.2 and also Sect. 6.1.4.3.

by the following equations according to the theory of absolute reaction rates (Glasstone *et al.*, 1941)

$$I^+ = n_i ai^{2/3}(kT/h)\exp\{[\tfrac{1}{2}(\Delta G_i - \Delta G_{i+1}) - \Delta G_\eta]/kT\} \tag{8}$$

$$I^- = n_{i+1} ai^{2/3}(kT/h)\exp\{[-\tfrac{1}{2}(\Delta G_i - \Delta G_{i+1}) - \Delta G_\eta]/kT\} \tag{9}$$

In these equations n_i and n_{i+1} represent the steady state number of β_i and β_{i+1} nuclei. The number of α molecules or molecular segments in contact with each β_i or β_{i+1} nucleus is $ai^{2/3}$. This value does not apply strictly to both nuclei; differences are, however, negligible for the larger nuclei. By assuming n_i and ΔG_i to be smooth functions of i so that their differences on varying i by 1 can be replaced by the respective differential quotients and by neglecting small quantities, Turnbull and Fisher (1949) derived† Eq. (5) for the net forward rate of nucleation $I^* = I^+ - I^-$.

Equation (5) can now be used to compute rates for all types of nucleation paths. At constant temperature the fastest nucleation rate is reached for the nucleus of lowest ΔG^*. The lowest free energy for a given size crystal is always coupled with the equilibrium shape.‡ ΔG^* for the critical nucleus can thus be computed by differentiation of Eq. (4) for the equilibrium morphology with respect to the size (see Fig. V.2)

$$d\,\Delta G_i/di = 0 \qquad \text{at} \qquad \Delta G_i = \Delta G^* \tag{10}$$

The change of ΔG_i with size for a nucleus of quadratic cross section (Fig. V.3a, $a = b$) is shown in Fig. V.7. ΔG_i has been computed using Eq. (4) in the form

$$\Delta G_i = -a^2 l\,\Delta g_f + 4al\gamma + 2a^2\gamma_e \tag{11}$$

where Δg_f is the bulk free enthalpy of 1 cm^3 of crystal without surface effects.§ The side and end surface free energies are designed by γ and γ_e, respectively.

The critical nucleus size can be computed from Eq. (11) by differentiation with respect to the dimensions a and l

† The quantities $[d\,\Delta G_i/(kT\,di)]$ and $[dn_i/(n_i\,di)]$ are assumed small relative to one, an approximation which is poor only for the smallest nuclei. Expanding the appropriate exponentials into series and neglecting small terms, one reaches the following differential equation:

$$-I^* = ai^{2/3}(kT/h)\exp(-\Delta G_\eta/kT)\{[n_i\,d\Delta G_i/(kT\,di)] + dn_i/di\}$$

which on solution leads to Eq. (5) for typical curves of $\Delta G_i/kT$ (see Fig. V.2., $\Delta G_i/kT = Ai^{2/3} - Bi$; see also the following).

‡ The equilibrium of a small particle is defined for the shape which minimizes the quantity $\sum \gamma A$. For a more detailed discussion of equilibrium morphology see Sect. 3.1.2.

§ Note that Δg_f is the free enthalpy of fusion which is equal to the negative of the free enthalpy of crystallization, this accounts for the negative sign of the first term on the right-hand side of Eq. (11).

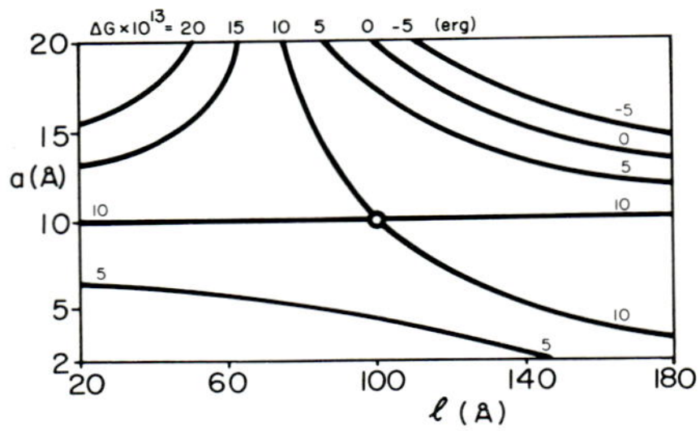

Fig. V.7 *Change of Free Enthalpy of Formation of a Nucleus.*
Dimensions: length l and quadratic cross section width a. Assumed parameters in Eq. (11) are $\gamma_e = 50$ erg/cm^2, $\gamma = 5$ erg/cm^2, $\Delta g_f = 2 \times 10^8$ erg/cm^3, the latter corresponds to a supercooling of 38.2°C in the case of the nucleation represented by Fig. V.5.

$$\partial\,\Delta G_i/\partial a = -2al\,\Delta g_f + 4l\gamma + 4a\gamma_e \tag{12}$$

$$\partial\,\Delta G_i/\partial l = -a^2\,\Delta g_f + 4a\gamma \tag{13}$$

The critical nucleus dimensions a^* and l^* are obtained by setting Eqs. (12) and (13) equal to zero

$$a^* = 4\gamma/\Delta g_f \tag{14}$$

$$l^* = 4\gamma_e/\Delta g_f \tag{15}$$

The free enthalpy change can be expressed furthermore by

$$\Delta g_f = \Delta h_f' - T\,\Delta s_f \tag{16}$$

As long as the degree of supercooling is not large, Δs_f the entropy of fusion per cm^3 can be taken to be temperature independent.† Then Δs_f can be computed from $\Delta h_f'$ the enthalpy of fusion per cm^3 at the equilibrium melting temperature T_m

$$\Delta s_f = \Delta h_f'/T_m \tag{17}$$

Noting further that the heat of fusion per gram of bulk crystal is

$$\Delta h_f = \Delta h_f'/\rho_c \tag{18}$$

† An approximation of the temperature dependence of Δg_f for large degrees of supercooling has been given by Hoffman (1958). It is based on a linear temperature dependence of Δh_f and leads to the expression $\Delta g_f = \Delta h_f\,T\Delta T/T_m^2$, usable down to the glass transition temperature. For polyethylene experimental data are even better represented by $\Delta g_f = 2\,\Delta h_f\,T\,\Delta T/(T_m^2 + TT_m)$ (see Hoffman and Weeks, 1962).

where ρ_c is the crystal density, we see that the critical nucleus dimensions can be expressed as

$$a^* = 4\gamma T_m/(\Delta h_f \Delta T \rho_c) \tag{19}$$

$$l^* = 4\gamma_e T_m/(\Delta h_f \Delta T \rho_c) \tag{20}$$

The supercooling $\Delta T = T_m - T$ and the surface free energies determine the size and shape of the nucleus. Figure V.7 exhibits a saddle point at the critical nucleus size. Table V.1 lists some typical dimensions for different super-

TABLE V.1

CRITICAL EQUILIBRIUM NUCLEUS DIMENSIONS[a,b]

ΔT (°C)	Surface free energy γ (erg/cm^2)						
	1	5	10	15	50	100	200
0.1	764.80	3824.0	7648.0	11472.0	38240.0	76480.0	152960
1.0	76.48	382.4	764.8	1147.2	3824.0	7648.0	15296
10	7.65	38.2	76.5	114.7	382.4	764.8	1530
25	3.06	15.3	30.1	45.4	153.0	305.9	612
50	1.53	7.6	15.3	22.9	76.5	153.0	306
100	0.76	3.8	7.6	11.5	38.2	76.5	153
150	0.51	2.5	5.1	7.6	25.5	51.0	102

[a] In Å as a function of supercooling and surface free energy.

[b] Δh_f is assumed to be 50 cal/g, $\rho_c = 1$ g/cm^3, and $T_m = 400$°K. These values are of the order of magnitude for typical polymer crystals. The dimensions are calculated using Eqs. (19) or (20) which are identical in form, so that all dimensions can be estimated whenever different surface free energies are given.

coolings and surface free energies. The greater the value γ_e is relative to γ, the more fibrous will be the shape of the nucleus. The greater a degree of supercooling, the smaller will be the size of the nucleus, at about 100°C of supercooling reaching the *ab* dimensions of the unit cell for the γ_e/γ ratio of Fig. V.7.

Nucleation rates can be estimated by substitution of a^* and l^* into the expression for ΔG_i (Eq. 11) to obtain an expression for ΔG^* and evaluation of Eq. (5)

$$\Delta G^* = 32\gamma^2\gamma_e T_m^2/(\Delta h_f \Delta T \rho_c)^2 \tag{21}$$

Figures V.4 and V.5 were computed using Eq. (21).

The treatment of nucleation just presented contains a number of limitations; the first and obvious is reached for large supercoolings. In the case where the critical nucleus is reduced to the order of magnitude of several unit cell sizes, opposing surfaces must interact giving rise to a surface free energy which is size dependent. Also, many of the simplifications included in the derivation of Eq. (5) apply only for reasonably large values of i.

A second limitation is the assumption of strictly crystallographic growth, permitting the use of Δh_f and Δs_f as measured from macroscopic melting experiments. The initially crystallized material does not have the initial perfection for these substitutions (see Sect. 6.1, Fig. VI.24 and Table VI.5).

Finally, the limitation of the nucleus to the equilibrium shape neglects other possible nucleation paths. Although these nonequilibrium shapes are less probable, their number may be large, and in the final analysis the multitude of nonequilibrium paths may be more important than the single equilibrium path.

Binsbergen (1972) has made an attempt to describe the nucleation process taking into account all possible paths. His calculations are based on a molecular model for a primitive cubic crystal ($\gamma = \gamma_e$) as proposed first by Kossel (1927). Limiting the lattice energy to only nearest-neighbor interaction, the single contact separation energy ϕ is related simply to the surface free energy by

$$\gamma = \phi/(2a^2) \tag{22}$$

as can be derived by computation of the surface energy needed for separation of a row or layer of molecules (Stranski and Kaischew, 1934). Figure V.8

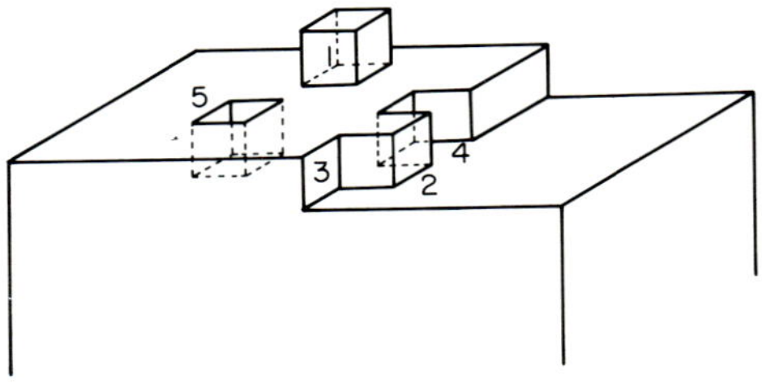

Fig. V.8 *Schematic View of a Kossel Crystal.* The numbers indicate the relative value of the separation enthalpy of a crystallizing unit at this location (1 and 2 drawn before removal, 3–5 drawn after removal).

gives a schematic view of the possible configurations at the surface of a Kossel crystal. The configurations are labeled by their respective separation energies which range from ϕ to 5ϕ. With help of a digital computer Binsbergen (1972) was able to follow the nucleation by permitting attachment and detachment of single units at any available site of the growing nucleus. The simulation was carried out in a 25 × 25 × 25 matrix. It led to a "simulated" activation free enthalpy of

$$\Delta G^* = 27.6\gamma^{15/4} T_m^2 / (\Delta h_f \Delta T \rho_c)^2 \qquad (23)$$

instead of the classical expression given in Eq. (21) ($\gamma = \gamma_e$). No critical size parameters could be given since the nuclei did not correspond to equilibrium shapes. Examples of typical shapes arising during nucleation are reproduced in Fig. V.9 for a two-dimensional case.

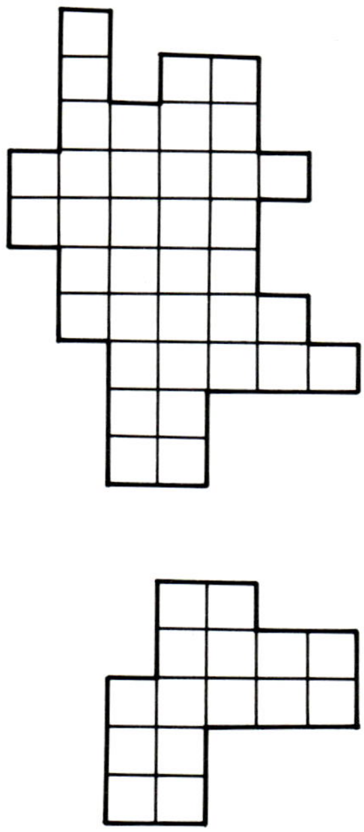

Fig. V.9 *Typical Nucleus Shapes during Growth in Two Dimensions.*
Drawn after computer simulations by Binsbergen (1970a).

A comparison of Eq. (23) with the classical Eq. (21) shows that the simulated activation enthalpy and its surface free energy dependence deviates perhaps less than one would expect. The shapes of Fig. V.9, in contrast, are considerably different from the equilibrium shape (square). The ΔT dependence shows no difference. The failure to include imperfections in the structure of the initial nucleus may not be a serious fault either since both the surface free energy and the bulk free enthalpy are often similarly affected and many of the changes may cancel out. As will be shown in the next section, classical nucleation describes many experiments adequately.

5.1.2 Homogeneous Nucleation

When estimating the volume of a homogeneous nucleus from the data of Table V.1, one finds that for typical degrees of supercooling ($\sim 50°C$) the nucleus is perhaps 10^4 Å^3. This volume is less than the volume of a macromolecule (10^5–10^7 Å^3 for 60 000–6 000 000 molecular weight). Thus homogeneous nucleation can involve only part of one macromolecule or even smaller portions of several macromolecules. On the other hand, when crystallization occurs during polymerization, the nucleus dimension in the molecular chain direction (~ 100 Å) is normally larger than the monomer length. In this case oligomers of appropriate length have to be polymerized first before nucleation can proceed. These two facts distinguish macromolecular crystal nucleation from nucleation of crystals of small molecules. Instead of one basic path possible for nucleation of crystals of small molecules, macromolecular crystals

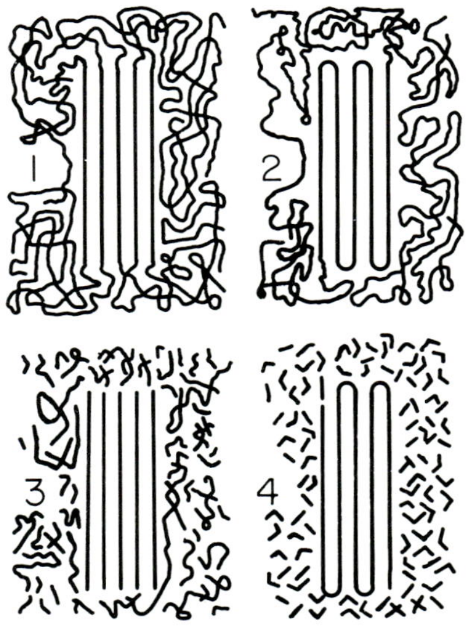

Fig. V.10 *Schematic Representation of Macromolecular Crystal Nuclei.*
1. Intermolecular fringed micelle nucleus; 2. intramolecular folded chain nucleus; 3. intermolecular oligomer nucleus; 4. intramolecular folded chain nucleus of a polymerizing molecule.

may follow any one of the four nucleation paths illustrated schematically in Fig. V.10.

In principle, the flexibility of linear macromolecules permits two different nucleation paths for crystallization from macromolecular melts or solutions. Figure V.10 shows in parts 1 and 2 a schematic representation of these two paths (Überreiter, 1962; Wunderlich, 1970). The first type of nucleation is intermolecular and will be called *fringed micelle* nucleation. The second type of nucleation is an intramolecular nucleation and will be called *chain folded* nucleation. The latter type of nucleation has received the most attention in the past 10 years.

Two completely different types of nucleation are of importance when the polymer crystal is grown directly from the monomer by crystallization during polymerization. Quite changed nucleation conditions exist in this case, as are shown in parts 3 and 4 of Fig. V.10. Only *oligomer* nucleation represented by part 3 has received any attention previously (Wunderlich, 1968). Part 4 of Fig. V.10 represents a *chain folded* nucleation, but this time of an actively growing molecule. Few details are known about its mechanism.

The following four sections deal with homogeneous nucleation of types 1–4. Much of the developed theory can also be used for the description of secondary nucleation, molecular nucleation, and crystal growth which are described later.

5.1.2.1 Fringed Micelle Nuclei

The first nucleation path of Fig. V.10 is called fringed micelle nucleation since it leads on incomplete crystallization to the fringed micelle type crystal morphology first proposed by Hermann *et al.* (1930).† The nucleus is thought to be a bundle of polymer chains with large sections of the molecules left uncrystallized. In case of sufficiently high molecular weight the molecules are even capable of entering other nuclei. The surface of the polymer crystal which is characterized by the surface free energy γ_e would represent the location of the transition from the crystalline to the amorphous area. Much of early nucleation theory centered around this crystallization type (Flory, 1949; Mandelkern, 1955).

As polymer crystal nucleation and growth was studied further, limitations to the fringed micelle structure were discovered. The first limitation to fringed micelle formation is that of size or shape. Flory (1962) showed that the strain

† Hermann *et al.* describe the fringed micelle (ausgefranste Micell) as follows: . . . "[the] crystallite . . . becomes towards its end even poorer, such that the lateral cohesion of the chain segments on both sides of the chains vanishes. The last segments of the chains (of variable length) lie randomly mixed . . .". A diagram was published by the same authors in *Naturwissenchaften* **18**, 755 (1930).

generated at the crystalline–amorphous interface by polymer molecules which traverse from the crystalline to the amorphous area must limit the nucleus or crystal dimensions.† For a strain-free surface the numbers of polymer chains (per cm^2), N_c and N_a which traverse planes parallel to the surface in the crystalline portion and in the amorphous region must be equal. The value of N_c can be computed from

$$N_c = (1/A_c)\sin\alpha \tag{24}$$

where the angle α represents the inclination of the surface to the molecular chain axis, and A_c represents the cross-sectional area of the polymer chain normal to the molecular chain axis. In the amorphous region all inclinations θ of the molecules relative to the reference plane are permitted so that the following average must be computed

$$N_a = \int_0^{\pi/2} N_\theta \, d\theta \tag{25}$$

The number of chains of angle θ relative to the reference plane (per cm^2) is given by N_θ. The cross section of the amorphous chain A_a can be estimated as V_a/l, where l is the segment length and V_a the amorphous volume per segment. $N_\theta \, d\theta$ is then given by

$$N_\theta \, d\theta = (1/A_a)\sin\theta \, d\sin\theta \tag{26}$$

Insertion of Eq. (26) into (25) and integration leads to

$$N_a = 1/(2A_a) \tag{27}$$

The condition for a strain free surface of large dimensions is thus

$$N_a/N_c = 1 = \tfrac{1}{2}(A_c/A_a)/\sin\alpha \tag{28}$$

Equation (28) shows that a large nucleus or crystal must have a surface inclined to the chain axis by about 30°. A surface normal to the chain axis is only permitted for rather small nuclei as, for example, in Fig. V.10, where the molecular chains can avoid crowding by splaying apart.

Another more serious limitation to fringed micelle nucleation involves the effect of the nucleus or crystal on the remaining amorphous portions of the molecules. The part of a molecule adding to a nucleus contributes to the free enthalpy of formation of the nucleus as described before [Eqs. (21) and (23)], but in addition, fixing a portion of the molecules reduces the entropy of its remaining amorphous portions. To compute the nucleation rate I^* for fringed micelle-type nucleation, a major modification of Eqs. (5) and (21) is necessary.

† For an early discussion see Frank (1958).

Following the description by Zachmann (1967, 1969), Eq. (21) needs to be changed to the following expression (see also Baranov *et al.*, 1972)

$$\Delta G^* = 32\gamma^2(\gamma_e + \gamma_e')/(\Delta g_f)^2 \tag{29}$$

The newly introduced term γ_e', represents the free energy contribution from the remaining amorphous portions of the molecules which are connected to 1 cm^2 of the appropriate surface, while γ_e is the normal surface free energy as defined earlier (see footnote p. 4). The term γ_e' does not represent a true surface free energy; however, it can be treated like a surface free energy because in a fringed micelle nucleus all crystallized chain sequences belong to different molecules, and it is assumed that all molecules traverse the nucleus, exiting on opposite *ab* surfaces. The total contribution of the amorphous portions of the molecules to ΔG^* is thus proportional to the *ab* surface area. The importance of γ_e' for molecular nucleation is discussed in Sect. 5.3.4.

Zachmann (1967, 1969) estimated the value of γ_e' by making use of a lattice model as shown schematically for two dimensions in Fig. V.11. [For details of lattice model calculations see Flory (1941, 1942); Huggins (1942); Prigogine (1957).] The entropy of the macromolecules in the melt can be calculated† from the number of different arrangements Ω

$$\Omega = (\tfrac{1}{2}x)^N[(z-1)/e]^{N(x-1)} \tag{30}$$

where x is the number of segments of a single macromolecule, N is the number of molecules in question, and z is the coordination number of the chosen lattice.

$$S_{\text{melt}} = k \ln \Omega = Nk[\ln(x/2) + (x-1)\ln[(z-1)/e]] \tag{31}$$

† According to Flory (1941, 1942) the first element of molecule $i+1$ can be placed in $[Nx - ix]$ ways onto the lattice. Nx represents the total number of available lattice sites of a melt (N molecules of x segments each). The previously placed i molecules occupied ix sites. Neglecting differences between the second and subsequent segments, we can place each further segment $[z-1][(Nx-ix)/Nx]$ ways, where the second factor corrects again for the fraction of sites occupied by previously placed molecules. The number of ways ν_{i+1} the molecule $i+1$ can be placed onto the lattice is

$$\nu_{i+1} = [Nx - ix]^x[(z-1)/Nx]^{x-1} \tag{a}$$

and the number of different arrangements of the total melt are

$$\Omega = \prod_{1}^{i=N} \tfrac{1}{2}\nu_i/N! \tag{b}$$

where the factor $\frac{1}{2}$ arises from an assumed identity of molecules placed onto the lattice starting with either end, which cuts the number of arrangements ν in half; $N!$ eliminates arrangements which differ only by exchange of the indistinguishable molecules. Carrying out operation (b) leads us to Eq. (30).

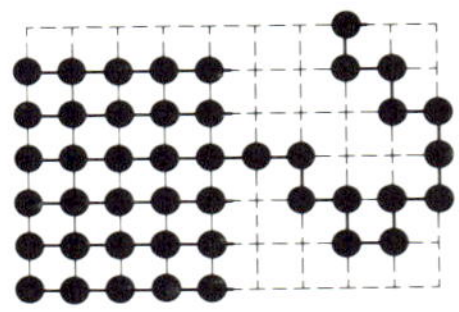

Fig. V.11 *Two-Dimensional Representation of the Finged Micelle Crystalline–Amorphous Interface on a Lattice.*
Only one chain is drawn in the amorphous region for clarity.

The first terms of the right-hand sides of Eqs. (30) and (31) originate from the free choice available for placement of the first segment of the molecule. The second term derives from the choices available for the continuing chain on the lattice ($z - 1$ for each segment) and the restrictions due to all other molecules ($1/e$ for each segment). Forcing the molecules to participate at a fixed location in a nucleus or crystal reduces Ω considerably.† First there is a loss of the factor $(x/2)^N$ due to the fixing of the chains by the nucleus. For a molecule of 2000 segments, this contributes approximately $7kT$ to the surface free energy per chain.

$$\gamma_1' = T\,\Delta s_1 = kT \ln 1000 \approx 7kT \tag{32}$$

This contribution must be divided between the two surface elements of the molecule traversing the nucleus. In addition, the conformations of the partially crystallized molecules are restricted due to the presence of the nucleus or crystal surface which does not permit any chain conformations in which the emerging amorphous chain crosses back over the surface. This conformational entropy effect Δs_k was calculated by Zachmann (1967) as shown in Fig. V.12. For approximately 1000 segments of a molecule emerging from the surface of the nucleus Fig. V.12 indicates a conformational entropy reduction of about $4k$, which leads to the following surface free energy contribution

$$\gamma_2' = T\,\Delta s_k \approx 4kT \tag{33}$$

† Only if all details of the boundary are predetermined is the change in Ω large. A much smaller change in Ω results if only the fraction of crystallized chains is fixed, not the position of the individual crystalline-amorphous boundaries. In this case, conformations must be counted which can only be reached by melting and reformation of the nucleus. For the present calculation of the stability of particular nuclei the first case applies; for the computation of ultimate thermodynamic equilibrium between partially crystalline polymers and the melt, the latter case applies (Zachmann, 1969, pp. 518–520, 526–527).

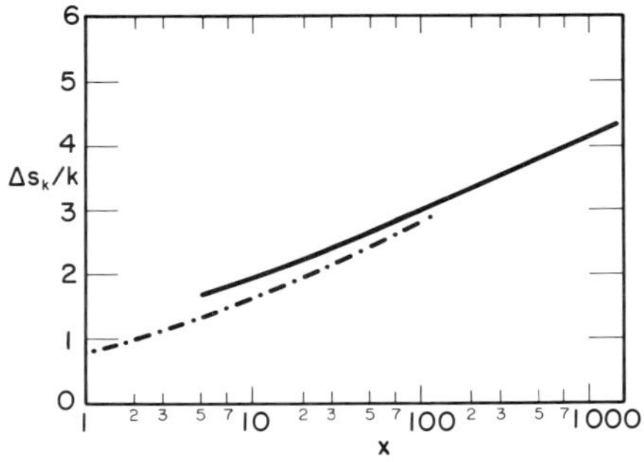

Fig. V.12 *Conformational Entropy Difference between a Molten Molecule aud a Segment of Equal Length Attached on One End to a Crystalline Surface.*
Drawn after Zachmann (1967). Broken curve: exact counting of conformations; solid curve: Gaussian approximation. Coordination number $z = 12$.

A final contribution comes from the fact that a partially crystallized molecule has a somewhat reduced volume for its vibrations. Per molecule this contribution has been estimated to contribute about k to the loss of entropy of a partially crystallized molecule. The corresponding surface free energy contribution is

$$\gamma'_3 \approx kT \tag{34}$$

Collecting all terms and assuming a molecule such as polyethylene which has a cross section of about 18 Å^2 per chain and nucleates at about 400°K, one finds the total contribution from entropy considerations

$$\gamma'_e = [(\gamma'_1/2) + \gamma'_2 + (\gamma'_3/2)]/18 \times 10^{-16} = 245 \text{ erg/cm}^2 \tag{35}$$

In case a molecule of sufficient length should in later stages of nucleation contribute in the formation of a second nucleus, both ends of the remaining amorphous molecular segment are fixed. In this case γ'_e is even larger, Zachmann (1967) estimates that under such conditions γ'_e may reach values up to 700 erg/cm^2 (see also Fig. V.15). A comparison with experimental values of the total surface free energies of folded chain nuclei with the range 50–200 erg/cm^2 for polyethylene (see Sect. 5.2.1.2) reveals the relative instability of fringed micelle type nuclei due to the additional free energy contribution arising from γ'_e.

The calculation of additional free energy based on entropy consideration is, however, only valid if nucleation is slow with respect to the time necessary to reach equilibrium among the different chain conformations. In the case of melt crystallization at temperatures approaching the glass temperature this may not be so, and nucleation may increasingly involve intermolecular nucleation rather than chain folding. Experimental evidence for homogeneous fringed micelle nuclei is scarce. Perhaps the best conditions for fringed micelle nucleation are given in cold crystallization (Dole, 1959), a crystallization carried out as closely as possible to the glass transition temperature.

5.1.2.2 Chain Folded Nuclei

Chain folded nuclei (Fig. V.10, type 2) are, according to the preceding discussion, more probable than fringed micelle nuclei whenever the segmental mobility is high. Particularly, crystallization from dilute solution should favor intramolecular nucleation mechanisms over intermolecular mechanisms. Practically all experiments yielding direct information on homogeneous nucleation from polymer melts or solution have been performed under such conditions of high chain mobility and/or dilution.

A direct proof of intramolecular folded chain nucleation was obtained by Bittiger and Husemann (1966), who crystallized cellulose tricarbanilates from dilute solution. In the case of sufficiently high molecular weight ($1 - 4 \times 10^6$) and dilute solution (0.01 g/liter) folded chain rods resulted which had about 550 Å thickness and lateral dimensions such that only one molecule could be contained per crystal. Nucleation had to be homogeneous and must have involved chain folding. The fold length was independent of molecular weight and corresponded to the thickness of lamellar crystals which could be obtained from lower molecular weight samples (molecular weights $< 5 \times 10^5$, see also Table III.3).

In an investigation of polychlorotrifluoroethylene Price (1952) found that from 190°C down to about 175°C the number of spherulites on crystallization remained constant with time, a possible indication of heterogeneous nucleation (see Sect. 5.1.3.3). This interpretation was further supported by multiple melting and crystallization experiments which showed that nucleation occurred repeatedly at fixed positions in the melt, presumably at the locations of the heterogeneities. Below 175°C the number of nuclei increased with time. At even lower temperature, however, another change in nucleation occurred. Catastrophic nucleation was observed. The resulting crystals were so small that the optical microscope was unable to resolve spherulites. This type of nucleation occurs at 154°C and was interpreted by Hoffman and Weeks (1962) to be due to homogeneous nucleation. The melting point of polychlorotrifluoroethylene is about 224°C. The homogeneous nucleation occurs thus at a supercooling of 70°C or higher. The increase in number of nuclei with time between 154°C and 175°C is most likely due to heterogeneous help of different degree in nucleus formation† (see Sect. 5.1.3.1).

One can recognize thus four distinct temperature regions in the nucleation of polymer melts: Immediately below the melting point there is a 10–30°C region where no crystal growth takes place. Even the addition of heterogeneous nuclei does not initiate large scale crystallization. Next, there is a region of ~20–40°C where crystal growth occurs on heterogeneous nuclei, but no homogeneous nucleation is possible. The number of heterogeneous nuclei remains constant throughout the crystallization experiment. The third region of temperature, of perhaps as much as 10–30°C, is one where in addition to the initial heterogeneous nuclei some smaller or less perfect heterogeneous nuclei can grow in time to critical size. These latter nuclei are also limited in number. Homogeneous nucleation finally becomes possible at a relatively well-defined temperature. The number of new nuclei in this region is usually so large that it does not lead to optically resolvable crystals when observed in bulk samples after completion of crystallization. With perhaps polychlorotrifluoroethylene the only exception, direct optical microscopy is usually only able to give information on the three regions of supercooling.

Another set of experiments to gain information on homogeneous nucleation was initially devised by Vonnegut (1948). He divided the melt into such small droplets that because of their number most could not contain heterogeneous nuclei. Homogeneous nucleation can then be followed by observation of the crystallization of the droplets under the optical microscope.‡ The crystallization rate is at the temperatures of homogeneous nucleation usually so large that a single nucleus crystallizes the whole droplet. Typical data generated by Cormia *et al.* (1962) on polyethylene are presented in Fig. V.13. The four regions of nucleation are clearly delineated. Only about 5% of the droplets solidified above 100°C. They usually represented the larger droplets (25–30 vol. %). More than 50% of the number of droplets crystallized in a critical temperature range between 85 and 87°C. Data on this critical temperature range of several polymers analyzed in the same fashion are collected in Table V.2. All experiments were carried out at about 0.1°C/min cooling rate and with droplet sizes of 1–3 μm diameter. The critical temperature of

† Sharples (1962) found additional support for this interpretation during the study of crystallization of poly(decamethylene terephthalate). He observed a limit in the final number of nuclei in the temperature region where the initial number of nuclei is time dependent. As this limit is reached long before crystallizable melt is exhausted, it must be caused by heterogeneous nuclei which in time develop into critical nuclei.

‡ Typical polymer crystallization experiments are carried out at 1–3 μm diameter droplet size using silicon oil or α-isooctylphenoxy-ω-hydroxypoly(ethylene oxide) [trade name Igepal CA 360] as dispersing agent.

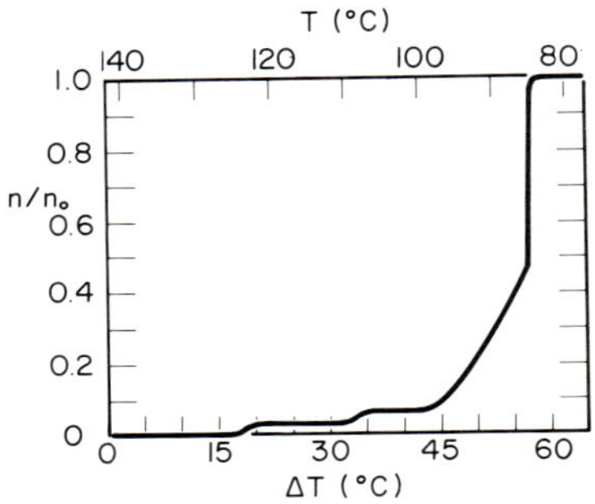

Fig. V.13 *Fraction of Solidified Polyethylene as a Function of Supercooling.* Data by Cormia, *et al.* (1962). Cooling rate 0.17°C/min. Region of no nucleation down to 125°C. Region of heterogenous nucleation down to 100°C. Region of increasing nucleation with help from heterogeneous nuclei perhaps between 100°C and 85°C. Homogeneous nucleation below 85°C.

TABLE V.2

CRITICAL TEMPERATURE OF SOLIDIFICATION IN DROPLET EXPERIMENTS

Polymer	Temp. (°C)	ΔT (°C)	$\gamma_e \gamma^2$ (erg²/cm⁶)	Reference
Polyethylene, linear	86	56	15 500	Cormia *et al.* (1962)
Polyethylene, linear	88	54	9 260	Koutsky *et al.* (1967)
Polyethylene, linear	87[a]	55	11 000	Gornick *et al.* (1967)
Polyethylene, branched	84	58	9 260	Koutsky *et al.* (1967)
Polypropylene, isotact.	79	107	25 800	Burns and Turnbull (1966)
Polypropylene, isotact.	85	101	11 500	Koutsky *et al.* (1967)
Poly-1-butene, isotact.	<0	>140	—	Burns and Turnbull (1968)
Polystyrene, isotact.	138	102	14 800	Koutsky *et al.* (1967)
Poly(ethylene oxide) MW 12 400	0	66	6 580	Koutsky *et al.* (1967)
Poly(ethylene oxide) MW 3×10^5	-4	70	6 580	Koutsky *et al.* (1967)
Polyoxymethylene (Delrin)	111	84	9 820	Koutsky *et al.* (1967)
Poly(3,3-bischloromethyl-oxacyclobutane) (Penton)	101	90	9 770	Koutsky *et al.* (1967)
Nylon 6	128	99	8 080	Koutsky *et al.* (1967)

[a] Estimated from isothermal data.

solidification due to homogeneous nucleation is for all analyzed polymers about 0.8 times the absolute melting temperature. The same relationship between temperature of homogeneous nucleation and melting was found by Jackson (1965) on analysis of droplet crystallization experiments on fifty metals, salts, and small organic molecules.

In order to form one nucleus in about 10 min in a volume of about 10^{-11} cm^3, the nucleation rate must be approximately 10^{10} nuclei/cm^3/min. Koutsky *et al.* (1967) used Eq. (5) to estimate on this basis the product of the surface free energies $\gamma_e \gamma^2$ by assuming in addition the WLF expression for ΔG_η [Eq. (6) and footnote p. 9] and Eq. (21) for ΔG^*. The temperature dependence of Δg_f was taken into account as suggested by Hoffman (see footnote p. 12). The computed values of $\gamma_e \gamma^2$ are listed also in Table V.2.

A more detailed analysis is possible for data from isothermal experiments. The fraction of droplets nucleated is given by

$$n/n_0 = 1 - \exp(-I^* v t) \tag{36}$$

where I^* is the nucleation rate of Eq. (5) and v is the volume of the droplet. Expressed in terms of half-life $t_{1/2}$ of an uncrystallized droplet at the crystallization temperature, Eq. (36) can be written

$$n/n_0 = 1 - \exp(-0.693 t_0/t_{1/2}) \tag{37}$$

The experimental half-life data of Gornick *et al.* (1967) on linear polyethylene can be summarized as follows:

86.28°C	2.34 min	87.40°C	27.49 min	
86.95°C	10.65 min	87.92°C	171.6 min	
87.12°C	14.01 min	88.45°C	488.1 min	(38)
87.24°C	14.94 min	88.76°C	715.0 min	
87.28°C	24.83 min	89.58°C	6710 min	

Regarding all slightly temperature dependent terms as a constant, we can evaluate the free enthalpy of nucleation ΔG^* [Eq. (21)] using Eqs. (36), (37) and (5) (see also Fig. V.4) to be

$$\ln t_{1/2} = \text{const.} + \Delta G^*/kT \tag{39}$$

Figure V.14 shows a plot of $\ln t_{1/2}$ versus $1/T\,\Delta T^2$ which, according to Eqs. (39) and (21), yields a straight line with a slope proportional to $\gamma_e \gamma^2$. With values of $\Delta h_f = 2.9 \times 10^9$ erg/g, $\rho = 1.00$ g/cm^3, $T_m = 415$°K, the product of surface free energies $\gamma_e \gamma^2$ becomes about 11 000 erg^3/cm^6.

This value compares well with the estimates of supercooling necessary for homogeneous nucleation on slow cooling listed in Table V.2. A slightly larger value (14 960 erg^3/cm^6) was derived from the same data by Gornick *et al.* (1967) by taking into account the more detailed temperature change of Δg_f

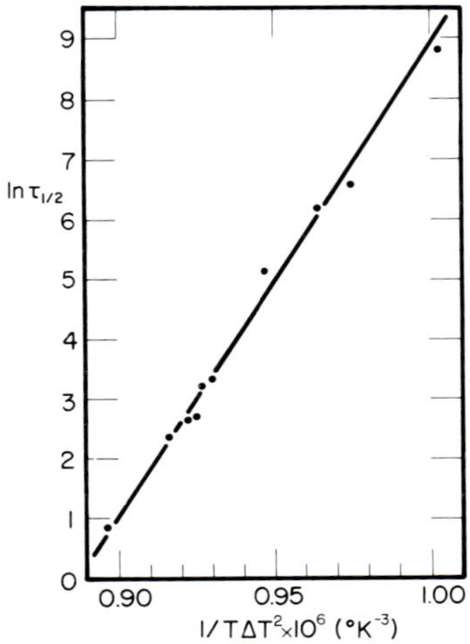

Fig. V.14 *Half-life of an Uncrystallized Polyethylene Droplet as a Function of Supercooling.*
Ordinate: natural logarithm of the half-life defined in Eq. (37); abscissa: $1/T\Delta T^2$ [see Eq. (21)]. From the slope 5.24×10^7 a surface free energy product $\gamma_e \gamma^2 = 11\,000$ erg^3/cm^6 can be computed from the data presented in Eq. (38).

(see footnote p. 12). Hoffman *et al.* (1969) used in another evaluation of these and additional data a somewhat higher T_m and arrived at 18 500 erg^3/cm^6 for $\gamma_e \gamma^2$. The quoted values for $\gamma_e \gamma^2$ by Turnbull *et al.* in Table V.2 were calculated from the measured supercoolings necessary to crystallize half of the droplets at different cooling rates. Cylindrical nucleus shapes were assumed for the calculation.

The data from the droplet experiments contain at least three uncertainties:

(1) Koutsky *et al.* (1967) suggested that except in polyethylene and polypropylene, nucleation might still have been heterogeneously helped by the droplet-dispersing liquid interface. This should be particularly true for the more polar polymers. The droplet experiment data would then represent only an upper temperature limit for homogeneous nucleation and a lower limit for $\gamma_e \gamma^2$.

(2) As is shown in Sect. 5.1.1, the classical expression for ΔG^* should be altered to Eq. (23) or its equivalent with a higher power of the dependence of ΔG^* on γ.

(3) The shape of the nucleus assumed is too simple according to Binsbergen (1970). Typical nuclei have more complicated shapes as is illustrated in Fig. V.9.

In the light of these comments all $\gamma_e \gamma^2$ data can be treated only as estimates. Some possible changes in magnitude of $\gamma_e \gamma^2$ based on nucleus shape are discussed by Burns and Turnbull (1966) who assume detailed nucleation paths.

Taking these uncertainties in the evaluated $\gamma_e \gamma^2$ product into account, one can nevertheless derive reasonable estimates of $\gamma_e \gamma^2$ and the homogeneous nucleus size. Assuming a side surface free energy of about 10 erg/cm^2, an end contribution γ_e between 50 to 200 erg/cm^2 can be calculated for typical polymers. The chosen side surface free energy can be estimated from paraffin data (Turnbull and Cormia, 1961) or via the empirical relation $\gamma \approx 0.1\, \Delta h_f b_0$, where b_0 is the growth plane separation.† The value of γ is most likely to be between 5 and 15 erg/cm^2 for all polymers listed in Table V.2. Such surface free energies coupled with a supercooling of 50–100°C lead to dimensions of a homogeneous polymer nucleus of 75–300 Å in the chain direction and 4–20 Å at right angles to the chain (see Table V.1). Typical separations between chains in crystals of the polymers listed in Tables V.2 are 4–6 Å, so that one homogeneous nucleus may contain about ten chain traverses.

Homogeneous nucleation from dilute solution of polyethylene (0.05% by weight) in toluene and xylene was concluded by Wunderlich and Mehta (1970) to occur at about 32°C supercooling. This estimate was made on the basis of the morphology of the resulting dendrites. Self-nucleation (see Sect. 5.1.4) was excluded by heating above the equilibrium dissolution temperature of the polymer before quenching to the crystallization temperature. The fraction of hedgehog dendrites grown alongside regular dendrites (see Sect. 3.6.2) was shown to reach infinity at the upper temperature of homogeneous nucleation (78°C). Hedgehog dendrites are known to grow on heterogeneous nuclei (see Sect. 5.1.3). The somewhat lower supercooling for homogeneous nucleation from solution may be caused by smaller nucleus dimensions or by a lower surface free energy.

For a folded chain nucleus the entropy contribution of the noncrystallized portions of the molecule to the surface free energy must also be considered. Following again the discussion by Zachmann (1967, 1969, see Sect. 5.1.2.1) one can see that only the one or two chain ends left after formation of the critical nucleus need to be taken into account; the overall effect of the uncrystallized chain ends on ΔG^* is thus much smaller than in an intermolecular fringed micelle-type nucleus.

† See Koutsky *et al.* (1967) for polymer calculations.

Adjacent and regular folding at the nucleus surface requires no extra entropy contribution γ_e' in Eq. (30). The magnitude of γ_e has been estimated under such circumstances to be 60–100 erg/cm^2 by Frank and Tosi (1961) and Petraccone *et al.* (1972) for polyethylene. The larger free energy of the fold surface results from the changed conformations necessary to make a fold, strain energy to achieve folding along the (110) or (200) plane, and a contribution similar to the normal surface free energy of magnitude γ.

Whenever folding at the surface is not regular and adjacent, a loop forms at the surface which will again add a free energy contribution as in the fringed micelle case because of the amorphous molecular chain portion in the loop. Zachmann (1967, 1969) calculated the appropriate entropy contribution as a function of separation h of the fixed exit positions of the chain on the end surface and the number of chain segments x in the loop. Figure V.15 illustrates the increase of the entropy contribution with longer loops and the

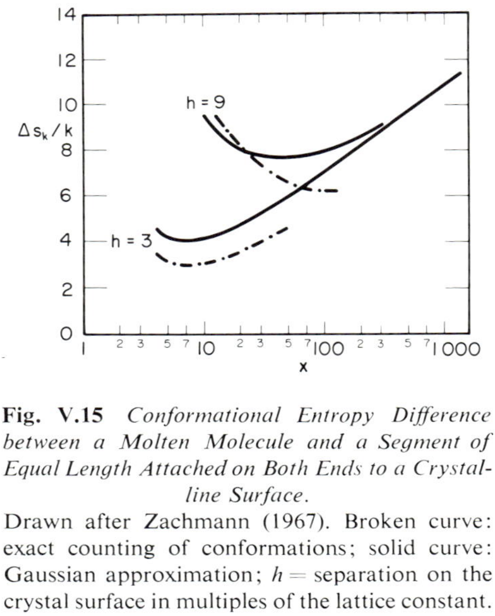

Fig. V.15 *Conformational Entropy Difference between a Molten Molecule and a Segment of Equal Length Attached on Both Ends to a Crystalline Surface.*
Drawn after Zachmann (1967). Broken curve: exact counting of conformations; solid curve: Gaussian approximation; $h =$ separation on the crystal surface in multiples of the lattice constant.

drastic increase with further separation of the endpoints of the noncrystallized chain portion. The limits of $\gamma_e + \gamma_e'$ established by the droplet experiment may accommodate some irregular folding, but not an excessive amount.

A homogeneous nucleus of a flexible linear macromolecule can thus be described as part of a molecule folded to a packet of perhaps 5–10 relatively regular traverses with a length of $\sim$150 Å.

5.1.2.3 Oligomer Nuclei

Oligomer nucleation is the intermolecular path on crystallizing during polymerization. The nucleus is schematically shown in Fig. V.10 as type 3. It is the analog to the fringed micelle nucleus of nucleation of an already polymerized molecule. The difficulties of fringed micelle nucleation discussed in Sect. 5.1.2.1 which are caused by the mobility and conformation restrictions of the noncrystallized portion of the polymer molecule are not existent for the oligomer nucleus if polymerization can occur directly at the nucleus surface (simultaneous polymerization and crystallization). The major problem of the oligomer nucleation lies in the initial formation of a sufficient oligomer concentration for nucleation. Since the monomer is normally much shorter than the critical nucleus dimensions listed in Table V.1, a certain oligomer concentration must first be polymerized by a homogeneous reaction step before nucleation can occur and crystallization during polymerization can commence. In analogy to standard polymerization kinetics, this initial homogeneous reaction may be a step or an addition reaction (see Sect. 1.2).

An initial homogeneous polymerization by step reaction kinetics leads to a relatively high oligomer concentration before higher molecular weights are formed. The weight fraction of oligomers of x repeating units as a function of degree of completion of the reaction p is shown in Fig. V.16. In case an oligomer of length $x = 10$ is needed for nucleation, a sufficient concentration is expected only after the polymerization reaction is about half completed.

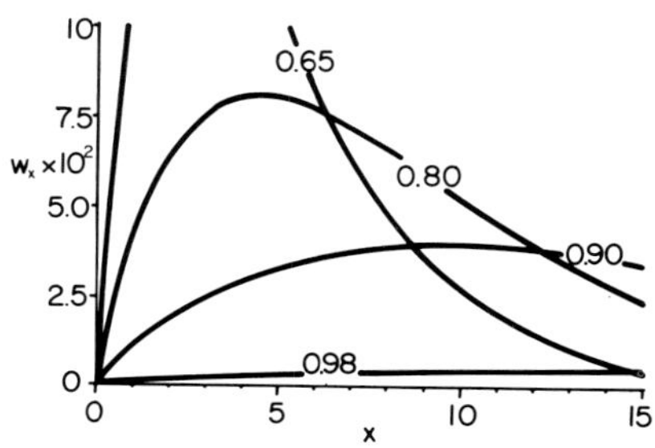

Fig. V.16 *Molecular Weight Distribution as a Function of Completion of Polymerization during a Homogeneous Step Reaction.* The weight fraction of molecules of length x is w_x. The completion of polymerization is expressed by the parameters p written next to each curve (0.65, 0.80, 0.90 and 0.98). Mathematically the curves are represented by $w_x = x(1-p)^2 p^{x-1}$ (see also Sect. 1.3).

After initial nucleation, the reaction mechanism must be able to change from a step reaction to an addition reaction at the active surface of the formed nuclei. This change in mechanism results in a change in the shape of the molecular weight distribution curve after nucleation (Wunderlich, 1968). The amorphous portion will not increase much further in molecular weight because of nucleation of further crystals or molecules on existing crystal surfaces. The molecular weight of the crystals will, in turn, not follow the increase dictated by Fig. V.16, rather it will be determined by crystal size and perfection.

An initial homogeneous polymerization by an addition reaction needs no change in type of mechanism after nucleation; the rate of reaction at the surface may, however, be largely different from the rate in solution. Another characteristic of the addition reactions is the frequent need to initiate the polymerization. After initiation, the growth of the molecule continues by monomer addition. The total number of growing molecules and intermediate oligomers for nucleation is thus fixed by the initiation conditions. The molecular weight distribution in the homogeneous phase is dependent on the initiation mechanism and the rate of removal of oligomers or polymers by nucleation and crystallization, while the final molecular weights of the crystals are also a function of the termination mechanisms which include the possibility of occlusion of the active chain end in the interior of the crystal or the blocking of further growth because of crystal morphology (see Fig. V.20).

Two examples of crystallization during polymerization with initial homogeneous polymerization by step reaction kinetics which have been analyzed in sufficient detail to draw conclusions about the primary nucleus size in the chain direction are the formation of polyphosphates and the formaldehyde polymerization (Wunderlich, 1968). Some additional information on other systems are given in connection with the description of crystallization during polymerization in Sect. 6.4. The polyanion of a metaphosphate can be represented by the formula

$$\mathrm{HO}(\overset{\displaystyle \mathrm{O} \atop \displaystyle \|}{\underset{\displaystyle | \atop \displaystyle \mathrm{O}^-}{\mathrm{P}}}\!-\!\mathrm{O}\!-\!)_x\mathrm{H} \qquad \text{or} \qquad (\mathrm{PO_3})_x^{x-} \cdot \mathrm{H_2O} \tag{40}$$

Its chain conformations in the crystalline state were discussed in Sect. 2.3.5. The formation of $(LiPO_3)_x \cdot H_2O$ by dehydration of LiH_2PO_4 is typical for all alkali metaphosphates and metaphosphoric acid. Thilo and Grunze (1955) have studied the polymerization by stepwise heating of LiH_2PO_4. The species present in the reaction mixtures were analyzed by paper chromatography. The reactions above 145°C can be represented schematically as follows

$$
\mathrm{LiH_2PO_4} \xrightarrow{>145^\circ\mathrm{C}}
\begin{array}{c}
\mathrm{Li_2H_2P_2O_7} \\ \text{(crystalline)} \\ \\ \mathrm{(LiPO_3)_x\,H_2O} \\ x = 2 \text{ to } 9 \\ \text{(amorphous)}
\end{array}
\xrightarrow{>210^\circ\mathrm{C}}
\begin{array}{c}
\mathrm{(LiPO_3)_x\,H_2O(l)} \\ \text{(crystalline polymer)} \\ \text{and} \\ \mathrm{(LiPO_3)_x\,H_2O} \\ \text{(amorphous oligomer)}
\end{array}
$$

$$
\xrightarrow{>350^\circ\mathrm{C}}
\begin{array}{c}
\mathrm{(LiPO_3)_x\,H_2O(h)} \\ \text{(crystalline polymer)}
\end{array}
\qquad (41)
$$

As soon as a little initial diphosphate is observed, amorphous oligophosphates are also found. The amount and chain length of the oligomers increases with temperature. At 170°C oligophosphates up to the nonaphosphate were isolated. Above 210°C, at which time 0.65 mole of H_2O have been lost per mole of LiH_2PO_4, the low temperature form of the linear high polymer $(LiPO_3)_x \cdot H_2O(l)$ crystallizes. It is likely that the nucleus of the crystalline polymer is thus formed by oligomers equal to or only little larger than $x = 9$. Considering that the melting point of the high temperature form of lithium metaphosphate is 630°C, this leads to a reasonable primary nucleus length. Analysis of the remaining amorphous material after initial nucleation revealed only traces of tri- and tetraphosphate at 270°C and no higher oligomer phosphates although 0.98 moles of H_2O were lost at this temperature per mole of LiH_2PO_4. It was shown by van Wazer and Callis (1962) that the molecular weight of polyphosphates is determined by the size of microcrystallites and internal crystal perfection. Malmgreen and Lamm (1944) have reported molecular weights up to several million for potassium metaphosphate which depended on time and temperature of reaction. The molecular weight was found to be quite uniform, increasing in uniformity with time of polymerization (annealing), quite in contrast to the distributions expected for homogeneous reactions (see Fig. V.16). The distribution predicted from Fig. V.16 could be observed as expected when polymerization was carried out entirely in the liquid phase at elevated temperature (Winkler and Thilo, 1959). Variations of this polymerization reaction are exhibited by potassium and sodium salts because of topochemical reasons and because of the greater stability of crystalline sodium diphosphate and sodium trimetaphosphate. The formation of the linear polyphosphate chain is thus a well-studied example of crystallization during polymerization and can be used as a model for similar organic and biological reactions. Similar inorganic polymerizations by dehydration involving As, Si, and possibly Ge have received much less attention, but seem to follow similar patterns of step reactions in the amorphous phase and crystallization during polymerization after nucleation of the crystalline polymer phase.

An analogous nucleation mechanism was proposed by Brown (1967) for the crystallization of polyoxymethylene from formaldehyde in aqueous or alcohol solution. The monomer in these solvents is present mainly as gem-diol or hemiformal. Since in this case the monomer double bond is replaced by two single bonds, the major driving force of polymerization ($\Delta G = -3.7$ kcal/mole) is reduced to little more than the driving force for crystallization ($\Delta G = -0.4$ kcal/mole). Brown (1967) describes the initial stages of the reaction as a step reaction

$$R{-}O{-}(CH_2{-}O{-})_x H + R{-}O{-}(CH_2{-}O{-})_y H \rightleftharpoons R{-}O{-}(CH_2{-}O{-})_{x+y} H + ROH$$

R=H or alkyl

As soon as a chain length of ten or slightly more monomer units is reached, spontaneous nucleation of solid crystalline polymer takes place following the type 3 path of Fig. V.10. As is to be expected, once the solid phase is present, further growth takes place on the crystal surface by the addition of a monomer [$y = 1$ in Eq. (42)]. The reaction probably follows a simultaneous polymerization and crystalline path. All polymer produced in this way is 100% crystalline. At a given temperature there is an equilibrium concentration of formaldehyde below which polyoxymethylene dissolves by depolymerization and above which formaldehyde adds to present crystals. Moreover, polyoxymethylene will precipitate spontaneously when a clear solution of formaldehyde reaches approximately 4–8 wt.% above the equilibrium concentration. Many of the polymer samples produced from aqueous formaldehyde are of low molecular weight because of inadequate control of nucleation. In particular, many occluded endgroups were discovered. By judicious choice of buffered systems, addition of inert diluent, and seeding, it was possible to reduce side reactions and occluded endgroups such that a reasonably high molecular weight product (38 000) could be made with about 80% of endgroups on the crystal surfaces (Brown, 1967).

Examples for initial homogeneous polymerization by an addition reaction are usually less well understood. The addition reactions have usually a much lower intermediate oligomer concentration for nucleation so that excessively long oligomers may form. In addition, polymerization and crystallization rates are frequently not matched leading to successive polymerization and crystallization. The resulting nuclei are then closer to type 1 fringed micellar nuclei. As the distance between polymerization site and crystallization site increases, the extra free energy contribution γ_e' to the ΔG^* of the nucleus [Eq. (29)] increases. Typical examples for the complications for nucleation caused by amorphous chain ends emerging from the nucleus surface are offered by the vinyl polymerizations involving Ziegler–Natta catalysts and the diazomethane polymerization catalyzed by boron trifluoride etherate (see also Sect. 6.4.2.1).

A Ziegler–Natta catalyst usually consists of the reaction product of an organometallic compound (like triethyl aluminum) and a transition metal compound (like titanium trichloride)† slurried in a dry, oxygen free, inert hydrocarbon solvent (like heptane) along with the vinyl monomer to be polymerized. The reaction can be written schematically as an insertion mechanism at the catalyst site:

$$\text{Catalyst}{-}CH_2{-}\underset{\displaystyle R}{\underset{|}{C}}HR' + xCH_2{=}CHR \longrightarrow \text{Catalyst}{-}(CH_2{-}\underset{\displaystyle R}{\underset{|}{C}}H{-})_xCH_2{-}\underset{\displaystyle R}{\underset{|}{C}}HR' \tag{43}$$

An insertion mechanism was also proposed for the diazomethane polymerization with boron trifluoride etherate [see Eqs. (120)–(123), Sect. 6.4.2.1].

Bulkiness of the catalyst in homogeneous catalysis (dissolved catalyst) or sparsity of reaction sites in the case of heterogeneous catalysis (solid catalyst) may preclude simultaneous polymerization and crystallization. Crystallization must then be successive or separate from polymerization (see Sect. 6.1.8).

The first portion of the polymer chain produced is in both cases amorphous because of oligomer solubility. Nucleation at a later stage must then be determined by quite unconventional parameters like the concentration and clustering of catalytic sites, the solubility and crystallization temperature of attached oligomer and polymer chains, and the reaction rate. For different conditions, analysis of the final crystalline macromolecules has yielded various morphologies that range from lamellar folded crystals to extended fibrillar crystals. Accordingly, nucleation is expected to vary greatly. One can imagine that nucleation at low catalytic site concentration and fast reaction rate may be retarded until the molecule is detached from the catalyst, resulting in standard type 2 folded chain nuclei. At high concentration of active sites and suitable reaction rate, nucleation may be of the type 3 oligomer nuclei, leading to an initially fibrillar nucleus with restricted crystal growth normal to the chain axis since new chains which are still attached to the active site are limited in number. As the chain length between catalyst site and nucleation site increases the type 3 nucleation becomes increasingly similar to type 1 fringed micellar nucleation. Finally, one can imagine type 4 nucleation of a still growing polymer chain which is long enough to form a folded chain intramolecular nucleus.

5.1.2.4 *Folded Chain Nuclei of Polymerizing Macromolecules*

The last type of nucleation shown in Fig. V.10 involves the formation of an intramolecular folded chain nucleus similar to the type 2 folded chain nucleus found frequently on crystallization of completed molecules. The difference of the present nucleus type lies in the fact that the molecules still polymerize while undergoing nucleation and crystal growth. Further growth of the nucleus may occur as in type 3 nucleation (Sect. 5.1.2.3) through simultaneous or successive polymerization and crystallization. The restrictions of

† See, for example, Reich and Schindler (1966).

crystal growth because of continued chain folding after nucleation discussed in Sect. 6.1.4 may be avoided after such nucleation if polymerization and crystallization sites are in close proximity. This type of nucleation is at present the least explored so that discussion of this nucleation mechanism is quite speculative. Since the folded chain nucleation carries the same increased surface free energy due to folding (see Sect. 5.1.2.2), one would expect type 4 nucleation only if the intermolecular path is not possible or the chain length between polymerization and crystallization sites is large so that type 3 nucleation would develop excessively large values of γ_e' (see Sect. 5.1.2.3).

As an intramolecular nucleation, type 4 nuclei are favored by low concentration of growing chains, a condition usually encountered in addition reactions (see Sect. 1.2). Particularly in free radical polymerizations, the concentration of growing chains must be kept low because of the termination reaction by the encounter of two free radicals. Typical levels of free radical concentrations in solution and melt crystallization are 10^{-8}–10^{-9} mole/liter. In this case, oligomer nucleation would be quite unthinkable.

A particularly interesting example is the dispersion polymerization of tetrafluoroethylene (Sperati and Starkweather, 1961). The polymerization is usually carried out at elevated pressures and in the presence of water. Persulfate, hydrogen peroxide, and redox initiators have been used. The resulting product, which is insoluble in water, is recovered as a granular powder or as a dispersion. The molecular weight is of the order of magnitude 10^6–10^8. The crystallinity is as high as 93–97%. Single dispersion particles give rise to discrete electron diffraction patterns. They have a diameter of 2×10^{-5} cm, or a weight of $\sim 2 \times 10^{-14}$ g. One molecule would weigh 10^{-18}–10^{-16} g, so that 200–20 000 molecules make up a single particle. The granular material seems to be an aggregate of many smaller particles. On the basis of this structural evidence, a nucleation mechanism has been proposed which involves a chain folded, partially polymerized molecule (Wunderlich, 1968). After nucleation, polymerization is most likely to occur close to the crystal surface as indicated schematically in Fig. V.10. Of interest for such a polymerization is the similarity to an emulsion polymerization process (see, for example, Lenz, 1967). Once a single molecule is nucleated, it may grow until another free radical in the form of an initiator or oligomer is adsorbed on its surface and terminates further growth by combination. A stable nucleus or crystal is, however, a preferred site for addition of another radical or oligomer which can restart the polymerization and crystallization process by epitaxial crystallization of the oligomer on the polymer crystal. Growth will continue until a fourth free radical terminates the second molecule. Through continued periods of growth and termination the final dispersion particle is produced. Since crystallization occurs from the monomer, there is no *a priori* reason for chain folding after nucleation, and indeed, the final morphology seems to be of an extended chain type as is discussed in Sect. 3.3.1.

5.1.2.5 Crystal Morphology and Nucleus Type

Several conclusions about the likely morphology of the resulting crystal can be drawn from the discussion of the different nucleation paths. A fringed micelle nucleus is expected to be elongated in the molecular chain direction due to the relatively large surface free energy ($\gamma_e + \gamma_e'$) at the end surface. Since its formation is most likely to be restricted to rather immobile molecules, the final crystal size and perfection should be low. The active growth face for further crystallization is the surface that is traversed by the molecules; it promotes a fiber-like morphology.

The folded chain nucleus has its active growth face parallel to the molecular chain direction. The limitation to the fold length set by the primary nucleation rate is changed for the subsequent growth (see Sect. 6.1.4), but the final lamellar morphology has a fold-length not much different to that of the primary nucleus l^*.

The oligomer nucleus should have an isometric shape. It has active polymerization and growth sites only at the chain ends. In the case where initiation of additional molecules on the nucleated crystal is inhibited a fibrillar morphology is possible. As the nucleus becomes more like the fringed micelle type (due to successive crystallization and polymerization), the tendency to fibrillar crystals growing from the nucleus must be additionally enhanced due to hindrance at the growth faces.

Finally, the folded chain nucleus of a continually polymerizing molecule may not have the same restrictions as a crystal grown after polymerization. The nucleus may thicken after its initial stages of growth, and further polymerization may lead to relatively isometric crystals. Lamellar morphologies are possible only when polymerization and crystallization sites are sufficiently separated. Some of these conclusions about crystal morphology have been applied to predict the primary nucleus type.

The types of nucleation just described are of necessity limiting cases and intermediate examples would be expected under appropriate conditions. One expects that most cases of fast crystallization of melts found in industrial processes are of the folded chain type with varying degrees of fringed micellar complications.

5.1.3 Heterogeneous Nucleation

A heterogeneous nucleation path makes use of foreign preexisting surfaces to reduce the free energy opposing primary nucleation. Following the classical nucleation theory outlined in Sect. 5.1.1, the critical nucleus is smaller if there is a reduction in free energy when the nucleus contacts a preexisting surface [see Eqs. (11)–(20)]. A lower overall free enthalpy of nucleation ΔG^* would then lead to faster nucleation rates. The droplet experiments described in Sect. 5.1.2.2 (see Table V.2 and Fig. V.13) indicate degrees of supercooling of 60°C to more than 100°C for homogeneous nucleation at about 0.1°C/min cooling rates. In contrast, typical crystallization experiments on macromolecular melts and solutions usually show only 10–30°C supercooling before crystallization. This suggests that most crystallization of macromolecules is initiated by heterogeneous nucleation.

Heterogenous nucleation ideas can be applied to all four nucleation paths outlined in Sect. 5.1.2 (Fig. V.10). The condition of low polymer chain mobility for fringed micelle type nucleation may, however, hinder a large effect of heterogeneous surfaces on crystallization because of the limited additional growth of each nucleus. Heterogeneous nucleation on crystallization during polymerization has been little analyzed to date. Most of the discussion in this section will thus deal with heterogeneous nucleation of polymer melts and solutions crystallizing to a chain folded macroconformation. First, general concepts of the classical nucleation theory described in Sect. 5.1.1 will be applied to heterogeneous nucleation. This is followed by a discussion of experimental data on nucleation by different surfaces. Finally, self-nucleation by nuclei out of the crystallizing material persisting in the melt and in solution will be treated in a separate section (Sect. 5.1.4). Overall, heterogeneous nucleation is still a process where much of the specific mechanism is unknown.

5.1.3.1 Heterogeneous Nucleation Concepts

A general discussion of heterogeneous nucleation was given by Binsbergen (1969, 1973). His review is largely followed here. Similar to Eq. (11), a rectangular heterogeneous nucleus of thickness b at right angles to a preexisting flat surface would show a free enthalpy of nucleation of

$$\Delta G_i = -abl\Delta g_f + al\Delta\gamma + 2bl\gamma + 2ab\gamma_e \tag{44}$$

where $\Delta\gamma$ is the specific interfacial free energy difference parameter accounting for one surface contacting the melt (of surface free energy $al\gamma$) and one surface contacting the heterogeneous nucleus [see Eq. (51)]. The other quantities are defined as in Eq. (11). Heterogeneous nucleation is preferred as long as $\Delta\gamma$ is less than 2γ.

The dimensions of the critical nucleus can be derived analogously to Eqs. (12)–(15). Dimensions a^* and l^* are again as given in Eqs. (19) and (20), while

$$b^* = 2\Delta\gamma/\Delta g_f = 2\Delta\gamma T_m/(\Delta h_f \Delta T \rho_c) \tag{45}$$

The free enthalpy of formation of the heterogeneous nucleus should be accordingly

$$\Delta G^* = 16\Delta\gamma\gamma\gamma_e T_m^2/(\Delta h_f \Delta T \rho_c)^2 \tag{46}$$

Thus ΔG^* remains proportional to $1/\Delta T^2$, the inverse square of the supercooling as in the case of homogeneous nucleation represented by Eq. (21).

For a sufficiently large ΔT, the dimension b^* approaches the molecular thickness b_0. In this case Eq. (44) changes to

$$\Delta G_i = -al(b_0 \Delta g_f - \Delta\gamma) + 2b_0(l\gamma + a\gamma_e) \tag{47}$$

The free enthalpy of formation of the critical nucleus is no longer proportional to $1/\Delta T^2$, but can be written

$$\Delta G^* = 4b_0 \gamma\gamma_e T_m/[\Delta h_f \Delta T \rho_c - (\Delta\gamma T_m/b_0)] \tag{48}$$

As $\Delta\gamma$ goes to zero, the substrate resembles in its action the polymer crystal itself and Eqs. (47) and (48) refer to secondary nucleation [see Sect. 5.2.2, Eq. (68)].

Heterogeneous nucleation on steps, cracks, holes, or other surface irregularities would be more advantageous than heterogeneous nucleation on a flat surface.† On such a rough surface the crystal nucleus could make use of a larger portion of the surface. Assuming an unlimited ledge on the substrate along direction l onto which the nucleus may form leads by the previous reasoning to

$$a^* = 2\Delta\gamma/\Delta g_f = 2\Delta\gamma T_m/(\Delta h_f \Delta T \rho_c) \tag{49}$$

$$\Delta G^* = 8(\Delta\gamma)^2 \gamma_e T_m^2/(\Delta h_f \Delta T \rho_c)^2 \tag{50}$$

[$4al\gamma$ in Eq. (11) is replaced by $2al\Delta\gamma$; again for a quandratic cross section, $a = b$.] At small supercooling (or large values for $\Delta\gamma$) the free energy of formation of the critical nucleus remains, as before, proportional to $1/\Delta T^2$. As supercooling increases, both dimensions a^* become monomolecular dimensions. In this case only a negligible dependence of heterogeneous nucleation on supercooling is left. If the ledge is limited in length, major enhancement of nucleation begins at the supercooling necessary to make l^* equal to the length of the ledge. At sufficiently higher supercooling, nucleation may again become supercooling independent.

The specific interfacial free energy difference parameter $\Delta\gamma$ may be written

$$\Delta\gamma = \gamma_{cm} + \gamma_{cs} - \gamma_{ms} \tag{51}$$

where γ_{cm} is the specific side surface free energy γ of the crystal as defined earlier, γ_{cs} is the specific crystal-substrate interfacial free energy, and γ_{ms} the specific melt–substrate surface free energy. Since most of the interactions of

† The general expressions were derived by Lacmann (1961) and discussed for crystallization of macromolecules by Binsbergen (1969, 1973) and Devoy and Mandelkern (1970).

the crystallizing macromolecules are due to dispersion forces (see Sect. 2.3), the interfacial free energies may be approximated according to Fowkes (1964) by

$$\gamma_{cm} = (\gamma_c^{1/2} - \gamma_m^{1/2})^2 + {}_o\gamma_c + {}_o\gamma_m \tag{52}$$

and analogous expressions for γ_{cs} and γ_{ms}. The terms γ_c, γ_m, and γ_s represent the contribution of the dispersion forces to the surface free energies of the phases indicated by the subscript. The ${}_o\gamma$ terms correct for other contributions. Substitution in Eq. (51) gives us

$$\Delta\gamma/2 = (\gamma_c^{1/2} - \gamma_s^{1/2})(\gamma_c^{1/2} - \gamma_m^{1/2}) + {}_o\gamma_c \tag{53}$$

Binsbergen (1969) introduced in addition an accommodation coefficient ζ $(0 < \zeta < 1)$ accounting for reduction of interfacial free energy by increased epitaxy (see Sect. 3.4) through crystal lattice parameter matching and other effects which leads to

$$\Delta\gamma/2 = (\gamma_c^{1/2} - \gamma_s^{1/2})(\gamma_c^{1/2} - \gamma_m^{1/2}) + (1 - \zeta)_o\gamma_c\,. \tag{54}$$

The two substrate dependent parameters of Eq. (54) that can enhance nucleation through reduction of $\Delta\gamma$ are γ_s and ζ. An effect is expected when γ_s, the dispersion contribution to surface free energy, is substantially larger than γ_c. In the case of epitaxy a large value of ζ can reduce $\Delta\gamma$.

Besides the effects of preexisting surfaces through their geometry and chemical structure on nucleation from molten or dissolved molecules, it was shown by Turnbull (1950) that for the nucleation of metals there may be a survival of crystallized material in cavities of a heterogeneous surface beyond the melting temperature. On cooling through the melting temperature these surviving small crystalline inclusions in the heterogeneous surface can act as nuclei for further growth. The size of the cavity determines the degree of supercooling necessary for growth, and also the degree of heating possible beyond the melting point before destruction. Figure V.17 shows schematically

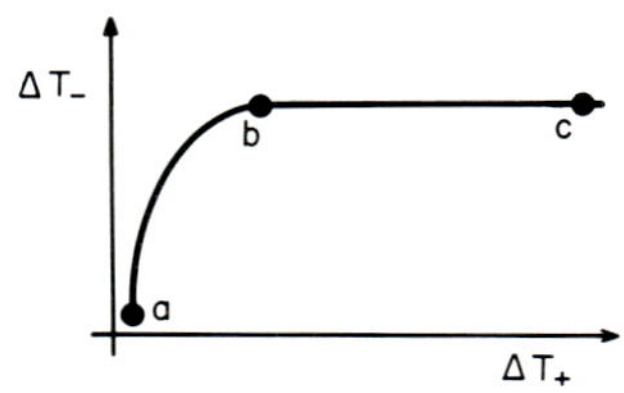

Fig. V.17 *Schematic Diagram of the Effect of Heating above the Melting Temperature ΔT_+ on the Supercooling ΔT_- Necessary to Achieve Nucleation in a Given Time.*

the dependence of supercooling below the melting temperature ΔT_- which is necessary for nucleation to occur in a given time as a function of the superheating above the melting temperature ΔT_+ prior to nucleation. For gallium and similar metals, increased nucleation was observed for ΔT_+ up to about 100°C when the heterogeneous surfaces were charcoal or powdered quartz (region *a–b*). Beyond this limit (region *b–c*) constant ΔT_- for nucleation was observed up to $\Delta T_+ = 1000$°C. Similar effects for linear macromolecules are described under the heading self-nucleation (see in particular Sect. 5.1.4.4).

5.1.3.2 Heterogeneous Nucleation from Solution

Crystallization from dilute solution yields well separated crystals so that morphological evidence for heterogeneous nucleation can be sought. The usual crystal morphology for small supercooling is the folded chain lamella (Sect. 3.3.2) and, for larger supercooling, the folded chain dendrite (Sect. 3.6). Fold lengths for both morphologies are normally between 50 and 200 Å (see Table III.3). Heterogeneous nuclei are usually larger in size and should be easily recognizable.

Figures V.18(a) and (c) show some typical crystal lamellae and dendrites of polyethylene grown from dilute solution. No evidence of foreign particles can be seen in these crystals (see also Figs. III.42–III.51). The general perfection of the whole crystal also requires the absence of a foreign particle in the center. On a substrate the crystal habit cannot develop fully because of the interference of the substrate surface. The crystals of Fig. V.18(a) and (c) must thus either be homogeneously nucleated or are self-nucleated (see Sect. 5.1.4). Most crystallizations from solution contain in addition some single crystal clusters or hedgehog dendrites. Figures V.18(b) and (d) illustrate examples of these single crystal clusters and dendrites, grown at somewhat higher concentration than the single crystals and dendrites in Figs. V.18(a) and (c). It was suggested by Wunderlich and Sullivan (1962) that the hedgehog dendrites are heterogeneously nucleated, in a similar way as some inorganic hedgehog dendrites. The separate parts of the single crystal cluster as well as the branches of the hedgehog dendrite are thought to nucleate independently of each other on the foreign particle surface which is large enough to nucleate many 100 Å thick lamellae. Wunderlich and Mehta (1970) showed by seeding with fluorescent nuclei that, indeed, fluorescence could be detected in the center of hedgehog dendrites, but never in the center of regular dendrites. Alfonso *et al.* (1973a) demonstrated that some pigments† are heterogeneous nuclei for 0.01% solution of polyethylene in xylene so that crystallization occurs preferentially on the pigment particles rather than on also present self-seeds (see Sect. 5.1.4.1). On melt crystallization, active solution heterogeneous nuclei

† Chromophthal GF Green and Chromophthal 8594 Yellow.

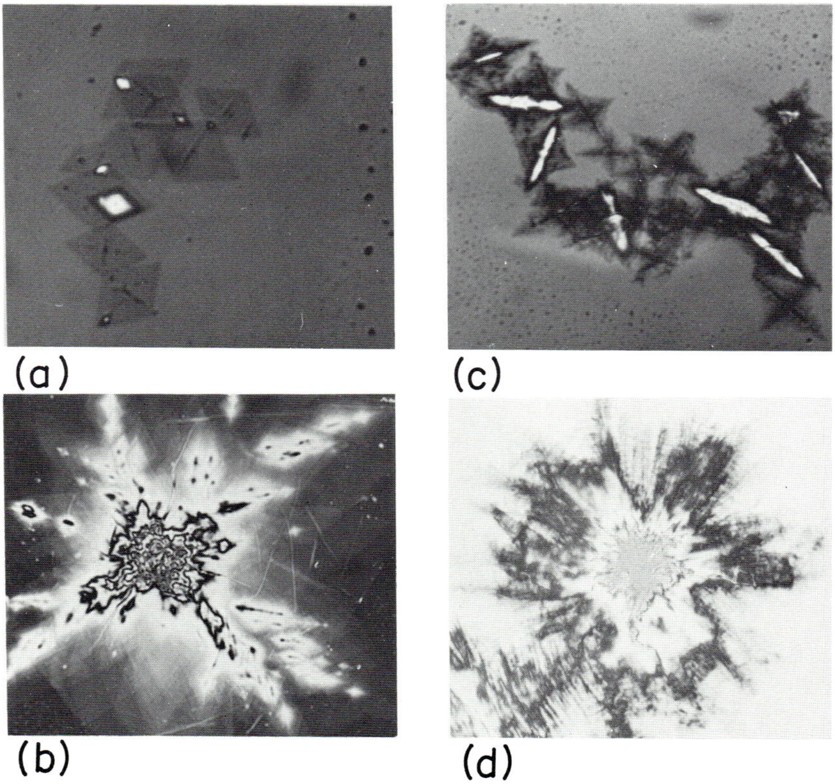

Fig. V.18 *Homogeneously and Heterogeneously Nucleated Polyethylene Single Crystals and Dendrites.*
(a) Single crystals grown isothermally at 91.2°C from a 0.04% solution in toluene. (b) Single crystal cluster grown onto a heterogeneous nucleus grown isothermally at 89.7°C from an 0.1% solution in *o*-chlorobenzene. (c) Dendrites grown on cooling from 100°C under a room temperature gradient from 0.006% solution in toluene. (d) Hedgehog dendrite grown onto a heterogeneous nucleus grown on cooling from 100°C under a room temperature gradient from 0.054% solution in toluene. Optical interference micrographs; (a), (c), (d), Wunderlich and Sullivan (1962); (b), Wunderlich, unpublished. Magnification about 500×.

enhanced the crystallization rate, while pigments inactive in solution crystallization showed also no effect on the melt crystallization rate.

Crystallization of polyamide solutions in the presence of large amounts of foreign material was shown by Yakhnin *et al.* (1969) to cause all crystallization to occur on foreign surfaces as long as the surfaces are hydrophilic (quartz, asbestos, barium sulfate, and oxides of zinc, iron, and chromium). Figure V.19 illustrates the change in crystallization on addition of quartz particles

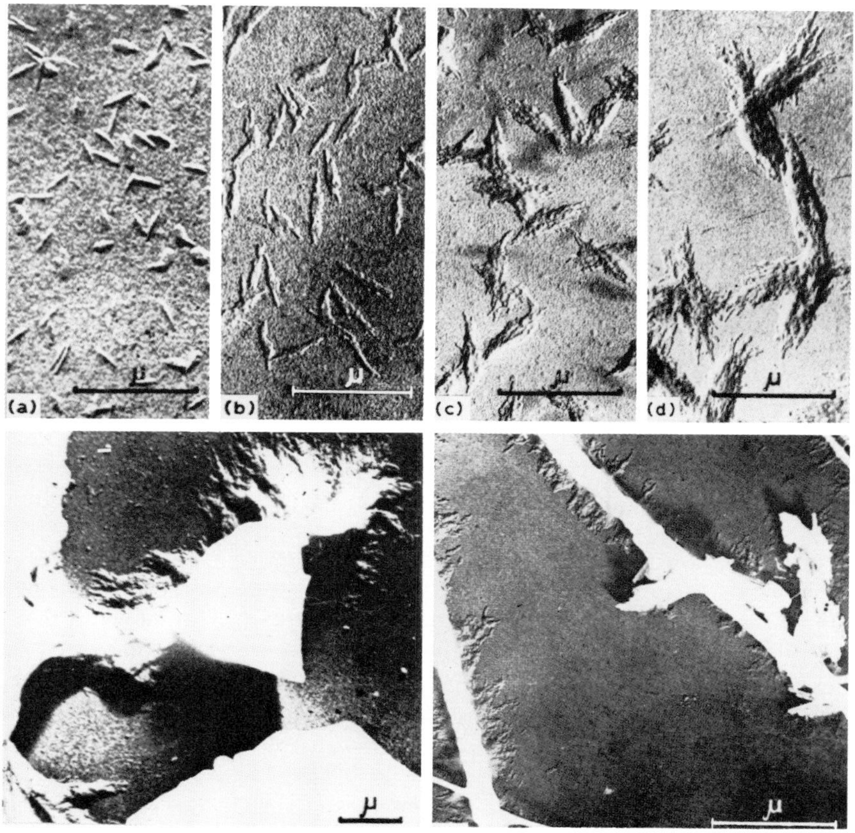

Fig. V.19 *Electron Micrographs of Homogeneously and Heterogeneously Nucleated Nylon.* (a)–(d): Nylon 6,6.6 copolymer crystallized for different lengths of time from 2% solution in ethyl alcohol without added nuclei at 30°C. Bottom micrographs: crystal growth from suspensions containing in addition about 20% by volume of quatrz particles and asbestos fibers, respectively. Practically all polymer deposits on the foreign surfaces. (Yakhnin *et al.*, 1969).

and asbestos fibers. The influence of the chemical nature of the foreign surface could be demonstrated by changing the hydrophilic quartz surface by coating with cetyldimethylbenzyl ammonium chloride, or Fe_2O_3 and $BaSO_4$ surfaces with stearic acid. The suspensions of the coated particles permitted considerable crystallization away from the particles, while the untreated particles did not.

The amount of inadvertent nuclei in solutions of macromolecules was shown by Lanceley and Sharples (1966) to be small. Little difference was found in the crystallization kinetics where solutions were first filtered. Similarly, Johnsen and Lehmann (1969) observed no effect on the kinetics when solutions were cleaned by initial partial crystallization and centrifugation which should remove heterogeneous nuclei. Together with the frequent morphological observation of well developed single crystals or dendrites these experiments indicate that homogeneous nucleation or self-nucleation are the normal nucleation modes from solution.

Some information about the early stages of crystal growth on foreign surfaces derives from the study of epitaxy (see Sect. 3.4). Carr *et al.* (1970) could show that initial crystal growth of polyethylene and polyoxymethylene (nucleation) occurred on NaCl surfaces at temperatures where further crystal growth was not possible. Polyethylene dissolved in xylene grows into larger crystals only below 97°C. Epitaxy of lamellae of a few hundred angstroms depth was observed, however, up to 109°C (Fig. III.81). At the higher temperatures increasingly larger molecular weights were found segregated in these crystal lamellae. The special nature of the foreign surface must have permitted such limited crystallization at a temperature where further growth does not occur because of limits imposed by molecular nucleation (see Sect. 5.3). The possible mechanism of the initial addition of polyethylene on NaCl has been discussed by Wellinghoff *et al.* (1974). After adsorption on the substrate, the CH_2-chains are aligned preferentially along the rows of sodium ions. A lower potential energy is expected whenever the permanent CH_2-dipole is perpendicular or parallel to the surface. The perpendicular arrangement is suitable for close side-by-side packing as in the triclinic polyethylene structure described in Fig. II.34 (t_T in the substrate surface, c along [110] of the NaCl). Only at some distance from the surface does the more stable orthorhombic structure develop which has the (110) plane parallel to the substrate. The crystal structure of the salt alone can, however, not be the sole basis of explanation of the nucleation since different grains of NaCl act differently (see Fig. III.81); some even do not support epitaxy despite proper crystallographic orientation. It has been speculated that impurities (to the order of 10^{-6} %) may give rise to a net electric charge to the surface which might increase the residence time of polymer segments close to the surface and thus increase the probability of interaction with the more specific forces of the surface ions. Quite clearly such possible effects cannot be described by the classical theory outlined in Sect. 5.1.3.1.

Additional difficulties not considered by the classical theory arise from the mechanism of adsorption of macromolecules on the surfaces. Only at low surface coverage was flat adsorption concluded from calorimetric experiments on poly(ethylene oxide) adsorbed from dilute solution on silica by Killmann and Eckart (1971). Joppien (1974) found high site coverage (80%) on the substrate (silica) for polyester because of strong interaction via hydrogen bonds. Because of space limitations the CH_2-portions had to form small loops away from the surface. At higher coverage one can predict from calculations (Hoeve, 1971) that only partial adsorption exists for any one molecule. A dense layer of polymer should exist close to the surface and a diffuse layer should extend into the bulk of the solution.† Little is known about the influence of such conformations of adsorbed layers and their possible change under conditions of heterogeneous nucleation (see also Sect. 5.1.3.3).

Heterogeneous nucleation in actively polymerizing formaldehyde solution was shown by Takahashi and Ogata (1971). Epitaxy occurred on drawn polyoxymethylene, but not on other substrates. It was observed, however, that the epitaxially grown polyoxymethylene had an appreciably lower melting point than solution-grown polymer crystals. This observation was explained by the authors assuming interference of the parallel nucleated lamellae at later stages of crystal growth. Figure V.20 illustrates that crystal growth can only occur until the spaces between the lamellae are filled. At later stages of growth all chain ends are occluded, leading to relatively low molecular weights and low melting temperatures.

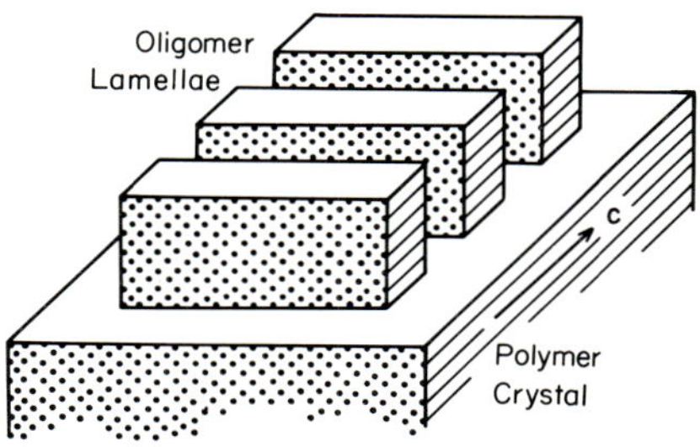

Fig. V.20 *Schematic Representation of Epitaxially Nucleated Polyoxymethylene Lamellae Grown during Polymerization.*
The parallel aligned lamellae hinder further growth by occluding chain ends. Drawn after Takahashi and Ogata (1971).

5.1.3.3 Heterogeneous Nucleation from the Melt

The region of heterogeneous nucleation from the melt was delineated in Sect. 5.1.2.2 through crystallization of small droplets. Figure V.13 shows a region of 30°C where crystallization could be enhanced by adding extra heterogeneous nuclei. Early interest in such experiments arose from the

† See, for example, Hoffman and Forsman (1970), also Schick (1971).

industrial efforts† to produce a fine grained spherulitic structure for improved optical and mechanical properties. It was described by Last (1959) that polyethylene and nylon 11 showed a marked reduction in spherulite size and increase in spherulite number without major change in crystallinity when finely divided solid foreign polymers were added to the melt. Polytetrafluoroethylene, isotactic polypropylene, nylons, a polyurethane, and poly(hexamethylene terephthalate) were active in nucleation of polyethylene melt; and nylon 6.10 and nylon 6.6 were active in nucleation of nylon 11. Inoue (1963) demonstrated that crystallization of nylon 6 could be speeded up considerably by addition of nylon 6.6, poly(ethylene terephthalate), titanium dioxide, lead phosphate, and sodium polyphosphate. Kargin *et al.* (1962, 1963, 1966, 1967) improved mechanical properties of polymers by the addition of heterogeneous, differently shaped nuclei as for example V_2O_5 to poly(vinyl alcohol). By adding coarsely subdivided foreign material which is visible under the microscope, they showed that solids, liquids, and even gas bubbles are able to nucleate polypropylene spherulites. The seed solids investigated were isotactic and atactic polystyrene, indigo, bismuth salicylate, titanium oxalate, cadmium benzoate, and zinc acetate; the seed liquids, glycerol and gallium; the seed gas, air. Slonimskii and Godovskii (1966) showed calorimetrically that addition of 1 % indigo to polypropylene leads to a higher crystallization temperature on cooling.

The connection between the experiments on epitaxy from solution described in Sect. 5.1.3.2 and melt droplets crystallization described in Sect. 5.1.2.2 was made by Koutsky *et al.* (1967b). They sprayed 5-μm droplets of a concentrated polyethylene solution onto glass and cleaved (001) alkali halide surfaces. After evaporation of the solvent and melting of the remaining polymer at 150°C, crystallization was induced by cooling at 0.1°C/min. Rodlike crystals were observed on the alkali halides, and spherulites of no preferred orientation on the surface of the glass. The orientation of the melt grown crystals on the (001) halide surface was identical to the solution grown polyethylene (see Table III.4). The *c*-axis of the polyethylene is parallel to the[110] and [$\bar{1}$10] directions of the halide crystal, with the (110) surface of the polyethylene contacting the (001) surface of the halide. Data of the supercooling ΔT_n of the droplets are tabulated in the following together with the unit cell

† See for example: E. I. duPont de Nemours, British patent 798,038 (1958); BASF, British patent 818,317 (1959); ICI, British patent 850,989 (1960); Wijga, P.O.W., U.S. patents 3,207,735; 3,207,736; 3,207,738 (1960). Monroe, G. C., Jr. and Vaughan, D. J., U.S. patent 2,991,264 (1961); Wijga, P. O. W. and Binsbergen, F. L., U.S. patent 3,299,029 (1961); Wales, M., U.S. patents 3,207,737; 3,207,739 (1961–1962); Binsbergen, F. L., U.S. patents 3,326,880; 3,327,020; 3,327,021 (1963); Greear, G. R., Schulken, Jr., R. M., U.S. patent 3,118,847 (1964); also British patents 951,158; 1,030,197; 1,095,427; and 1,001,709; and U.S. patents 2,958,098; 3,327,020; 3,367,926; and 3,359,346.

lengths a and an estimate of the overall surface free energy of the particular alkali halide.

Crystal	a (Å)	γ_{NaCL} (erg/cm²)	ΔT_n (°C)
LiF	4.01	343	15.2
NaCl	5.63	173	16.0
KCl	6.28	110	18.1
KI	7.05	80	19.2

A comparison with Table V.2 shows that the heterogeneously nucleated droplets need only about $\frac{1}{3}$ the supercooling as the homogeneously nucleated ones. Assuming the nucleation can be described through the effect of a flat substrate (Eq. 46), a comparison of the homogeneous and heterogeneous nucleation rates via Eq. (5) can be used to estimate $\Delta\gamma$. Koutsky *et al.* (1967b)† computed values of 2.9 to 4.6 erg/cm² for the alkali halides in the sequence listed. Furthermore, Eq. (54) may be used to see whether reasonable estimates for the dispersion contribution to the surface free energy of the alkali halides (γ_s) can be obtained. Neglecting $(1 - \zeta)_o\gamma_c$ and using the values $\gamma_c = 70$ erg/cm² and $\gamma_m = 36$ erg/cm² as proposed by Schonhorn (1968) leads to γ_s-values decreasing from 60 to 55 erg/cm² in the order listed above. Some increase in γ_s is possible due to the term $(1 - \zeta)_o\gamma_c$. The order of magnitude of the γ_s values appears reasonable, but too many uncertainties in the use of Eq. (46) and the parameters in Eq. (54) preclude further discussion.

Studies of transcrystalline morphology (see Sect. 3.4.3, Fig. III.84) by Schonhorn (1967, 1968) and Fitchmun and Newman (1970) proved that selected foreign surfaces could substantially enhance nucleation. Only a large number of polymer lamellae which nucleate on the surface before the bulk of the polymer melt nucleates further away from the surface can lead to transcrystalline morphology. Gold, for example, permits transcrystalline morphology on crystallization in contact with polyethylene, nylon 6.6, polychlorotrifluoroethylene, and polypropylene. Hobbs (1971) grew polypropylene crystals from a melt in contact with graphite fibers. He could show that only graphite fibers with crystals of the size of the fold length of the polypropylene lamellae were active nuclei, producing transcrystallization. A match of the helical period of polypropylene with the H-adsorption sites of graphite was suggested to enhance the nucleation. The polymer surface free energy determined from the contact angle with various liquids is much larger

† Note that $\Delta\gamma = 2\sigma_s - \delta$ in the notation of Koutsky *et al.* (1967b).

for a transcrystallized surface (measured after removal from the foreign surface) than for surfaces grown in contact with inactive foreign surfaces. Schonhorn (1967, 1968) interpreted these data as indicating a much higher surface free energy of the crystals than previously assumed (70 erg/cm^2 dispersion contribution for polyethylene). In contact with surfaces which do not initiate transcrystallization (usually low energy surfaces), the polymer surface layer has after crystallization still a surface free energy which indicates a structure close to amorphous (36 erg/cm^2 for polyethylene).

Heterogeneous nucleation by addition of foreign seed material to the bulk polymer melt was extensively studied for polypropylene. Beck and Ledbetter (1965) characterized the relative quality of different seeds by thermal analysis. They determined the temperature for fastest rate of crystallization on cooling

TABLE V.3

CRYSTALLIZATION TEMPERATURES[a] OF SEEDED[b] POLYPROPYLENE[c]

Seed	Cryst. temp. (°C)	Seed	Cryst. temp. (°C)
No seed	106	Na octanoate	113
$AgNO_3$	107	Na *p*-nitrobenzoate	114
Na 1-naphthoate	108	Na_2 sebacate	114
Pb $(benzoate)_2$	108	Na_2 oxalate	115
Na cyclopropanecarboxalate	108	Na cyclobutanecarboxalate	115
LiCl	109	Na cyclopentanecarboxalate	115
CaO	110	Na m-hydroxybenzoate	116
$MgSO_4$	110	Ca $(benzoate)_2$	116
Na_2SO_4	110	Mg $(benzoate)_2$	117
Na 1-naphthylacetate	110	Na *m*-methylbenzoate	118
TiO_2	111	Cu $(benzoate)_2$	118
Al_2O_3	111	Na caproate	118
Na *o*-nitrobenzoate	111	Na *o*-methylbenzoate	119
Na *m*-nitrobenzoate	111	Ba $(benzoate)_2$	121
Na *p*-aminobenzoate	112	Na *p*-hydroxylbenzoate	122
Zn $(benzoate)_2$	112	Li benzoate	125
Na_2 adipate	112	Na cycloheptanecarboxalate	125
Na acetate	112	Na 2-naphthoate	127
Na_2 terephthalate	112.5	Na_2 1,2-cyclohexanedicarboxalate	128
Pyrex glass	113	Na cyclohexanecarboxalate	130
Na_2 malonate	113	Na benzoate	131
Na formate	113	Na *p*-methylbenzoate	132

[a] Peak temperature as measured on cooling at 12.5°C/min from 200°C by DTA.
[b] 0.25% by weight, milled at 150°C for 5 min.
[c] Selected from data by Beck (1967).

the polymer–seed mixture. A selection from data on some 200 different seeds is shown in Table V.3. To be good nucleating agents an organic and a polar group has to be present. The most ideal nucleating agent found was sodium benzoate.

Binsbergen (1970b) tested 2000 substances for their possible nucleating effect on the crystallization of polyolefins. Most substances were only tested in polypropylene, but nuclei active in polypropylene also had activity in polyolefins like polyethylene, poly(4-methyl-1-pentene) and polystyrene. Qualitatively the nucleating effect was judged by the decrease in size and increase in number of the spherulites when compared to the pure polymer, and by the possible increase in crystallization temperature at constant cooling rate (100°C/min from 240°C, see also Table V.3). A good nucleating agent was noted when the spherulite diameter had decreased to $\frac{1}{5}$–$\frac{1}{10}$ of the value for the unseeded polymer. The following groups of substances were found to be good nucleating agents:

(1) Aluminum salts of:
aromatic and cyclo aliphatic mono- and dicarboxylic acids, aliphatic α,ω-dicarboxylic acids, aliphatic monocarboxylic acids bearing aromatic or cyclo aliphatic substituents, aromatic phosphonic acids, phosphoric and phosphorous acid (partially esterified to aromatic esters).

The hydrocarbon part of the molecule generally did not carry any polar substituents other than the acid group(s) mentioned. If in the hydrocarbon part one or more hydrogens had been replaced by polar groups such as $-NH_2$, $-OH$, $-NO_2$, $-O-CH_3$, $-(CO)-O-CH_3$, etc., the nucleating effect was often negligible. Exceptions to this rule were the aluminim salts of, e.g., *p*-hydroxy- and *p*-nitrobenzoic acid.

(2) Alkali salts of:
all the acids mentioned under (1), together with various of these acids carrying one polar substituent, short, branched-chain aliphatic monocarboxylic acids, aromatic, cycloaliphatic, and branched-chain aliphatic sulfonic acids.

(3) Several salts of divalent metals, such as Ca, Ba, Cu, and Co, of trivalent metals, Ga and In; and of tetravalent metals, Ti and V (in the form of the titanyl and vanadyl ions, resp.) with a number of the acids mentioned above.

(4) Various organic pigments, such as
Cu phthalocyanine and chlorinated Cu phthalocyanine, anthraquinone type pigments of condensed aromatic structure, such as flavanthrone, quinacridones, perylene type pigments, some azo pigments carrying Na, Ca, or Ba sulfonate groups.

(5) Colloidal silver and gold.

(6) Hydrazones.

The only nonsalt, nonpigment type organic substances found to show nucleating activity were hydrazones, made by reaction of salicyl hydrazide with either benzaldehyde or salicylaldehyde.

Binsbergen (1970b) reported that inorganic salts and oxides were found inactive, contrary to patent claims and epitaxy observed from solution and melt as well as transcrystallization. The degree of supercooling necessary to cause large scale crystallization in such seeded melts was larger than for the substances listed above and often similar to the base polymer without additives (see also Table V.3). The good nucleating agents were shown to be a fine dispersion of small crystals or a thin crystalline layer adsorbed on another substance. Epitaxy based on lattice match could be excluded because of activity toward several polyolefins tested. Good nucleating agents were insoluble in the polymer or crystallized on cooling before the polymer did. The nature of many crystals common for good nucleation seemed to be the probable existence of alternating rows of polar and nonpolar groups at the seed crystal surface.

An estimate of the number of nuclei and their development in time is also possible by interpretation of the overall development of crystallinity. Since it is experimentally established that the spherulitic radius increases at constant temperature with the linear rate v, any one spherulite has at time t the volume

$$V = \tfrac{4}{3}(vt)^3\pi \tag{55}$$

Assuming N heterogeneous nuclei per unit volume lead to a simultaneous start of N spherulites, the development of the volume fraction of crystallinity v^c [see Eq. (8) Sect. 4.1.1] can be expressed after taking into account impingement of neighboring spherulites by (see Sect. 6.1.3)

$$v^c = 1 - \exp(-K_3 t^3) \tag{56}$$

with

$$K_3 = \tfrac{4}{3}Nv^3\pi \tag{57}$$

Measurement of v^c as a function of time allows a check on the exponent of t which should be three for the case assumed (see Table VI.2 for other exponents). Furthermore, additional measurement of the spherulitic growth rate v permits an estimate of the number of active nuclei N.

Dilatometry and depolarization methods (see Sect. 4.1) were used by Binsbergen and DeLange (1970) to follow the development of crystallinity in seeded polypropylene. Slonimskii and Godovskii (1966) and Johnsen *et al.* (1970) used DTA for similar, but less complete measurements. The crystallinity development in polypropylene with various amounts of sodium *p*-tertiary butylbenzoate and monohydroxyl aluminum *p*-tertiary butylbenzoate, two of the most active seeding materials, were analyzed by Binsbergen and DeLange (1970) over the temperature range of 130–155°C. The data could be represented by Eq. (56) with the exponent three. Nucleation densities derived from experimental values of K_3 [Eq. (57)] are plotted in Figs. V.21 and V.22. The number of nuclei are seen to be strongly crystallization temperature dependent. The number increases by a factor of ten for every 4° of

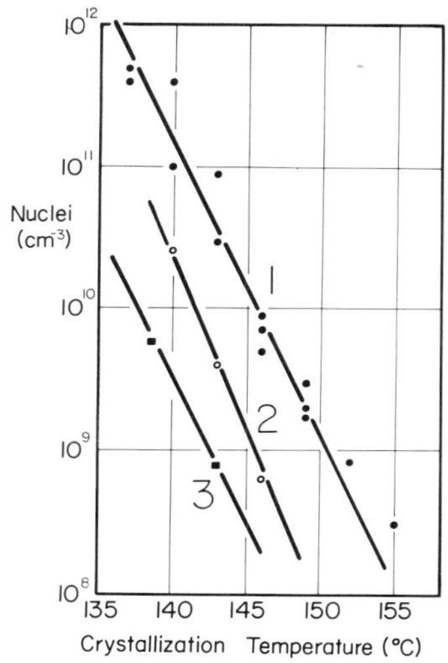

Fig. V.21 *Number of Nuclei as a Function of Temperature and Concentration.*
Seeds of sodium *p*-tertiary butylbenzoate were precipitated by evaporation from solution on finely divided polypropylene, followed by milling at 180°C and pressing at 270°C. Curve 1, 0.3 wt.% seeds; Curve 2, 0.1 wt.% seeds; Curve 3, 0.03 wt.% seed. Data by Binsbergen and DeLange (1970).

increase in supercooling. Up to the limit of about 1 wt. % seeds, the number of nuclei increases with increasing seed weight, however, not linearly. Variation of the temperature to which the nucleated samples where heated before crystallization did not affect the number of nuclei or the kinetics in the range of 175–280°C. A brief cooling to a lower temperature and reheating to crystallization temperature before noticeable development of crystallinity would

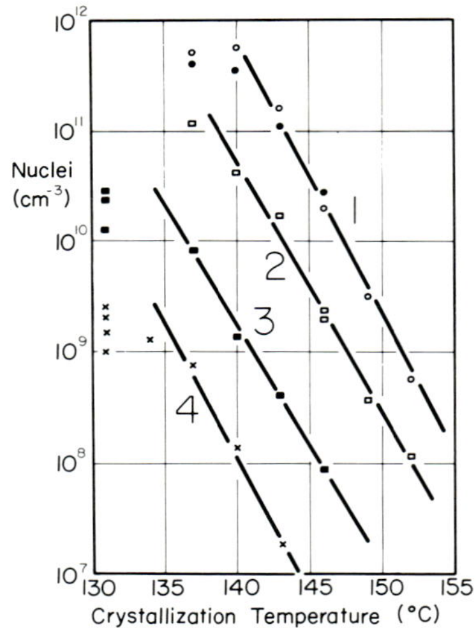

Fig. V.22 *Number of Nuclei as a Function of Temperature and Concentration.* Seeds of monohydroxyl aluminum *p*-tertiary butylbenzoate were precipitated by aqueous metathesis in a slurry of finely divided polypropylene. The polymer containing the salt was filtered off, washed, dried, milled at 180°C and pressed at 270°C. Curve 1, 1.0 wt.% seeds (filled circles) and 3.0 wt.% seeds (open circles); Curve 2, 0.3 wt.% seeds; Curve 3, 0.1 wt.% seeds; Curve 4, 0.03 wt.% seeds. Data by Binsbergen and DeLange (1970).

always lead to a number of nuclei characteristic of the lower temperature. Nucleation was thus practically instantaneous. Counting the spherulites using electron microscopy gave values for N in agreement with the computed ones using K_3 values of Fig. V.22 [Eq. (57)].

An attempt to reconcile the extensive data on polypropylene with the principles outlined in Sect. 5.1.3.1 did not give a completely satisfactory answer (Binsbergen 1969, 1973). Lattice matching type epitaxy as the major mechanism can be excluded because of the large variety of foreign surfaces showing activity and the variety of polymers being seeded by the same foreign surface. Flat surface nucleation [Eqs. (44)–(48)] is difficult to reconcile with the instantaneous activity of the seeds. Self-nucleation by residual polypropylene crystals on the surface, in cracks, or in holes is excluded because of the constant number of active centers when heating prior to crystallization to 175–280°C. Formally, nucleation on limited step lengths [Eqs. (49)–(50)] may be able to explain the data if a proper distribution of step length is assumed and if $\Delta\gamma$ is sufficiently small.

The interfacial free energy difference parameter $\Delta\gamma$ may be low for two reasons, a high energy foreign surface (large γ_s) or a high accommodation coefficient ζ [see Eq. (54)]. Apparently a high γ_s is not the reason for nucleation in polypropylene since the best nucleation agents as listed are expected to have a low surface free energy† (see also Table V.3). Most active nuclei seem to have a γ_s which is close to the surface free energy of the polymer crystal γ_c and can at best cancel the first term in Eq. (54). The ${}_0\gamma_c$ value has been estimated to be about 8 ergs/cm^2 so that an accommodation coefficient ζ close to 1 is the only solution for an explanation of heterogeneous nucleation in this framework. Binsbergen (1969, 1973) proposes that the probable existence of alternating rows of polar and nonpolar groups on the surface may be the cause of the large accommodation coefficient. These rows of nonpolar material are even more effective than the rows of positive charges proposed for the heterogeneous nucleation of polyethylene on alkali halides. Alkali halides are quite inactive in nucleating polypropylene, perhaps because of the close distances of the rows of positive ions.

Several less detailed studies of other polymer melts nucleated by addition of foreign materials have been reported and are briefly summarized in the following. Polychlorotrifluoroethylene was nucleated by Rodchenko and Belyi (1970) with cadium, cadium oxide, and molybdenum disulfide. The spherulites appeared on cooling at a 4–6° higher temperature than without heterogeneous nuclei. Because of the higher crystallization temperature, the crystals had also a somewhat higher melting temperature than the unseeded polymer.

Poly(ethylene terephthalate) mixed with 0.2–0.3 vol. % of talc, kaolin, silion dioxide, or titanium dioxide of about 1 μm particle size were crystallized from the glassy state by Groeninckx *et al.* (1974). The crystallization rate was found to vary with concentration, size distribution, and nucleation ability of the nucleating agent. Talc and titanium dioxide were most effective. Transcrystallinity was observed on the effective nucleation particles. The coefficient of t in Eq. (56) was fractional, but constant for different crystallization temperatures 100–113°C for the same nucleating agent (3.1–3.9). The authors assumed heterogeneous nucleation with time dependent numbers of nuclei.

Poly(2,6-dimethylphenylene oxide) melts were nucleated with 0.005 to

† Graphite, $BaSO_4$, and Fe_2O_3 for which the high dispersion contribution to the surface free energy of 120, 77, and 107 erg/cm^2, respectively, has been measured (Schonhorn, 1968) are not good nucleating agents.

0.05 wt. % of indigotin, anthraquinone, or 2-mercaptobenzimidazole by Packter and Sharif (1971). Without nucleating agents no crystallization could be achieved in the temperature region of 280–180°C. With seeds, crystal growth took place to a crystallinity of about 0.5.

Polydimethylsiloxane was shown by Yim and St. Pierre (1970) to crystallize on cooling at rates between 1.5 and 12°C/min only after seeding with silica (specific surface area 200 m^2/g). The DTA peak temperature for crystallization on cooling of the polymer–seed mixture was dependent on the amount of silica added (for 9 wt. % crystallization occurred at −70°C, for 28 wt. %, at −65°C). Reducing the surface free energy of the silica by coating the surface increased the crystallization temperature.

Isotactic polystyrene was crystallized by Kargin *et al.* (1967) with and without 1% of indigo at 160 and 184°C. Addition of indigo sharply increased the development of crystallinity and the ultimate level of crystallinity. The linear growth rate of spherulites did not change on adding the heterogeneous nuclei (15.6 μm/hr at 160°C).

Polyethylene containing organic and inorganic pigments was crystallized by Turturro *et al.* (1973). It was shown that the heterogeneous nuclei did not only speed up crystallization, but also decreased the coefficient of t in Eq. (56).

5.1.4 Self-Nucleation

The term self-nucleation or self-seeding has been coined by Blundell, Keller, and Kovacs to describe nucleation of folded chain crystals grown from solution on high molecular weight remnants of its own crystals which remained through the dissolution step (Blundell *et al.*, 1966). Self-nucleation will be used here as the general term describing nucleation of a macromolecular melt or solution by its own crystals grown previously. Macromolecules are particularly well suited for self-nucleation because of a large temperature range where crystals do not melt, but even nucleated melts do not crystallize. The droplet experiments illustrated by Fig. V.13 (Sect. 5.1.2.2) indicated for polyethylene such a temperature range between 125 and 142°C. An explanation for this special behavior of macromolecules is given by the molecular nucleation phenomenon (Sect. 5.3). Crystals for self-nucleation can also be created on deformation which shifts the melting point of macromolecules to higher temperatures because of rubber-elastic effects. Finally, it was proposed by Turnbull (1950) that crystal fragments grown in small cracks of a foreign surface may have an elevated melting point and thus survive the initial dissolution or melting of the bulk of the polymer and serve as nuclei on subsequent cooling.

In all cases of self-nucleation it is found experimentally that the supercooling ΔT_- necessary for crystal growth to occur in a given time scale is, over a certain temperature range, critically dependent on ΔT_+, the degree of heating above the melting or dissolution temperature of the bulk of the material. A schematic representation of ΔT_- as a function of ΔT_+ is shown in Fig. V.17. In the region a–b increasing numbers of the nuclei are destroyed; in the region b–c none remain, and nucleation is either truly heterogeneous on foreign surfaces or homogeneous.

Early evidence of the decrease of the number of spherulites nucleated with increasing temperature to which the melt was heated and also the time at the high temperature before cooling to the crystallization temperature was cited by Hawkins and Richards (1949) and Rabesiaka and Kovacs (1961) for polyethylene, Price (1952) and Hoffman *et al.* (1959) for polychlorotrifluoroethylene, Hartley *et al.* (1954) for poly(ethylene terephthalate), von Falkai (1960) for polypropylene, Magill (1962) for nylon 6, Banks and Sharples (1963) for poly(ethylene oxide), and McLaren (1964) for nylon 6.6. Various explanations such as crystal maintained in cracks, order in the melt above the melting temperature, and stress were advanced for the observed effects, but could not be proven conclusively. Self-nucleation by residual high molecular weight crystals as described in the following can possibly account for many of the observations, even for the occasional reports of little or no effect. In the latter cases the high molecular weight portion particularly active in self-nucleation may have been missing in the analyzed sample, or the temperature range chosen was outside the range a–b in Fig. V.17.

5.1.4.1 Self-Nucleation from Dilute Solution

The first detailed analysis of self-nucleation by incomplete destruction of crystals of high molecular weight macromolecules on disordering was done on solution crystallization.† The typical experiment involves precrystallization followed by heating to a temperature T_s above the recognizable clearing temperature where practically all polymer dissolves. Between the clearing temperature (97°C for polyethylene in *o*-xylene) and a maximum temperature T_f (110°C for polyethylene in *o*-xylene) which is close to the dissolution temperature for extended chain crystals, various numbers of nuclei survive. On subsequent cooling these self-nucleate crystal growth. Such seeding can under

† Earlier work which may be relevant is that of Holland and Lindenmeyer (1962) who crystallized fractions of polyethylene from dilute solution in xylene to single layer lamellae of practically uniform size. Higher molecular weights yielded smaller crystals and thus contained more nuclei. Lanceley and Sharples (1966) crystallized poly(decamethylene terephthalate) from solution in 1,2-dichloroethane to constant size aggregates. In both cases, however, the solutions were initially heated so high above the dissolution temperature that no conclusion about the nature of the nuclei can be made. Self-nucleation may have been accidental.

proper conditions lead to uniform size, single layered crystals as shown in Fig. V.23. Detailed conditions for such self-nucleation are described by Blundell and Keller (1968b). Some upper temperatures for self-nucleation of solutions and melts are listed in Table V.4. The factors influencing self-nucleation are according to Blundell and Keller (1968a) as follows:

1. The number of nuclei under otherwise similar conditions is determined mainly by the dissolution temperature T_s. Figure V.24 shows a typical dependence of the number of nuclei computed from the number of subsequently grown crystals as a function of T_s. The temperature of rapid decrease of nuclei

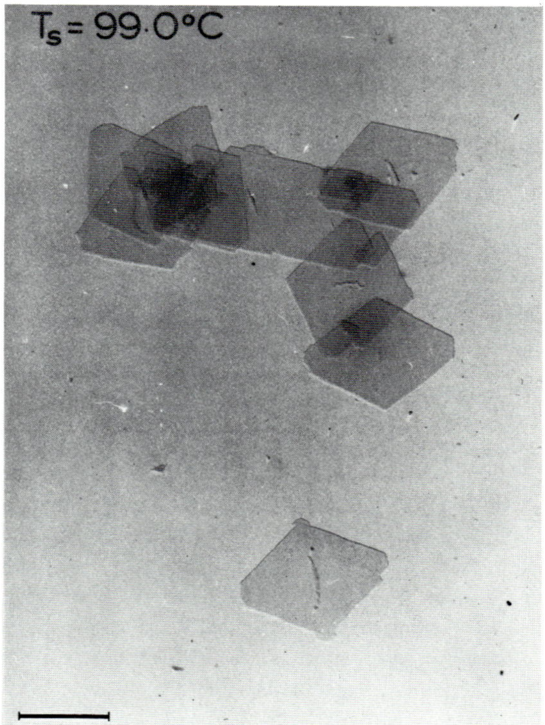

Fig. V.23 *Self-Nucleated Crystals of Polyethylene.* Electron micrographs of monolayer crystals grown at 80°C from 0.1% solution in *o*-xylene. The solution was formed by heating at 10°C/hr to 99°C (T_s). Compare with Fig. III.51b for crystals heated to 101.5°C. Blundell *et al.* (1966). Scale bar 1 μm.

TABLE V.4

MAXIMUM TEMPERATURES FOR SELF-NUCLEATION

Polymer	Temp. (°C)	Reference
A. Solutions		
Polyethylene (xylene, 1%)	110	Blundell and Keller (1968)
Poly(ethylene oxide) (ethyl benzene) Block copolymerized with polystyrene	44	Kovacs and Manson (1966)
Poly(3,3-bischloromethyloxa-cyclobutane) (Penton)	143	Johnsen and Lehmann (1969)
B. Melts		
Polyethylene	138.5	Vidotto *et al.* (1969)
Poly-1-butene form I	141	Vidotto *et al.* (1969)
Poly-1-butene form II	130	Vidotto *et al.* (1969)
Poly(ethylene oxide)	(69)100	Vidotto *et al.* (1969)
Polychlorotrifluoroethylene	305	Hoffman *et al.* (1959)
Poly(ethylene terephthalate)	~290	Hartley *et al.* (1954)
Nylon 6	~260	Magill (1962)
Polystyrene	230	Boon *et al.* (1968)

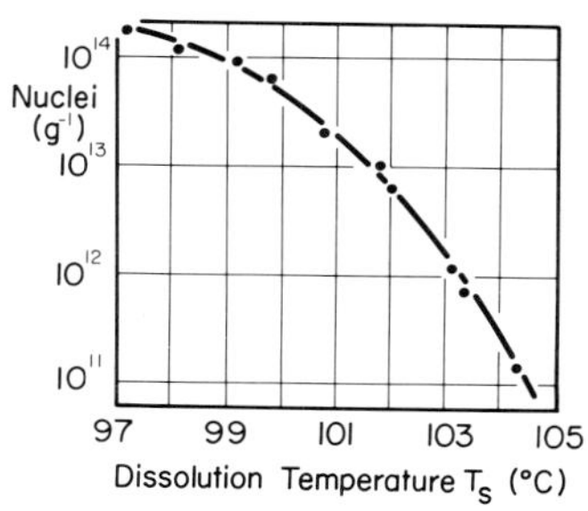

Fig. V.24 *Nuclei of Polyethylene in Xylene as a Function of Dissolution Temperature T_s.* Initial crystallization at 80°C on cooling from a boiling solution with 0.01 wt.% polymer. The initial crystals gave a clearing temperature at 96–97°C on heating. Drawn after Blundell and Keller (1968a).

in the case of polyethylene in *o*-xylene (~110°C) is close to the dissolution temperature of extended chain crystals of similar polyethylene in *o*-xylene (110.4°C, Cormier and Wunderlich, 1966), but less than the equilibrium dissolution temperature for large molecular weight samples (~113°C). The holding time at T_s has little influence on the number of nuclei (see also Keller and Willmouth, 1970).

2. The final crystallization temperature T_c has little effect on the number of active nuclei, quite in contrast to heterogeneous nucleation on foreign surfaces where the number of nuclei was strongly T_c dependent (see Figs. V.21 and V.22, Sect. 5.1.3.3). A 5°C increase in T_c from 80 to 85°C decreased the number of nuclei only by about 20%.

3. The high molecular weight fraction of the sample is particularly active in self-nucleation. The temperatures T_s which give for the indicated molecular weight fractions of polyethylene ($M_w/M_n = 1.1$) the same number of nuclei (2.4×10^{12} g^{-1}) in xylene are

M_n	T_s (°C)
2.6×10^6	105.5
2.4×10^5	102
7.6×10^4	98.5
6.8×10^3	95

Addition of increasing amounts of a polymer with a particularly large high molecular weight fraction to the solution increases linearly the number of nuclei. (Keller and Willmouth, 1970).

4. The rate of heating to T_s after initial crystallization, the concentration, and the temperature of initial crystallization T_1 are of some importance. The number of nuclei per gram of polymer increases with decreasing heating rate, increasing concentration, and decreasing initial crystallization temperature T_1. These effects are quite similar to those found for the ease of reorganization observed for solution grown crystals on annealing (see Sect. 7.2).

5. The concentration of the final crystallization at T_c had little effect on the number of nuclei per gram of polymer when checked over the range 0.01–1.0 wt. % polymer.

The nature of these nuclei was analyzed through electron microscopy and light scattering by Blundell and Keller (1968a). Figure V.25 shows some typical self-seeded polyethylene single crystals. The central protuberances visible in the lamellae are the nuclei.† A larger magnification of such central portion of a self-nucleated crystal is shown in Fig. V.26. The nucleus diameter is about 200–400 Å. Light scattering studies on solutions with self-nuclei revealed the presence of particles of single molecules with a molecular weight of over 10^7 at a T_s of about 100°C. These observations and experiments with

† Such seeds were previously observed by Geil (1963) and Price (1966), however, these authors interpreted them as homogeneous nuclei.

narrow fractions (Keller and Sadler, 1970) make it likely that the nuclei consist in this case of only one molecule. The molecular weight of the residual nuclei

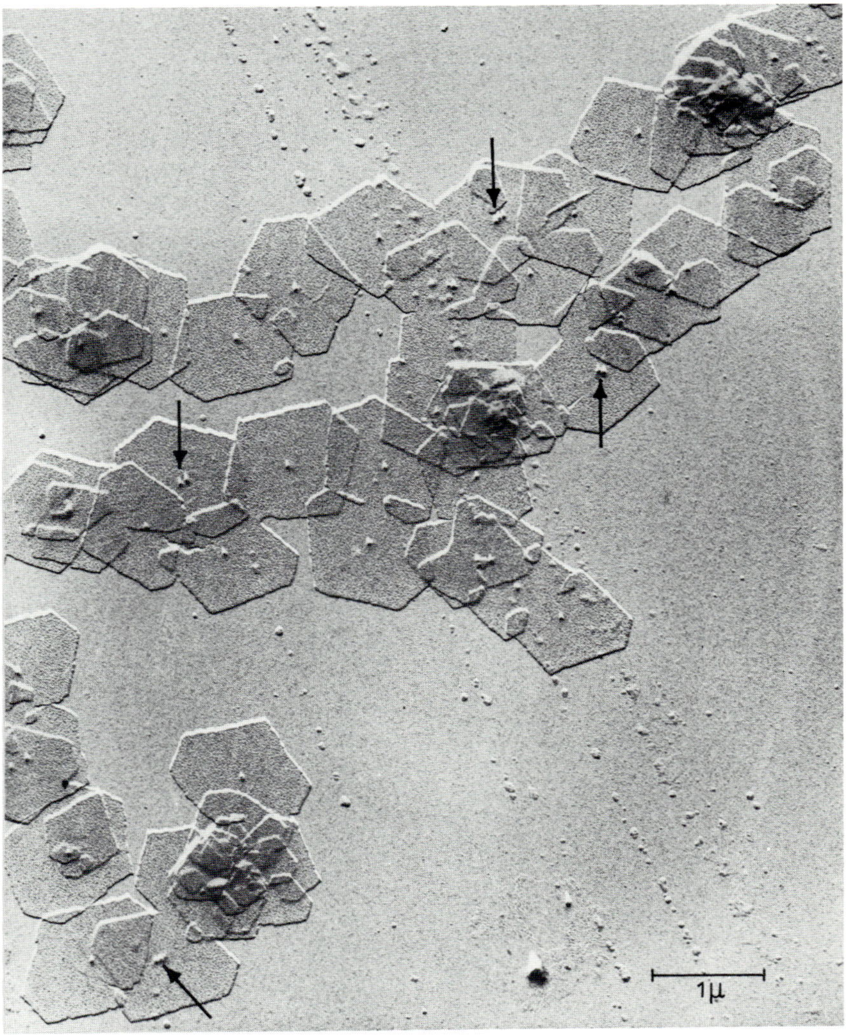

Fig. V.25 *Typical Self-Nucleated Monolayer Single Crystals of Polyethylene Grown from Xylene Solution at* 80°*C.*

The crystals show clear evidence of the nuclei in the center of the crystals. The arrows point at some multiple nuclei. Transmission electron micrograph by Keller and Willmouth (1970) (see also Fig. III.95).

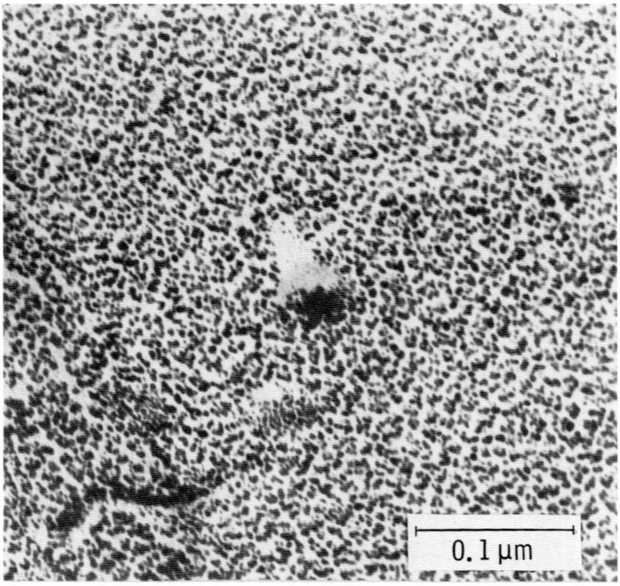

Fig. V.26 *Electron Micrograph of the Center of a Self-Nucleatea Crystal of Polyethylene.*
The crystal has been shadowed with Pd/Au at an angle of about 22° to the surface of the crystal. Blundell and Keller (1968a).

increases for increasing T_s. Self-nucleation by sufficiently large macromolecules rests thus with the special ability of single molecules to anneal on slow heating to a size which shows only a somewhat higher free enthalpy than extended chain crystals of many molecules, but much less free enthalpy than folded chain crystals of many molecules. Once formed, the nucleus is active until it is dissolved by heating above its dissolution temperature. Low molecular weight solutions which lack high molecular weight fractions may perhaps show self-nucleation by several molecules as was suggested by Alfonso *et al.* (1973b). A 3250 molecular weight polyethylene was self-nucleated from xylene at temperatures T_s of 94–99°C.

Practically all high molecular weight molecules are contributing to nuclei when heated sufficiently slowly. This was shown by Carr *et al.* (1970). They found strongly reduced epitaxy of polyethylene in xylene solution on halide crystal surfaces in the high temperature range between clearing of the crystal suspension (97°C) and the limit of epitaxy (109°C) up to which high molecular weight molecules can crystallize to a limited degree on the foreign surface (Sect. 5.1.3.2, Fig. III.81). Practically all high molecular weight polymer was involved in self-nuclei and could not participate in epitaxy. The ability for epitaxy was restored on cooling from a temperature where all nuclei are dissolved. Partial dissolution of the nuclei at 102°C led, for example, to some epitaxy at 100°C. By self-nucleation, the high molecular weight can thus be assessed with high precision (Keller and Sadler, 1970). On the other hand, it is also possible by high temperature epitaxy to remove these high molecular weight molecules expediently (Carr *et al.*, 1970).

5.1.4.2 Self-Nucleation from the Melt

Self-nucleation is naturally also of importance for crystallization from the melt. Most of the phenomena of decreasing overall crystallization on heating to higher temperatures close to the melting point which were discovered earlier are probably caused by self-nucleation. Polyethylene, Poly(ethylene oxide), and poly-1-butene have been studied in detail by Vidotto *et al.* (1969). Quite similar to self-nucleation from solution, the number of surviving nuclei was found to depend on the temperature and time of original crystallization, the heating rate, or more generally the time spent at temperatures above the original crystallization temperature and the ultimate melting temperature T_s before cooling to the crystallization temperature.

Figures V.27 and V.28 show the crystallization kinetics of polyethylene and poly-1-butene. Polyethylene under the chosen condition is, except for the nuclei, practically completely molten at or above 136.8°C. The exponent of t in Eq. (56) which describes the crystallization is 3. The number of nuclei decreases rapidly as shown in Fig. V.29 until at about 138.5°C a limit of about

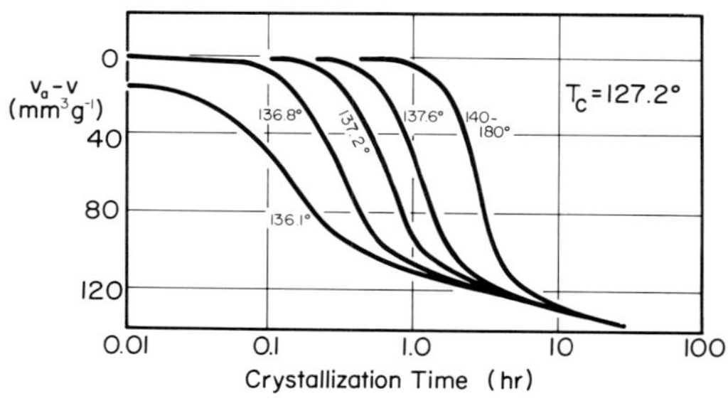

Fig. V.27 *Crystallization Kinetics of Polyethylene at 127.2°C Using Self-Nucleation Techniques.*
The difference in specific volume between the melt (v_a) and the partially crystalline sample at various times (v) is plotted against time. The temperature T_s to which the samples were heated before quenching to T_c are listed for each isotherm. Drawn after data by Vidotto *et al.* (1969). Molecular weight 1.2×10^5 ($M_w/M_n \approx 2$).

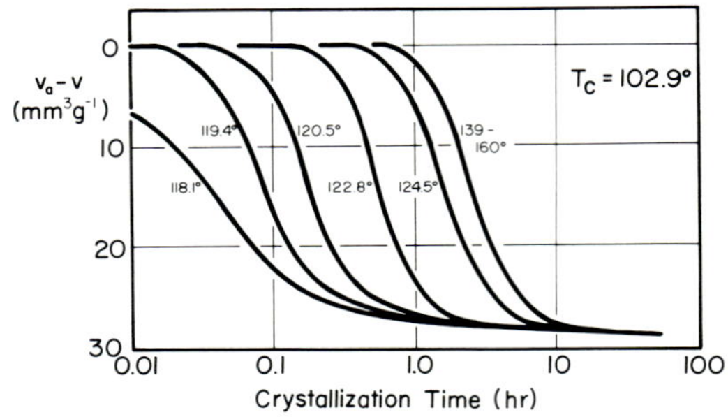

Fig. V.28 *Crystallization Kinetics of Poly-*1*-butene* II *at* 102.9°C *Using Self-Nucleation Techniques.*
The difference in specific volume between the melt (v_a) and the partially crystalline sample at various times (v) is plotted against time. The temperatures T_s to which the samples were heated before quenching to T_c are listed for each isotherm. Drawn after data by Vidotto *et al.* (1969).

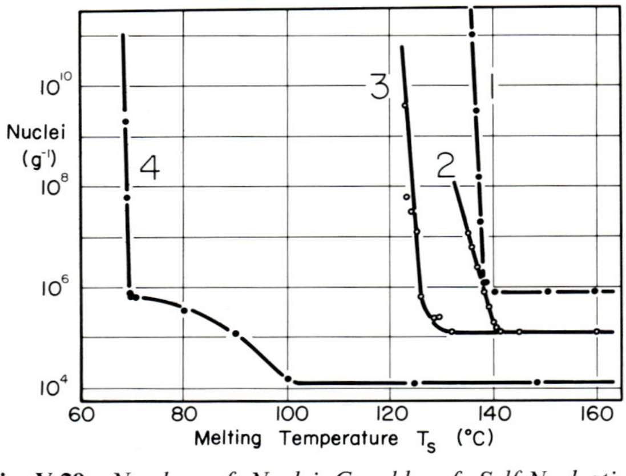

Fig. V.29 *Number of Nuclei Capable of Self-Nucleation from the Melt Heated to the Temperature* T_s.
Curve 1, polyethylene $T_c = 127$°C; Curve 2, poly-1-butene Form I, $T_c = 110$°C; Curve 3, poly-1-butene Form II, $T_c = 110$°C; Curve 4, poly(ethylene oxide), $T_c = 58.6$°C. Drawn after data by Vidotto *et al.* (1969).

10^6 nuclei/g is reached. These remaining permanent nuclei seem to be foreign, heterogeneous nuclei. The region of temperature to which a melt of polyethylene can be heated for self-nucleation is smaller than that of a solution.† The

† Compare Fig. V.24 with Fig. V.29, but see also Figs. VII.10 and VII.11 which indicate how from solution a larger level of self-nucleation is achieved by poorer initial crystallization.

reason for the narrower region of temperature of stability of self-nuclei can be found in the larger fold length and perhaps faster annealing of the whole sample when crystallized from the melt. Typical solution-grown polyethylene lamellae grown at 80°C have a thickness of 120 Å, while polyethylene crystals grown at 127°C from the melt have a fold thickness of about 200 Å (see Figs. III.12 and III.13). The lowering of the melting point corresponding to these thicknesses assuming a fold surface energy of about 83 erg/cm^2 is approximately 20°C and 13°C, respectively, in good correspondence with the change in the regions of self-nucleation.

A special problem arises in the melt crystallization of poly-1-butene. This polymer crystallizes in two crystal forms. Form I (trigonal, see Table II.10 and Fig. II.42A) is the stable crystal form, but only Form II (tetragonal, see Table II.11) crystallizes at high temperature from the melt. The conversion of Form II to Form I occurs with maximum rate at about 25°C. Practically no conversion occurs above 80°C. The crystallization kinetics for Fig. V.28 applies only to Form II, i.e., the sample was never below 80°C. In this case the t-exponent of Eq. (56) is 3, as expected for heterogeneously nucleated spherulites. Converting the initially crystallized samples first to Form I by holding it before heating to T_s for 200 hr at 25°C changes the kinetics drastically. Between 130°C and 138°C, Form I melts only partially, but the molten portion recrystallizes, being self-nucleated by the remaining Form I crystals as Form II. The limit of self-nucleation of Form II crystals on seeds of Form I is finally reached at 141°C in contrast to Form II nuclei which reach their limit at 131°C (compare curves 2 and 3 in Fig. V.29).

Data on poly(ethylene oxide) give also a somewhat different behavior. The crystallization process is clearly separated into two processes (see Sect. 6.3.1.3). Initially, the t exponent in Eq. (56) in the self-seeding range is again 3, but changes at longer crystallization times to a second process of lower exponent. The numbers of nuclei calculated from the initial process are shown in curve 4 of Fig. V.29. Above a T_s of 68.8°C, the second process of crystallization is superimposed on the first, instead of being consecutive. Two critical temperatures can be extrapolated from the data of Fig. V.29, one at about 69°C, close to the melting temperature, and the other at about 100°C. Both temperatures are observed independently of the final crystallization temperature (55–60°C). The 100°C critical temperature is thought to be possibly the limiting dissolution temperature for crystal fragments stabilized by specific heterogeneous nuclei of the sample.

5.1.4.3 Self-Nucleation at Low Temperature

A somewhat different type of self-nucleation has been described by Boon *et al.* (1968). It has been observed on macromolecules which crystallize slowly enough so that the melt can be supercooled below the region where crystallization can occur. Isotactic polystyrene is a typical example of such a polymer. Nuclei can in this case be created at low temperature without large scale crystallization. Figure V.30 illustrates the data by Boon *et al.* (1968). Between 140°C and 80°C increasing numbers of nuclei are created on cooling (curve 1).

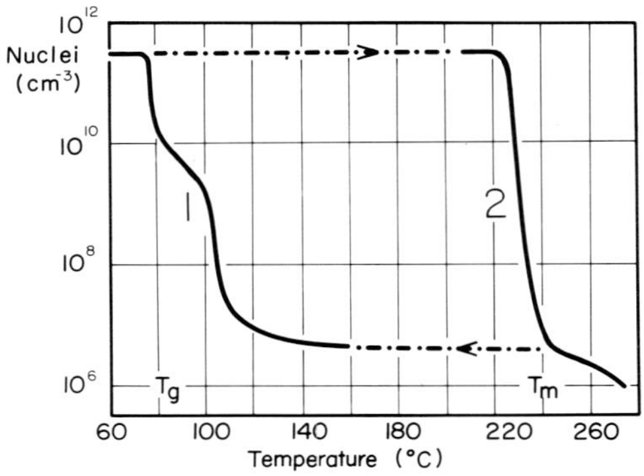

Fig. V.30 *Number of Polystyrene Nuclei Capable of Self-Nucleation from the Melt as a Function of Quench Temperature (Curve 1) and Subsequent Heating (Curve 2).* The number of nuclei was evaluated through Eq. (56) on crystallization at 160°C after quenching for 5 min (curve 1) or after quenching for 5 min at 20°C followed by heating to the indicated temperature (curve 2). Data by Boon *et al.* (1968).

These can be measured by crystallization on heating to temperatures between 140°C and 200°C. Even at temperatures close to the glass transition most nuclei are already created within 5 min. A limiting value seems to be reached after several hours. On quick heating from the low temperature to a higher temperature, the number of nuclei stays constant up to the melting region (curve 2). The number of nuclei were assessed by following the crystallization kinetics at 140°C or 160°C and evaluating the number of nuclei through Eq. (56). Again, the exponent of t was close to 3.

These nuclei are certainly not large, annealed single-molecule crystals as just described since close to the glass transition temperature, practically all molecules could form stable crystals. The mobility should, however, be insufficient for the required rearrangements. One might suspect that the nuclei might even be of the fringed micelle type (see Sect. 5.1.2.1 and Fig. V.10); details are not available at present. An effort to remove possible foreign particles by centrifugation led to a reduction of the number of nuclei. Particularly those counted at the lowest crystallization temperatures decreased in numbers. This observation, together with the indication of a limited number of nuclei at larger nucleation times mentioned above, point to a possible help of the nucleation by foreign surfaces.

Nucleation of crystals in network forming (inorganic) glasses has been studied extensively. In this field most of the available experimental evidence seems also to point to heterogeneous rather than homogeneous nucleation (Hammel 1969).

5.1.4.4 Self-Nucleation from above the Melting Temperature

At first glance, quenching a melt or solution from above its melting or dissolution temperature should be the perfect process to produce conditions of homogeneous or heterogeneous nucleation, not influenced by crystalline structures of the starting material. Two effects can, however, be imagined causing residual structures capable of self-nucleation. A small crystal properly grown at low temperatures in a crack of a foreign substance may survive temperatures higher than its own melting point; and nuclei of subcritical size (embryos) created by steady state fluctuations at the higher temperature might become supercritical or stable on quenching to a lower crystallization temperature. Experimental evidence for both processes is not conclusive at present and must be carefully analyzed because of the chance of superheating of macromolecular crystals above their equilibrium melting temperature (see Chapter IX) and because of the possibility of strain retention in the melt for some time after deformation, leading to self-nucleation caused by orientation (see Sect. 5.1.4.5).

The possibility of retained crystals in cracks was explored by Turnbull (1950). For example, embryos are retained above the melting temperature in cylindrical cavities of sufficient depth which have a radius smaller than

$$r_c = (2\gamma_{cm} \cos \theta)/\Delta g_c \tag{58}$$

A sufficiently negative surface free energy contribution to the free enthalpy of nucleus formation ΔG results in a negative ΔG.† The specific surface free energy γ_{cm} refers to the crystal–melt interface, Δg_c is the free enthalpy of crystallization per cm^3, and θ is the (macroscopic equilibrium) contact angle between the foreign surface and the crystal. Once the embryos are molten by a further raising of the temperature, they will regrow only at lower temperature. Their growth would have to start on one side of their cavities that would act similar to a flat surface and this cannot lead to a negative surface free energy.

† The interfacial free energy between melt and substrate γ_{ms} must naturally be larger than that between crystal and substrate γ_{cs} ($\gamma_{ms} - \gamma_{cs} = \gamma_{cm} \cos \theta$).

To estimate the critical size r_c, one may assume γ_{cm} to be perhaps 5–15 erg/cm^2 (see Sect. 5.1.2.2), Δg_c for 10°C above the melting temperature is for typical macromolecules 3–10 × 10^7 ergs/cm^3 so that r_c should be in the range of 10–100 Å. This estimate makes it unlikely that such effect extends to temperatures far above melting. In fact, the true equilibrium melting temperature of macromolecules is often not known with sufficient certainty to distinguish surviving nuclei in cracks from surviving extended chain nuclei.

A complication for nucleation by crystals surviving in cracks arises from the special nature of macromolecules. One might expect that in order to have maximum stabilization and permit initial crystallization of the embryo, the molecular chains should be oriented more or less parallel to the assumed cylindrical cavity axis. In this case the cavity opening would be inactive for further crystal growth because it shows only chain folds or chain ends which terminate the retained crystal.

The retention of subcritical nuclei on quenching seems more probable, as they are inactive at high temperature, but can cause immediate crystallization through self-nucleation at lower temperature. The steady state distribution of subcritical nuclei (embryos) $Z(i)$ of size i can be drawn schematically as shown in Fig. V.31. Mathematically the curve is represented by

$$Z(i) = \text{const.} \exp(-\Delta G_i/kT) \tag{59}$$

On cooling, the critical size is shifted from a to the left. If the shift is far into the region of reasonable concentration, self-nucleation may be observed (Zachmann 1964). Experimentally such self-nucleation by cooling should lead to time-independent initiation of crystals.

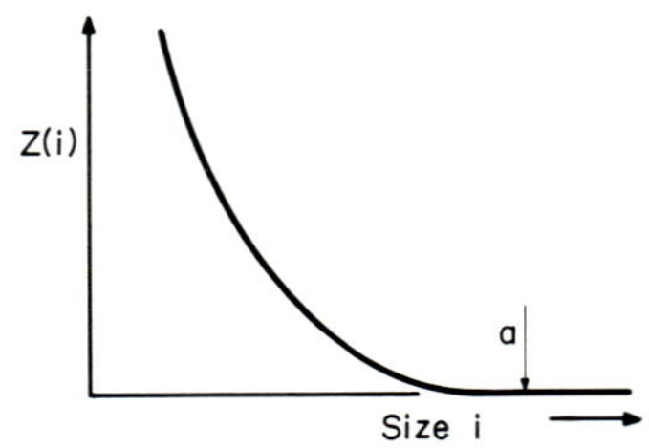

Fig. V.31 *Embryo Distribution as a Function of Size at a Given Temperature.*
The distribution can be computed using Eq. (59). The critical size is marked by a. On cooling to a lower temperature the critical size shifts into the region of appreciable concentration leading to self-nucleation.

In case the critical nucleus size is not reached on cooling, there should still be an effect of the initial embryo size on the time needed to reach the steady state of nucleation I^* [Eq. (5)]. The time to reach steady state is frequently called the *induction time* τ. Figure V.32 shows schematically the initially slow increase in number of critical nuclei which leads into the steady state linear increase described by I^* and the extrapolation to evaluate τ experimentally. Mathematically τ was evaluated by Frisch (1957) using

$$\tau = \int_0^{i_{crit}} [(Z_\infty(i) - Z_0(i))/I^*]\, di \qquad (60)$$

where $Z_\infty(i)$ is the steady state distribution at the ultimate crystallization temperature and $Z_0(i)$ is the steady state distribution at time zero. On quick quenching from a temperature T_s the distribution $Z_0(i)$ would correspond to the steady state distribution at that temperature.

Foreign surfaces may play a role in the production of these subcritical nuclei; in particular, if the surfaces are capable of adsorbing the polymer molecules in a somewhat more extended chain conformation as suggested for active heterogeneous nuclei (Sect. 5.1.3.3).

An effect of heating far above the melting temperature on nucleation was observed by Schultz (1969) on polyethylene. Samples of polyethylene melt were cooled under similar conditions from 150°C and 180°C. Both temperatures are far enough above the melting temperature so that no crystalline remnants should be present (holding time before cooling: 15–120 min; the 150°C sample was also heated to 180°C before holding at 150°C). On cooling at 30°C/min, the sample cooled from 150°C crystallized about 2°C earlier. On isothermal crystallization, quenching from the lower melt temperature gave rise to larger spherulites and a higher degree of crystallinity with no

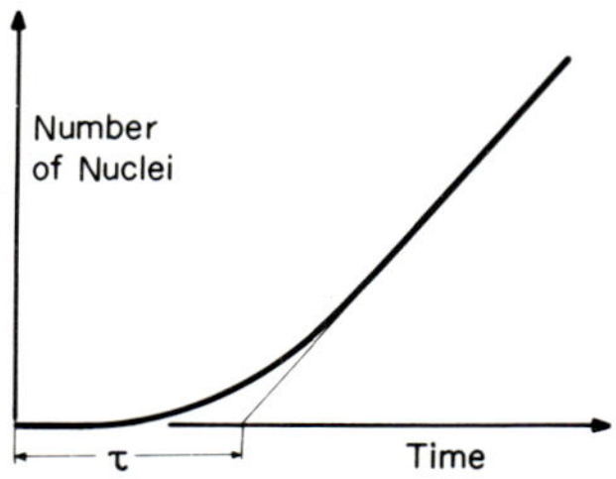

Fig. 32 *Number of Nuclei as a Function of Time.*
The induction time τ is caused by the need to establish the steady state distribution of subcritical nuclei $Z_\infty(i)$ after cooling from a higher temperature.

change in fold period. None of the two previous suggestions can explain these data satisfactorily. The larger number of embryos at 150°C could on cooling start crystallization earlier; however, on isothermal crystallization they should lead to more and not less spherulites. The suggestion of Schultz (1969) that perhaps the overall order in the melt is affected, changing ΔG of crystallization, will need further study.

5.1.4.5 Self-Nucleation by Orientation

All types of self-nucleation discussed above were concerned with nucleation processes involving macromolecules in an unstrained solution or melt. The entropy of the molecules in the random phase was assumed to be at a maximum. Such state is best described by a random coil conformation. Deformation of the macromolecule to a more extended conformation can lead to enhanced nucleation and crystallization. A more extended chain would have a lower entropy, but change only little in enthalpy, so that the melting temperature, expressed as

$$T_m = \Delta H/\Delta S \tag{61}$$

increases. At a given temperature a deformed macromolecule may have thus a larger supercooling and greater driving force towards nucleation and crystallization (see also Sects. 6.2.2 and 6.3.2). On chain extension, the conformation approaches the final macroconformation in the crystalline state, so that one might expect in addition an enhanced rate of nucleation because of kinetic considerations.

The morphology of macromolecular crystals grown under conditions of shear is basically fibrillar (shish-kebobs, Figs. III.79, III.80, and III.136). Detailed analysis revealed a compound structure. The innermost portion is a more or less macroscopically smooth thin extended chain fiber, which suggests itself to be closely related to the initial nucleus. It is usually covered with irregularly folded material, to be followed, when crystallized at sufficiently low temperature, with a further layer of more regularly folded lamellae which give the aggregate the shish-kebob appearance (for details see Sect. 3.8.2).

Early evidence that even residual strain caused by preorientation can enhance nucleation was found through the study of memory effects. Rabesiaka and Kovacs (1961) studied polyethylene, Frolova *et al.* (1963, 1965) measured deformation effects on poly(ethylene terephthalate), isotactic polystyrene, polyethylene, and polycarbonate, and Mitchell (1967) analyzed polybutadiene. The enhanced nucleation could be shown to persist through several meltings. Electron microscopy on thin films of stretched polyisoprene (natural rubber) by Andrews (1964) showed typical nucleation lines in direction of the stress (see Fig. III.80C). Such nucleation can easily be demonstrated by deformation of polymer melt under an optical microscope followed by crystallization as shown in Fig. V.33. Andrews (1964) could show that the number of nucleation lines increased with the stress applied. More detail about the nature of the nucleation lines was gained by the study of crystal growth from strained, lightly cross-linked melts, within impinging jets of solutions, and from stirred solutions. The results from such experiments are summarized next.

To extend a macromolecule, purely extensional flow is necessary. In simple shear processes, as produced in ordinary viscometers, the fluid elements rotate as well as experience shear. The strain rate will in this case change its sign periodically. Before large elongation can be achieved by the molecule, the deformation mode is reversed. Under conditions of uniaxial extension, as found in fiber spinning, or pure shear, as produced by impinging jets, there is no rotational component and the magnitude of the change in molecular conformation is simply $\dot{\varepsilon}\tau_1$ where $\dot{\varepsilon}$ is the rate of strain and τ_1 is the slowest relaxation time for the recovery of the macromolecules to their statistical random coil conformation (for a more detailed discussion see Frank, 1970). Since the relaxation time τ_1 is proportional to the $\frac{3}{2}$ to $\frac{4}{2}$ power of the molecular weight, it is possible to achieve much larger elongations with higher

Fig. V.33 *Optical Micrograph of Polyethylene Melt, Sheared before Crystallization.*
Observed between crossed polarizers after partial crystallization, the nucleation lines are clearly visible due to subsequent crystallization. G. Czornyj unpublished, magnification about 500×.

molecular weights. A small fraction of a polymer of molecular weight 15 times the average would, for example, extend 100 times as much as the average molecule. A frequent situation at moderate extensional flow may thus be that most shorter molecules are barely affected, while a few large molecules are almost fully aligned and able to nucleate crystallization at sufficiently low temperature (Frank, 1970).

Experiments on the extension of macromolecules in the pure shear field set up by directly opposing jets of polyethylene solutions were carried out by Frank *et al.* (1971) (see also Sect. 6.3.2, Fig. VI.56). The temperature for the growth of (fibrous) crystals was increased from the normal 95–97°C for unstrained solutions, to 108°C for a 5 wt. % sample of 202 000 weight average molecular weight in xylene, and 112°C for a similar sample of 1.5×10^6 weight average molecular weight. Birefringence persisted above the temperature where crystallization was possible, an indication of the extension of the polymer molecules. The reduction in birefringence with decreasing concentration was, however more abrupt than would be expected from the change in number of molecules, so that one must conclude that the major extension is perhaps achieved by clusters of molecules rather than single molecules. These observations and qualitative estimates make it likely that in a deformed sample of macromolecules, the high molecular weight fraction is particularly active for nucleation, permitting later additional growth of lower molecular weight species on its surface. Finally, growth of relaxed molecules at sufficiently low temperature leads to the typical shish-kebob morphology.

Some morphological information on the early stages of crystallization of deformed macromolecular melts has been given by Hill and Keller (1969). By selective dissolution and by nitric acid etching (Sect. 4.1.4) of a stress crystallized film of polyethylene, fibers are revealed in the direction of stress which are particularly resistant to dissolution or oxidation as is shown in Fig. V.34. The observed fibers are thought to be related to the nuclei. X-ray diffraction during the course of crystallization in 300% strained films of lightly cross-linked polyethylene (to prevent flowing) established that, indeed, the fibrous material grows first. Later the fibers are covered at sufficiently low temperature (133°C or lower) with folded chain, twisted lamellar overgrowth (see Sect. 3.8.2 and Figs. III.80A,B). The fibrous material could be grown at temperature as high as 135°C, which compares to crystallization without stress at about 129°C in the same time scale (several hours). A strong amorphous halo in the X-ray diffraction diagram of the stretched sample material indicates that the fibers makes up only a small portion of the overall sample. By removal of the stress, the initial fibers detected by X-ray diffraction melt again down to temperatures as low as 132.5°C as long as no lamellar overgrowth had occurred. With lamellar overgrowth, the whole aggregate could be molten after stress removal at about 135°C; with stress not removed, the lamellar overgrowth melted selectively between 136°C and 138°C, while the fibrous material persisted at least for some time up to 139.8°C. The stress

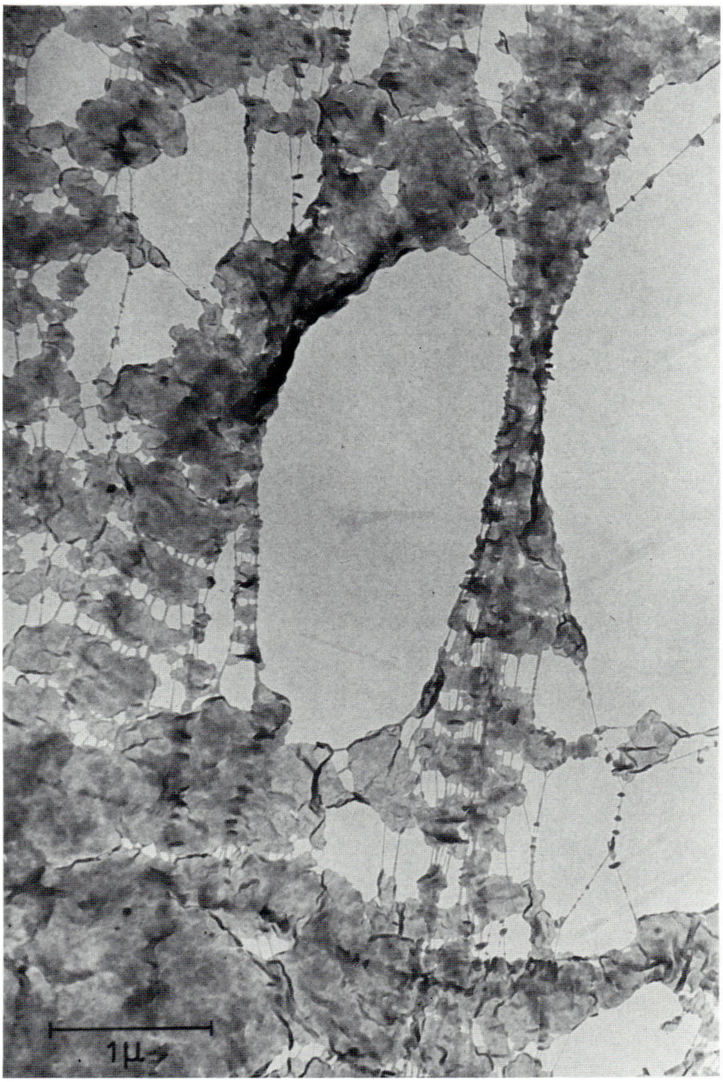

Fig. V.34 *Electron Micrograph of Stress Crystallized Polyethylene Selectively Treated with Xylene at 80°C.*
The lamellar overgrowth of fibrous structure has been selectively dissolved. Hill and Keller (1969).

must thus be supported mainly by the small fraction of the molecules of the fibrous crystals which melt at higher temperatures when strained.

After all traces of X-ray crystallinity were molten in the relaxed stress crystallized samples, orientation could be recovered on recrystallization at lower temperature. This indicates that self-nucleation leads under such conditions to oriented nuclei. Such oriented nuclei were detected after melting of the initial fibers grown by stress crystallization at 132.5–135°C through stress relaxation, or after heating of the relaxed, fully stress crystallized sample to 139.8°C. The nuclei are thus related to the fibers visible by electron microscopy (see Fig. V.34) and studied by X-ray diffraction, but must by themselves be much smaller since their presence cannot be detected by X-ray diffraction.

Shish-kebob type crystals grown from stirred solution showed similar results (Pennings *et al.*, 1970, see also Sect. 6.2.2). The maximum crystallization temperature of linear polyethylene from xylene solution could be increased as much as 20°C by means of stirring. The key condition for the increased nucleation and growth was the presence of Taylor vortices or turbulent flow in the solution. Simple shear, even of considerable magnitude, did not lead to nucleation. Evidently the extensional flow connected with the vortices or turbulence causes the enhanced nucleation. Pennings *et al.* (1970) suggest that in this case, clusters of molecules are particularly effective in nucleus formation. Once nucleated, further growth can be enhanced by the stirring as described in Sect. 6.2.2 (see also Sect. 3.8.2).

5.1.5 General Conclusions about Primary Nucleation

This section on primary nucleation has provided some clear trends and left a series of open questions. Both will be summarized here because of the importance of the understanding of primary nucleation for the discussion of experimental evidence on secondary and molecular nucleation and in preparation for the description of crystallization in Chapter VI.

Primary nucleation is an experimental fact. By inspection of the morphology of the grown crystals, either crystals of all equal size or of varying sizes are observed. The former indicates that all crystals started growing at the same time, which is termed *athermal nucleation*; the latter indicates that new crystals started growing throughout the crystallization, which is termed *thermal nucleation*. While the homogeneous nucleation of Sect. 5.1.2 is connected with thermal nucleation, the converse is not true. Heterogeneous nucleation may be either athermal or thermal (Sect. 5.1.3). Since the number of heterogeneous nuclei is limited, however, the thermal nucleation of heterogeneous origin must level off after all nuclei are exhausted. Usually, self-nucleation seems to be practically athermal nucleation (Sect. 5.1.4). Any sample may now contain all three types of nuclei, but clean separation on the basis of observation of athermal and thermal nucleation alone is not possible. Only if a more detailed morphology is visible, as for example in Fig. V.18, can additional conclusions about homogeneous or self-nucleation on the one hand and heterogeneous nucleation on the other be made.

Classical nucleation theory as outlined in Sects. 5.1.1 and 5.1.3.1 can account at least qualitatively for the observed athermal and thermal nucleation. Quantitative agreement of nucleation rates can often be achieved by choosing appropriate parameters in Eq. (5). Equation (5) has been derived for the steady state of nucleation [see Eqs. (7)–(9)]. Thus in the initial phase of nucleation one expects an induction period τ until steady state is reached as illustrated in Fig. V.32 (see Sect. 5.1.4.4).

A more detailed look at the primary nucleation has shown up serious flaws in the classical theory. The equilibrium shape was shown by computer simulation to be a poor approximation to the probable critical nucleus shape (Sect. 5.1.1). This was found to be true despite the fact that the expression for ΔG^* seemed to be acceptable [see Eq. (23)].

The fringed micelle-type nucleus (Sect. 5.1.2.1) can be described perhaps formally by classical nucleation. However, the meaning of the surface free energy term is altered because of the influence of the remaining connection between crystalline and amorphous portions of the molecule.

An estimate of primary homogeneous nuclei was possible for chain folded nuclei, showing a diameter of 5–10 Å and a length of ~150 Å (Sect. 5.1.2.2). Greater precision is not likely because of the uncertainty in Δh_f and γ, which must be derived from macroscopic measurements unrelated to nucleation.

Oligomer nuclei have not been studied to any extent at present, but further studies on them may lead to a better estimate of the chain length in a primary nucleus by molecular weight measurement of the surrounding melt at the nucleation time (Sect. 5.1.2.3).

Heterogeneous nucleation described by the classical concepts of Sect. 5.1.3.1 contains similar problems. The strong temperature dependence of the number of active nuclei (see Figs. V.21 and V.22) and the frequently observed athermal mode of nucleation are only explicable by rather special nucleus size distributions with special surface step structures. Most experiments seem to have been carried out with μm-size nuclei which should have given rise to many simultaneous lamellae on their surface (see Figs. III.81 and III.82). Each lamella should be capable of initiating a spherulite (see Sect. 3.7.1). In this case, the temperature dependence should have been less important (note that the much smaller self-nuclei have a smaller temperature dependence).

The indication of special lamellar growth on alkali halide surfaces at temperatures where self-nuclei do not grow leads to an additional problem: What is the mechanism of formation of macromolecular lamellae at temperatures where nuclei cannot grow? (Sect. 5.1.3.2). The suggestion of possible influence of ordered surface forces on the conformation of the adsorbed molecule may be one explanation, but it must receive more experimental verification.

Self-nucleation described in Sect. 5.1.5 is perhaps best understood; however, the questions of the detailed structure of the nucleus and how it arises initially on annealing of poorer crystals or by extensional flow remain only partially answered. Finally, the difference in molecular macroconformation of molecules in the melt and in solution have not been brought to bear on the nucleation problem (Sect. 5.1).

5.2 Secondary and Tertiary Nucleation on Crystal Surfaces

The problem to be discussed in this section is the need for further nucleation of crystal growth after the initial crystal nucleus is established through homogeneous, heterogeneous or self-nucleation (Sect. 5.1). Experimental evidence for nucleation control of crystal growth is given in Sect. 5.2.1. This is followed by a description of the principles of secondary and tertiary nucleation (Sect. 5.2.2). The general conclusion and some direct experiments for secondary nucleation are given in Sect. 5.2.3.

5.2.1 Nucleation Control of Crystal Growth

The growth rate of many crystals has been observed either directly through microscopy, or indirectly through dilatometry or thermal analysis (see Chapter VI). Growth of a crystal by simple addition of crystallizable particles without creating new surface is understandable within the framework of the theory of absolute reaction rates used for the description of primary nucleation [Sect. 5.1.1, Eqs. (8) and (9)]. The molten and crystallized state of a crystallizable particle can be represented by Fig. V.6 with the molten state characterized by the free enthalpy difference $\Delta G_i = 0$ and the crystalized state characterized by $\Delta G_{i+1} = \Delta G_c$. Below the melting temperature T_m the free enthalpy of crystallization ΔG_c is negative, and the net rate of crystallization v' can be represented by

$$v' = \text{const.}\ \exp[-\Delta G_\eta/kT](\exp[-\Delta G_c/2kT] - \exp[+\Delta G_c/2kT]) \quad (62)$$

As before, ΔG_η represents the free enthalpy of activation which governs the short distance diffusion of the crystallizing unit across the phase boundary. Approximating the free enthalpy of crystallization as before [Eqs. (16)–(18)] leads us to

$$\Delta G_c = -\Delta H_f \cdot \rho_c \cdot \Delta T/T_m \quad (63)$$

At high supercooling ΔG_c is numerically much larger than $2kT$ and negative so that the second term in Eq. (62) can be neglected. At low supercooling the two exponentials can be developed into a series. Using only the first two terms of the series results in a simple linear supercooling dependence of the crystallization rate at low supercooling.

$$v' = \text{const.} \exp[-\Delta G_\eta/kT](\Delta H_f \cdot \rho_c \cdot \Delta T/kTT_m) \tag{62a}$$

Such growth rate behavior has not been observed experimentally. Even in crystal growth from the gas phase one finds that the growth rate drops more rapidly to zero when approaching the melting temperature (see also Sect. 6.1.2.2).

In an effort to seek an explanation for the different observed behavior we might reason that only additional growth to the corner of a crystal can fulfill the growth condition without increasing surface area (position 3 in Fig. V.8). We could thus conclude that the initially irregular surfaces of the nucleus (as indicated in Fig. V.9) grow faster to molecular smoothness than fluctuations can start new layers of growth. However, a molecularly smooth surface can support growth only after secondary nucleation (see Fig. V.3b). Secondary nucleation is a slow process and should under these circumstances govern crystal growth.

When screw dislocations intersect the surface, the slow-down of crystallization due to the need of secondary nucleation can be avoided. A screw dislocation represents a continuous, indestructible secondary nucleus in the surface (see Sect. 4.3.3 and Fig. IV.43). The rate of advancement of crystal growth on a surface with a screw dislocation is governed by the rate of generation of fresh turns of the growth spiral multiplied with the thickness of a single layer b_0 (Frank, 1949). The crystal growth rate can in this way be expressed as [see also the discussion of Eq. (20), Chapter VI]

$$v = b_0 v_\infty/(2a^*\pi) \tag{64}$$

where v_∞ is the rate of advance of a nucleated surface layer, calculated using an expression like Eq. (62a), and a^* is the diameter of the critical secondary nucleus whose dimension is fixed in the center of the screw dislocation. The rate of advance of the surface layer v_∞ is dependent on ΔT as shown by Eq. (62a); the critical secondary nucleus dimension a^* is proportional to $1/\Delta T$ as shown by Eq. (66). This makes v proportional to ΔT^2. Many instances of crystal growth of small molecules from the vapor or solution at low supercooling (1% supersaturatation or less) seem to be adequately described by these assumptions.† Chernov (1961) lists literature data on water, glycerol, tin, phosphorous and salol‡ for which the supercooling dependence of the

† For a more detailed discussion see, for example, Strickland-Constable (1968) or Laudise (1970). For a more detailed derivation of Eqs. (62a) and (64) see Chernov (1961).

‡ Phenyl ester of 2-hydroxybenzoic acid.

growth rate varies between $\Delta T^{1.7}$ and $\Delta T^{2.9}$. One expects that crystallization during polymerization of gaseous or dissolved monomers may behave similarly under appropriate circumstances.

The crystallization of macromolecules presents additional difficulties. Knowing the morphology of the crystals, one must distinguish between two basically different directions of crystal growth, one of which is in the direction of the macromolecular chain axis, the other being normal to the molecular

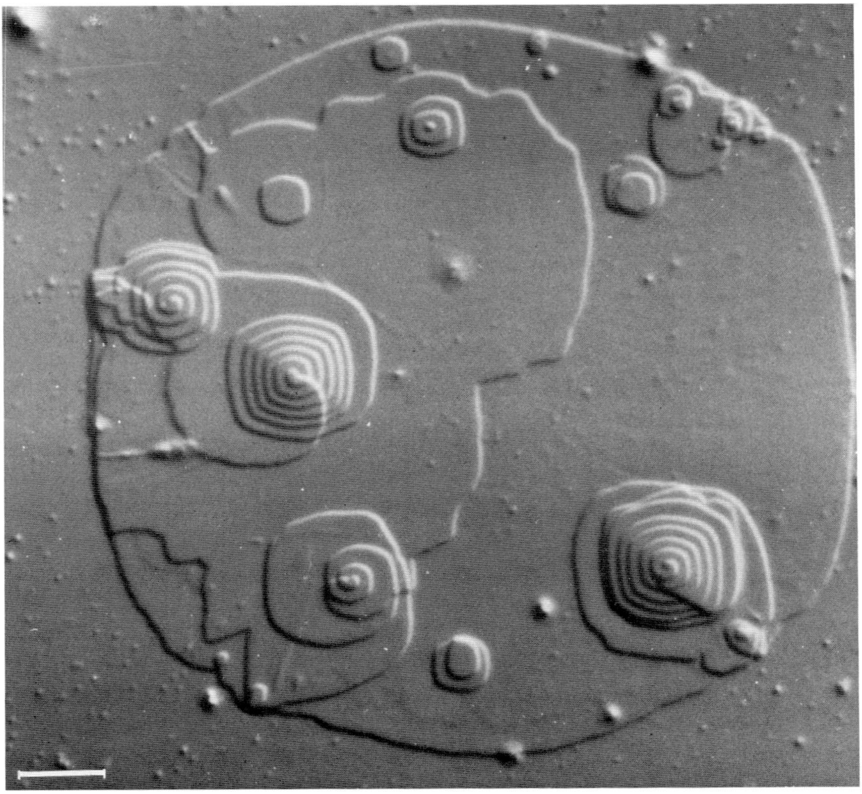

Fig. V.35 *Interference Micrograph of Poly(ethylene oxide) Growth Spirals.* Grown from the melt (6000 molecular wieght) at 59.5°C. Normarski interference optics. Left and right-handed growth spirals and terraces (caused by pairs of close screw dislocations) are visible. The molecular chains are normal to the plane of the figure. The Burgers vector is equal to the fold length. Photograph courtesy of A. Kovacs. Scale bar 10 μm.

chain axis. The difference of fast crystal growth is usually close to normal to the chain axis. Chain folding or chain ends provide a rather special surface for further growth in the direction of the chain axis. In fact, it has been shown for polytetrafluoroethylene that new crystals can be initiated on the (001) surfaces in such an orientation that they are not in crystallographic register with the base lamella, but rather branch at an angle close to 90° (Melillo and Wunderlich, 1972). Folded chain lamellar dendrites of isotactic polypropylene show a similarly large-angle branching which leads to a quite regular network of laths (see Figs. III.113 and III.114).

Screw dislocations with a Burgers vector of the size of the lamellar fold thickness provide the only mode to grow multilayer lamellae close to crystallographic register (see Sect. 3.3.3). Figure V.35 illustrates such screw dislocations and the resulting growth spirals for poly(ethylene oxide) (see also Sect. 4.3.3). The smallest radius of curvature of growth spirals of polyethylene, polyoxymethylene, and poly(ethylene oxide) was analyzed by Barnes and Price (1964) as a function of supercooling. They found significantly larger than critical nucleus dimensions governing the growth rate [Eq. (64)]. The rate of advance of a nucleated surface layer v_∞, in turn, is not proportional to ΔT as we will show. Equation (64) does thus not apply to folded chain polymer growth spirals of Burgers vectors of the lamellar thickness size.

Crystal growth rates under situations which require larger degrees of supercooling were analyzed first by Volmer and Marder (1931). Only by assuming that the secondary nucleation of new layers is the rate-determining step could the authors achieve reasonable agreement between experimental and calculated growth rates. The crystal growth rate was found to be proportional to $\exp(-1/\Delta T)$. Such dependence on ΔT leads to a much larger degree of supercooling before appreciable growth can be observed, and the increase in crystallization rate on increasing supercooling is much steeper. Figure V.36 illustrates the growth rate of a crystal which is governed by secondary nucleation. Also shown for comparison is the homogeneous nucleation rate as expressed by Eq. (5). The overall diagram indicates that under the chosen condition there should be an approximately 7° range of supercooling (shown shaded) where the melt is metastable. Even addition of seed crystals could not start large scale crystallization in this temperature region. This is followed by 40° or more where crystal growth will not start through homogeneous nucleation, but once started by heterogeneous nucleation, will continue through repeated secondary nucleation. Finally, close to the glass transition temperature, both primary and secondary nucleation rates drop to very small values. A melt which can be quenched into this temperature range without crystallization will again be metastable (again shown shaded in Fig. V.36).

Crystal growth rates of a number of linear macromolecules from the melt as a function of temperature are shown in Fig. V.37. It is obvious from these

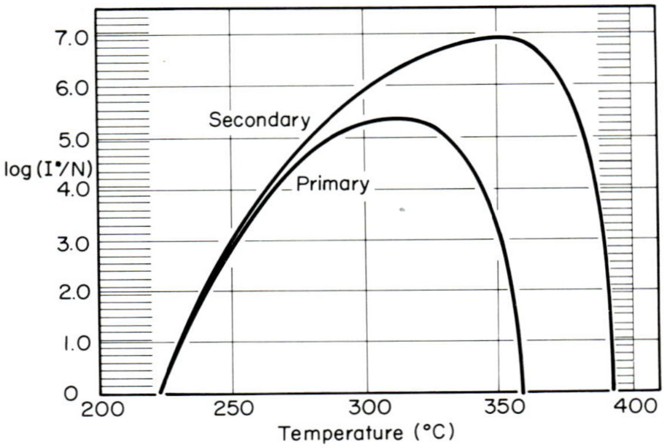

Fig. V36 *Rates of Nucleation I^* in sec^{-1}.* Calculated using Eq. (5) to express the temperature dependence. For secondary nucleation Eq. (68) was used for the calculation of ΔG^* (to obtain the crystal growth rate I^* must be multiplied with the layer thickness b_0). For primary nucleation Eq. (21) was used for the calculation of ΔG^*. Other parameters were chosen to fit a typical polymer: $\Delta G_n/kT$ was calculated using the WLF expression with $Tg = 200°K$ (see footnote on page 9); $\gamma = 5$ erg/cm^2, $\gamma_e = 50$ erg/cm^2, $T_m = 400°K$, $\Delta h_f = 50$ cal/g, $\rho_c = 1.00$ g/cm^3 (as in Figs. V.4, V.5, V.7, and V.39). The shaded areas indicate metastable regions of no crystal growth.

curves that growth of macromolecular crystals from the melt is governed by a nucleation process with a temperature dependence expected for secondary nucleation (Sect. 5.2.2). In Sect. 5.3 it is shown, however, that for linear macromolecules a molecular nucleation process with similar supercooling dependence must exist in addition to secondary nucleation of a new crystal layer. Both secondary nucleation or molecular nucleation may govern crystal growth.

Direct evidence for the need of nucleation beyond primary nucleation was brought by Wunderlich and Cormier (1966) for linear polyethylene. A polyethylene melt was seeded with polyethylene extended chain crystals. About 80 cm^2 of crystal surface area per gram of melt was added, and the development of crystallinity was observed. Practically no crystallization occurred on the seeds in 48–120 hr when the crystallization temperature was 131.2°C, which is about 7.5°C below the melting temperature of the extended chain crystals used. At 128°C crystallization is virtually complete within 48 hr. A comparison with Fig. V.36 shows that only nucleation could account for such observations.

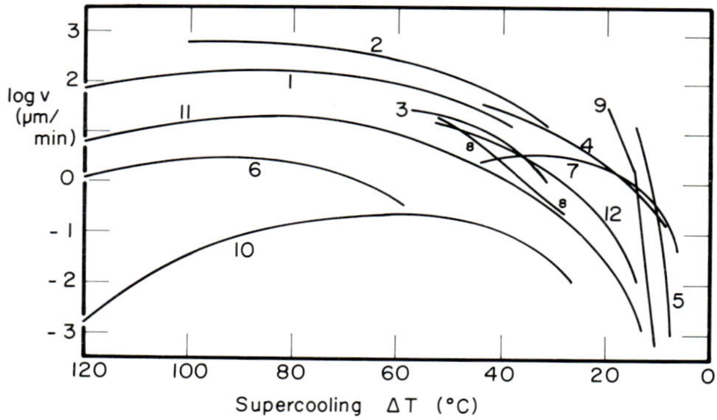

Fig. V.37 *Logarithm of Crystal Growth Rates as a Function of Supercooling.*

The growth rate v is expressed in μm/min. The curves are for the following macromolecules: 1. nylon 6 [Burnett and McDevit, (1957) *J. Appl. Phys.* **28**, 1101]. 2. nylon 6.6 [see curve 1]. 3. polychlorotrifluoroethylene (Price, 1952). 4. nylon 11 [Kahle and Stuart, (1957). *Z. Elektrochemie*, **61**, 1318]. 5. poly(decamethylene sebacate) [Flory and McIntyre, (1955). *J. Polymer Sci.* **18**, 592]. 6. poly(ethylene terephthalate) [Antwerpen, (1971). Thesis, Delft, Technical University]. 7. poly(ethylene adipate) [Takayanagi, (1975). *Mem. Fac. Eng., Kyushu Univ.* **16** (3), 111]. 8 polypropylene [see Table VI.9]. 9. polyethylene [see Fig. VI.53]. 10. isotactic polystyrene [Suzuki and Kovacs, (1970). *Polymer J.* **1**, 82]. 11. poly(tetramethyl-*p*-silphenylene siloxane) [Magill, (1964). *J. Appl. Phy.* **35**, 3249; (1967). *J. Polymer Sci. Part A2*, **5**, 89]. 12. poly(propylene oxide) [Magill, (1965). *Makromol. Chemie* **86**, 283].

The growth of folded chain crystals of polyethylene on an extended chain crystal of polyethylene was studied even more directly by Wunderlich *et al.* (1967) by electron microscopy. Fracture surfaces of extended chain crystals of polyethylene are rough with surface striations which run parallel to the molecular chain direction (see Fig. III.32). On heating such lamellar surfaces to a temperature between 100 and 130°C much of the roughness melts and recrystallizes from the molten state in form of epitaxial, folded chain lamellae (see Sect. 3.4.1, Figs. III.74 and III.75). Figure V.38 is a typical example of such folded chain lamellae grown from the melt on an extended chain crystal subsubstrate. The folded chain lamellae are several micrometers in length and show a characteristic average separation which increases with temperature (200, 250, 330, and 670 Å at 100, 115, 120, and 130°C, respectively). Their fold lengths are identical to the lengths found on crystallization from the melt at the respective temperatures without the help of nuclei (see Fig. III.13). The

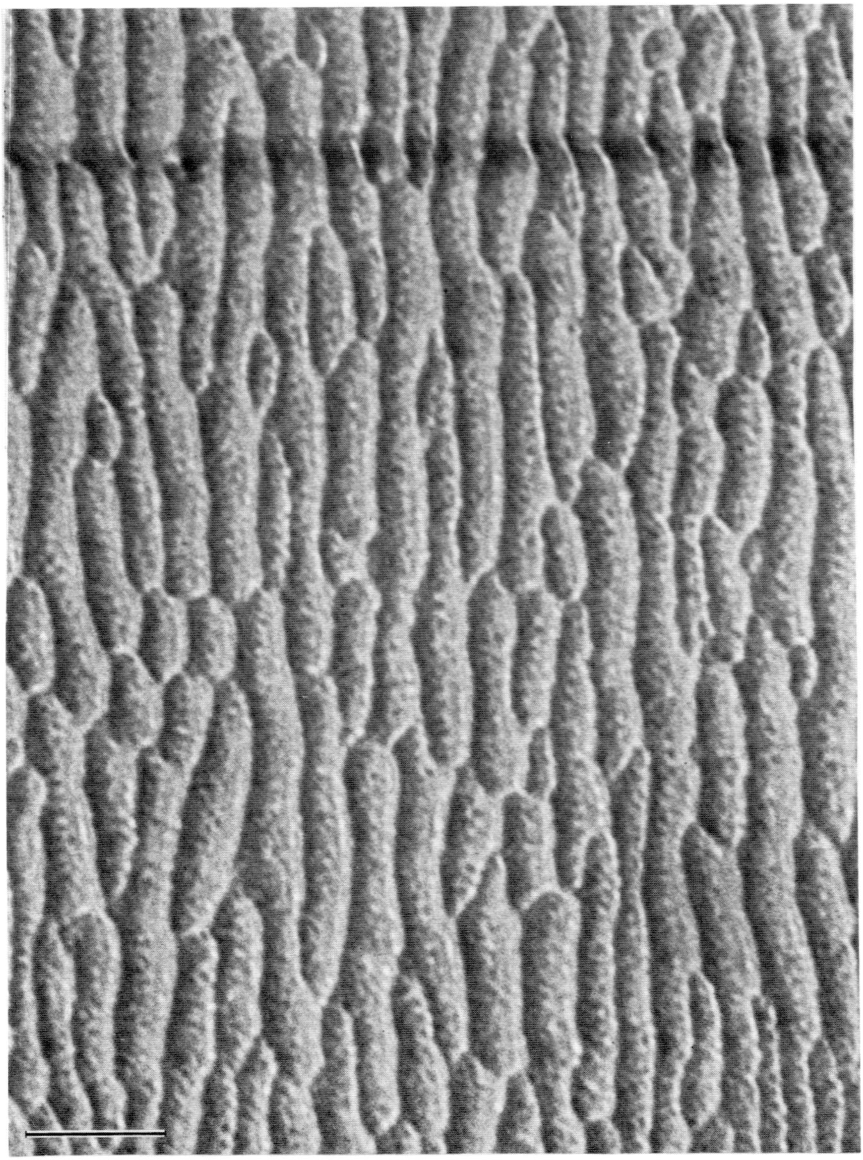

Fig. V.38 *High Magnification Electron Micrograph of an Extended Chain Fracture Surface of Polyethylene Covered with Epitaxially Grown Folded Chain Lamellae.* Replica, showing the need of secondary nucleation. Crystallographic *c*-axis horizontal. (Wunderlich *et al.*, 1967). Scale 1000 Å.

only interpretation possible for these observations is that the start of a folded chain lamellae even on a perfect crystalline substrate is a rare event; growth, in contrast, is faster. Figure V.38 shows also that it is easier for polymer molecules to diffuse 200–700 Å (depending on temperature) than to start a new lamella by secondary nucleation.

5.2.2 Secondary and Tertiary Nucleation Concepts

The secondary nucleation concept as applied to macromolecular crystals needs to be treated for two very different cases. The first occurs on crystallization during polymerization and involves the formation of oligomer nuclei which grow further by addition of monomer molecules similar to low molecular weight compounds. The treatment of secondary nucleation is in this case similar to the crystallization of such low molecular weight materials as paraffins. The second case requires the detailed accounting for the macromolecular nature of the crystallizing species. It applies to crystallization after polymerization. A folded chain or fringed micelle secondary nucleus is formed.

In addition to considering the nucleation of a new crystal layer, it may also be necessary to consider the nucleation of a new molecule on the crystal surface. Generally, this nucleation, called molecular nucleation, has been neglected in the past. It will be treated separately in Sect. 5.3.

5.2.2.1 Secondary and Tertiary Nucleation of Crystals Grown from Small Molecules

Secondary nucleation of a new layer of crystallizing material on a molecularly smooth surface is related to the heterogeneous nucleation case described by Eq. (44) with the condition that $\Delta\gamma$ is zero. From Fig. V.3 it can be seen that the free enthalpy of formation of the surface nucleus is

$$\Delta G_i = -ab_0 l\Delta g_f + 2b_0 l\gamma + 2ab_0\gamma_e \tag{65}$$

The dimension b has been assumed to be of molecular dimension b_0 so that the crystal would grow one layer after the other. The critical nucleus dimensions are obtained as before (Sects. 5.1.1 and 5.1.3.1) by differention of ΔG_i with respect to the variables l and a and setting the slopes equal to zero.

$$a^* = 2\gamma/\Delta g_f = 2\gamma T_m/(\Delta h_f \rho_c \Delta T) \tag{66}$$

$$l^* = 2\gamma_e/\Delta g_f = 2\gamma_e T_m/(\Delta h_f \rho_c \Delta T) \tag{67}$$

Insertion of Eqs. (66) and (67) into Eq. (65) leads us to the free enthalpy of formation of the critical nucleus

$$\Delta G^* = 4\gamma\gamma_e b_0 T_m/(\Delta h_f \rho_c \Delta T) \tag{68}$$

which can be compared to Eqs. (21) and (46) and needs to be inserted in Eq. (5) to yield the nucleation rate I^*. Figure V.36 illustrates the magnitude of the nucleation rates I^* for the secondary nucleation (Eq. 68) and also homogeneous primary nucleation (Eq. 21). Both rates increase many orders of magnitude over a small temperature interval after being practically zero for a well-defined region of supercooling. The reduction of rates at lower temperature is caused by an increasing ΔG_η in Eq. (5) and is similar for both cases.

For a detailed discussion of Eq. (65) it is useful to replace the surface free energies by the critical nucleus dimensions through Eqs. (66) and (67). The following simple expression results for the change of free enthalpy during secondary nucleation (Price, 1959, 1969)

$$\Delta G = \Delta G^* - b_0 \, \Delta g_f (a - a^*)(l - l^*) \tag{69}$$

Figure V.39 is a plot of this equation which represents a hyperbolic parabloid surface. For any constant a or l, ΔG is represented by a straight line. For $a = a^*$ and $l = l^*$ horizontal lines result, dividing the surface into four areas with the saddle point $\Delta G = G^*$ at their intersection. In the lower left quadrant any increase in a or l yields an increase in ΔG so that secondary nuclei of this size are embryos (see Fig. V.2). The critical nucleus at the saddle point has the lowest possible free enthalpy of formation. The upper right quadrant shows decreasing ΔG with increasing a as well as l and leads finally to the stable, completed crystal layer of overall negative ΔG.

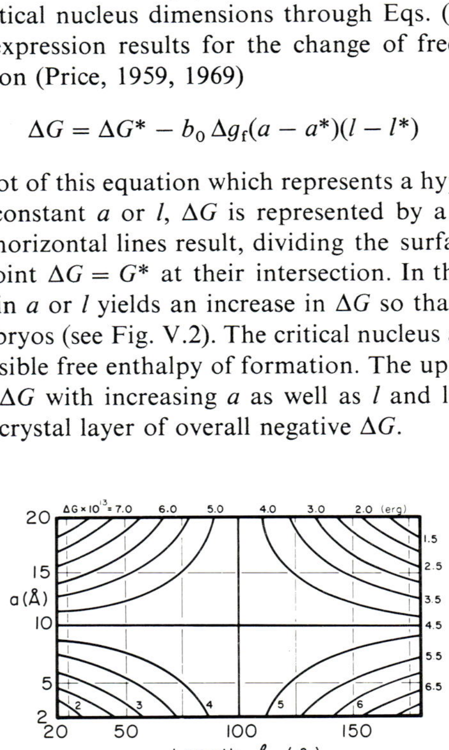

Fig. V.39 *Free Enthalpy of Formation of a Secondary Nucleus.*
Dimensions: Length l, width a, single layer thickness $b_0 = 4.5 \times 10^{-8}$ cm. Assumed parameters for Eq. (69) are $\gamma_e = 50$ erg/cm², $\gamma = 5$ erg/cm², $\Delta g_f = 10^8$ erg/cm³; the latter corresponds to a supercooling of about 19°C in the case of nucleation represented by Fig. V.36 (see also Fig. V.7 for a similar plot for primary nucleation at about double the supercooling).

The elementary growth steps of a crystal in our classical model must be the addition of one crystallizing unit after the other onto a location where addition causes no increase in surface free enthalpy, namely addition to a corner position (3 in Fig. V.8). Once a row of new crystallization is started, the corner can propagate by further addition without change in configuration, permitting a growth as described by Eq. (62). The first creation of a new set of corners by addition of the crystallizing unit to a smooth ledge may be called tertiary nucleation (see Fig. V.3c). Its free enthalpy of formation may be written in analogy to secondary and primary nucleation [Eqs. (65) and (11)]

$$\Delta G_i = -a_0 b_0 l \Delta g_f + 2a_0 b_0 \gamma_e \tag{70}$$

Both the a and b dimensions have the fixed molecular dimensions a_0 and b_0. Equation (70) indicates that ΔG_i has its maximum value for the smallest possible value of l and decreases continuously with increasing l to reach stable size ($\Delta G_i = 0$) for

$$l^\circ = 2\gamma_e/\Delta g_f = 2\gamma_e T_m/(\Delta h_f \rho_c \Delta T) \tag{71}$$

the length of the critical secondary nucleus given in Eq. (67).

Discussion of crystal growth from the gas phase (Sect. 6.1.2.2, Eq. 6) seems to indicate that there is always a large number of kinks in each surface ledge obviating the need of tertiary nucleation. The computer simulation of nucleation described in Sect. 5.1.2.2 is in accord with this finding. The typical nuclei shown in Fig. V.9 show considerable numbers of kinks which generate at a sufficient rate so that crystal growth from small molecules is probably not tertiary nucleation limited. Binsbergen (1970a, 1973) suggests that it may even be possible that secondary nucleation is also a fictitous concept because of the shapes generated by computer simulation of crystal growth. Equilibrium habits with molecularly smooth surfaces which would be in need of secondary nucleation for further growth were just not found.

5.2.2.2 *Secondary and Tertiary Nucleation of Crystals Grown from Large Molecules*

Secondary nuclei formed by linear macromolecules on a large substrate are illustrated in Fig. V.40. The two limiting types of secondary nuclei are, as in the case of primary nucleation, fringed micelle intermolecular nuclei and folded chain intramolecular nuclei (see Sect. 5.1.2; Fig. V.10). Even more than in the primary nucleation, the secondary nucleus size is much smaller than the size of the molecule.

A comparison of these nuclei with nuclei of small molecules reveals two differences. The first difference is connected with the fact that the nature of the macromolecule tends to make the growing edge of a nucleus or crystal

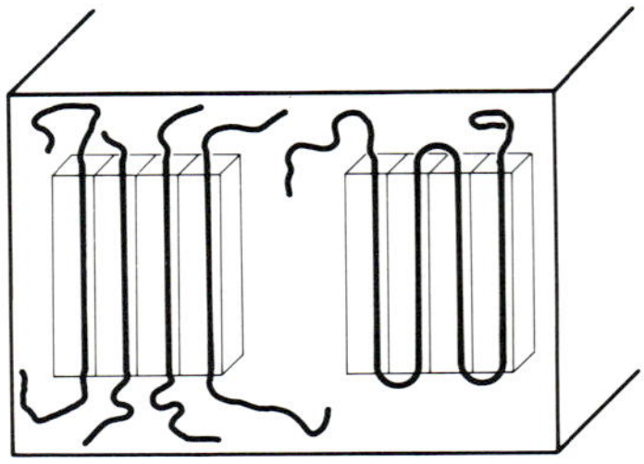

Fig. V.40 *Schematic Drawing of Secondary Nuclei of Macromolecules.* A. Fringed micelle type; B. folded chain type.

straight rather than kinked. The introduction of kinks or jogs in a polymer takes considerably more energy than in small molecules as discussed in Sect. 4.2.3. In addition to the increased surface area, kinks and jogs have an increased conformational energy due to the deformation of the backbone chain. A newly added chain portion as shown in Fig. V.40 will thus rather quickly straighten as many kinks and jogs as possible to reach a more stable macroconformation. If a larger number of kinks remains, it is likely that the chain will remelt or redissolve.

The second difference to nuclei of small molecules is connected with the nature of the upper and lower surface of nuclei of chain molecules. In case of chain folding, further growth in the chain direction is not possible by addition of new material to the fold surface. Refolding through a solid state diffusion process or complete remelting and new crystallization are the only paths to greater fold length. In case of the fringe-surface in fringed micelle nuclei, a large free energy effect is introduced at the amorphous–crystalline interface as discussed in Sect. 5.1.2.1.

Disregarding these difficulties for the moment, and staying in the framework of classical nucleation theory (Sect. 5.1.1), we can obtain the free enthalpy of formation of a secondary nucleus of rectangular shape by Eq. (69) and Fig. V.39. A different interpretation must be given to the specific surface free energy γ_e. When folding takes place, the fold-free energy must be added to γ_e (see Sect. 5.1.2.2); for fringed micelles, the effect of the noncrystalline portion of molecules must be taken into account by the extra contribution γ_e' as discussed in Sect. 5.1.2.1 [Eq. (29)]. Using the parameters evaluated from primary nucleation of folded chain nuclei (Sect. 5.1.2.2), we should find typical values for γ and γ_e to be 5 and 50 erg/cm^2, and this is assumed for the computations leading to Figs. V.36 and 39. For the 19°C supercooling assumed in Fig. V. 39, the fold length of the secondary nucleus should be 100 Å, the width, 10 Å (2 to 3 chain width). This fold-length is of the order of magnitude of the final fold-length observed on crystallization from dilute solution (see Sect. 3.2.2.2 and Fig. III.12).

The coincidence of reasonable secondary nucleus length l^* and observed fold-length on crystallization from dilute solution directed the attention to nucleation as a reason for chain folding. Price (1959) showed that a relatively narrow distribution of fold-length can be derived from Eqs. (67)–(69) if one assumes that the lamellar thickness on later growth remains close to the initially set critical fold-length of the secondary nucleus. By finding the probability for crossing the critical nucleus size in Fig. V.39 at different values for $l > l^*$ it was shown that the average should be about $1.12l^*$.

Besides establishing a possible reason for chain folding, this explanation leads to a prediction of the variation of fold-length with temparature. Figure V.41 shows a plot of the experimental data of several laboratories for the thickness of polyethylene lamellae grown from dilute solution of xylene (for data points see Fig. III.12). The resulting crystal thickness is approximately proportional to $1/\Delta T$ over the whole temperature range analyzed (55 to 95°C) as would be predicted from Eq. (67). The practically straight line of Fig. V.41 is represented by the equation

$$l = 70 + 1650/\Delta T \quad \text{Å} \tag{72}$$

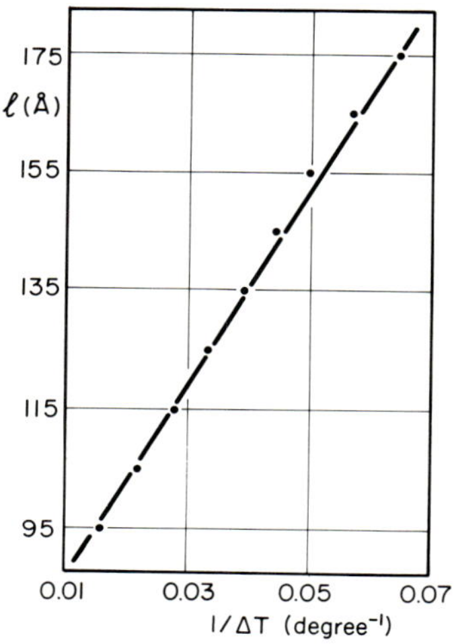

Fig. V. 41 *Thickness of Polyethylene Lamellae Grown from Dilute Xylene Solution.*
Averaged data of Fig. III.12, plotted as function of $1/\Delta T$.

and is well within the experimental error. Assuming l^* to be 0.89 times the final lamellar thickness, one could compute from the slope of the curve that γ_e is about 50 erg/cm^2, which is in good agreement with primary nucleation. The constant 70 Å fold-length for large ΔT shows the trend of the lamellar thickness to level off at large supercooling rather than to decrease further as predicted by Eq. (67). This indicates that such a simple model is incomplete.

The next step to connect chain folding to nucleation was presented by Lauritzen and Hoffman (1960).† The detailed nucleation path chosen is in this case the growth of a folded chain lamella on a substrate in steps of one chain width with fixed fold-length. In Fig. V.39 this would correspond to an increase of a in steps of $a_0 = 4.5$ Å. Schematically the free enthalpy of this path is illustrated in Fig. V.42. From Eq. (69) it can be seen that ΔG decreases linearly with increasing a as soon as $l > l^*$. The first chain length laid down experiences the large nucleation barrier. From Fig. V.39 it can be seen that this nucleation barrier is larger than ΔG^* of Eq. (68) as long as a^* at the given supercooling is different from a_0. Fixing the possible increase in thickness to one chain width of constant fold-length gives the process the character of tertiary nucleation as described by Eq. (70). It reduces the two-dimensional nucleation problem to a one-dimensional tertiary nucleation problem.‡ The decrease in ΔG for each successive new fold is computed by differentiating Eq. (69) at constant l, expressing a in multiples of the unit length $a_0 (a = a_0 \cdot \nu)$, to obtain

$$d\,\Delta G_l/d\nu = -a_0\, b_0\, \Delta g_f(l - l^*) \tag{73}$$

The greater the value l is given above l^*, the more negative Eq. (73) becomes, and the faster is the expected growth of such a lamella. Initial nucleation, however, expressed by

$$\Delta G_l = 2b_0\, l\gamma + a_0\, b_0\, \nu(2\gamma_e - l\,\Delta g_f) \tag{74}$$

shows increasingly larger positive values of ΔG_l for increasing values of l as

† For a summary of this work see, for example, the reviews by Price (1969), Gornick and Hoffman (1966), and Hoffman *et al.* (1969).

‡ Note that in this case l is fixed and a is variable, in contrast to Eq. (70) which takes l as the variable.

---

§ Derivation of Eq. (75): Designating the occupancies of the various states of Fig. V.42 by N_ν, with ν increasing from zero in steps of one, the process of growth of the lamella can be expressed in terms of A_0, the rate of generating the first strip on the substrate, B_1 its reverse, and the subsequent forward rates A_ν and their reverses $B_{\nu+1}$. For steady state one assumes that

$$A_1 = A_2 = A_3 \cdots = A \tag{a}$$

and

$$B_2 = B_3 \cdots = B \tag{b}$$

long as $2\gamma/a_0\,\nu > \Delta g_f$. Although growing faster, lamellae of larger fold-length would be nucleated less frequently. Based on these assumptions, a steady state expression for lamellar growth $S(l)$ has been derived (see also Frank and Tosi, 1960, and Gornick and Hoffman, 1966).§

$$S(l) = N_0\left(\frac{kT}{h}\right)\left[1 - \exp\left[\frac{-(d\,\Delta G_l/d\nu)}{kT}\right]\right]\exp\left[\frac{-\Delta G\eta}{kT}\right]\exp\left[\frac{-\Delta G_1}{kT}\right] \tag{75}$$

The overall rates of change are then given by

$$dN_1/dt = N_0 A_0 + N_2 B - N_1(B_1 + A) \tag{c}$$

and

$$dN_\nu/dt = N_{\nu-1}A + N_{\nu+1}B - N_\nu(B + A) \qquad \text{for} \quad \nu > 1 \tag{d}$$

Since at steady state $dN_\nu/dt = 0$ for all values of ν, the overall flux S can be written

$$S = N_0 A_0 - N_1 B_1 = N_1 A - N_2 B \tag{e}$$

and furthermore, at any point along the growth in Fig. V.42 the flux must be

$$S = N_\nu A - N_{\nu+1} B \tag{f}$$

From Eq. (f) a recursion formula for the expression of $N_{\nu+1}$ in terms of lower occupancies can easily be derived

$$N_{\nu+1} = (A/B)N_\nu - (S/B) \tag{g}$$

$$N_{\nu+1} = (A/B)^\nu N_1 - \sum_0^{\nu-1} (A/B)^\nu (S/B) \tag{h}$$

Summing up Eq. (h) leads us to

$$N_{\nu+1} = (A/B)^\nu N_1 - (S/B)[(A/B)^\nu - 1]/[(A/B) - 1] \tag{i}$$

$$N_{\nu+1} = (A/B)^\nu (N_1 - S/(A - B)) + [S/(A - B)] \tag{j}$$

For steady state crystallization $A/B > 1$, and ν can reach large values. In order to keep Eq. (j) bounded the coefficient of $(A/B)^\nu$ must vanish so that for the flux the following condition arises

$$S = N_1(A - B) \tag{k}$$

Eliminating N_1 between Eqs. (k) and (e) leads to the overall flux equation

$$S = N_0 A_0(A - B)/(A - B + B_1) \tag{75a}$$

For many purposes B_1 is also set equal to B to yield the even simplier equation

$$S = N_0 A_0[1 - (B/A)] \tag{75b}$$

Equation (75), above, is obtained as a function of fold-length l by setting [see also Eqs. (7) and (8)]

$A_0 = (kT/h)\exp[-G_\eta/kT]\exp[-\Delta G_1/kT]$ (see Eq. (74) for ΔG_1)

$B/A = \exp[-(d\,\Delta G_l/d\nu)/kT]$ (see Eq. (73) for $d\,\Delta G/d\nu$)

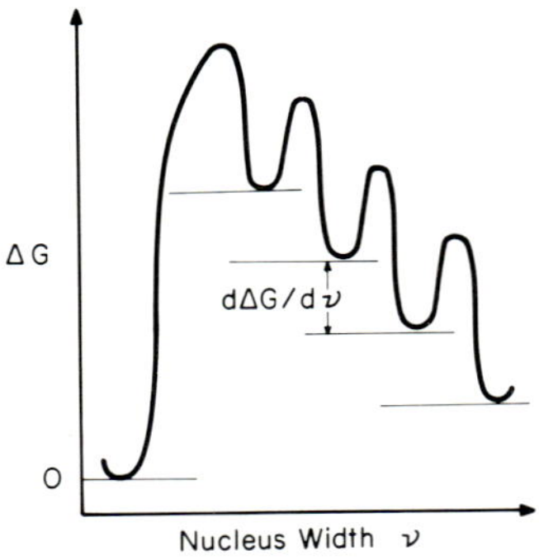

Fig. V.42 *Free Enthalpy Diagram of Nucleation and Growth of a Folded Chain Lamella.* ΔG is given by Eqs. (74), $d\ \Delta G/d\nu$ by Eq. (73).

$S(l)$ is thus the rate of appearance of new nuclei of fold-length l per mole of surface units per second at steady state which lead to new layers of chain folded material. The average thickness of the resulting single crystals from all possible lengths was computed by Gornick and Hoffman (1966) by evaluating

$$\langle l \rangle = l_g{}^* = \int_{l*}^{\infty} S(l) l\, dl \Big/ \int_{l*}^{\infty} S(l)\, dl \tag{76}$$

The final expression for $l_g{}^*$ can be shown to be

$$l_g{}^* = \frac{2\gamma_e}{\Delta g_f} + \frac{kT}{2b_0\gamma} \left[\frac{(4\gamma/a_0) - \Delta g_f}{(2\gamma/a_0) - \Delta g_f} \right] \tag{77}$$

The first term on the right-hand side of Eq. (77) contains the major $1/\Delta T$ dependence at low supercooling, it is equal to the critical fold-length of the secondary nucleus given in Eq. (67). The factor $kT/2b_0\gamma$ is about 10 Å in the temperature region of interest for the standard parameters chosen for Figs. V.36 and V.39. The factor in brackets is about two at small supercooling ($\Delta T < 10°C$) and increases slowly at larger supercooling.

At large supercooling the term in brackets in Eq. (77) increases rapidly, to approach infinity when $2\gamma/a_0$ becomes equal to Δg_f (in our sample calculation at $\Delta T = 42°C$). At low supercooling Eq. (77) has thus the proper form to represent the experimental data given in Eq. (72); however, the quantitative fit is not perfect. The rapid increase in l_g^* predicted by Eq. (77) has not been observed. This problem which concerns crystal growth is discussed in more detail in Sect. 6.1.4.

Computer simulations of secondary nucleation of polymer chains have been carried out by Binsbergen (1970c). His results could be represented by

$$\Delta G^* = C + 3.0(\gamma\gamma_e)^{1.25}(b_0^6\, a_0\, c_0)^{0.25}/[\Delta g_f(kT)^{0.5}] \tag{78}$$

with C representing a ΔT independent (but T and $\gamma\gamma_e$ dependent) term. The constants a_0, b_0, and c_0 are the dimensions of the unit of the chain added per fundamental step of crystallization. Equation (78) shows some similarity to Eq. (68) [and even more to Eq. (82) which describes molecular nucleation]. The ΔT dependence remains the same; however, the exponents of the surface free energy are somewhat larger. Checking the shape of typical computer generated secondary nuclei reveals for us, as for primary nucleation (Sect. 5.1.1), large deviation from the shapes assumed in the classical theory. The ΔT constant, additional term C in Eq. (78) may also show up in simulated expressions for l^* and bring better agreement with the experiment [Eq. (72)].

Thus, the final observed fold-length depends not only on the classical secondary nucleus length, but is altered by fluctuations, which must continue all through later crystal growth. Some fold-length fluctuation calculations have been attempted and are described in Sect. 6.1.4.3.

Fold-lengths of crystals grown from the melt, as displayed in Fig. III.13 for polyethylene, show deviations from the secondary nucleation theory also at low supercooling. The fold-lengths increase much faster than predicted by a $1/\Delta T$ dependence. In this case evidence points to a rearrangement of the folds to greater length after initial crystallization. Thus in some polymers, extended chain crystals can be obtained by secondary chain extension (see Sects. 3.3.1, 6.3.3, and 7.1.3).

5.2.3 General Conclusions about Secondary and Tertiary Nucleation

There is direct morphological evidence that secondary nucleation is of importance for macromolecular crystal growth (Fig. V.38). Crystallization rates from macromolecular melts (Fig. V.37) indicate a temperature dependence as predicted by classical secondary nucleation theory [Sect. 5.2.2.1, Eq. (68)]. It is shown, however, in Sect. 5.3 that molecular nucleation has the same temperature dependence and may be the crystallization rate determining step (see also Sect. 6.1). Only for crystallization during polymerization would one expect at low supercooling a crystal growth governed by nondestructable growth spirals as found in the crystallization of small molecules (Sect. 5.2.2.1). The tertiary nucleation concept is of little importance for crystallization of small molecules. Its formalism has found application in the description of simplified models of chain-folded crystallization [Eqs. (73) and (74)].

The greatest interest in secondary nucleation was created by the observation that the final fold-length of solution grown crystals [Eq. (72)] is close to the predicted average nucleus dimension [Eq. (77)]. The supercooling dependence could also be fitted, but only over a limited temperature range (see Sect. 6.1.4).

The failure of quantitative agreement points to oversimplifications in the application of the classical secondary nucleation model. It may even be possible that in some cases no secondary nucleation is necessary because of large deviations of the growth forms of the crystal on a microscopic scale from the ideal equilibrium shape (Sect. 5.2.2.1).

5.3 Molecular Nucleation

Molecular nucleation is the term applied to the process which establishes the first part of a macromolecule in the crystalline phase. It is of particular importance for the crystallization of already polymerized molecules. Molecular nucleation is to be distinguished from nucleation of the crystal, which is the main topic dealt with in Sects. 5.1 and 5.2 and is concerned with the start of a new crystal or a new crystal layer, usually irrespective of the parts of macromolecules contributing to the growth of the nucleus.

In some cases molecular nucleation is intimately connected with crystal nucleation. The primary nucleus of a crystal grown from macromolecules is normally smaller than a single molecule (see Sect. 5.1.2); it must thus involve both molecular and crystal nucleation simultaneously. As a first approximation for a folded chain primary nucleus (Sect. 5.1.2.2), it was possible to neglect the effect of the noncrystallized chain ends. One can interpret the success of this approach as an indication that crystal nucleation is the dominant feature in primary nucleation of folded chain crystals.

In contrast, for fringed micelle primary nucleation the molecular nucleation dominates (see Sect. 5.1.2.1). Each new chain traverse in the nucleus belongs to a new molecule. The molecular nucleation effect is taken into account through the extra free energy term γ_e' in Eq. (29). A change in nucleation mechanism from folded chain to fringed micelle type is only possible if the mobility (and flexibility) of the macromolecule is restricted to reduce the magnitude of γ_e'. One can interpret this result as an indication that the need of multiple molecular nucleation retards the fringed micelle formation relative to the folded chain nucleus.

Analyzing secondary nucleation (see Sect. 5.2), one predicts a smaller nucleus than for primary nucleation. Thus, molecular nucleation should be of greater importance. Even in the case of folded chain secondary nucleation, it may be necessary to consider molecular nucleation. Again, as in the case of primary nucleation, since the molecular nucleation term would be added to the surface free energy [see Eq. (29)], there seems to be no easy way of experimental verification as long as independent estimates of the surface free energy of the fold surface are unavailable.

After primary and secondary nucleation, molecular nucleation must still take place every time a new molecule is to be added to the crystal. The key experimental observation indicating molecular nucleation is the rejection of molecules of lower molecular weights from the growing crystal surfaces, despite the fact that they are below their respective melting temperature. Other indications for molecular nucleation arose from the discussion of molecular structure on crystal growth rate [Andrews (1974); Sect. 6.1.5] and the detailed description of linear crystal growth rates and its concentration dependence (Sect. 6.1.4.5). Qualitative experimental evidence for molecular nucleation from molecular weight segregation will be described first (Sect. 5.3.1), to be followed by a description of quantitative data of segregation of molecular weights on crystallization of polyethylene from solution and from the melt (Sect. 5.3.2 and 5.3.3). This is followed by a discussion of the initial theory of molecular nucleation.

5.3.1 Some Qualitative Experimental Evidence for Molecular Nucleation

Qualitative evidence for molecular nucleation can be derived from the observation of segregation of molecular weights into different crystals. For polyethylene crystallized slowly from the melt at temperatures between 125 to 130°C, the final folded chain crystals are between 200 Å and 300 Å thick (see Sect. 3.2.2.2 and Fig. III.13). Molecular weight segregation occurs up to about 12 000 molecular weight as was shown by Anderson (1964). The low molecular weights could be found as fully extended, separate crystals, while the higher molecular weight was contained in folded chain lamellae of mixed molecular weights. Similar results were observed on crystallization of molten polyethylene at similar supercooling but at elevated pressure; in this case the mixed crystals were of the extended chain type (Prime and Wunderlich, 1969a). An analysis of these observations shows clearly the need for a reversible nucleation step involving each molecule, as is shown next.

Figure V.43 is a plot of the equilibrium melting points T_m^0 of fully extended chain paraffin and polyethylene crystals as established experimentally by Prime and Wunderlich (1969b) (see also Chapter VIII). Also shown is the temperature T_m of high molecular weight polyethylene crystals of fold-length L at which their free enthalpy is equal to the free enthalpy of the melt (zero enthalpy production melting temperature, see also Chapter IX). The folded chain crystal is more stable than the lower molecular weight fully extended chain crystal of molecular length equal to the fold-length. At about 250 Å

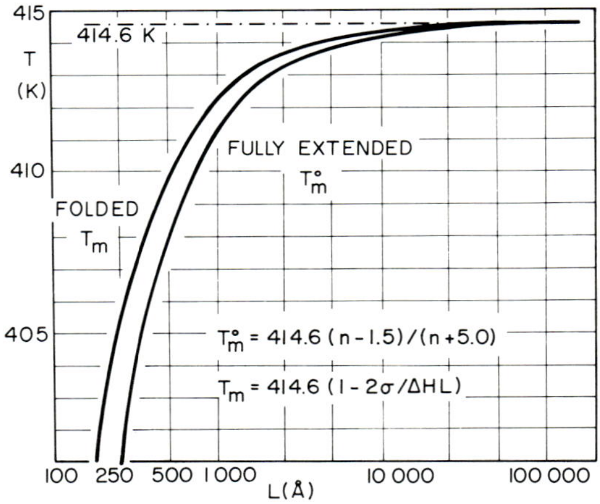

Fig. V.43 *Melting Temperatures of Polyethylene Crystals as a Function of Lamellar Thickness. L*
The curve for folded chain crystals is calculated for high molecular weight, assuming melting occurs whenever the free enthalpy of the crystal of given L is identical to the free enthalpy of the supercooled melt (zero entropy production melting, see Chapter IX). The curve for fully extended chain crystals represents the true equilibrium melting temperatures T_m^0. For data see Prime and Wunderlich (1969b).

fold-length, or about 2750 molecular weight, the difference between the two curves is 3.2°C. At about 1000 Å fold-length (11 000 molecular weight), the difference between the two curves is only 0.9°C and decreases rapidly to zero for higher molecular weights. Based on equilibrium considerations, crystallization at 130°C could at best explain segregation of fully extended chain crystals of up to 3200 molecular weight (290 Å length). Somewhat larger molecular weight segregation may come about because of the melting point becoming lower due to the solvent effect of the lower molecular weights. However, this cannot reach the 8°C lowering of the melting temperature necessary to make 1000 Å long molecules stay molten. (Typical values are 0.5 to 3°C; see Chapter VIII and Sect. 5.3.4).

A solution to the problem of segregation is the assumption that every molecule must go through a molecular nucleation step. Only when the molecule is long enough to produce a nucleus of sufficient size to overcome the fold surface and chain end effects, can crystal growth occur.

Molecular nucleation would then explain the irreversibility of melting and crystallization. As long as the volume of a molecule which remains attached to a crystal is larger than the molecular nucleus size, its crystallization and melting should be reversible. As soon as the molecule is removed from the crystal surface, however, new nucleation is necessary which requires increased supercooling. Similarly, segregation on crystallization is not governed by the equilibrium length representing the lower limit of crystallization, but by the larger size of the molecular nucleus under the given condition of temperature, concentration, and surface geometry.

5.3.2 Segregation on Crystallization from Solution

Early work on segregation on crystallization from solution was done in connection with the analysis of fractionation of molecular weight distributions (see for example Johnson *et al.*, 1967 and Konigsveld, 1970). It was observed that while liquid–liquid solution separation was rapid and reversible, separation of crystallizing macromolecules was a poorly reversible process showing large supercooling effects. It was quickly realized that equilibrium considerations should permit fractionation (lower curve, Fig. V.43), but that kinetic effects may prevent at low supercooling the precipitation of larger molecular weights (Flory, 1953). Thus, whenever possible, the crystalline phase was avoided for fractionation by choosing poor solvents which permitted liquid–liquid fractionation above or a little below the crystal dissolution temperature. These difficulties were generally accepted to preclude fractionation on crystallization, despite the fact that Richards (1946) had demonstrated that at least in the low molecular weight region polyethylene could be fractionated by cooling a 2.5% solution of the polymer in benzene to room temperature. The lower the average molecular weight of the sample, the sharper was the fraction of the polymer which remained dissolved.

Fractionation is obviously even more effective when, in addition to the molecular weight varying from molecule to molecule, the crystallization ability of some molecules is affected by branching (polyethylene, Hawkins and Smith, 1958), stereoblock copolymerization (polypropylene, Natta *et al.* 1958; poly(propylene oxide) Allen *et al.*, 1964), or other defects which cannot be accommodated in the host crystal (see Sects. 2.4.8 and 4.3.4). Booth and Price (1966) could show that the fractionation effect of stereoregularity is larger than the molecular weight effect particularly for high molecular weights. They compared the fractionation of stereoregularity affected poly(propylene oxide) with a broad molecular weight distribution of poly(ethylene oxide) which has no stereoisomers. The latter polymer showed over rather narrow crystallization temperature intervals (about 1–5°C) only segregation of the lowest molecular weights in various solvents (in the most favorable case up to 40 000 molecular weight out of an average molecular weight of 90 000).

Studies of segregation on crystallization from dilute solution in conjunction with the analysis of the crystallization process were made by Konigsveld and Pennings (1964), Kawai and Keller (1964), and Peterlin and Meinel (1964a,b). Crystals and filtrates of polyethylene and polypropylene were analyzed by these authors with respect to the average molecular weights after isothermal crystallization. The precipitated crystals clearly contained the larger molecules, while the filtrates were enriched in shorter molecules. In the lower molecular weight range, quite sharp separation could be achieved. Crystallization of linear polyethylene from 0.5% solution in xylene at 85°C left 12% uncrystallized polymer. Redissolution of the crystals and recrystallization at 85°C left only 0.5% in solution, and no dissolved polymer was observed on a further redissolution and crystallization. Koningsveld and Pennings (1964) concluded from these observations that sharper separations with regard to molecular weight are possible on crystallization than on liquid–liquid separations under similar conditions.

Detailed molecular weight distributions of precipitated polyethylene crystals from xylene solution and of the supernatent solution were finally carried out by Prime and Wunderlich (1969a) and Sadler (1971). Figure V.44 illustrates the cumulative moleculecular weight distributions of crystals and supernatent liquid at 85°C (0.7 wt % polymer in *p*-xylene). In Table V.5 the weight and number average molecular weights are listed for polyethylene left uncrystallized at the given crystallization temperature. Average molecular weights of 10^4 can easily be separated by crystallization. Repeated crystallization at increasing temperatures as suggested by Koningsveld and Pennings

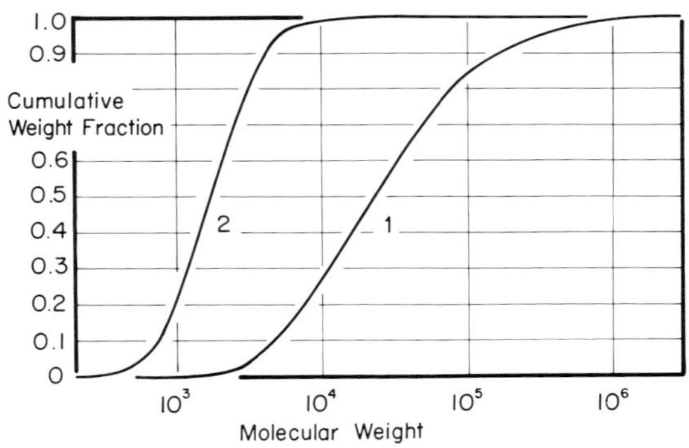

Fig. V.44 *Cumulative Molecular Weight Distributions of Polyethylene Precipitate (Curve 1) and Filtrate (Curve 2).* Concentration at the start of crystallization, 0.7 wt.% in *p*-xylene, with crystallization temperature, 85°C. Data from Prime and Wunderlich (1969a).

TABLE V.5

AVERAGE MOLECULAR WEIGHTS OF POLYETHYLENE LEFT DISSOLVED AFTER CRYSTALLIZATION FROM XYLENE[a]

Crystal. temp. (°C)	Weight av. mol. wt	No. av. mol. wt	Left in sol. (wt %)
92.8[b]	11 000	5 030	23
90.0[b]	4 300	2 400	—
87.6[b]	4 280	2 600	10
83.5[c]	2 580	1 960	—
75.1[c]	2 050	1 670	—
72.6[b]	1 900	1 530	1.2
70.0[b]	1 430	1 220	—
67.0[b]	1 400	1 240	—
65.0[b]	1 208	1 062	—
50.0[b]	1 050	1 030	0.32
23.0[b]	460	420	—

[a] Data by Sadler (1971); after 48 hr crystallization (except at 92.8°C crystalization temperature where crystallization was allowed for 150 hr).
[b] 0.75 % initial concentration.
[c] 0.1 % initial concentration.

(1964) permits further segregation of the higher molecular weights. Initial crystallization of broad molecular weight distribution polyethylene from *p*-xylene at 85°C and redissolution of the precipitate, followed by recrystallization at 90°C and again redissolution of the precipitate and final recrystallization at 92.5°C led to the following fractions (Prime and Wunderlich 1969a):

Samples	Weight average, $\overline{M}_w$	Number average, $\overline{M}_n$	Ratio $\overline{M}_w/\overline{M}_n$
Starting polymer	138 000	8 880	15.5
85°C solution	2 550	1 640	1.6
90°C solution	24 100	6 300	3.8
92.5°C solution	101 000	29 200	3.5

Separation of higher molecular weights in this manner is not possible since crystallization times become unreasonably large. Special techniques have been devised by Pennings (1967) for separation of the higher molecular weights by using stirring. In this case fibrous crystals of polyethylene and polypropylene were produced. Higher molecular weights are preferentially included in the inner portions of the fibers (see Sects. 3.8.2, 5.1.4.5, and 6.2.2). The extremely high molecular weight portion can be removed from the solution by high temperature epitaxy (see Sect. 5.1.4.1). It is thus possible to segregate the various fractions of different molecular weight by judiciously adjusting crystallization conditions. This fractionation is, however, largely based on kinetic rather than on equilibrium effects.

5.3.3 Segregation on Crystallization from the Melt

The earliest suggestion of segregation on crystallization from the macromolecular melt came from morphological studies. Extended chain crystals of different thickness in the molecular chain direction were observed by electron microscopy of melt grown crystals of polytetrafluoroethylene and it was speculated by Bunn *et al.* (1958) that perhaps "The molecules may be to some extent locally segregated into regions of roughly uniform molecular length" (see Fig. III.11). It turned out later that the molecular weights in the polytetrafluoroethylene sample were too large to correspond to the observed lamellar thickness. The same explanation of segregation according to molecular weight held, however, for the lower molecular weight tail of slowly crystallized polyethylene (Anderson, 1964; and Geil *et al.*, 1964; also see Sects. 5.3.1 and 3.3.1).

More direct evidence of segregation on crystallization of polyethylene from the melt (as well as from solution) was provided by Bank and Krimm (1969, 1970) through analysis of the infrared spectrum of mixed crystals of polyethylene and deuteropolyethylene. At the higher temperature of crystallization some preferential crystallization of the higher molecular weight deuteropolyethylene occurred under rejection of the lower molecular weight polyethylene species. Less or no such segregation occurred at 121°C. More evidence for rejection is the accumulation of tritium-labeled low molecular weight polyethylene ($M_n = 320$) on the crystal surfaces as observed by Moyer and Ochs (1963) (see also Sect. 4.1.7).

Kardos *et al.* (1971) crystallized mixtures of narrow fractions of polyethylene of 11 260 ($\overline{M}_w/\overline{M}_n = 1.2$) and 100 000 ($\overline{M}_w/\overline{M}_n = 1.5$) under elevated pressure so that both fractions crystallized when pure in the form of fully extended chain crystals of about 1000 and 9000 Å lamellar thickness, respectively (up to 3000 bar pressure, see also Sect. 3.3.1). On mixing melts of both fractions in different ratios and crystallizing such that mainly fully extended chain crystals resulted, two distinct lamellar thicknesses could be observed, corresponding to the two components of the mixture. The morphology suggested that the high molecular weight lamellae grew first, rejecting the lower molecular weight fraction which crystallized later.

Quantitative evaluation of the morphology of broad molecular weight distributions of polyethylene crystallized under elevated pressure to extended chain crystals revealed a molecular weight distribution and crystal thickness

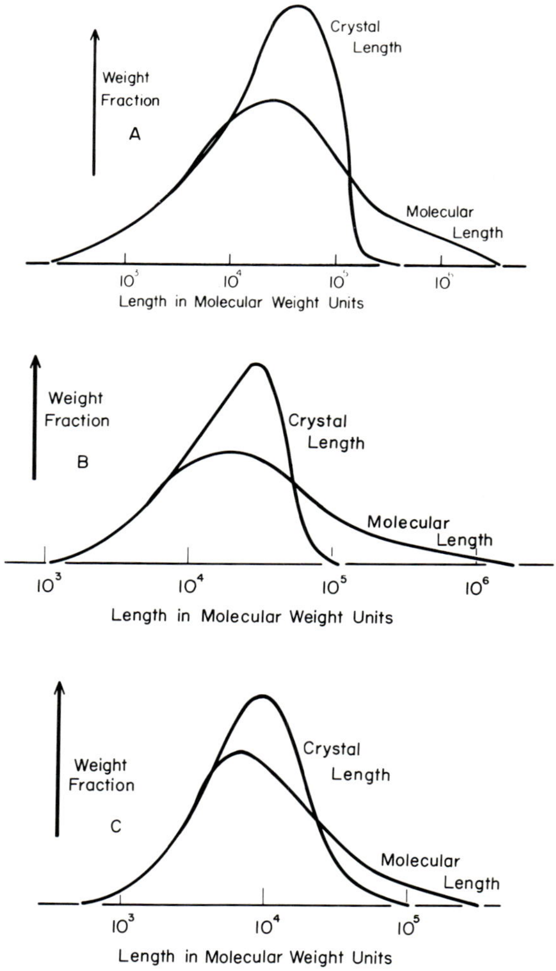

Fig. V.45 *Molecular Weight and Crystal Size Distributions of Polyethylene.*

Crystallized at 4800 bar and 215–225°C, followed by annealing to yield crystals of high perfection and chain extension. Note that 1 Å crystal thickness corresponds to about 11 molecular weight units. Measurements on 300 to 1000 randomly chosen lamellae for each sample. The differential distribution curves were normalized and replotted on a logarithmic scale. Curve A: broad molecular weight sample; Curve B: narrower molecular weight sample; Curve C: lower molecular weight sample. Data from Prime and Wunderlich (1969a).

distribution match up to about 10 000 molecular weight (Prime and Wunderlich (1969a). Figure V.45 illustrates the molecular weight and crystal length distributions of three different samples. A sample with even lower molecular weight than distribution A in Fig. V.45 led to fully extended chain crystals, a sample with no low molecular weight fraction showed no crystals of length smaller than would correspond to 10 000 molecular weight (see Fig. III.37). Melting experiments which can detect the true free enthalpy of the crystals grown were in agreement with the morphological evidence (see Chapter VIII and Prime and Wunderlich, 1969b).

Mehta and Wunderlich (1974) developed an extraction method to separate supernatant melt and crystals of polyethylene after completion of crystal growth. After crystallization, the sample was quenched in ice water to poorly crystallize the supernatant melt. The resulting sample showed a two-peaked differential thermal analysis melting curve (see Chapter IX) amenable to quantitative analysis. The low temperature melting peak was due to the poor

TABLE V.6

CRITICAL MOLECULAR WEIGHTS OF POLYETHYLENE LEFT MOLTEN AFTER CRYSTALLIZATION FROM THE MELT[a]

Cryst. temp. (°C)	Av. critical mol. wt[b]	Wt % not crystallized
100	(600)[c,d]	—
110	(1 000)[c,d]	—
115	3 100[d]	8
120	3 700[d]	10
125	4 400[e]	12
127	7 300[e]	16
129	22 300[e]	20
238[f]	30 300[e]	34

[a] Data by Mehta and Wunderlich (1974), broad molecular weight sample, $\overline{M}_w = 153\ 000$, $\overline{M}_n = 8530$.

[b] Determined after extended time crystallization (1 day up to 120°C crystallization temperature, 2 days at 125°C, 10 days at 127°C, 21 days at 129°C, 24 hours at 238°C at 5 kbar pressure).

[c] Data from a low molecular weight sample since not enough polymer was left uncrystallized for analysis of the sample described above.

[d] Data from thermal analysis.

[e] Data from molecular weight analysis of extracted low melting polymer (see text).

[f] Sample crystallized at elevated pressure (5 kbar), followed by slow cooling. The critical molecular weight corresponds approximately to a crystallization at 129.6°C at atmospheric pressure.

crystals grown from the supernatant melt on quick cooling, the high temperature melting peak was due to the crystals grown at the crystallization temperature. Fractional dissolution was used for complete separation of the two fractions. Table V.6 illustrates the change in critical molecular weight with crystallization temperature. The critical molecular weight is chosen such that 90% of the crystallized molecules are higher in molecular weight and 90% of the molecules in the supernatant melt are lower in molecular weight. Figure V.46 shows the decreasing segregation on increasing supercooling through the double peaked differential thermal analysis curves. The low temperature melting peak corresponds to the low molecular weight polymer crystallized only after quenching in ice water. Figure V.47 illustrates the sensitivity of the methods by showing the decreasing low temperature melting peak with decreasing content of low molecular weight in the sample. The data on segregation from the melt are similar to those derived from solution (see Sect. 5.3.2). Both sets of data are analyzed in more detail in Sect. 5.3.4 in the light of molecular nucleation.

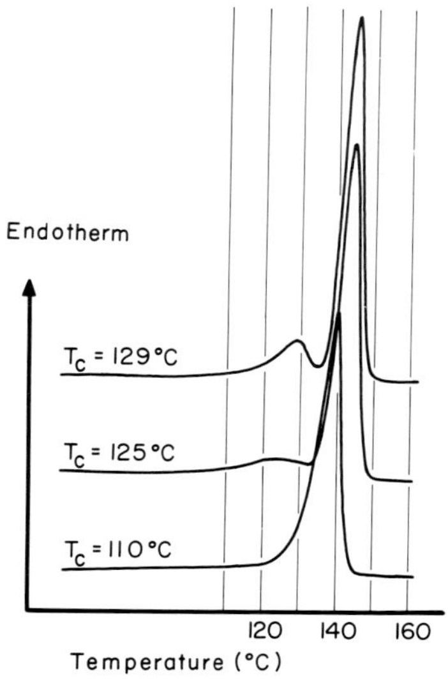

Fig. V.46 *Melting Curves of Polyethylene Crystallized at Different Temperatures.*
The low molecular weight fraction not crystallized at the given crystallization temperature was crystallized poorly by quenching in ice water, giving rise to the low temperature melting peak. Polymer as in Fig. V.45A. Data from Mehta and Wunderlich (1974).

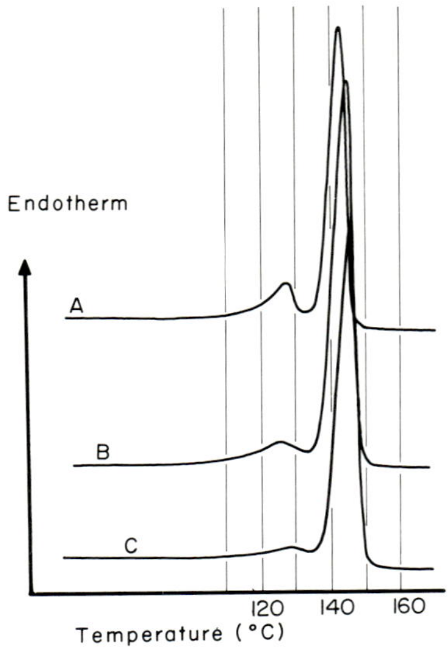

Fig. V.47 *Melting Curves of Polyethylene of Different Molecular Weight Distributions Crystallized at 127°C for 10 Days.*
The decreasing low temperature melting peak is parallel to the decreasing amount of low molecular weight polymer. Molecular weights: Curve A, $\overline{M}_n = 5530$, $\overline{M}_w = 92\,500$; Curve B, $\overline{M}_n = 8530$, $\overline{M}_w = 153\,000$; Curve C, $\overline{M}_n = 21\,000$, $\overline{M}_w = 311\,000$. Data from Mehta and Wunderlich (1974).

5.3.4 Concepts of Molecular Nucleation

The data of the previous two sections reveal quantitative evidence that segregation of molecular weights does occur on crystallization of linear macromolecules from the melt and from solution. It can also be seen over a considerable range of temperature that the segregated species are larger than the crystals are thick in the molecular chain direction (compare Tables V.5 and V.6 with Figs. III.12 and III.13). Furthermore, data as presented in Fig. V.44 illustrate that the segregation is relatively sharp, i.e., in the example a critical length of molecule can be found (400 Å) so that at least 90% by weight of the molecules in the supernatant solution are shorter than this length and that at least 90% by weight of the molecules in the crystallized polymer are longer than this critical length. The fold-length for these crystals is only 135 Å; a 400 Å long molecule must thus make three traverses through the crystal. In case the molecule is slightly shorter, it is most likely redissolved after initial crystallization. Primary and secondary nucleation of the crystal cannot account for this segregation since these occur too infrequently to have 90% efficiency. The solution to this problem is molecular nucleation.

The lower limit of segregation is set by the hypothetical equilibrium of crystallization. Irrespective of growth kinetics, it is assumed for this purpose that dynamic equilibrium between fully extended chain crystals and the surrounding solution can be achieved. At equilibrium the molecular length of the crystallizable species would correspond rather sharply to the lamellar thickness in the molecular chain direction. Molecules too short or too long would introduce defects and increase the overall possible free enthalpy and thus would be rejected by the crystal. The melting and crystallization temperature T_m^0 of a melt of a given well-defined molecular weight is represented by the lower curve of Fig. V.43. The effect of the different species in the melt or in solution can be computed through the Flory–Huggins expression (see Chapter VIII). Overall the equilibrium has been represented by Prime and Wunderlich (1969b) as

$$(1/T_m) - (1/T_m^0) = (R/\Delta H_p)[-\ln v_p + (\bar{x} - 1)(1 - v_p) - \bar{x}\chi(1 - v_p)^2] \quad (79)$$

where T_m is the temperature of the melting–crystallization equilibrium considered; T_m^0 is the equilibrium melting–crystallization temperature of the pure species of the molecular weight considered (given by the lower curve of Fig. V.43); v_p is the volume fraction of the considered molecular weight species in the melt or in solution; ΔH_p is the molar heat of fusion of the considered molecular weight species; χ is an interaction parameter; and $\bar{x}$ is the appropriate average molecular volume ratio of the crystallizing species to all other species i in the melt

$$\bar{x} = \sum v_i x_i / \sum v_i \quad (80)$$

A typical set of phase diagrams for different molecular weights assuming $\chi = 0$ (which may be acceptable for melts) is shown in Fig. V.48.† With the help of similar phase diagrams for all fractions present in a given initial molecular weight distribution, it is possible to establish the equilibrium concentrations in the melt at any chosen temperature. The calculations are rather

† Closely related expressions to Eq. (79) were also used by Kawai (1965, 1967), Sadler (1971), and Kawai *et al.* (1971) for the description of fractionation as a cause for changing lamellar thickness (see, however, Mehta and Wunderlich 1972); Kilian (1965) and Asbach *et al.* (1967) used a similar expression with an additional surface free energy term for the description of pseudo-eutectic homopolymer and copolymer melting.

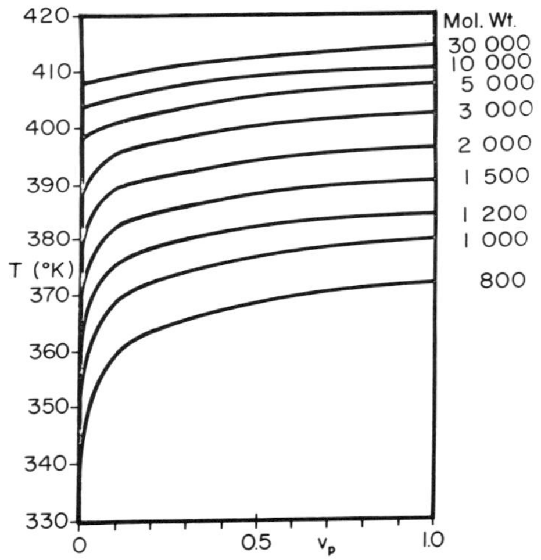

Fig. V.48 *Phase Diagrams for Polyethylene at Melting–Crystallization Equilibrium.*
The noncrystallizing molecules of fraction (1-v_p) are assumed to have a molecular weight of 2500. The molecular weights of the crystallizing–melting species are listed in the figure. Curves calculated using Eq. (79) and the lower curve of Fig. V.43. Prime and Wunderlich (1969b).

tedious and must be carried out by successive iterations on a digital computer (Prime, 1968). The resulting molecular weight distribution curves at various temperatures in the crystallization region are similar to the experimental segregation curves shown in Fig. V.44. Critical molecular weights (which refer in this case to the molecular weight fraction that remains at the given temperatures up to 90% crystalline) have been extracted from these curves. Data are listed in Table V.7 for six different molecular weight distributions, ranging in polydispersity from 18 to 1.6. Despite the large variations in the assumed molecular weights and distributions, the critical molecular weights do not change much from sample to sample. At 120°C, for example, an equilibrium melt of polyethylene should have practically all molecules longer than 2000–3000 molecular weight (180–270 Å) crystallized and nearly all shorter molecules molten.

Figure V.49 illustrates in curve 1 the equilibrium critical molecular length for a typical broad molecular weight distribution. Also shown in Fig. V.49 are the experimental data described in Sects. 5.3.2 and 5.3.3 (curves 3 and 2). They approach the equilibrium curve at large supercooling, but show significant deviations at lower supercooling. Much higher molecular weight is segregated at low supercooling. The difference between the three curves is to be

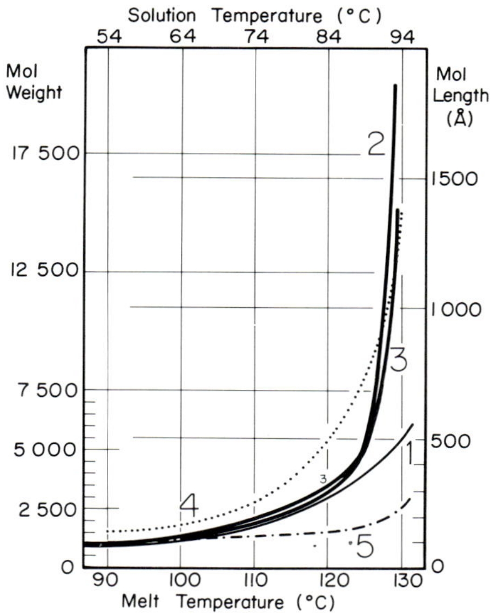

Fig. V.49 *Critical Molecular Weight (Length) of Polyethylene for Nucleation as a Function of Crystallization Temperature.*

Curve 1, lower limit set by equilibrium (Eq. (79), Table V.7, type M polymer). Curve 2, from selective extraction and thermal analysis data for melt crystallized type M polymer (see Table V.6). Curve 3, dilute solution data from the supernatent polymer molecular weight on crystallization of fractions and whole polymer of 0.5 to 0.7 wt.% in zylene (data by Konigsveld and Pennings, 1964; Prime and Wunderlich, 1969a; and Sadler, 1971; curve 3 has been shifted by 36°C to achieve coincidence with Curves 1 and 2. The upper temperature scale should be used for curve 3). Curve 4, calculated using Eq. (83). Curve 5, initial crystal thickness as found experimentally mainly from solution crystallization [Eq. (72), Fig. V.41]. Mehta and Wunderlich (1974).

however, that at least on an extended chain substrate, secondary nucleation is a rarer event than molecular nucleation. Solutions to this problem are at present only speculative (Mehta and Wunderlich, 1974). It may well be that the initial creation of a folded-chain secondary nucleus on an extended chain crystal surface takes a higher free enthalpy because of an initially irregular fold surface area. This could be accounted for in Eq. (68) by a considerably higher value of γ_e (see also Fig. V.9). Once regularized through reorganization after the nucleation step, the first folded chain layers form a "fold directing" substrate for further growth by eliminating large folds which would be the main cause of the excessive surface area.

Another point of interest in studying the results of Fig. V.49 is the 36°C shift of the solution data necessary to agree with the melt data. Dissolution and melting experiments on extended chain crystals of polyethylene have revealed a shift of the melting temperature of only about 28°C (Cormier and Wunderlich, 1966; Arakawa and Wunderlich, 1967). The added shift of 8°C could only be explained by a larger surface free energy of the molecular nucleus in contact with the solvent (xylene).

Lindenmeyer and Peterson (1968) and Lindenmeyer (1969) made an attempt to explain segregation on crystallization to be due to the increase in crystal free enthalpy on the inclusion or exclusion of chain ends in folded chain crystals. They could show that in case included chain ends cause an increase in free enthalpy, the overall free enthalpy would be lowered on segregation. Mechanistically it is, however, difficult to see in this case a path for segregation since the defect would not act as such on the surface. It would become active only after additional polymer layers have been deposited (see Sect. 4.3.4, Fig. IV.58). After deposition of additional layers it would be difficult to reverse the crystallizations process. In principle, the assumption of excluded chain ends is not different from the just discussed kinetic picture.

5.3.5 General Conclusions about Molecular Nucleation

The discussion of molecular nucleation rounds out the description of nucleation processes on crystallization of macromolecules. Experimentally the molecular nucleation effect is based on the theory of molecular weight segregation on crystallization beyond the equilibrium prediction (see Sects. 5.3.1 and 5.3.4 and Fig. V.49). It may well be that secondary nucleation of the crystal is only of importance on large substrates as shown in Figs. V.38 and III.81, and that the reason for secondary nucleation is a more irregular initial fold surface. Molecular nucleation, in turn, is the key to understanding the growth features which have been explained by secondary crystal nucleation (see Sects 5.2.1 and 6.1.2.1).

Formally, molecular nucleation has a larger free enthalpy barrier than secondary nucleation at equal nucleus size, and should thus be rate determining [Eqs. (82) and (68)]. The question of the meaning of the side surface free energy γ is left open. In order to show the $(1/\Delta T)^2$ dependence of the rejected molecular weight, a surface free energy parameter γ must be included [Eq. (83)]. Thus a new molecule would not be added onto a ledge of the prior existing crystal as assumed in secondary nucleation. Instead, one may expect each new molecule to start as a new, isolated patch on the surface, affected mainly by the underlying (perhaps already regularized) fold surface of given thickness. The detailed structure of the growth face seems only of secondary importance.

As already pointed out in Sects. 5.1.3 and 5.2.3, nucleation can be described in outline form by the classical nucleation theory by assuming lowest free enthalpy shapes. For more quantitative answers a more detailed discussion of the actual steps in going from the melt or from solution to the crystal is necessary. A large number of questions remain open, particularly for molecular nucleation. There must be some effect of the surface structure on the molecular nucleus broadening the molecular weight segregation. The term γ_e' is not well understood (see Sect. 5.1.2.1). It should be strongly length dependent, an effect difficult to assess. Our assuming a constant value in Eqs. (80)–(82) and basing the segregation on the size of the nucleus thus calculated, must result in an oversimplification since a matching molecule in a folded chain crystal should show γ_e' close to zero. That such matching can occur in sharp molecular weight fractions is illustrated in Fig. III.17. Thus details of the variation of γ_e' still need to be explored in connection with molecular nucleation. Differences from polymer to polymer have not been studied yet. Finally, one must raise the question whether polymer nucleation can be described at all without taking into account the cooperative nature of the process (see Sect. 6.1.4). At present one uses the side surfaces created on crystallizing a single chain on the surface as the reason for slowing nucleation to observed time scales. In reality there should be also a slowing due to the need to crystallize one unit after the other in sequence.

References

Alfonso, G. C., Olivero, L., Turturro, A., and Pedemonte, E. (1973a). Crystallization of high density polyethylene from dilute solution: Heterogeneous mucleation. *Brit. Polym. J.* **5**, 141.

Alfonso, G. C., Saccone, A., Valenti, B., and Pedemonte, E. (1973b). Crystal nuclei in the formation of low molecular weight polyethylene single crystals. *Chim. Ind. (Milan)* **55**, 941.

Allen, G., Booth, C., and Jones, M. N. (1964). Polypropylene oxide II-Dilute solution properties and tacticity. *Polymer* **5**, 257.

Amrhein, E. M., and Frischkorn, H. (1973). Das Absorptionsverhalten von Polyäthylen und anderen teilkristallinen und amorphen Polymeren zwischen 20 cm^{-1} und 600 cm^{-1}. *Kolloid Z. Z. Polym.* **251**, 369.

Anderson, F. R. (1964). Morphology of isothermally bulk-crystallized linear polyethylene. *J. Appl. Phys.* **35**, 64. See also (1962). Internal morphology of bulk crystallized polyethylene. *Int. Congr. Electron Microsc.*, **5***th*, *Philadelphia* 1362; (1965). Bulk fractionation of linear polyethylene. *J. Polym. Sci. Part C* **8**, 275.

Andrews, E. H. (1964). Crystalline morphology in thin films of natural rubber. II. Crystallization under strain. *Proc. Roy. Soc.* (*London*) **A277**, 562.

Andrews, E. H. (1974). Structure-property relationships in a polymer. *Angew. Chem. Int. Ed.* **13**, 113 (*Angew. Chem.* **86**, 151).

Arakawa, T., and Wunderlich, B. (1967). Thermodynamic properties of extended chain polymethylene single crystals. *J. Polym. Sci. Part C* **16**, 653.

Asbach, G. I., Kilian, H. G., and Müller, F. H. (1967). Fractions of linear polyethylene as molecularly dispersed eutectic systems. *J. Polym. Sci.* **C18**, 133.

Ballard, D. G. H., Wignall, G. D., and Schelten, J. (1973). Measurement of molecular dimensions of polystyrene in the bulk polymer by low angle neutron diffraction. *Eur. Polym. J.* **9**, 965.

Bank, M. I., and Krimm, S. (1969). Mixed crystal infrared study of chain folding in crystalline polyethylene. *J. Polym. Sci. Part A2* **7**, 1785.

Bank, M. I., and Krimm, S. (1970). Mixed crystal infrared study of chain segregation in polyethylene. *J. Polym. Sci. Part B* **8**, 143.

Banks, W., and Sharples, A. (1963). The effect of melt temperature on nucleation in crystallizing polymers. *Macromol. Chem.* **64**, 42.

Baranov, V. G., Zurabyan, R. S., and Frenkel, S. Ya. (1971). Nucleus formation during crystallization of polymeric systems. *Fiz. Tverd. Tela* (*Leningrad*) **13**, 2336 [*English transl.*: *Soviet Physics.—Solid State* **13**, 1957 (1972)].

Barnes, W. J., and Price, F. P. (1964). Morphology of polymer crystals: Screw dislocations in polyethylene, polymethylene oxide and polyethylene oxide. *Polymer* **5**, 283.

Beck, H. N. (1967). Heterogeneous nucleating agents for polypropylene crystallization. *J. Appl. Polym. Sci.* **11**, 673.

Beck, H. N., and Ledbetter, H. D. (1965). DTA study of heterogeneous nucleation of crystallization in polypropylene. *J. Appl. Polym. Sci.* **9**, 2131.

Benoit, H. (1974). Determination of polymer chain conformation in amorphous polymers. *Polym. Preprints Div. Polym. Chem. Inc. Amer. Chem. Soc.* **15** (2), 324.

Benoit, H., Decker, D., Higgins, J. S., Picot, C., Cotton, J. P., Farnoux, B., Jannink, G., and Ober, R. (1973). Dimensions of a flexible polymer chain in the bulk and in solution. *Nature Phys. Sci.* **245**, 13.

Binsbergen, F. L. (1969). Nucleation in the Crystallization of Polyolefins. Thesis, Rijksuniversiteit te Groningen.

Binsbergen, F. L. (1970a). A revision of some concepts in nucleation theory. *Kolloid Z. Z. Polym.* **237**, 289.

Binsbergen, F. L. (1970b). Heterogeneous nucleation in the crystallization of polyolefins: Part I. Chemical and physical nature of nucleating agents. *Polymer* **11**, 253.

Binsbergen, F. L. (1970c). A reconsideration of polymer crystallization theory. *Kolloid Z. Z. Polym.* **238**, 389.

Binsbergen, F. L. (1972). Computer simulation of nucleation and crystal growth. *J. Crystal Growth* **16**, 249; see also **13–14**, 44 and Binsbergen (1970a).

Binsbergen, F. L. (1973). Heterogeneous nucleation in the crystallization of polyolefins III. Theory and mechanism. *J. Polym. Sci. Polym. Phys. Ed.* **11**, 117.

Binsbergen, F. L., and DeLange, B. G. M. (1970). Heterogeneous nucleation in the crystallization of polyolefins. 2. Kinetics of crystallization of nucleated polypropylene. *Polymer* **11**, 309.

Bittiger, H., and Husemann, E. (1966). Elektronenmikroskopische Untersuchungen über die Kristallisation von Makromolekülen aus verdünnter Lösung. 3. Über den Einfluss von Molekulargewicht und Konzentration auf Keimbildung und Kristallwachstum bei Cellulosetricarbanilaten. *Makromol. Chem.* **96**, 92.

Blundell, D. J., and Keller, A. (1968a). Nature of self-seeding polyethylene crystal nuclei. *J. Macromol. Sci.* **B2**, 301.

Blundell, D. J., and Keller, A. (1968b). Controlled crystal-growing procedures in polyethylene involving self-seeding: Some novel twinning habits. *J. Macromol. Sci.* **B2**, 337.

Blundell, D. J., Keller, A., and Kovacs, A. J. (1966). A new self-nucleation phenomenon and its application to the growing of polymer crystals from solution. *J. Polym. Sci. Part B* **4**, 481.

Boon, J., Challa, G., and van Krevelen, D. W. (1968). Crystallization kinetics of isotactic polystyrene. II. Influence of thermal history on number of nuclei. *J. Polym. Sci. Part A2* **6**, 1835.

Booth, C., and Price, C. (1966). The fractional crystallization of poly(ethylene oxide). *Polymer* **7**, 85.

Brown, N. (1967). Polymerization of formaldehyde. *J. Macromol. Sci.* **A1**, 209.

Bunn, C. W., Cobbold, A. J., and Palmer, R. P. (1958). The fine structure of polytetrafluoroethylene. *J. Polym. Sci.* **28**, 365.

Burns, J. R., and Turnbull, D. (1966). Kinetics of crystal nucleation in molten isotactic polypropylene. *J. Appl. Phys.* **37**, 402.

Burns, J. R., and Turnbull, D. (1968). Nucleation of crystallization in molten isotactic polybutene-1. *J. Polym. Sci. Part A2* **6**, 775.

Carr, S. H., Keller, A., and Baer, E. (1970). Relationship between self-seeded and epitaxial crystallization from polymer solutions: A potentially new method for molecular weight separation and a new decoration method for alkali halides. *J. Polym. Sci. Part A2* **8**, 1467.

Chernov, A. A. (1961). The spiral growth of crystals. *Usp. Fiz. Nauk* **73**, 277 [*English Transl. Sov. Phys.–Usp.* **4**, 116].

Cormia, R. L., Price, F. P., and Turnbull, D. (1962). Kinetics of crystal nucleation in polyethylene. *J. Chem. Phys.* **37**, 1333.

Cormier, C. M., and Wunderlich, B. (1966). Equilibrium dissolution temperatures of polyethylene and the surface free energy of folded-chain single crystals. *J. Polym. Sci. Part A2* **4**, 666.

Devoy, C., and Mandelkern, L. (1970). On the heterogeneous nucleation of long-chain molecules. *J. Chem. Phys.* **52**, 3827.

Dole, M. (1959). Thermodynamic properties of high polymers as a function of their pretreatment as determined by specific heat measurements. *Kolloid Z.* **165**, 40.

Fischer, E. W., Wendorff, J. H., Dettenmaier, M., Lieser, G., and Voigt-Martin, I. (1974). Chain conformation and structure in amorphous polymers as revealed by X-ray, neutron-, light- and electron scattering. *Polym. Preprints, Div. Polym. Chem. Inc. Amer. Chem. Soc.* **15** (2), 8.

Fitchmun, D. R., and Newman, S. (1970). Surface crystallization of polypropylene. *J. Polym. Sci. Part A2* **8**, 1545.

Flory, P. J. (1941). Thermodynamics of high polymer solutions. *J. Chem. Phys.* **9**, 660.

Flory, P. J. (1942). Thermodynamics of high polymer solutions. *J. Chem. Phys.* **10**, 51.

Flory, P. J. (1949). Thermodynamics of crystallization in high polymers. IV. A theory of crystalline states and fusion in polymers, copolymers and their mixtures with diluents. *J. Chem. Phys.* **17**, 223.

Flory, P. J. (1953). "Principles of Polymer Chemistry." Cornell Univ. Press, Ithaca, New York.

Flory, P. J. (1962). On the morphology of the crystalline state in polymers. *J. Amer. Chem. Soc.* **84**, 2857.

Fowkes, F. M. (1964). Attractive forces at interfaces. *Ind. Eng. Chem.* **56** (12), 40.

Frank, F. C. (1949). The influence of dislocations on crystal growth. *Discuss. Faraday Soc.* **5**, 48; see also discussion p. 67.

Frank, F. C. (1958). "Growth and Perfection of Crystals" (R. H. Doremus, B. W. Roberts, D. Turnbull, eds.), pp. 529–530. Wiley, New York.

Frank, F. C. (1970). The strength and stiffness of polymers. *Proc. Roy. Soc. London* **A319**, 127.

Frank, F. C., Keller, A., and Mackley, M. R. (1971). Polymer chain extension produced by impinging jets and its effect on polyethylene solution. *Polymer* **12**, 467.

Frank, F. C., and Tosi, M. (1960). On the theory of polymer crystallization. *Proc. Roy. Soc.* **263A**, 323.

Frisch, H. L. (1957). Time lag in nucleation. *J. Chem. Phys.* **27**, 90.

Frolova, A. A., and Kozlov, P. V. (1963). Acceleration of crystallization process in poly-(ethylene terephthalate) under the influence of mechanical force. *Dokl. Akad. Nauk* SSSR **149**, 1390 [*English transl.*: *Dokl. Phys. Chem.* **149**, 372].

Frolova, A. A., and Kozlov, P. V. (1965). Effect of deformation rate on structural transformation in crystallizing polymers. *Vysokomol. Soedin.* **7**, 432 [*English Transl.*: *Polym. Sci. USSR* **7**, 479].

Frolova, A. A., Kozlov, P. V., and Kargin, V. A. (1963). Effect of mechanical action on the rate of crystallization of isotactic polystyrene. *Dokl. Akad. Nauk SSSR* **153**, 349.

Geil, P. H. (1963). Primary nucleation of poly-3,3-bischloromethyloxacylobutane single crystals. *Polymer* **4**, 404.

Geil, P. H., Anderson, F. R., Wunderlich, B., and Arakawa, T. (1964). Morphology of polyethylene crystallized from the melt under pressure. *J. Polym. Sci. Part A* **2**, 3707.

Gibbs, J. W. (1878). On the equilibrium of heterogeneous substances. *Trans. Conn. Acad.* **III**, 343; see also "The Scientific Work of J. Willard Gibbs," Vol. I, p. 219. Longmans, Green, New York. 1906 and later.

Glasstone, S., Laidler, K. J., and Eyring, H. (1941). "The Theory of Rate Processes." McGraw-Hill, New York.

Gornick, F., and Hoffman, J. D. (1966). Nucleation in polymers. *Ind. Eng. Chem.* **58** (2), 41.

Gornick, F., Ross, G. S., and Frolen, L. J. (1967). Crystal nucleation in polyethylene: The droplet experiment. *J. Polym. Sci. Part C* **18**, 79.

Groeninckx, G., Berghmans, H., Overbergh, N., and Smets, G. (1974). Crystallization of poly(ethylene terephthalate) induced by inorganic compounds. I. Crystallization behavior from the glassy state in a low temperature region. *J. Polym. Sci. Polym. Phys. Ed.* **12**, 303.

Hammel, J. J. (1969). Nucleation in glass-forming materials; *In* "Nucleation" (A. C. Zettlemoyer, ed.). Dekker, New York.

Hartley, F. D., Lord, F. W., and Morgan, L. B. (1954). Crystallization phenomena in polymers. III. Effect of melt conditions and the temperature of crystallization on the course of the crystallization in polyethylene terephthalate. *Phil. Trans. Ser. A* **247**, 23.

Hawkins, S. W., and Richards, R. B. (1949). Light transmission and the formation and decay of spherulites in polythene. *J. Polym. Sci.* **4**, 515.

Hawkins, S. W., and Smith, H. (1958). The fractionation of polyethylene. *J. Polym. Sci.* **28**, 341.

Hermann, K., Gerngross, O., and Abitz, W. (1930). Zur röntgenographischen Strukturforschung des Gelatinemicells. *Z. Phys. Chem.* **B10**, 371.

Herring, C. (1953). The use of classical macroscopic concepts in surface-energy problems; *In* "Structure and Properties of Solid Surfaces" (R. Gomer and C. S. Smith, Eds.). Univ. of Chicago Press, Chicago, Illinois.

Hill, M. J., and Keller, A. (1969). Direct evidence for distinctive, stress induced nucleus crystals in the crystallization of oriented polymer melts. *J. Macromol. Sci.* **B3**, 153.

Hobbs, S. Y. (1971). Row nucleation of isotactic polypropylene on graphite fibers. *Nature (London) Phys. Sci.* **234**, 12.

Hoeve, C. A. J. (1971). Theory of polymer adsorption at interfaces. *J. Polym. Sci. Part C* **34**, 1.

Hoffman, J. D. (1958). Thermodynamic driving force in nucleation and growth processes. *J. Chem. Phys.* **28**, 1192.

Hoffman, J. D. (1964). Theoretical aspects of polymer crystallization with chain folds: Bulk polymers. *SPE Trans.* **4**, 315.

Hoffman, J. D., and Weeks, J. J. (1962). Rate of spherulitic crystallization with chain folds in polychlorotrifluoroethylene. *J. Chem. Phys.* **37**, 1723.

Hoffman, J. D., Weeks, J. J., and Murphy, W. M. (1959). Experimental and theoretical study of kinetics of bulk crystallization in poly(chlorotrifluoroethylene). *J. Res. Nat. Bur. St.* **63A**, 67.

Hoffman, J. D., Lauritzen, J. I., Jr., Passaglia, E., Ross, G. S., Frolen, L. J., and Weeks, J. J. (1969). Kinetics of polymer crystallization from solution and the melt. *Kolloid Z. Z. Polym.* **231**, 564.

Hoffman, R. F., and Forsman, W. C. (1970). Use of numerical methods in the prediction of polymer adsorption. *J. Polym. Sci. Part A-2* **8**, 1847.

Holland, V. F., and Lindenmeyer, P. H. (1962). Morphology and crystal growth rate of polyethylene crystalline complexes. *J. Polym. Sci.* **57**, 589.

Huggins, M. L. (1942). Some properties of long-chain compounds. *J. Phys. Chem.* **46**, 151.

Inoue, M. (1963). Nucleation effect on the kinetics of crystallization and the spherulites of nylon 6. *J. Polym. Sci. Part A* **1**, 2013.

Jackson, K. A. (1965). Nucleation from the melt. *Ind. Eng. Chem.* **57** (12), 29.

Johnsen, U., and Lehmann, J. (1969). Die Kinetik der Kristallisation des Poly-3,3-bis-(chloromethyl)-oxacylobutan (Penton) aus verdünnter m-Xylol-Lösung. *Kolloid Z. Z. Polym.* **230**, 317

Johnsen, U., Spilgies, G., and Zachmann, H. G. (1970). Abhängigkeit der heterogenen Keimbildung in Polypropylen von der Kristallisationstemperatur und von der Art der Beimengung der Fremdsubstanz. *Kolloid Z. Z. Polym.* **240**, 762.

Johnson, J. F., Cantow, M. J. R., and Porter, R. S. (1967). Fractionation. *Encyclop. Polym. Sci. Technol.* **7**, 231

Joppien, G. R. (1974). Strukturuntersuchungen an Adsorptionsschichten makromolekularer Stoffe, 1. Haftstellen von Linearen Polyestern und Polystyrol an Aerosiloberflächen. *Makromol. Chem.* **175**, 1931.

Kardos, J. L., Li, H. -M., and Huckshold, K. A. (1971). Fractionation of linear polyethylene during bulk crystallization under high pressure. *J. Polym. Sci. Part A2* **9**, 2061.

Kargin, V. A., Kitaigorodskii, A. I., and Slonimskii, G. L. (1957). Structure of Linear polymers. *Kolloid Zh.* **19**, 131 [*English transl.*: *Colloid J.* (*U.S.S.R.*) **19**, 141); see also Kargin, V. A. (1958). Structure and phase state of polymers. *J. Polym. Sci.* **30**, 247.

Kargin, V. A., Sogolova, T. I., and Meleskaya, T. K. (1962). The effect of fillers with anisodiametric particles on the properties of polymers. I. *Vysokomol. Soedin.* **4**, 601.

Kargin, V. A., Sogolova, T. I., and Shaposhnikova, T. K. (1963). The effect of fillers with anisodiametric particles on the properties of polymers. II. *Vysokomol. Soedin.* **5**, 921 [*English transl.*: *Polym. Sci. USSR*, **4**, 1645].

Kargin, V. A., Sogolova, T. I., and Kurbanova, I. I. (1966). Effect of concentration and phase condition of nuclei on the spherulite dimension and the properties of isotactic polypropylene. *Vysokomol Soedin.* **8**, 2104 [*English transl.*: *Polym. Sci. USSR* **8**, 2325].

Kargin, V. A., Sogolova, T. I., Rapoport, N. Ya, and Kurbanova, I. I. (1967). Effect of artificial nucleating agents on polymer structure and properites. *J. Polym. Sci. Part C* **16**, 1609.

Kawai, T. (1965). A note on fractionation of crystalline polymers. *J. Polym. Sci. Part B* **3**, 83; Molecular fractionation and thickening of lamellar crystals during isothermal crystallization. *Makromol. Chem.* **84**, 290.

Kawai, T. (1967). Fractionation of polymers from the crystalline state. *Makromol. Chem.* **102**, 125.

Kawai, T., and Keller, A. (1964). Some effects on molecular fractionation during single crystal growth of polyethylene. *J. Polym. Sci. Part B* **2**, 333.

Kawai, T., Hosoi, M., and Kamide, K. (1971). Thermal analysis of lamellar thickening during isothermal crystallization. *Makromol. Chem.* **146**, 55.

Keller, A., and Sadler, D. M. (1970). Self-seeded crystallization and its potential for molecular weight characterization. II. Experiments on narrow fractions. *J. Polym. Sci. Part A2* **8**, 1457.

Keller, A., and Willmouth, F. M. (1970). Self-seeded crystallization and its potential for molecular weight characterization. I. Experiments on broad distributions. *J. Polym. Sci. Part A2* **8**, 1443.

Kilian, H. G. (1965). Partiell kristallisierende Copolymere als pseudoeutektische Systeme. *Kolloid Z.* **203**, 97.

Killmann, E., and Eckart, R. (1971). Kalorimetrische Untersuchungen zur Adsorption von Makromolekülen aus verdünnten Lösungen. *Makromol. Chem.* **144**, 45.

Kirste, R. G., Kruse, W. A., and Schelten, J. (1973). Die Bestimmung des Trägheitsradius von Polymethylmethacrylat im Glaszustand durch Neutronenbeugung. *Makromol. Chem.* **162**, 299.

Konigsveld, R. (1970). Preparative and analytical aspects of polymer fractionation. *Advan. Polym. Sci.* **7**, 1.

Konigsveld, R., and Pennings, A. J. (1964). Crystallization fractionation of polymers. *Rec. Trav. Chim.* **83**, 552.

Kossel, W. (1927). Zur Theorie des Krystallwachstums. *Nachr. Ges. Wiss. Göttingen, Math.-Phys.* **Kl.** 135; (1928). *Phys. Z.* **29**, 553.

Koutsky, J. A., Walton, A. G., and Baer, E. (1967a). Nucleation of polymer droplets. *J. Appl. Phys.* **38**, 1832.

Koutsky, J. A., Walton, A. G., and Baer, E. (1967b). Heterogeneous nucleation of polyethylene melts on cleaved surfaces of alkali halides. *J. Polym. Sci.* **B5**, 185.

Kovacs, A. J., and Manson, J. A. (1966). Propriétés des copolymères bisequencés polyoxyéthylène polystyrène. III. Cinétique decroissance des monocristaux en solution. *Kolloid Z. Z. Polym.* **214**, 1.

Lacmann, R. (1961). Über die Kristallisation auf Unterlagen. Die Keimbildungsarbeiten. *Z. Krist.* **116**, 13.

Lanceley, H. A., and Sharples, A. (1966). A kinetic study of polymer crystallization from solution. *Makromol. Chem.* **94**, 30.

Last, A. G. M. (1959). Nucleation effects in high polymers. *J. Polym. Sci.* **39**, 543.

Laudise, R. A. (1970). "The growth of Single Crystals." Prentice-Hall, Englewood Cliffs, New Jersey.

Lauritzen, J. I., Jr., and Hoffman, J. D. (1960). Theory of formation of polymer crystals with folded chains in dilute solution. *J. Res. Nat. Bur. St.* **64A**, 73.

Lenz, R. W. (1967). "Organic Chemistry of Synthetic High Polymers." Wiley (Interscience), New York.

Lindenmeyer, P. H. (1969). Imperfections in polymer crystals. *Kolloid Z. Z. Polym.* **231**, 593.

Lindenmeyer, P. H. (1973). Molecular confirmation in molten and glassy polymers. *J. Macromol Sci.* **B8**, 361.

Lindenmeyer, P. H., and Peterson, J. M. (1968). Chain-folding and molecular-species segregation in the crystallization of linear high polymers. *J. Appl. Phys.* **39**, 4929.

Magill, J. H. (1962). Melting behavior and spherulitic crystallization of polycaproamide (nylon 6). *Polymer* **3**, 43.

Malmgreen, H., and Lamm, O. (1944). Dispersionsmessungen an hochpolymeren Kaliummetaphosphaten. *Z. Anorg. Allg. Chem.* **252**, 256

Mandelkern, L. (1955). Crystallization kinets of high polymers II. Polymer-diluent mixtures. *J. Appl. Phys.* **26**, 443.

McLaren, J. V. (1963). A kinetic study of the isothermal spherulitic crystallization of polyhexamethylene adipamide. *Polymer* **4**, 175.

Mehta, A., and Wunderlich, B. (1972). Molecular fractionation during crystallization and its application to the study of lamellar thickening during isothermal crystallization. *Macromol. Chem.* **153**, 327.

Mehta, A., and Wunderlich, B. (1974). Macromolecular Nucleation. *J. Polym. Sci. Polym. Phys. Ed.* **12**, 255; see also (1975). A study of molecular fractionation during crystallization of polymers. *Colloid and Polym. Sci.* **253**, 193.

Melillo, L., and Wunderlich, B. (1972). Extended-chain crystals VIII. Morphology of polytetrafluoroethylene. *Kolloid Z. Z. Polym.* **250**, 417.

Michaels, A. S. (ed.) (1966). "Nucleation Phenomena." Amer. Chem. Soc. Publ., Washington, D.C.

Mitchell, J. C. (1967). Mechanical history effects in the crystallization of cis-1,4-polybutadiene. *Polymer* **8**, 369.

Morawetz. H. (1965). "Macromolecules in Solution." Wiley (Interscience), New York.

Moyer, J. D., and Ochs, R. J. (1963). Microstructure of polymers by tritium autoradiography. *Science* **142**, 1316.

Natta, G., Pegoraro, M., and Peraldo, M. (1958). Chromatographic fractionation of stereoblock polymers. *Ricerca Sci.* **28**, 1473.

Ovchinnikov, Yu. K., Markova, G. S., and Kargin, V. A. (1969). Electron diffraction study of the structure of polymer melts. *Vysokomol. Soedin.* **A11**, 329 [*English transl.*: *Polym. Sci. USSR* **11**, 369].

Packter, A., and Sharif, K. A. (1971). Heterogeneous nucleation and crystallization of poly(2,6-dimethyl)phenylene oxide melts. *J. Polym. Sci. Part B* **9**, 435.

Pechhold, W. (1968). Molekülbewegung in Polymeren. I. Teil: Konzept einer Festkörperphysik makromolekularer Stoffe. *Kolloid Z. Z. Polym.* **228**, 1.

Pechhold, W., and Blasenbrey, S. (1967). Kooperative Rotationsisomerie in Polymeren. I. Schmelztheorie und Kinkenkonzentration. *Kolloid Z. Z. Polym.* **216–217**, 235.

Pechhold, W., and Blasenbrey, S. (1970). Molekülbewegung in Polymeren. III. Teil. Mikrostruktur und mechanische Eigenschaften. *Kolloid Z. Z. Polym.* **241**, 955.

Pechhold, W., Hauber, M. E. T., and Liska, E. (1973). The meander model of condensed polymers. *Kolloid Z. Z. Polym.* **251**, 818.

Pennings, A. J. (1967). Fractionation of polymers by crystallization from solutions. II. *J. Polym. Sci., Part C* **16**, 1799.

Pennings, A. J., van der Mark, J. M. A. A., and Booij, H. C. (1970). Hydrodynamically induced crystallization of polymers from solution. II. The effect of secondary flow. *Kolloid Z. Z. Polym.* **236**, 99.

Peterlin, A., and Meinel, G. (1964a). Investigation of polyethylene single crystal suspensions by light scattering. *J. Appl. Phys.* **35**, 3221.

Peterlin, A., and Meinel, G. (1964b). Light scattering by polyethylene single crystal suspensions. *J. Polym. Sci. Part B* **2**, 751.

Petermann, J., and Gleiter, H. (1973). The molecular structure of molten polyethylene films. *Phil. Mag.* **28**, 271.

Petraccone, V., Allegra, G., and Corradini, P. (1972). Calculations of minimum potential energy of folds and kinks in polyethylene crystals. *J. Polym. Sci. Part C* **38**, 419.

Price, F. P. (1952). The development of crystallinity in polychlorotrifluoroethylene. *J. Amer. Chem. Soc.* **74**, 311.

Price, F. P. (1959). Growth habit of single polymer crystals. *J. Chem. Phys.* **31**, 1679.

Price, F. P. (1966). Nucleation and condensation in polymeric systems. *In* "Fundamental Phenomena in the Materials Sciences" (L. J. Bonis, P. L. deBruyn, and J. J. Duga, eds.), Vol. 3, p. 85, Plenum Press, New York.

Price, F. P. (1969). Nucleation in polymer crystallization. *In* "Nucleation" (A. E. Zettlemoyer, ed.), Dekker, New York.

Prigogine, I. (1957). "The Molecular Theory of Solutions," p. 323. North-Holland Publ., Amsterdam.

Prime, R. B. (1968). The Equilibrium Melting of Homopolyers. Thesis, Dept. of Chem., Rensselaer Polytechnic Inst., Troy, New York.

Prime, R. B., and Wunderlich, B. (1969a). Extended-chain crystals. III. Size distribution of polyethylene crystals grown under elevated pressure. *J. Polym. Sci. Part A2* **7**, 2061.

Prime, R. B., and Wunderlich, B. (1969b). Extended-chain crystals. IV. Melting under equilibrium conditions. *J. Polym. Sci. Part A2* **7**, 2073.

Rabesiaka, J., and Kovacs, A. J. (1961). Isothermal crystallization kinetics of polyethylene. III. Influence of the sample preparation. *J. Appl. Phys.* **32**, 2314

Reich, L., and Schindler, A. (1966). "Polymerization by Organometallic Compounds." Wiley (Interscience), New York.

Richards, R. B. (1946). The phase equilibria between a crystalline polymer and solvents. I. The effect of polymer chain length on the solubility and swelling of polythene. *Trans. Faraday Soc.* **42**, 10.

Robertson, R. E. (1965). Polymer order and polymer density. *J. Phys. Chem.* **69**, 1575.

Rodchenko, D. A., and Belyi, V. A. (1970). Melting and crystallization of polychlorotrifluoroethylene containing synthetic nucleating agents. *Dokl. Akad. Nauk Beloruss. SSR* **19**, 425.

Sadler, D. M. (1971). Fractionation during crystallization. *J. Polym. Sci. Part A2* **9**, 779.

Scherr, H., Pechhold, W., and Blasenbrey, S. (1970). Molekülbewegung in Polymeren. II. Teil. Kräftedipol, Volumeneffekt und Relaxationsverhalten der 2gl-Kinke in PÄ-Kristall. *Kolloid Z. Z. Polym.* **238**, 396.

Schick, M. J. (ed.) (1971). Polymers at interfaces. *J. Polym. Sci. Part C* **34**.

Schonhorn, H. (1967). Heterogeneous nucleation of polymer melts on surfaces. I. Influence of substrates on wettability. *J. Polym. Sci. Part B* **5**, 919.

Schonhorn, H. (1968). Heterogeneous nucleation of polymer melts on high-energy surfaces. II. Effect of substrate on morphology and wettability. *Macromolecules* **1**, 145.

Schultz, J. M. (1969). Transient effects in the crystallization of polyethylene. *J. Polym. Sci. Part A2* **7**, 821.

Sharples, A. (1962). The formation of nuclei in crystalline polymers. *Polymer* **3**, 250.

Slonimskii, G. L., and Godovskii, Y. K. (1966). Calorimetric melting and crystallization study of polypropylene using artificial nuclei. *Vysokomol. Soedin.* **8**, 718 [*English transl.*: *Polym. Sci. USSR* **8**, 787].

Sperati, C. A., and Starkweather, Jr., H. W. (1961). Fluorine containing polymers. II. Polytetrafluoroethylene. *Advan. Polym. Sci.* **2**, 465.

Stranski, I. N., and Kaischew, R. (1934). Über den Mechanismus des Gleichgewichtes kleiner Kriställchen. *Z. Phys. Chem.* (*B*) **26**, 100, 114, 312.

Strickland-Constable, R. F. (1968). "Kinetics and Mechanism of Crystallization." Academic Press, New York.

Stuart, H. A. (1943). Form und Beweglichkeit von Fadenmolekülen im Modellversuch. *Naturwissenschaften*, **31**, 123

Takahashi, T., and Ogata, N. (1971). Epitaxial crystallization of polyoxymethylene in the course of polymerization in aqueous solution. *J. Polym. Sci. Part B* **9**, 895; see also Takahashi, T., Ogata, N., Okazaki, H., and Tsujimoto, I. (1972). *Kobunshi Kagaku* **29**, 331.

Tammann, G. (1903). "Kristallisieren und Schmelzen," 2nd ed., 1923 under the title "Aggregatzustände," Voss, Leipzig.

hilo, E., and Grunze, H. (1955). Zur Chemie der kondensierten Phosphate und Arsenate. Teil XIII. *Z. Anorg. Allg. Chem.* **281**, 262.

Turnbull, D. (1950). Kinetics of heterogeneous nucleation. *J. Chem. Phys.* **18**, 198.

Turnbull, D., and Cormia, R. L. (1961). Kinetics of crystal nucleation of some n-alkanes. *J. Chem. Phys.* **34**, 820.

Turnbull, D., and Fisher, J. C. (1949). Rate of nucleation in condensed systems. *J. Chem. Phys.* **17**, 71.

Turturro, A., Olivero, L., Pedemonte, E., and Alfonso, G. C. (1973). Crystallization kinetics and morphology of high density polyethylene containing organic and inorganic pigments. *Brit. Polym. J.* **5**, 129.

Überreiter, K. (1962). Kristallisieren, Kristallzustand, und Schmelzen; *In* "Kunststoffe" (R. Nietsche and K. A. Wolf, eds.), Vol. 1, p. 189, Springer Verlag, Berlin.

van Wazer, J. R., and Callis, C. F. (1962). Phosphorous-based macromolecules. *In* "Inorganic Polymers" (F. A. G. Stone and W. A. G. Graham, eds.). Chapter 2. Academic Press, New York.

Vidotto, G., Lévy, D., and Kovacs, A. J. (1969). Crystallisation et fusion des polymerès autoensemencés. I. Polybutène-1, polyéthylène et polyoxyéthylène de haute masse moléculaire. *Kolloid Z. Z. Polym.* **230**, 289.

Volmer, M. (1939). "Kinetik der Phasenbildung." Dresden: Steinkopff.

Volmer, M., and Marder, M. (1931). Zur Theorie der linearen Kristallisations-geschwindigkeit unterkühlter Schmelzen und unterkühlter fester Modifikationen. *Z. Phys. Chem. A* **154**, 97.

von Falkai, B. (1960). Schmelz- und Kristallisationserscheinungen bei makromolekularen Substanzen. I. Kristallisationskinetische Untersuchungen an isotaktischem Polypropylen. *Makromol. Chem.* **41**, 86.

Vonnegut, B. (1948). Variation with temperature of the nucleation rate of super-cooled liquid tin and water drops. *J. Colloid Sci.* **3**, 563.

Wellinghoff, S., Rybnikar, F., and Baer, E. (1974). Epitaxial crystallization of polyethylene. *J. Macromol. Sci.* **B10**, 1.

Williams, M. L., Landel, R. F., and Ferry, J. D. (1955). The temperature dependence of relaxation mechanisms in amorphous polymers and other glass-forming liquids. *J. Amer. Chem. Soc.* **77**, 3701.

Winkler, A., and Thilo, E. (1959). Zur Konstitution des Grahamschen Salzes II. *Z. Anorg. Allg. Chem.* **298**, 302.

Wunderlich, B. (1968). Crystallization during polymerization. *Advan. Polym. Sci.* **5**, 568–619; A shorter summary was published in (1968). *Angew. Chem.* **80**, 1009; (1968). *Int. Ed. English* **7**, 912.

Wunderlich, B. (1970). Das Kristallisieren und Schmelzen von Hochpolymeren. *Ber. Bunsenges. Phys. Chem.* **74**, 768.

Wunderlich, B., and Cormier, C. M. (1966). Seeding of supercooled polyethylene with extended chain crystals. *J. Phys. Chem.* **70**, 1844.

Wunderlich, B., and Cormier, C. M. (1967). Heat of fusion of polyethylene. *J. Polym. Sci. Part A2* **5**, 987.

Wunderlich, B., and Mehta, A. (1970). The nucleation of polyethylene dendrites from solution. *J. Mater. Sci.* **5**, 248.

Wunderlich, B., and Sullivan, P. (1962). Solution-grown polyethylene dendrites. *J. Polym. Sci.* **61**, 195.

Wunderlich, B., Melillo, L., Cormier, C. M., Davidson, T., and Snyder, G. (1967). Surface melting and crystallization of polyethylene. *J. Macromol. Sci.* **B1**, 485.

Yakhnin, E. D., Egorova, Yu. V., Evko, E. I., Taubmann, A. B., and Luk'yanovich, V. M. (1969). Crystallization of polyamide in alcohol suspensions of mineral particles. *J. Crystal Growth* **5**, 184; see also *Vysokomol. Soedin. Ser. A* **12**, 771 (1970).

Yeh, G. S. Y. (1972). A structural model for the amorphous state of polymers: Folded-chain fringed micellar grain model. *J. Macromol Sci.* **B6**, 462.

Yim, A., and St. Pierre, L. E. (1970). The effect of interfacial energy on heterogeneous nucleation in the crystallization of polydimethylsiloxane. *J. Polym. Sci. Part B* **8**, 241.

Zachmann, H. G. (1964). Das Kristallisations- und Schmelzverhalten hochpolymerer Stoffe. *Fortschr. Hochpolym. Forsch.* **3**, 591.

Zachmann, H. G. (1967). Der Einfluss der Konfigurationsentropie auf das Kristallisations- und Schmelzverhalten von hochpolymeren Stoffen. *Kolloid Z. Z. Polym.* **216–217**, 180.

Zachmann, H. G. (1969). Statistische Thermodynamik des Kristallisierens und Schmelzens von hochpolymeren Stoffen. *Kolloid Z. Z. Polym.* **231**, 504.

Zettlemoyer, A. C. (1969). "Nucleation." Dekker, New York.

CHAPTER VI

The Growth of Crystals

6.1 General Description of Crystal Growth

In the present chapter crystal growth subsequent to the three nucleation steps, primary, secondary, and molecular nucleation is described. After a brief summary of the historical development of knowledge on crystallization, Sect. 6.1 deals with the general description of crystal growth, subdivided into the linear growth of a single crystal and the resulting overall growth of an agglomerate of a polycrystalline sample. Special effects in macromolecular systems as they arise due to chain folding, changes in molecular weight and structure, secondary crystallization, crystal perfection, growth without chain folding, and crystallization during polymerization are discussed next. Sections 6.2 and 6.3 contain a larger selection of examples of solution and melt crystallization, analyzed in some detail. The final section of this chapter presents an effort to bring together information on the crystallization during polymerization. This last section builds on the initial review of this subject by Wunderlich (1968).

6.1.1 History

The microscopic structure and the morphology of crystals has been shown in Chapters II and III to have been the subject of early speculation and theory. The actual growth of crystals, on the other hand, is a more difficult subject. A detailed theory requires from the outset some knowledge of the crystal itself and its constituent particles. Although Nicolaus Steno in the seventeenth century disproved the idea of minerals possessing life (see Sect. 3.1.1), there was still support of the thesis throughout the eighteenth century that nature is uniform in the treatment of all her productions, and consequently there should be uniform generation of animal, vegetable, and mineral beings (see for example Burke, 1966). Robinet (1761), for example, stated: "It seems to me that one cannot give a more satisfactory answer other than to admit the existence of fossil germs ... ; stones engender stones, metals produce metals, as animals engender their kind, as plants produce plants, by seeds, germs, or eggs, because all these words are synonymous."

One observation properly accounted for by this panpsychism is the usual need of a seed before crystal growth can commence. Steno (1669) admittedly could not offer any explanation how or why the initial seed crystal is produced. He proved only that further growth after this initial step occurred by the external addition of particles on the existing surfaces of the crystal.

Detailed description of the crystallization process comes from Newton (1730): "When any saline liquor is evaporated to a cuticle and let cool, the salt concretes in regular figures; which argues that the particles of the salt, before they concreted, floated in the liquor at equal distances in rank and file, and by consequence that they acted upon one another by some power which at equal distances is equal, at unequal distances unequal. For by such a power they will range themselves uniformly, and without it they will float irregularly, and come together as irregularly. And since the particles of island crystal act all the same way upon the rays of light for causing the unusual refraction, may it not be supposed that, in the formation of this crystal, the particles not only ranged themselves in rank and file for concreting in regular figures, but also by some kind of polar virtue turned their homogeneal sides the same way." Quite clearly in this statement Newton recognizes the need for regular, sequential addition of properly oriented particles, but he also places much of the ordering of particles in the solution before crystallization, caused by the balance of forces between the particles. This idea is not completely foreign to some modern thoughts on crystallization of macromolecules where much of the ordering in solution or melt is presumed to be caused by geometic requirements in the melt or in solution before crystallization (see Figs. V.1 and VI.32–VI.37 and Sects. 5.1 and 6.1.7).

The first quantitative theory of crystal growth was suggested by Gibbs (1878) on the basis of his thermodynamics: "The following view of the molecular state of a crystal when in equilibrium with respect to growth or dissolution appears as probable as any. Since the molecules at the corners and edges of a perfect crystal would be less firmly held in their places than those in the middle of a side, we may suppose that when the condition of theoretical equilibrium is satisfied several of the outermost layers molecules on each side of the crystal are incomplete toward the edges. The boundaries of these imperfect layers probably fluctuate, as individual molecules attach themselves to the crystal or detach themselves .... Now a continued growth on any side of a crystal is impossible unless new layers can be formed. This will require a value of μ_1'' [the potential of the crystal] which may exceed that given by equation (665) [for equilibrium] by a finite quantity. Since the difficulty in the formation of a new layer is at or near the commencement of the formation, the necessary value of μ_1'' may be independent of the area of the side .... It does not appear that any part of the operation of removing a layer of molecules presents any special difficulty so marked as that of commencing a new layer; ...." This statement clearly contains the facts that the actual crystallization step is reversible up to the stage of what is called in Chapter V the secondary nucleation that has to overcome a higher free energy barrier, nonexistent for the dissolution step. The two stages in the formation of a crystal, nucleation and growth, were by this time given a precise, thermodynamic meaning.

These early investigations were followed by a large number of increasingly more detailed studies of the crystallization process. Much of this early work met frustration. Wells (1946), who gives a review of the earlier work quotes that after fifteen years (since the last review) it is still true "that it is almost impossible to draw any certain conclusions from the mass of information that has been collected on the growth and solution of crystals, because very few of the aspects have been investigated from all sides, and insufficient systematic work has been carried out." The recent advances in crystal growth mechanisms of small molecules have been summarized by Parker (1970) who characterizes the progress as follows: "It has become customary to refer to crystallization as more an art than a science. The rate at which the art is changing to science has greatly increased during the past 15 years .... There still appears, however, to be a certain lack of connectedness between the various parts of the subject, a certain inability to treat crystallization as an overall, coupled process."

The various factors influencing crystallization rate are discussed in Sect. 6.1.2. The enormous detail which must be known for a full description of crystal growth can be judged by looking at the large variety of different morphologies of crystals which in the final analysis must be derived from a theory of crystal growth. Ice crystals are perhaps the most intensely studied. Bentley and Humphreys (1931) illustrate over 2000 different crystals in their catalog of snow flakes. There seems little hope to give an accounting of the reasons for the appearance of one shape or the other.

The study of crystallization of macromolecular materials was initially carried out much along the lines of already known works on low molecular weight systems. In particular, the development of crystallinity with time was studied intensely and correlated with the growth of the spherulite structure (see, for example, Jenckel, 1955, and Mandelkern, 1964a). The next observation was the chain folding principle of flexible, mobile macromolecules on crystallization (Sect. 3.2.2.1) and the discovery of a large amount of disorder, even in outwardly seemingly perfect lamellae (see Chapter IV). It was rather quickly concluded that chain folding and imperfect crystallization had to be kinetic phenomena since the resulting crystals were greatly metastable (see Chapter IX). The increasingly more complex kinetic theories of crystal growth of macromolecules from the melt and from solution which included chain folding as a central feature had their beginning with the work of Lauritzen and Hoffman (1959) and Price (1959), just two years after the initial recognition of the importance of chain folding in polymer crystallization (see also Sect. 3.2.2). The details have been reviewed as they developed by Price (1964), Zachmann (1964), Hoffman *et al.* (1969), and Sanchez (1974). Molecular nucleation described in Sect. 5.3 finally brings the understanding of crystallization of macromolecules from solution and from the melt up to date by providing an explanation for the fractionation of different molecular weight species on crystallization. Very little is known about the reasons for a particular morphology found under given conditions of crystallization beyond the general outlines of the causes for dendritic and spherulitic growth (see Sects. 3.6.1. and 3.7.1). No knowledge exists about the reasons for the resulting crystal morphologies as they are represented by their different fold sectors, twin laws, etc. (see Chapter III). Also, the processes involved in crystallization under stress are not well understood (see Sects. 6.2.2 and 6.3.2). Even less is known about the direct crystallization of macromolecular crystals from the monomer. In this case even kinetic data describing the simultaneous or successive chemical reaction and crystallization are scarce (see Sects. 6.1.8 and 6.4). One can therefore conclude that for the field of crystallization of macromolecules the rate at which art is changing to science has also increased in the last fifteen years; the problems are, however, much more involved than for small molecules. Many of the problems require the quantitative connection of several parts of the subject which are presently only qualitatively understood. Key problems are concerned with the understanding of the conformation, interaction, and motion of molecules in the melt and solution, the processes involved in adsorption before crystallization and rearrangements after initial crystallization, the effect of macroscopic stresses on the conformation and interaction between macromolecules, and the influence of chemical reactions on crystallization.

6.1.2 Linear Growth

6.1.2.1 Experimental Observations

The macroscopic advance of a crystal front is termed its linear growth rate. It can be observed directly by optical or electron microscopy, or indirectly as is shown in Sect. 6.1.3 by noting the overall increase in crystalline content by any of the methods discussed in Sect. 4.1. Macromolecular crystals are usually so small that direct observation of linear growth rate is only possible under special conditions. Frequently crystallization must be followed by observation of the increase of the radius of the spherulitic superstructure as shown for example in Fig. VI.1 which illustrates schematically the different

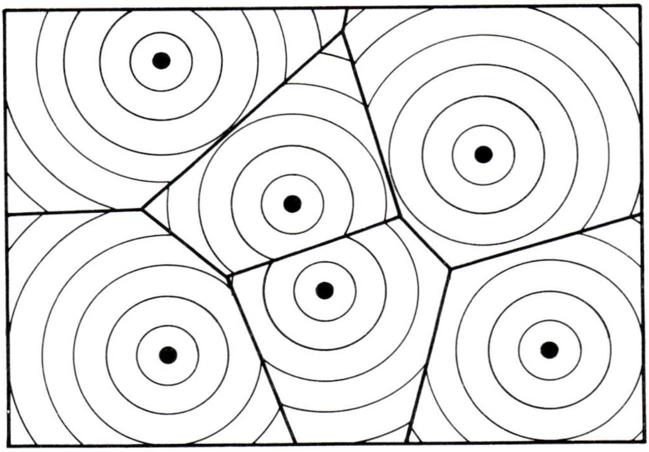

Fig. VI.1 *Schematic Drawing of the Change in Spherulite Radius during Crystallization.*
For the drawing, athermal nucleation was assumed at the marked centers at time zero. Consecutive circles mark the size of the growing spherulite at constant time intervals. Note the straight lines of intersection between neighboring spherulites on impingement which suggest athermal nucleation and constant growth rate (Brenschede, 1949). Compare with Fig. III.116.

stages of spherulitic growth, terminated for each spherulite by impingement on neighboring spherulites. The final picture should be compared with Fig. III.116. An example of direct observation of linear crystal growth is shown in Fig. VI.2. The time dependence of the crystal size is in this case measured by quenching parts of the crystallizing solution at certain intervals. Electron microscopic observation of the compound crystals, grown at two temperatures, permits the measurements of the increase of the diameter of the

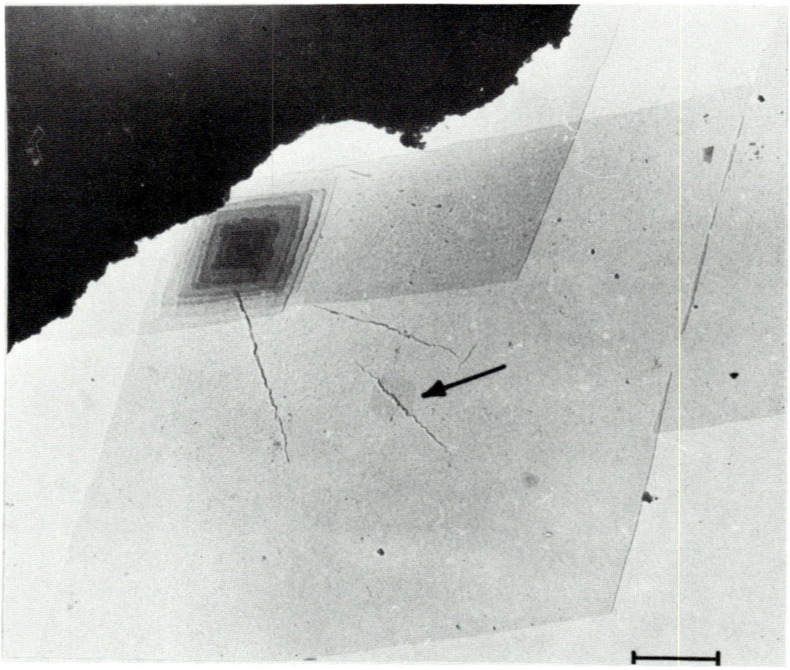

Fig. VI.2 *Electron Micrograph of Initial Crystal Growth of Polyethylene from Dilute Solution in Xylene Followed by Low Temperature Crystallization after Quenching.*
The arrow indicates the small, six-sided, high temperature crystal of larger thickness (greater contrast) grown at the higher crystallization temperature. The four-sided low temperature crystal grown around it crystallized at a later time after quenching (compare with Fig. III.51). The change in dimension of the six-sided crystal as a function of time gives crystallization rate information. Keller and Pedemonte (1973). Scale bar 1 μm.

inner, thicker crystal. In Fig. VI.2 the latter also shows the high temperature morphology with {100} growth faces in contrast to the low temperature morphology with only {110} growth faces that grew after quenching (see Fig. III.51).

Early observations of spherulite growth indicated that the linear growth rate is constant with time (see, for example, Zachmann, 1964). Similarly, it was shown by Holland and Lindenmeyer (1962) that the linear growth rate of crystals from solution is constant with time until close to exhaustion of the solution. Figure VI.3 illustrates the change in crystal size with time for polyethylene grown from dilute solution in xylene at two temperatures. Besides

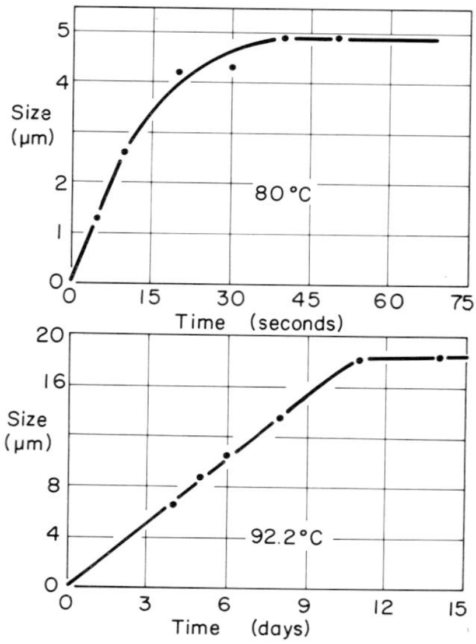

Fig. VI.3 *Increase in Crystal Size with Crystallization Time for Polyethylene Grown from Dilute Solution in Xylene.*
$M_n = 120\,000$, 0.01 wt. %, at the given crystallization temperatures. Note the factor of 1.7×10^4 between the two time scales. Data by Holland and Lindenmayer (1962).

illustrating the initially constant crystal growth rate, Fig. VI.3 shows also the enormous increase in crystallization rate with decreasing temperature (see the following). A sizeable slowdown of the spherulitic growth rate was observed by Price and Kilb (1962) and Barnes *et al.* (1961) when two spherulites of poly(ethylene oxide) approached each other to distances of 0.15 mm or less. It was suggested that this slowdown before impingement is caused by the temperature increase expected in the vicinity of the crystal growth front; it may, however, also result from, or be enhanced by, accumulation of non-crystallizable material before the growing crystal as discussed for dendrite formation (Sect. 3.6.1) and segregation of molecular weights on crystallization (Sects. 5.3.2 and 5.3.3).

Data on the linear growth rate variation with the degree of supercooling are presented in Fig. V.37. This figure is taken as definite evidence that the linear crystal growth rate of macromolecules from the melt is nucleation controlled (see Sect. 5.2.1). Similar data are shown in Sect. 6.2.1 for the change of crystal growth rate from solution (see Fig. VI.42). Again the growth rate is nucleation controlled. From the study of fractionation of broad molecular weight distributions, it became clear that the rate determining nucleation step is not secondary crystal nucleation, but molecular nucleation (see Sect. 5.3.4). Little is known about the temperature dependence of crystal growth on crystallization during polymerization.

6.1.2.2 Crystal Growth from the Vapor Phase (Surface Kinetics Governed Crystallization)

The first detailed attempt to describe the kinetics of crystal growth was made on the model of a one-component perfect crystal built of small atoms growing from the vapor phase. Schematically the assumed process is represented by Fig. VI.4. The basic idea for the model comes from the work of

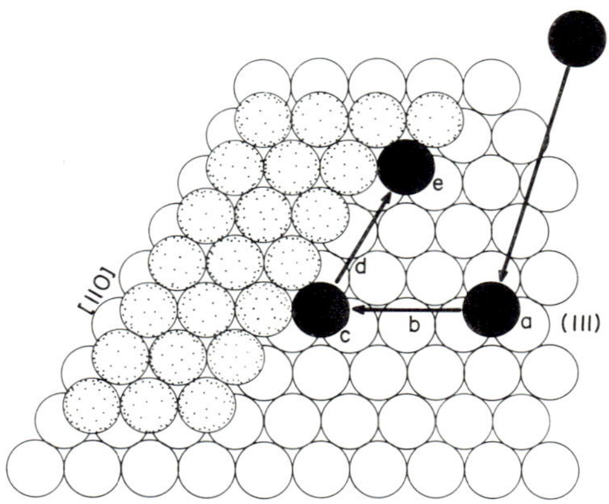

Fig. VI.4 *Schmeatic Drawing of the Crystallization of an Atom by Adsorption and Diffusion on the Surface.*
Ledge in direction [110]. Close packed cubic face centered crystal, (111) face (see Fig. II.13).

Gibbs (see Sect. 6.1.1). Tammann (1903, 1923, 1925), Volmer (1939), Kossel (1927), Stranski (1928), Becker and Döring (1935), Frenkel (1946), and Burton and Cabrera (1949) developed the necessary formalism for the description of the macroscopically observable growth rate in terms of the microscopic assumed process. In its simplest form such crystallization involves the following steps: (a) adsorption to the surface; (b) two-dimensional diffusion to a growth ledge; (c) attachment to the ledge; (d) one-dimensional diffusion to a corner; and (e) crystallization. The reverse of the five steps are the stages of evaporation. At equilibrium, evaporation and crystallization rates are equal;

for crystallization, the outlined sequence a → e must be preferred over the reverse process.

The detailed structure of a monomolecular layer on a perfect crystal surface such as is shown in Fig. VI.4 has been assessed by Burton and Cabrera (1949) using the energetics of the atoms in different positions (see Fig. V.8). The heat of evaporation Δh_v is related to the nearest-neighbor pair interaction ϕ ($\phi/2$ per atom). For a close-packed face centered cubic crystal of coordination number 12, Δh_v is given by

$$\Delta h_v = 6\phi \tag{1}$$

At the absolute zero of temperature the close-packed surface energy is $3\phi/(2a^2)$. As the temperature increases, one expects increasing surface roughness to cause an increase in the surface free energy.

The state of a surface layer can be assessed through the model of a two-dimensional square lattice gas developed by Lee and Yang (1952). The energy of a single atom on the (100) surface of a simple cubic crystal is raised by 2ϕ over that of an atom in the surface (4 times $\phi/2$, see position 1 in Fig. V.8). Two atoms in adjacent positions reduce their combined potential energy by ϕ. For two atoms which occupy the same position, the potential energy is assumed to be infinity. Atoms in nonadjacent positions show no interaction. Such a square lattice gas can be described by the two-dimensional Ising model for which an exact partition function has been given in a complex derivation by Onsager (1944). The result of interest to the surface roughness problem is the existence of a critical temperature T_c^I given by

$$\exp[-\phi/(2kT_c^I)] = \sqrt{2} - 1 \tag{2}$$

below which absorbed surface atoms are in equilibrium with practically densely packed surface layers. The phase diagram of such a two-dimensional lattice gas is shown in Fig. VI.5. The critical temperature can be expressed in terms of the heat of evaporation using the analog of Eq. (1) for the simple cubic lattice ($\Delta h_v = 3\phi$):

$$T_c^I = 0.095\,\Delta h_v \qquad (\Delta h_v \text{ in cal/mole, } T_c^I \text{ in °K}) \tag{3}$$

The critical temperature T_c^I is typically far above the bulk melting temperature of the solid,† which means that the crystal contains densely packed surface layers in equilibrium with a dilute gas of adsorbed atoms (see Fig. VI.5).

Exciting a single atom out of a smooth surface to the single atom adsorbed position creates in the simple cubic lattice four unpaired bonds. Its probability of formation is thus $\exp(-2\phi/kT)$. Pairs of such adsorbed atoms

† Note that Trouton's rule states $\Delta h_v = 21T_b$, where T_b is the boiling temperature.

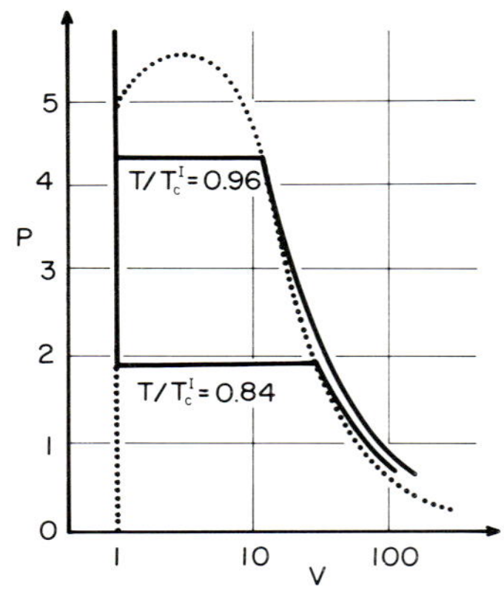

Fig. VI.5 *Pressure–Volume Isotherms of a Two-Dimensional Lattice Gas.*
At low temperature ($<0.96T_c^I$) the gas is in equilibrium with a practically close packed surface patch. Dotted curve: Coexistence curve. Solid Curves: Isotherms. Calculations by Lee and Yang (1952).

would have analogously a probability of $\exp(-3\phi/kT)$. The energetics of a surface would accordingly call for typical temperatures of interest ($T \ll T_c^I$) for some single atoms absorbed on the surface, but few pairs and practically no clusters.

The treatment of the surface roughness just given does not consider the medium adjacent to the surface or effects arising from crystal layers other than the perfect crystal substrate and a single, rough layer. Attempts to include such effects have been reviewed by Parker (1970), but add little to the description of crystallization from the gas phase.

Before discussion of the growth kinetics, it is also useful to look at the structure of a ledge surrounding a condensed surface patch. Once created, such a ledge could develop corners by adding or losing atoms. At the absolute zero of temperature the ledge should naturally be straight in order to have minimum free energy. The probabilities of having at higher temperature a positive or a negative corner at any position along a ledge are n_+ and n_-. They are governed by the loss of nearest-neighbor interaction at the corner atoms. For a cubic close packed structure,† a ledge along the [110] direction

† The simple cubic crystal leads to the same expression for ledges along [100], see Fig. V.8.

that has the probability q of not having a corner at the chosen position, the following holds (see Fig. VI.4):

$$n_+ = n_- = q \exp[-\phi/(2kT)], \qquad n_+ + n_- + q = 1 \tag{4}$$

The mean separation between corners x_0 has been shown by Frenkel (1946) and Burton and Cabrera (1949) to be simply $a/(n_+ + n_-)$ or

$$x_0 = (a/2)\{\exp[\phi/(2kT)] + 2\} \tag{5}$$

where a is the interatomic distance. The average melting temperature of 69 metallic elements occurs at 0.5 times the boiling temperature T_b so that Eq. (5) yields with Trouton's rule ($\Delta h_v = 21T_b$) a mean separation of only

$$x_0 \approx 4a \tag{6}$$

between corners at the melting temperature. At typical crystallization temperatures a ledge is thus interrupted frequently by positive and negative corners (see also Fig. V.9).

The next point of interest is the number of single atoms adsorbed on the surface (adatoms, see position a, Fig. VI.4) and their average lifetime before evaporation. The number of adatoms n_s is governed by the energy of evaporation from a corner position to a surface position (3ϕ)

$$n_s = n_0 \exp[-3\phi/(kT)] \tag{7}$$

where n_0 is the number of sites on the surface. The lifetime τ_2 of the adatom is determined by the frequency of evaporation which is governed by the atomic vibration frequency ν_2 ($\sim 10^{12}$–10^{13} Hz) and the evaporation energy from the surface to the gas phase which, in the present case of a face centered cubic (111) face is again 3ϕ

$$1/\tau_2 = \nu_2 \exp[-3\phi/(kT)] \tag{8}$$

The motion of the adatom in the surface is governed by a lateral activation energy U_s, so that the overall energetics can be schematically represented as shown in Fig. VI.6. The time τ_1 required for a lateral jump of interatomic distance is

$$1/\tau_1 = \nu_1 \exp[-U_s/(kT)] \tag{9}$$

The frequency ν_1 is again the appropriate atomic vibration frequency. The surface diffusion coefficient D can be calculated from Eq. (9) using the Einstein relation for the random walk surface diffusion

$$D = a^2/\tau_1 \tag{10}$$

The mean displacement x_s before evaporation of the adatom can then be calculated from

$$x_s^2 = D\tau_2 \tag{11}$$

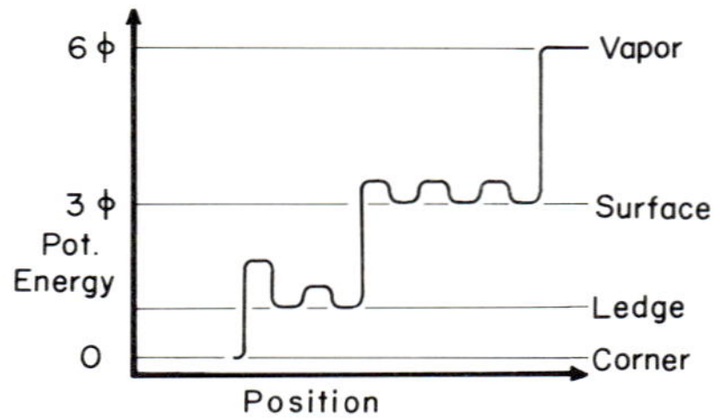

Fig. VI.6 *Schematic Drawing of the Change in Potential Energy along the Crystallization Path in Fig. VI 4.*
From a face centered cubic (111) surface the heat of evaporation of a single atom is 3ϕ, the heat of transfer from a corner to the surface is also 3ϕ, so that the overall heat of evaporation is 6ϕ [see Eq. (1)]. U_s may be less than $\phi/3$.

Combining Eq. (11) with Eqs. (8)–(10) yields us

$$x_s = a(v_1/v_2)^{1/2} \exp[(3\phi - U_s)/(2kT)] \tag{12}$$

where the ratio v_1/v_2 should be close to 1. Assuming furthermore that U_s is small relative to ϕ, x_s can be estimated to be about $200a$ or about 500 Å at the melting point when using the same quantities as for the estimate Eq. (6).

With these preliminary calculations, it is now possible to discuss the growth process of a single ledge. The direct condensation of an atom into the ledge is rather unlikely as long as the ledge concentration is small. The corner density, on the other hand, was shown to be large [Eq. (6)], so that the diffusion along the ledge is not rate determining and can be ignored (step d and e can be combined). Under equilibrium conditions the number of atoms arriving from the gas phase and the number leaving is equal. The rate of departure depends on temperature through Eq. (8); the rate of arrival is proportional to the vapor pressure p. Increasing the vapor pressure from the equilibrium value p_0 to p will cause more atoms to join corners and advance the ledge. The supersaturation can be expressed by

$$\sigma = (p - p_0)/p_0 \tag{13}$$

Far away from a ledge, the supersaturation increases the surface fraction of adatoms to

$$(n_s/n_0)(1 + \sigma) \tag{14}$$

Close to a ledge the old equilibrium fraction (n_s/n_0) is maintained due to increased crystallization. A reasonable estimate of the distance from a ledge which has a reduced surface atom concentration is x_s, the mean distance traveled by an adatom before reevaporation [Eq. (12)]. The concentration gradient can thus be estimated by

$$n_s \sigma/(n_0 x_s) \tag{15}$$

The rate of advance v_∞ of the ledge at right angles to its average direction is now a^2 times† the number of atoms arriving at a 1 cm length of the ledge per second or

$$v_\infty = 2a^2 D n_s \sigma/(n_0 x_s) \tag{16}$$

where the factor 2 arises because atoms may diffuse to a growing ledge from both directions. D is expressed by Eq. (11) and together with Eq. (8) for τ_2 and Eq. (7) for n_s/n_0, Eq. (16) becomes

$$v_\infty = 2\sigma x_s \nu_2 \exp[-6\phi/(kT)] \tag{17}$$

where 6ϕ is the total heat of evaporation [Eq. (1)]. Qualitatively this means that all excess molecules which hit a diffusion zone of width $2x_s$ about the ledge will reach the advancing step and, because of the large corner concentration, crystallize. Thus, the rate of advancement of a ledge at constant temperature, is dependent only on the supersaturation σ. This conclusion is also reached in Sect. 5.2.1 using a less detailed approach based on the formalism developed for the discussion of nucleation.

A crystal surface containing ledges will grow in contact with its supersaturated vapor until the partial layers are completed. The speed v_∞ is proportional to the degree of supersaturation. This means that all ledges which are initially present will quickly disappear. Further crystal growth then will rely on the generation of new ledges which is possible through secondary nucleation (see Sect. 5.2). Since secondary nucleation needs a much larger supersaturation for observable rates (see Fig. V.36), it may become the rate determining step. Crystal growth rates governed by secondary as well as molecular nucleation (Sect. 5.3.4) have an $\exp[1/(T\Delta T)]$ dependence, a rate observed for linear growth rates from macromolecular melts (see Fig. V.37) and solutions (see Fig. VI.42).

Defect crystals may contain screw dislocations (Sect. 4.3.3) which produce indestructable ledges (see Fig. IV.43) in the surface. The calculation of the crystal growth rate under these conditions has been described by Burton *et al.* (1951) (see also Parker, 1970). An approximation to the shape of a growth spiral is obtained from the equation of the spiral of Archimedes (see Fig. V.35)

$$r = a^*\theta \tag{18}$$

† Each atom advances the front a cm, but covers, however, only a cm of the 1 cm reference length.

where r and θ are the polar coordinates of the spiral and a^* is the diameter of the critical secondary nucleus, the smallest permissible nucleus diameter as described by Eq. (67), Chapter V. The angular velocity of the spiral then should be

$$\omega = v_{\infty}/a^* \tag{19}$$

and the rate of crystal growth v along the dislocation vector (normal to the crystal surface) is the number of times a fixed point is passed by the spiral per second multiplied with the step height b_0, the Burgers vector magnitude

$$v = b_0 v_{\infty}/(2a^*\pi) \tag{20}$$

The growth rate v is, as is pointed out in Sect. 5.2.1 in the discussion of Eq. (64), Chapter V, proportional to ΔT^2 (or the square of the supercooling σ).

Adatoms landing within $\pm x_s$ from the growth spiral ledge may contribute to the crystal growth, but all others tend to evaporate before reaching the growth spiral ledge. When successive ledges of the growth spiral come closer together than $2x_s$, Eq. (20) does not apply (see for example Parker, 1970). With decreasing ledge separation the supercooling dependence of the crystal growth rate v approaches the first power. For the more general situation when a crystal face contains several growth spirals, special interactions arise between them. The crystal growth rate v changes, however, never to a lower than linear σ dependence, nor does it increase beyond the quadratic dependence as given by Eq. (20) (as long as the dislocation vectors are further apart than a^*).

Experimental verification of the crystal growth rate law expressed by Eq. (20) can be seen in Fig. VI.7. More detailed reviews of experimental data have been given by Hirth and Pound (1963) and Heyer (1966).

The just outlined mechanism of crystal growth also has implications concerning the final Tracht of the crystal (see Sect. 3.1.2 and Fig. III.1). The principle of minimum surface free energy (Sects. 3.1.2 and 5.1) requires surfaces with high Miller indices (vicinal faces) to break up into close packed, stepped surfaces of low index as shown in Fig. VI.8. Under growing conditions all the ledges will be completed to the limits of the crystal, causing a disappearance of the vicinal faces and leaving only the faces with the largest lattice spacings as required by the Bravais–Friedel law (see Sect. 3.1.2). Complications arise when the bonding is not similar in all directions of space, a condition particularly germane to macromolecular crystals. For example, for crystals of polymeric tellurium and selenium crystallized and polymerized from the gas phase, the (001) face was not observed even on slow crystallization (Straumanis, 1935, 1940). Some work towards the illumination of the

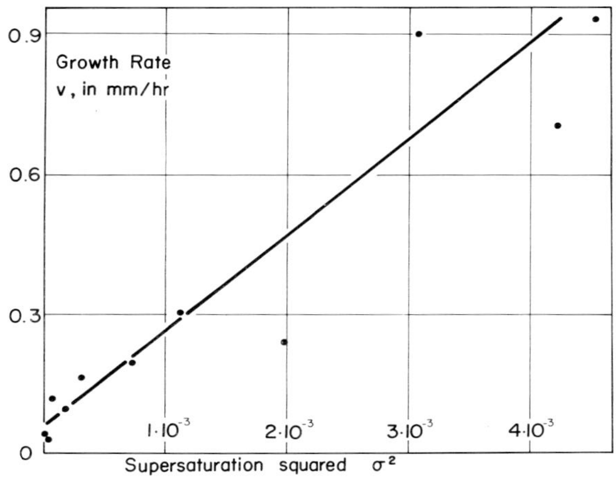

Fig. VI.7 *Rate of Iodine Crystal Growth from the Vapor Phase at 25°C as a Function of* σ^2.
Data by Bradley and Drury (1959). *Trans. Faraday Soc.* **55**, 1848.

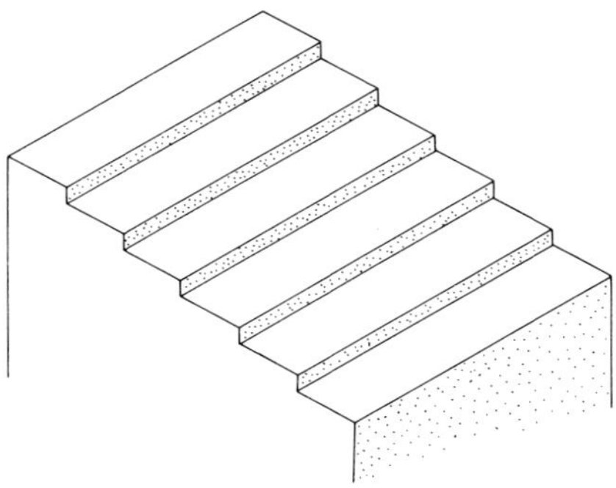

Fig. VI.8 *High Index Surface Made Up of Stepped Low Index Planes.*

morphology of such crystals was carried out by Hartman and Perdok (1955). On the basis of the comparison of many crystal morphologies and bonding within the crystals, they concluded that strong bonds should form more easily than weak bonds and lead in the extreme cases as in SiS_2 to fibrous crystals (see also Hartman, 1963). It is shown in Sect. 6.4 and also in Sects. 3.8, 5.1.2.5, and 6.1.4 that this trend to faster crystal growth in the direction of strong bonds is even enhanced by possible nucleation difficulties for growth at right angles to the molecular chain direction.

6.1.2.3 Crystal Growth Governed by Transport Processes

Crystallization from the gas phase under conditions governed by the surface kinetics is shown in Sect. 6.1.2.2 to be understood in some detail. The nucleation control of crystallization is discussed in Sect. 5.2.1. A third limiting type of crystallization is the case where the transport process alone determines the crystallization. The extended flat faces observed above on crystals from the gas phase, and under proper conditions also from the melt or solution, are evidence that in these cases growth is not governed by a diffusion field since a diffusion field leads to higher supersaturation near edges and corners (see Sect. 3.6.1).

Under proper conditions it is always possible to make the transport process to the crystal interface the dominant factor. In melt crystal growth it may be possible that the conduction of heat through the crystal or through the supercooled melt is the rate determining process for crystallization. From solution, crystallization may be limited by diffusion of crystallizable molecules to the interface.

Transport processes can usually be treated as macroscopic problems. The general group of transport problems which involves the diffusion differential equation with a moving or free boundary are called Stefan problems. Stefan (1891) studied the rate of growth of the thickness of the ice layer of the polar seas. The heat of crystallization has in this case to be conducted through the ice and carried away by the cold air. The temperature of the liquid T_l is in this case above the freezing temperature, and the heat of crystallization is liberated at the liquid–solid interface. The problem is to find the position of the interface n_i as a function of time. As the layer of crystal increases in thickness, the conduction of heat becomes less efficient and the crystallization slows down. The conduction of heat in an isotropic medium is given in analogy to the diffusion of matter (Sect. 3.6.1) by

$$(\partial T_s/\partial t) = \kappa_s(\partial^2 T_s/\partial n^2), \qquad (\partial T_l/\partial t) = \kappa_l(\partial^2 T_l/\partial n^2) \tag{21}$$

where κ is the thermal diffusivity in cm^2/sec, n is the normal distance from the cold surface of the crystal and the subscripts s and l denote the solid and liquid phases, respectively. At the interface position n_i the temperatures† are equal to the melting temperature

$$T_l = T_s = T_m \tag{22}$$

In order to preserve continuity of heat flow at the interface, the following conditions must apply

$$K_s(\partial T_s/\partial n) - K_l(\partial T_l/\partial n) = \Delta h_f(\partial n_i/\partial t) \tag{23}$$

† Temperatures are expressed relative to the constant cold surface of the crystal layer.

where K represents the thermal conductivity in cal/(cm sec deg) and Δh_f is the heat of fusion/cm^3. The solution of the problem leads to a dependency of the boundary position n_i on the square root of time

$$n_i = 2\lambda_1(\kappa_s t)^{1/2} \tag{24}$$

where λ_1 is the "growth constant" given in terms of physical properties of the solid and liquid and the superheating of the water. The detailed solution of the problem is given, for example, by Parker (1970) or Carslaw and Jaeger (1959). Figure VI.9 illustrates the increase in thickness as a function of time

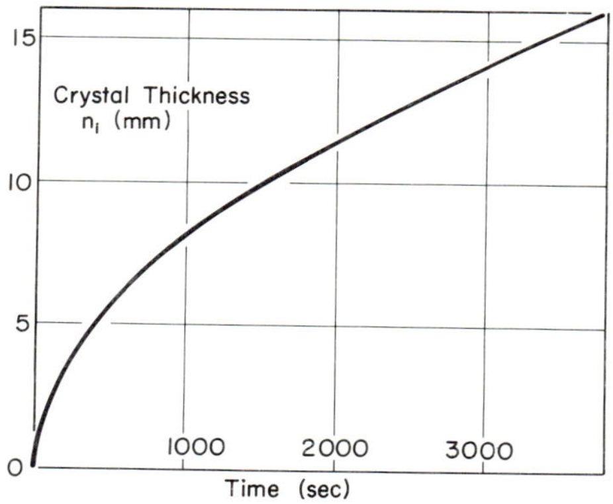

Fig. VI.9 *Increase in Crystal Thickness with Time as Predicted by the Simple Stefan Problem.* Crystal growth is only heat flow dependent; see text. Parameters: Cold crystal surface at $n = 0$ at 5°C below T_m. Liquid bulk temperature far from the crystal–liquid interface 2°C above T_m. $\kappa_s =$ 0.0115 cm^2/sec and $\lambda_1 = 0.121$ (quantities chosen for ice-water).

for the example of ice. A macromolecular application of such crystal growth rate would be the crystallization of a melt on injection molding with crystallization occurring by transcrystallization (see Sect. 3.4.3) whenever nucleation control is not the limiting factor for crystal growth.

The many variations of the Stefan problem and complications are discussed by Parker (1970).

6.1.2.4 *Conclusions about Linear Growth of Crystals*

The linear growth rate of crystals can only be described if the detailed mechanism of crystallization is understood. Three limiting cases can be recognized and are treated in some detail. The case most important for the isothermal crystallization of macromolecules is the nucleation controlled growth (Sects. 5.2.1 and 5.3.4) where one assumes that all processes except nucleation are fast. The case of crystal growth governed by surface kinetics is treated in Sect. 6.1.2.2 by the example of crystal growth from the gas phase, but may also have applications to solution and possibly also melt growth. For macromolecular crystals, such growth kinetics would apply mainly to the little studied case of crystallization during polymerization (see Sects. 6.1.8 and 6.4). Finally, the transport governed crystal growth can be attempted to be fitted to one of the many forms of the macroscopic Stefan problem (Sect. 6.1.2.3). Each of the limiting cases showed considerable difficulties and for any intermediate case the difficulties are compounded. Special effects on the linear growth rate caused by chain folding of macromolecules are described in Sect. 6.1.4.5.

6.1.3 Overall Crystallization

6.1.3.1 The Basic Avrami Equation

The knowledge of the linear growth rate and the primary nucleation rate should be sufficient to calculate the overall crystallization rate, an important step in the correlation of experimental data with microscopic models of crystal growth. Many measurements of crystallization involve the macroscopic determination of crystallinity (see Sect. 4.1) as a function of time. The first effort to describe quantitatively macroscopic development of crystallinity in terms of nucleation and linear crystal growth was made by Kolmogoroff (1937), Johnson and Mehl (1939) and Avrami (1939, 1940, 1941).

Evans (1945), seemingly without knowledge of prior work, based similar descriptions on the old problem of expanding waves created by raindrops on a pond, first solved by Poisson (1837). This problem may be stated as follows: If raindrops fall randomly on a surface of water creating one leading expanding circular wave each, what is the chance that the number of waves which pass over a representative point P up to time t is exactly n? The solution to the problem is given by the Poisson distribution†

$$P_n(t) = e^{-E}(E^n/n!) \tag{25}$$

where E is the expected number of waves, available for example by evaluation of the average number of counts of a large number of samples or by computation. For the rate of rain of I^* droplets per unit time and area, the number of

† See any standard text on theory of probability: If events occur in such a way that the probability of their occurrence is independent of how many times the event occurred before, and of other time intervals, the number of events in equal finite intervals of time follows the Poisson distribution. The binomial distribution $(q + p)^n$ approaches the Poisson distribution when $p \to 0$ and $n \to \infty$ such that $E = np$, the mean, remains constant.

waves having reached a point from a distance of r at time t since the beginning of the rain is

$$dE = I^*[t - (r/v)]2\pi r\, dr \tag{26}$$

where v is the rate of travel of the waves, so that if t is counted from the beginning of the rain, every element of the ring of distance r about P can contribute to the waves arriving at P for a time interval $t - (r/v)$. In order to find E, Eq. (26) must be integrated over all possible values of r (from 0 to vt).

$$E = 2\pi I^* \int_0^{vt} [t - (r/v)]r\, dr = \pi I^* v^2 t^3/3 \tag{27A}$$

Equation (27A) can now be inserted into Eq. (25) for final computation.

To treat Poisson's problem as an analog for crystallization one may assume a thin layer of melt crystallizes by formation of two-dimensional spherulites. One would be interested in the chance that a randomly chosen point P is *not* reached by any of the circularly spreading crystal aggregates ($n = 0$), i.e., P remains amorphous. Equation (25) becomes

$$P_0(t) = e^{-E} = 1 - v^c \tag{28}$$

where v^c is the volume fraction of the material crystallized [Eqs. (6)–(9), Chapter IV].†

Figure VI.1 illustrates the development of crystallinity through two-dimensional spherulites. The filled center circles represent athermal nuclei, active from the beginning of crystallization (Sect. 5.1.5). The increasing circles represent the spherulite sizes at constant, successive time intervals. Finally, the spherulites impinge upon each other and terminate further crystal growth. The schematic drawing of Fig. VI.1 should be compared with Fig. III.116 which represents an actual spherulite grown in a thin film of melt between microscope slide and cover glass.

The calculation of the expected number of spherulites which start growing at time zero and reach a representative point P is analogous to Poisson's rain drops problem. It leads to

$$E = 2\pi N \int_0^{vt} r\, dr = \pi N v^2 t^2 \tag{27B}$$

where N represents the number of nuclei per unit area and v is the linear rate of crystal growth. The probability that point P is not reached by a crystal and

† In the present case v^c represents also the area fraction because of constant layer thickness assumed. The question of equivalence of $P_0(t)$ to the volume or weight fraction of uncrystallized material is discussed in Sect. 6.1.3.3.

remains amorphous is given by Eq. (28) and the corresponding crystalline fraction is

$$v^{c} = 1 - \exp(-\pi N v^{2} t^{2}) \tag{29}$$

Another situation arises in the case of thermal nucleation, meaning that the nuclei are created sporadically in time and space. For computation of E, the nucleation rate I^* must be known. The conclusion of the discussion of primary nucleation (Sect. 5.1.5) was that for thermal nucleation it is frequently permissible to assume a time-constant, steady state nucleation rate after a certain induction time which is necessary to establish the initial steady state population of crystal embryos. The derivation of the expression for two-dimensional spherulitic crystal growth with constant nucleation rate is similar to the rain drops problem just solved. Equations (27A) and (28) describe the change in crystallinity.

A derivation that takes the phase transformation geometry explicitly into account was given by Tobin (1974) for the two-dimensional case with thermal nucleation. The overall nucleation rate is reduced in this treatment in the time interval dt as given by

$$dI^* = I^*[A_0 - A(\tau)]\, d\tau \tag{30}$$

where A_0 is the initial total area and $A(\tau)$ is the total transformed area at time τ. Impingement is corrected for by assuming that each crystal has its area at time t reduced by the fraction $[A_0 - A(t)]/A_0 = 1 - v^c$. The transformation kinetics takes then the form of a nonlinear Volterra integral equation

$$\begin{aligned} A(t) &= \pi I^* v^2 \int_0^t (1 - v^c)(t - \tau)^2 [A_0 - A(\tau)]\, d\tau \\ v^c/(1 - v^c) &= \pi I^* v^2 \int_0^t (t - \tau)^2 [1 - v^c(\tau)]\, d\tau \\ v^c/(1 - v^c) &= (\pi I^* v^2 t^3/3) - \int_0^t (t - \tau)^2 v^c(\tau)\, d\tau \end{aligned} \tag{31}$$

The zeroth-order solution $v^c/(1 - v^c) = \pi I^* v^2 t^3/3$ is identical to the first two terms of the expansion of the exponential of the combined Eqs. (27A) and (28). It also represents $(1 - v^c)$ times the unimpeded free growth equation [see also the discussion of Eq. (33)]. In addition, it was found to be of a type that sometimes empirically fits crystallization data over a longer crystallization time than the expressions based on Eq. (28) (Rabesiaka and Kovacs, 1961). This is because Eqs. (31) show at longer crystallization times a slower increase in crystallinity, a fact frequently observed, but usually attributed to secondary crystallization and crystal perfection (see Sect. 6.1.6).

In three dimensions, spherulitic crystal growth can be treated similarly to the rain drops problem. For the cases of athermal and thermal nucleation, the following expected crystal fronts reaching point P can be calculated.

Athermal nucleation:

$$E = 4\pi N \int_0^{vt} r^2\,dr = 4\pi N v^3 t^3/3 \tag{32A}$$

Thermal nucleation:

$$E = 4\pi I^* \int_0^{vt} [t - (r/v)]r^2\,dr = \pi I^* v^3 t^4/3 \tag{32B}$$

In general, the development of crystallinity based on Eq. (28) can thus be represented by the equation

$$1 - v^c = \exp[-Kt^n] \tag{33}$$

which is commonly called the Avrami equation. The meaning of K and n can be deduced from Eqs. (27), (32), and (28). For the treated cases n should be a whole number, characteristic of the nucleation type and the crystal growth geometry. Figure VI.10 shows a general plot of the development of crystallinity v^c as a function of the logarithm of time for values of n varying from 1 to 6.

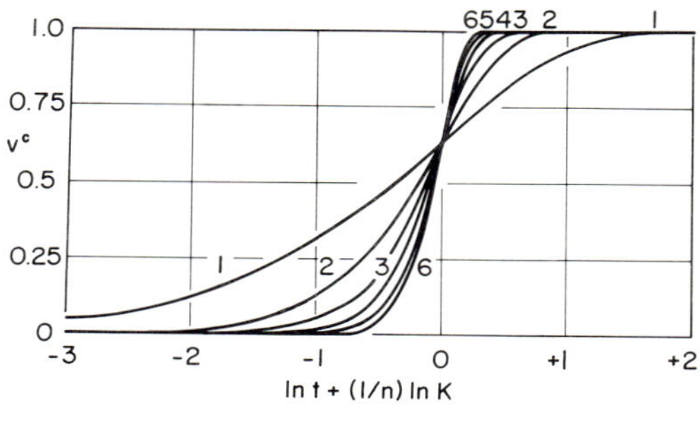

Fig. VI.10 *Increase in Volume Crystallinity as Given by the Avrami Equation.*
The different values of n in Eq. (33) are listed in the figure.

In the early stages of crystallization there is only little restriction of crystallization due to impingement. Equation (33) can then be $v^c = Kt^n$. In this form the equation represents unimpeded crystal growth (Göler *et al.*, 1932):

$$(\exp[-Kt^n] = 1 - Kt^n + \cdots).$$

6.1.3.2 Experimental Evaluation

Dilatometry and calorimetry are easily adapted to measure crystallinity as a function of time. Calorimetric experimental data on crystallization of a copolymer of ethylene terephthalate (80 mole %) and ethylene sebacate (20 mole %) are shown in Fig. VI.11. The data (left) show the required

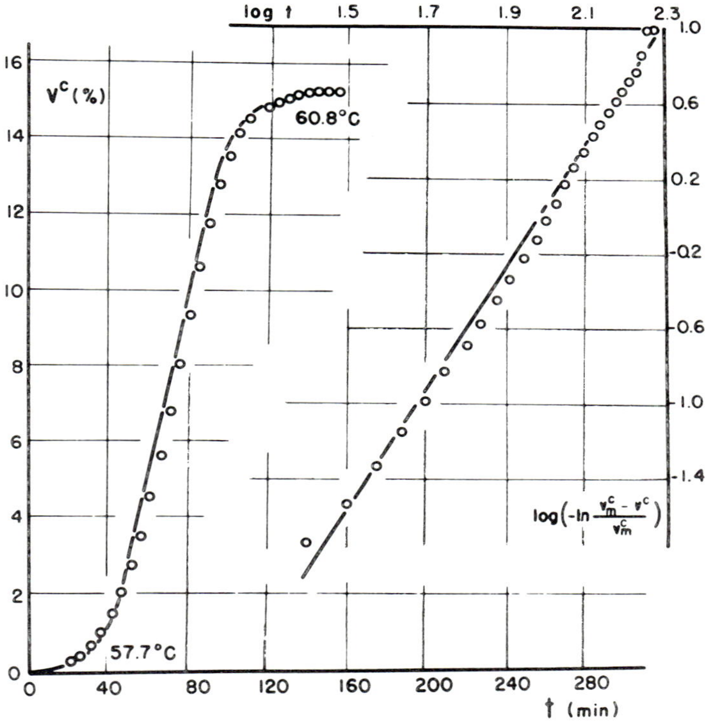

Fig. VI.11 *Experimental Data for the Crystallization of a Macromolecular Melt.*
Data for a random copolymer of ethylene terephthalate and ethylene sebacate (80/20). (○) Experimental points. Calorimetric data by Wunderlich, unpublished.

sigmoidal shape for an Avrami type analysis; however, the sample does not reach complete crystallization as required for the model. To continue an analysis along the lines of Eq. (33), one assumes that complete crystallization is reached at the level v_m^c (=0.152). In the framework of the model, this means that the growing spherulites are only partially crystallized within themselves, but have crystallized to the extent v_m^c at the instant the crystal–amorphous interface has passed. To evaluate n, the right-hand side of Fig. VI.11 shows a plot of the equation

$$\log(-\ln[(v_m^c - v^c)/v_m^c]) = \log K + n \log t \tag{34}$$

The slope represents n and was in the chosen example 3.2. Since the experiment was not strictly isothermal,† perhaps it is close enough to support the case of three-dimensional spherulitic growth with athermal nucleation conditions, a picture which fits the known crystallization habit of the copolymer. Additional data are discussed in Sect. 6.3.1.

The use of depolarization of linearly polarized light measured photometrically through a polarizing microscope as a measure of progress of crystallization (see Sect. 4.1.7) causes some complications. The optical retardation Δ is proportional to the sample thickness [Eq. (8), Chapter III] the depolarization (light intensity) however, is proportional to $\sin^2(\Delta/2)$. The Avrami coefficient n observed in this manner should thus be greater by 1 when expressed in terms of the increase in depolarization rather than in terms of increase in crystallinity. Binsbergen (1970) calculated the necessary average values for the depolarization of plane polarized light going through randomly oriented uniaxially birefringent crystallites. He found, indeed, that a t^3 dependence of the initial increase in crystalline volume results in an initial t^4 dependent increase in depolarization. After impingement of the crystals this leads to Avrami-type equations with $n = 4$ instead of $n = 3$, which would have been observed by measuing the crystallinity directly.

To check on the overall crystallization progress, direct data on nucleation rate and linear growth velocity are desirable. Measurements are frequently possible by optical microscopy. Particularly for metals, but in principle also for macromolecular materials, it is easiest to cut specimens after various crystallization time, polish, and count the nuclei and measure the size of the grains. The problem of converting the random-cut planar distribution to a spacial distribution was already treated by Huber (1935). Macromolecular samples are usually observed directly under the microscope on a hotstage. In this case the melt is restrained between cover glass and slide to a layer which is often thinner than the freely grown spherulite. The resulting crystallinity development should thus change with decreasing layer thickness from the true three-dimensional cases [Eqs. (32)] to the two-dimensional cases [Eqs. (27)]. Stein and Powers (1962) derived the equation for an intermediate thickness d by the introduction of a correction factor f for dE to account for

† The calorimetric experiment involved close to adiabatic conditions so that the evolved heat of crystallization heated the sample from 57.7°C to 60.8°C.

the truncation of the volume out of which a spherulite could reach point P due to restraining coverglass and slide planes.† For the general case of i nonoverlapping truncating surfaces it can be easily seen that

$$f = 1 - (1/2) \sum (1 - a_i/r) \qquad \text{for } r \geq a_i \tag{35}$$

where a_i is the right angle distance from the point P to the ith truncation. Averaging over all choices of point P within the layer of thickness d

$$\bar{f} = 1 - (r/2d) \qquad \text{for } r \leq d \tag{36}$$

and

$$\bar{f} = d/2r \qquad \text{for } r \geq d \tag{37}$$

After integration of the expressions for dE corrected with Eq. (37), Eqs. (32) change to

$$E = \pi N\, dv^2\, t^2 \tag{38}$$

and

$$E = \pi I^*\, dv^2\, t^3 \tag{39}$$

as expected from the two-dimensional cases of Eqs. (27). Thus, for sufficiently large spherulites it may be possible to have a discrepancy of 1 in the Avrami exponent n when comparing thin films with bulk samples.

The general problem of interpretation of experimental data using the Avrami equation is that fractional exponents from data fitting is more usual than exceptional (see Sect. 6.3.1). Also, the experimental data deviate frequently at long crystallization times, showing slower increases in the crystallinity than expected from the initial data. Both observations are clear indications that the simple cases of crystallization described in Sect. 6.1.3.1 are not fully applicable to the case in hand. Some effort will be made in Sect. 6.1.3.3 to fit the model more closely to the actual macromolecular crystallization cases.

At this point one may start to wonder, however, why the Avrami equation usually fits the data so well. Binsbergen and DeLange (1970) proved that the frequent good fit rests with the inherent strong correlation between t_0, the starting time of crystallization, and K and n of Eq. (33). By nonlinear regression analysis the influence of variations in the different parameters describing the experimental results was assessed using typical data. After some correlated variation of δK, δn, and δt_0 of the parameters, the experimental data v^c and t were still reasonably well represented by Eq. (33). A special consequence of

† Stein and Powers (1962) actually calculated the reduction in rate of appearance of crystallinity rather than the rate of disappearance of unconverted volume used above.

this inherent correlation of the parameters of the Avrami equation is the difficulty of estimating the induction time for nucleation (see Sect. 5.1.5). Binsbergen and DeLange (1970) estimate that the induction time must be more than 0.3 times the crystallization half-time before it can be safely detected though the fitting of the Avrami equation to experimental data.

6.1.3.3 Complications in the Avrami Analysis

The derivation of the Avrami expression for the overall change in crystallinity contains a number of simplifications which do not necessarily apply to the crystallization of macromolecules.

1. The volume does not stay constant on crystallization as assumed (see Table IV.1).
2. The linear growth rate is under certain circumstances not constant with time (see Sect. 6.1.2.3).
3. The number of nuclei may not increase continuously, but reach a limiting level after exhaustion of heterogeneous nuclei (see Sect. 5.1.3).
4. The crystal morphology does not conform exactly to a spherical habit (or circular for the two-dimensional approximation) (see Chapter III).
5. The actual crystalline substructure frequently branches (see Sects. 3.3.3, 3.6, and 3.7.1).
6. There can be a two-stage crystallization (see Sect. 6.1.6).
7. Finally, there can be a perfection of the crystals after crystallization (see Sect. 6.1.6 and Chapter VII).

These complications of the Avrami analysis will be discussed next. The large number of complications makes it quite clear that more then just macroscopic observation of the increase in crystallinity is necessary to draw definite conclusions about the microscopic mechanism of crystal growth. Without morphological information, the Avrami analysis provides little more than a convenient representation of the macroscopic data.

The effect of the volume change on crystallization has been discussed by Price (1965a). E, the expected number of crystals reaching a representative point P up to time t for the case of changing sample volume is given by Price (1965a) to be (for randomly centered spheres)

$$E = \sum V_i/V \tag{40}$$

where V_i is the volume at time t of the crystalline sphere i which would have grown from nucleus i had there been no impingement, and V is the total sample volume at time t. The sum is to be taken over all spheres i initiated in the sample.† The condition of constant mass during the crystallization leads to the relationship

$$V_0 \rho_l = V v^c \rho_c + V(1 - v^c)\rho_l \tag{41}$$

where V_0 is the initial liquid volume, ρ_l and ρ_c are the liquid and crystalline densities, and v^c is the volume fraction crystallinity (see Sect. 4.1.1). The total volume at a given crystallinity v^c is thus

$$V = V_0 \rho_l / [v^c \rho_c + (1 - v^c)\rho_l] \tag{42}$$

Defining the fractional density change on crystallization‡ δ as

$$\delta = (\rho_c - \rho_l)/\rho_l \tag{43}$$

the expression for E (Eq. 40) changes to

$$E = \sum (V_i/V_0)(1 + v^c\delta) \tag{44}$$

Substitution of Eq. (44) into Eq. (28) does not give us an explicit solution of v^c except for equal densities of crystal and liquid ($\delta = 0$). The general Eq. (33) for the three-dimensional case assumes the form

$$(1 - v^c) = \exp[-K(1 + v^c\delta)t^n] \tag{45}$$

or, in the double logarithmic form used for the analysis of experimental data (Eq. 34):

$$\log \left[\frac{-\ln(1 - v^c)}{(1 + v^c\delta)} \right] = \log K + n \log t \tag{46}$$

† The derivation can be summarized as follows: Consider a fixed mass of supercooled liquid and mark the random positions of nuclei present or to be generated in time. At a given time t the volume has changed from V_0 to V. The crystallized spheres V_i are now injected, one at a time, into the volume V. The probability that point P is not inside any of the spheres of total volume $\sum V_i$ is $1 - v^c$ and is approximated by E of Eq. (40) if there are many crystals. For athermal nucleation $\sum V_i = 4\pi N V_0 v^3 t^3/3$, where N is the number of nuclei in one cubic centimeter. For thermal nucleation it can easily be derived that $\sum V_i = \pi I^* V_0 v^3 t^4/3$, where I^* is the rate of nucleation in nuclei/(cm^3 sec) [compare with Eqs. (32)]. [But note also the discussion of Rohleder and Stuart (1960) who came to the different conclusion that w^c should be substituted for v^c in Eq. (33) and the correction of crystal growth rates for sample shrinkage could not compensate the volume change. See also Tobin (1974) for an attempt at explicit calculation of overall crystal growth, described for two-dimension growth in Sect. 6.1.3.1, Eq. (31).]

‡ This definition of δ deviates from that of Price (1965a) who set $\delta = \rho_l/\rho_c$. With the present definition the discussion simplifies, especially since $P_0(t)$ has been already set equal to the volume fraction amorphous instead of the weight fraction as is frequently usual (see Eq. 28).

A comparison with Eq. (34) reveals an almost insignificant change† in the parameters K and n which, in addition, are strongly correlated (see Sect. 6.1.3.2). A larger error is introduced if Eq. (33) is written using the weight fraction crystallinity without correction [$w^c = (\rho_c/\rho)v^c$]

$$w^c = 1 - \exp[-Kt^n] \tag{47}$$

In this case Price (1965a) states that errors of 0.15 to 0.25 in n are easily obtainable.

The variation of the linear growth rate of the crystals with time would affect the evaluation of E through Eqs. (27) and (32). The case of crystallization wholly governed by transport processes is discussed in Sect. 6.1.2.3. For planar and spherical interfaces it leads to a $t^{1/2}$-dependence of the crystal dimension. Insertion of the appropriate temperature dependence leads us to the following exponents n of Eq. (33):

Two-dimensional crystal growth:
$n = 1$ for athermal nucleation (Eq. 27A)
$n = 2$ for thermal nucleation (Eq. 27B)
Three-dimensional growth:
$n = \frac{3}{2}$ for athermal nucleation (Eq. 32A)
$n = \frac{5}{2}$ for thermal nucleation (Eq. 32B)

The decrease in nucleation rate as observed in thermal heterogeneous nucleation when the nuclei are being exhausted should not be described by a single Avrami type expression. Rather, at a certain time the exponent n of Eq. (33) should decrease by 1.

For crystal growth in a three-dimensional sample, Eq. (33) applies strictly to spherical morphology. Spherulitic crystallization approximates spherical morphology reasonably well as long as the spherulite is well developed and crystallization is completed close to the spherulite boundary (see below). Many polymeric samples crystallize, however, with a simple lamellar (Sect. 3.3) or fibrillar (Sect. 3.8) morphology. Morgan (1954) discussed the case of fibrillar growth. Using the method of the derivations of Eqs. (32A and B) one would assume that fibrils of diameter d from the spherical shell of volume $4\pi r^2\,dr$ could approach point P with rate v. However, their chance of reaching it is not one, as in the case of a sphere growing, but it is reduced by the ratio of twice the fibril‡ cross section $d^2\pi/2$ to the sphere surface $4\pi r^2$. Thus, for

† For example, assuming equal K and a one-decade half-time of crystallization, the increase in n of Eq. (34) over Eq. (46) is only 0.02 for $\delta = 0.1$ and decreasing at longer times.

‡ Assumed to grow in both directions and to keep a constant diameter d.

athermal nucleation, the expected number of crystals reaching P is [compare with Eq. (32A)]

$$E = \tfrac{1}{2}\pi N d^2 \int_0^{vt} dr = Nd^2vt/2 \tag{48}$$

Analogously for thermal nucleation [compare with Eq. (32B)]

$$E = \tfrac{1}{2}\pi I^* d^2 \int_0^{vt} [t - (r/v)]\, dr = \pi I^* d^2 vt^2/4 \tag{49}$$

Equations (48) and (49) are incorrect to the degree that some fibers which were calculated to be able to reach point P may not do so since their path is blocked by other fibers. This restriction becomes less serious, however, if one assumes, and this is usually the case, that the fibers will grow around each other or are only deflected. Price and Thornton (1973) tried to assess the effect of termination of fibers by computer modeling. The exponent n was found to decrease; the biggest effect, however, was a large decrease in K of Eq. (33). Equation (49) has an additional restriction: it permits new nucleation in areas which are actually already occupied by fibers. Thus, the equation would only apply to the early stage of crystallization and n, the exponent of t, would decrease somewhat below 2 with time.

For lamellar growth, similar equations arise. In this case the lamellar thickness is assumed to be fixed at d (the fold-length) and the overall shape is taken to be that of a disk of side surface $2dr\pi$. For athermal nucleation, the expected number of crystals reaching P at time t is

$$E = 2\pi N d \int_0^{vt} r\, dr = \pi N d v^2 t^2 \tag{50}$$

and again, for thermal nucleation,

$$E = 2\pi I^* d \int_0^{vt} [t - (r/v)] r\, dr = \pi I^* d v^2 t^3/3 \tag{51}$$

The same restrictions hold for these equations that hold for the fibrillar cases. Some evidence for interpenetration of lamellae can be gained from Figs. III.41 and III.77B.

The blocking of crystals and reduction in nucleation frequency lowered the temperature exponents in Eqs. (48)–(51). The opposite effect, an increase in exponent n, is expected for branching crystals (Morgan, 1954). Macromolecular crystals are known frequently to have an inherent or a supercooling caused branching mechanism (see Sects. 3.3.3, 3.6.1, and 3.7.1). Thus, it is of interest to study the influence of branching of crystals on the overall crystallization rate. A simple model was developed by Booth and Hay (1972).

It assumes that a once-nucleated fibrillar crystal of cross section $d^2\pi/4$ branches into two fibers after b cm of growth. Both daughter branches have the same cross-sectional area and branch again after b cm. Overall, one expects from such growth a sheaflike structure and finally a spherulitic development as illustrated in Fig. III.117B. The total cross-sectional area A of the branching fibrils growing again in both directions is

$$A = (d^2\pi/2)(1 + 1 + 2 + 4 + \cdots 2^{(r/b)-1}) \tag{52}$$

$$A = d^2\pi 2^{(r/b)-1} \tag{53}$$

Using now the reasoning for fibrillar growth of Eq. (48), we can derive† E for athermal nucleation as

$$E = \tfrac{1}{2}\pi N d^2 \int_0^{vt} 2^{r/b}\, dr = \pi N d^2 b(2^{vt/b} - 1)/(2 \ln 2) \tag{54}$$

This expression has the same limitations as Eq. (48), but to a larger degree. The branching fibers will not only be blocked by other sheafs of fibers from reaching point P, but they will also block each other. The termination of fibers would reduce the exponent of t when analyzing Eq. (54) in terms of the simple expression Eq. (33). By developing Eq. (54) into a series, it can be written

$$E = (\pi N d^2 vt/2)\{1 + [\ln 2/(2b)](vt) + [(\ln 2)^2/(6b^2)](vt)^2 + \cdots\} \tag{55}$$

The first term is identical to Eq. (48) which was developed for the growing of a single fiber. With increasing branching (decreasing b) and increasing growth (vt), exponents with increasing values must be included in the analysis. Booth and Hay (1972) showed that data computed from Eq. (54) when analyzed with Eq. (33) show an increasing exponent as growth (branching) continues, reaching values as high as 14 for the development of 10^6 branches/cm ($b = 100$ Å). This is to be compared with the analysis by Morgan (1954) who assumed the branching to be efficient enough to produce a solid sheaf around auxiliary circles of radius a as shown in Fig. VI.12. He estimated‡ the initial decrease in amorphous content to follow the expresions

$$1 - v^c = \exp[-Kt^5] \tag{56}$$

for athermal nucleation and to follow the expression

$$1 - v^c = \exp[-Kt^6] \tag{57}$$

† Except for the ln 2 term, Eq. (54) is identical to Eq. (2) of Booth and Hay (1972).

‡ The volume of the sheaf is $(\pi v^5 t^5/10a^2)$ $[1 - (1/42)$ $(vt/a)^2 + (1/3024)$ $(vt/a)^4 \cdots]$. From this Morgan (1954) estimated that the sheaf could be approximated by a sphere of average radius $(\text{volume})^{1/3}$. Eqs. (56) and (57) follow then on neglecting higher terms of the series.

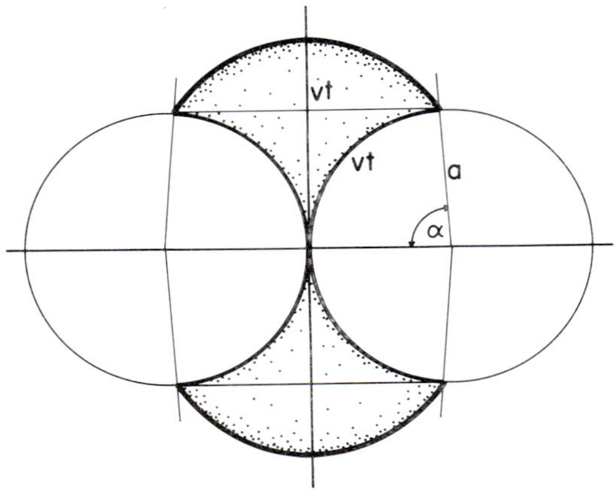

Fig. VI.12 *Schematic Drawing of a Solid Sheaf Developing around Circles of Radius a.*
Compare with Figs III.117B and III.118. The drawing shows a cross section through the sheaf which is cylindrically symmetric about the axis.

for thermal nucleation. Both exponents of t also to increase at later times. Iguchi (1972) made a similar effort to assess the effect of branching using model experiments and only slightly different calculations.† He also made an attempt to correct for the additional blocking of fibrils by inter- and intrasheaf collision beyond the assumptions of Eq. (48) by the introduction of an arbitrary shape factor.

Two-stage spherulitic crystallization as a complication of the overall crystallization rate‡ has been considered by Price (1965b), Hillier (1965), and Hoshino *et al.* (1965). The three treatments are similar, but here the ideas of Hillier (1965) will be followed. After the spherulite boundary has passed over a particular point at time τ, a second crystallization step is assumed to be initiated, following also an Avrami-type expression, but with the parameter $n_2 = 1$ (for convenience and because it is known that the interior of a macromolecular spherulite is often fibrillar) [see Eq. (33)]. This step increases the

† Iguchi (1972) used a constant branching rate s $(=v/b$ in the present nomenclature) rather than successive step branching. Their final expression for E on athermal nucleation is [compare to Eq. (54)].

$$E = \pi N d^2 b(e^{vt/b} - 1)/4$$

‡ Baur (1968) has shown by thermodynamic considerations that the mixing effect in multicomponent polymers may result in nonmonotonic crystallization of various molecular weight components. This leads also to a delay of crystallization analogous to secondary crystallization.

crystallinity from an initial maximum v^c_{am} by an additional amount v^c_{bm}. This process is not to be confused with the perfection of once-crystallized material discussed in Sect. 6.1.6 and which is frequently described by a crystallinity dependence on the logarithm of time. The initial crystallinity (a) is thus

$$v^c(a, t) = v^c_{am}(1 - \exp[-K_1 t^{n_1}]) \tag{58}$$

with n_1 being either 3 or 4 depending on the type of nucleation [see Eqs. (32A and B)]. This is followed for each crystallized element from time τ by the secondary process (b)

$$v^c(b, t - \tau) = v^c_{bm}(1 - \exp[-K_2(t - \tau)]) \tag{59}$$

The overall macroscopicially observable development of crystallinity is then

$$v^c(t) = v^c(a, t) + \int_t^0 \frac{v^c(a, \tau)}{v^c_{am}} \frac{d}{d\tau} v^c(b, t - \tau)\, d\tau \tag{60}$$

or, after insertion of Eqs. (58) and (59),

$$v^c = v^c_{am}(1 - \exp[-K_1 t^{n_1}]) + v^c_{bm} K_2 \int_0^t [(1 - \exp[-K_1 \tau^{n_1}])\exp[-K_2(t - \tau)]\, d\tau \tag{61}$$

The final overall crystallinity at long times t is thus $v^c_{am} + v^c_{bm}$. Table VI.1 illustrates how a hypothetical crystallization following Eq. (61) would appear on analysis with the simple Avrami expression Eq. (33) using only the parameters K and n. The last column indicates the standard deviations of n

TABLE VI.1

AVRAMI PARAMETERS FOR A TWO-STAGE CRYSTALLIZATION ANALYZED BY EQ. (33)[a]

K_1	K_2	n_1	v^c_{am}	v^c_{bm}	K	n	$\sigma(\%)$
10^{-6}	10^{-1}	3	0.7	0.3	7.47×10^{-7}	3.04	0.08
10^{-6}	10^{-2}	3	0.7	0.3	1.46×10^{-5}	2.32	2.71
10^{-6}	10^{-3}	3	0.7	0.3	3.04×10^{-5}	2.18	4.62
10^{-6}	10^{-2}	3	0.5	0.5	2.29×10^{-5}	2.16	2.79
10^{-6}	10^{-2}	3	0.9	0.1	3.00×10^{-6}	2.73	1.43
5×10^{-9}	10^{-1}	4	0.7	0.3	5.01×10^{-8}	3.50	1.84
5×10^{-9}	10^{-2}	4	0.7	0.3	2.87×10^{-7}	3.06	2.90
5×10^{-9}	10^{-3}	4	0.7	0.3	4.26×10^{-7}	2.99	4.83
5×10^{-9}	10^{-2}	4	0.5	0.5	3.74×10^{-8}	3.39	4.20
5×10^{-9}	10^{-2}	4	0.9	0.1	4.70×10^{-8}	3.50	1.56

[a] Data by Hillier (1965).

which show a systematic trend with time. For all considered cases the overall standard deviation is less than 5%, a fluctuation tolerable for many experimental data. Fractional exponents have resulted whenever the half lives of the two crystallizations are comparable. Making the second process much slower than the first leads us to the description which may be useful for the case of changing perfection of crystals after initial crystallization (see below and Sect. 6.1.6). The application of Eq. (61) may be able to describe the overall crystallization process better, but at the cost of more adjustable parameters. The interpretation of the derived parameters is only simple for the first stage of crystallization; the second stage crystallization is not well understood, neither does Eq. (59) have an exact correspondence to a simple model [see also the discussion of Eq. (48)].

The final complication of the Avrami analysis mentioned at the beginning of this section is the effect of increasing crystal perfection after the initial growth front has passed. Macroscopically this is recognized as a delayed slow increase in crystallinity. This effect can again be taken into account only by a second process as in Eqs. (58)–(61). The perfection of crystals is a solid state process and has often been described using a logarithmic time dependence. There is little reason, however, to use any form of the Avrami expression Eq. (33) for this increase in crystallinity. Such slowing down of the crystallization at long times is not only characteristic for many macromolecular systems, but has been observed also on the crystallization of metals (see Avrami, 1940). More details about this perfection process are given in Sect. 6.1.6 and Chapter VII.

6.1.3.4 Conclusions about the Avrami Equation

The treatment of the overall crystallization led to several stages of approximation. Exact solutions are possible for the one-dimensional crystallization along a line, the two-dimensional crystallization forming a circle, and the three-dimensional crystallization leading to spheres. Other morphologies have only been treated approximately. The equations are usually only valid at low conversion as long as the impingement is not serious. More insight is expected from computer based model experiments. Table VI.2 shows a collection of results obtained by analyzing the overall crystallization of given microscopic models with the general Avrami equation (33). The exponent of temperature n may show values ranging from below 1 to far above 6. Any one value, whole or fractional, however, is not uniquely fixed to any one set of conditions. Additional information on nucleation, morphology, and possibly even mechanism is necessary to fully interpret the exponent n. Special caution is necessary due to the strong correlation between K, t, and n

TABLE VI.2

EXPONENTS OF TIME IN THE AVRAMI EQUATION (33)

Type of crystallization	Nucleation	n	Remarks
A. Linear problem:			
Line	Athermal	1	Exact, analogous to Eq. (27B)
Line	Thermal	2	Exact, analogous to Eq. (27A)
B. Two-dimensional problem:			
Ribbon	Athermal	≤ 1	Limiting, analogous to Eq. (48)
Ribbon	Thermal	≤ 2	Limiting, analogous to Eq. (49)
Circular	Athermal	2	Exact, Eq. (27B)
Circular	Thermal	3	Exact, Eq. (27A)
Circular, diffusion control	Athermal	1	Exact, see tabulation p. 141
Circular, diffusion control	Thermal	2	Exact, see tabulation p. 141
Circular	Thermal, exhaustion	$3 \rightarrow 2$	Successive, p. 141
C. Three-Dimension Problem:			
Filbrillar	Athermal	≤ 1	Limiting, small v^c, Eq. (48)
Fibrillar	Thermal	≤ 2	Limiting, small v^c, Eq. (49)
Circular lamellar	Athermal	≤ 2	Limiting, small v^c, Eq. (50)
Circular lamellar	Thermal	≤ 3	Limiting, small v^c, Eq. (51)
Spherical	Athermal	3	Exact, Eq. (32A)
Spherical	Thermal	4	Exact, Eq. (32B)
Spherical, diffusion control	Athermal	3/2	Exact, see tabulation p. 141
Spherical, diffusion control	Thermal	5/2	Exact, see tabulation p. 141
Spherical	Thermal, exhaustion	$4 \rightarrow 3$	Successive, p. 141
Two-stage	Athermal/thermal	fractional	Limiting, Eq. (61)
Branching fibrillar	Athermal/thermal	1, $2 \rightarrow$ large	Limiting, Eq. (55)
Solid sheaf	Athermal	≥ 5	Limiting, Eq. (56)
Solid sheaf	Thermal	≥ 6	Limiting, Eq. (57)
Truncated sphere	Athermal	2–3	Exact, Eqs. (35)–(39)
Truncated sphere	Thermal	3–4	Exact, Eqs. (35)–(39)
Volume decrease on cryst.	Athermal/thermal	fractional increase	See Eqs. (40)–(47)
Perfection after initial cryst.	Athermal/thermal	decrease	See Sect. 6.1.6.

(see Sect. 6.1.3.2), which may hide changes in n by correlated changes in K and t_0.

Without the parallel knowledge of the microscopic, independently proven mechanism, the macroscopic, experimentally derived Avrami equation and the Avrami parameters are only a convenient means to represent empirical data of crystallization.

6.1.4 CRYSTAL GROWTH AND CHAIN FOLDING

In this and the following sections, points of crystal growth unique to the crystallization of linear macromolecules are discussed. The central question about crystallization of already formed macromolecules is: How can one understand and describe chain folding of the type discussed in Sect. 3.2.2?

6.1.4.1 The Basic Problem

The treatment of primary and secondary crystal nucleation and the molecular nucleation of the macromolecules has revealed that all these nuclei possess a critical size which is normally smaller in volume than a single molecule and larger in width than the molecular diameter (Chapter V). A direct consequence of the higher free enthalpy of the remaining uncrystallized portions of the partially crystallized macromolecules, expressed by γ' (see Sect. 5.3.4), is nucleation by single molecules as long as γ' is large (mobile molecules). A single molecule must then fold in order to reach nucleus dimensions of larger than single chain widths. Since each molecule goes through the molecular nucleation step (Sect. 5.3), the question of chain folding can be restated as: Why does a once-nucleated molecule continue to fold?

6.1.4.2 Nucleation Theories

The kinetic theory of chain folded crystal growth of macromolecules has had many refinements since its initial suggestion by Lauritzen and Hoffman (1959, 1960) and Price (1959, 1960). More recent reviews and comparisons of the various aspects of the theory have been given by Hoffman *et al.* (1969) and Sanchez (1974). A summary of the key contributions to the kinetic theory of chain folding and its major problems is given next.

The simplest assumption about chain folding is that the critical secondary or molecular nucleus size as expressed by Eq. (67), Chapter V, is maintained for the rest of the crystal or molecule. By adjustment of the surface free energy γ_e, reasonable agreement with the experimental fold-length can be achieved for not-too-large a supercooling (see Sect. 5.2.2.2). The temperature dependence of the fold-length $1/\Delta T$ is reproduced properly in this way. However, for large supercooling, the practically constant fold-length observed experimentally cannot be fitted [compare Eqs. (67) and (72), Chapter V]. In addition, one would expect from the free enthalpy diagram of Fig. V.39 that crystallization after nucleation should increase the fold-length in order to decrease the overall free enthalpy of the crystal. Even if one postulates that the once-folded crystallized portions cannot rearrange, further crystallization of each molecule slowly should increase the subsequent fold-lengths and lead to a nonuniform fold surface and a strong molecular weight dependence of the fold-length, neither of which seem to be in agreement with the experimental findings (see Sect. 3.2.2.2 and 3.2.2.4).

An improvement in the description of the fold-length of the resulting crystals is achieved by assuming that all nuclei of a fold-length l larger than the critical fold length l^* [Eq. (67), Chapter V] would contribute to the crystal by growing to completion, but still without change in fold-length. The thicker folded chain layers would nucleate more slowly, but would complete growth faster. The average fold-length of the resulting crystal is given under these conditions by Eq. (76), Chapter V. This equation can also be written as

$$\langle l \rangle = \frac{2\gamma_e T_m}{\Delta h_f \rho_c \Delta T} + \delta l \tag{62}$$

with the usual meanings: γ_e, specific fold surface free energy; T_m, melting or dissolution temperature; Δh_f, heat of fusion; ρ_c, density of the crystal; and ΔT, supercooling. The term δl is $kT/b_0\gamma$ at the melting temperature [see Eq. (76), Chapter V] and increases initially slowly with supercooling (b_0, single layer thickness; γ, specific side surface free energy). Equation (62) has at first sight the proper mathematical form to describe the experimental data as described by Figs. III.12 and V.41 and Eq. (72), Chapter V. Inspection of Eq. (76), Chapter V, shows, however, that there are no additional widely adjustable parameters since γ is known, for example, from nucleation experiments (see Sect. 5.1.2.2). Reasonable values for γ of polyethylene range between 5 and 10 erg/cm^2 which would lead at a temperature of 350°K to a δl range from about 20 to 10 Å, which is considerably smaller than the experimental value of 70 Å. The poor fit of theory and experiment could be removed making γ_e strongly temperature dependent, a result obtained when considering fluctuations [see Eq. (67)].

There is an additional difficulty when Eq. (62) is used to predict fold-lengths at high supercooling. At a certain temperature at which Δg_f the free enthalpy of fusion approaches $2\gamma/a_0$, δl approaches infinity [see Eq. (77), Chapter V]. This unrealistic increase in predicted fold-length can be understood from the detailed study of the assumptions which went into the derivation of Eq. (77), Chapter V. It was assumed that the first step, the establishment of the initial new traverse on the substrate, would be increasingly slower with increasing length because of the higher free enthalpy level ΔG_1 which would have to be overcome [see Fig. V.42, Eq. (74), Chapter V]. At a supercooling where Δg_f reaches $2\gamma/a_0$, this assumption no longer holds. From Eq. (74), Chapter V, it can be seen that under this condition ΔG_1 is independent of l.

$$\Delta G_1 = 2\gamma_e a_0 b_0 \qquad (\text{for } \Delta g_f = 2\gamma/a_0) \tag{63}$$

Thus, all fold-lengths have the same nucleation barrier. It was shown before that, once established, larger fold-lengths show a faster decrease in free enthalpy for subsequent fold segments which should lead to faster growth [see Eq. (73) and Fig. (39), Chapter V]. Thus, the average fold-length must tend to infinity. Figure VI.13 illustrates experimental data on isotactic

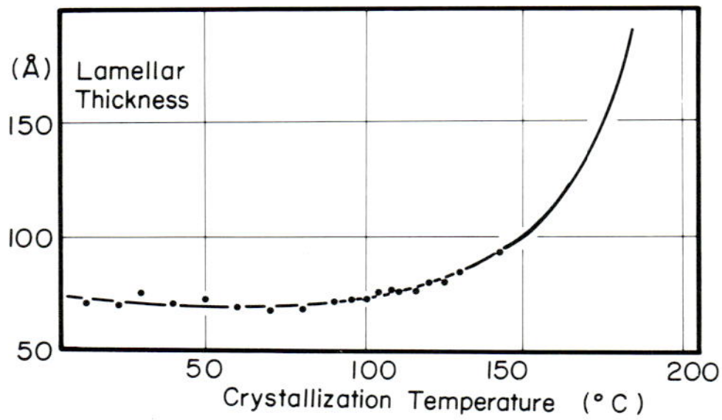

Fig. VI.13 *Lamellar Thickness l of Isotactic Polystyrene as a Function of Supercooling* ΔT. Crystallization from 0.1% dilute solution of dimethylphthalate, estimated dissolution temperature of extended chain crystals 205°C. (Data by Jones *et al.*, 1973).

polystyrene, a macromolecule which crystallizes slowly enough to reach a supercooling substantially larger than found in polyethylene. Little or none of the predicted upturn in thickness can be seen.† The use of molecular nucleation arguments does not improve the situation since in this case ΔG_1 is only larger by an approximately constant amount $2a_0 b_0 \gamma'_e$; again, ΔG_1 has no fold-length dependence left when Δg_f reaches $2\gamma/a_0$ [see Eq. (81), Chapter V, $\nu = 1$]. Thus, empirically, one may say that only a ΔG_1 which increases with l over the whole range of supercooling can be fitted satisfactorily to the experimental data within the framework of a chain fold-length governed by classical nucleation theory based on constant fold-length for subsequent folds. Lauritzen and Hoffman (1973) tried to satisfy this condition by assuming that the initial step involves mainly the creation of surface free energy by weak absorption of the macromolecule ($\Delta G_1 = 2b_0 l\gamma$) which naturally would be length dependent.

6.1.4.3 Fluctuation Theories

Perhaps the most unrealistic assumption made previously is that once the nucleation step is over, the fold-length is kept constant. One would expect that fluctuations to larger or smaller fold-lengths would be possible. Initial efforts to describe the influence of fluctuations of the fold-length after nucleation were made by Frank and Tosi (1961) and Price (1961). Frank and Tosi

† Similar invariance of fold-length at larger supercooling has been observed for melt crystallized poly(tetramethyl-*p*-silphenylene siloxane) by Pollack and Magill (1969), *J. Polym. Sci. Part A*-2, 7, 551; for nylon 6.6 crystallized from solution by Dreyfuss and Keller (1973), *J. Polym. Sci. Polym. Phys. Ed.* **11**, 193; and for melt crystallized poly(ethylene terephthalate) by Wlochowicz and Przygocki (1973), *J. Appl. Polym. Sci.* **17**, 1197.

(1961) permitted single fluctuations after nucleation such that a too-small nucleus for fastest growth showed a tendency to fluctuate (once) towards a larger fold-length l^*. Similarly, a too-large nucleus tended towards l^* by a decrease in fold-length. In this way l^* defines a stable mean single layer fold-length. Similarly, the subsequent layers growing onto the first are thought to converge to a stable average layer fold-length l^{**} by a single fluctuation. The final observed fold-length should thus be l^{**}. Figure VI.14 indicates that

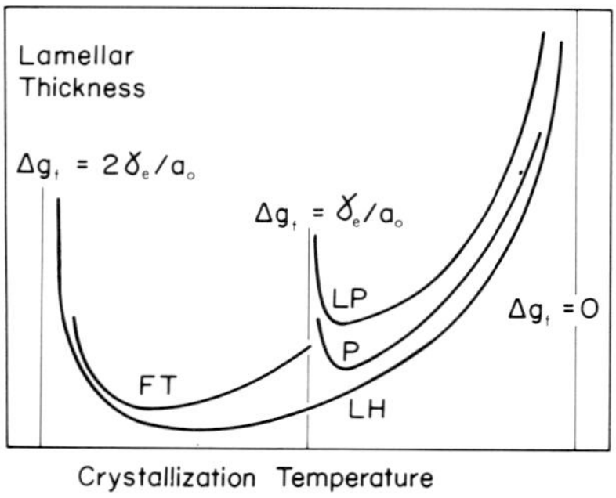

Fig. VI.14 *Lamellar Thickness l as Predicted by Various Theories of Chain Folding.* Parameters chosen to approximate a macromolecule like polyethylene, drawn after Hoffman *et al.* (1969). LH: nucleation theory by Lauritzen and Hoffman (1960), see Sect. 5.2.2. FT: single fluctuation theory by Frank and Tosi (1961). P: unlimited fluctuation theory using the Markov chain model by Price (1961). LP: general fluctuation model by Lauritzen and Passaglia (1967).

again acceptable parameters can be found to fit the theory at low supercooling†; at high supercooling fluctuations are also expected to tend to unlimited l^{**}. The reasons for this behavior are similar to the Lauritzen and Hoffman (1960) nucleation approach. Fluctuations to larger sizes need no increase in free enthalpy when the condition $\Delta g_f = 2\gamma/a_0$ is reached. In this case, the largest fluctuation leads naturally to the fastest subsequent growth.

Price (1961) based the evaluation of the most probable fold-length on a study of successive additions of layers of molecules on a substrate of given,

† However, at a supercooling of less than $\Delta T = \gamma T_m / a_0 \, \Delta h_f \, \rho_c$ this treatment predicts a fluctuation to the minimum thickness of a stable crystal so that crystal growth should cease unless a fold-length increase can occur in a secondary step.

limited thickness l_s in the chain direction (for example the primary nucleus fold-length). Figure VI.15 illustrates the appearance of such a new folded

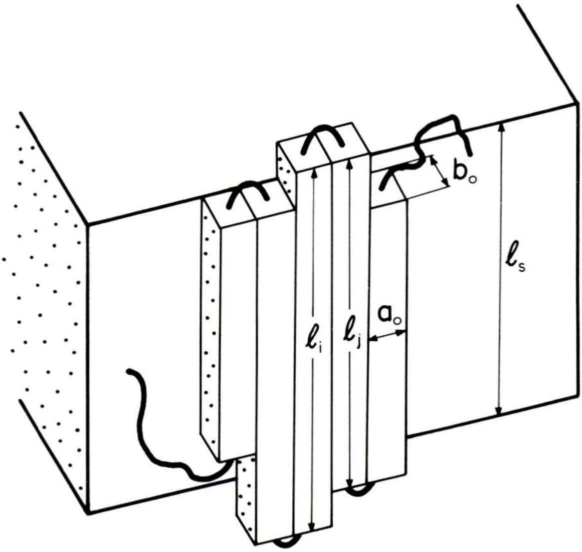

Fig. VI.15 *Schematic Drawing of a Macromolecule Growing on a Smooth, Finite Substrate with Fluctuating Fold-Length.*

chain layer on the smooth substrate. A restricted equilibrium is assumed to determine the distribution of fold-lengths. Unlimited numbers of fluctuation are permitted within each layer, but it is required that the average free enthalpy within the crystallized layer must be a minimum (i.e., the layer is in equilibrium within itself relative to the substrate thickness). A crystal is now built up of many such layers, each previous layer serving as a substrate for the next after smoothing to the average fold-length for computation. As soon as two successive layers reach some average fold-length, the final crystal fold-length has been found. The actual calculations are based on the Markov chain formalism developed by Mullins (1959). As can be imagined from the previous discussion, as soon as the bulk free enthalpy gain on crystallization of a chain segment is equal to the restricting effect of the surface created by fluctuations, the final fold-length tends again to infinity. Figure VI.14 illustrates a typical fold-length curve; the unlimited fluctuation supercooling is reached as soon as $\Delta g_f = \gamma/b_0$ ($b_0 \approx a_0$), earlier than in the above cases. The stoppage of crystal growth due to fluctuation of the critical nucleus size observed in the case of single fluctuations was avoided by prohibiting fold-lengths shorter or equal to the critical nucleus fold-length.

The most general treatment to date of fluctuations of fold-lengths during the crystallization of macromolecules has been given by Lauritzen and Passaglia (1967) using a general, one-dimensional growth rate theory of multicomponent chains (Lauritzen *et al.* 1966). This treatment allows for fluctuations at every stage of the crystallization process on a finite substrate under kinetic conditions. Figure VI.15 is used as a model to establish the necessary kinetics expressions. While the treatment of Price (1961) relies on equilibrium for each crystallizing layer, which may be approached only at small supercooling, it is possible to check the changes which occur when kinetics governs the fluctuations. The rate constant for the addition of a species j in length to a species i in length is A^{ij} and its reverse is B^{ij}. At equilibrium their ratio should be (see Fig. VI.15)

$$A^{ij}/B^{ij} = \exp[a_0 b_0 l_j \Delta g_f - 2a_0 b_0 \gamma_e - 2b_0 U(l_j - l_i)\gamma - 2a_0 U(l_j - l_s)\gamma]/kT \tag{64}$$

where $U(l_j - l_i)$ is equal to $l_j - l_i$ as long as $j > i$ and zero when $j \leq 0$; $U(l_j - l_s)$ is to be evaluated similarly. For the first segment added to a substrate the respective ratio of rate constants is (see Fig. VI.15)

$$A_0^j/B_i^j = \exp[a_0 b_0 l_j \Delta g_f - 2b_0 l_j \gamma - 2a_0 U(l_j - l_s)\gamma]/kT \tag{65}$$

Under kinetic conditions the assumption is made that the ratios of Eqs. (64) and (65) still hold and computations are made empirically, apportioning the free enthalpy changes variously to the forward and reverse rate constants.† Total flux could now be calculated analogously to the derivation of Eq. (75), Chapter V, for example. In order to circumvent the cessation of growth at an upper temperature limit below the melting temperature, the free enthalpy of the system (consisting of crystals of various thicknesses, each with layers of various thicknesses and lengths) was minimized with respect to its variables. The rate constant ratios which minimize the free enthalpy contain the terms

$$a_0 |l_j - l_s| \gamma \qquad \text{instead of} \qquad 2a_0 U(l_j - l_s)\gamma \tag{66}$$

in Eqs. (64) and (65), i.e., the loss of free enthalpy due to a short fold-length is also considered. To arrive at Eq. (66), we see that the distribution of l_j must of necessity be symmetric. With the substitutions indicated in Eq. (66), the rate constant ratios of Eqs. (64) and (65) lead to crystallization processes which proceed continuously to the melting temperature. The reason for the need for this substitution (which removes the wrongly predicted cessation of

† Note that sometimes all free enthalpy change is assigned to the less favorable step (see Frank and Tosi 1961, Lauritzen and Passaglia 1967, and also Lauritzen and Hoffman 1973). A 50 : 50 apportioning is assumed in the derivation of Eq. (5), Chapter V, and in the calculations for Figs. VI. 16–19.

crystal growth at temperatures below melting) lies in the basically one-dimensional nature of the theory which is applied to a two-dimensional problem. A plot of the average fold-length thus calculated is also drawn in Fig. VI.14. Mathematically this curve is represented by

$$l = (2\gamma_{\text{eff}}\, T_m)/(\Delta h_f\, \rho_c\, \Delta T) + \delta l \tag{67}$$

an expression quite analogous to Eq. (62), except that the meaning of γ_{eff} and δl is somewhat different. The fold free energy γ_{eff} is an "effective" value which is strongly temperature dependent. The difference between γ_e and γ_{eff} can be interpreted from the different nature of the two surface free energy terms. The term γ_{eff} arises from a rough surface which has extra exposed lateral surfaces (and also contains mixing entropy contributions). The detailed equations of the Lauritzen and Passaglia (1967) treatment permit calculation of these extra contributions. Figure VI.16 illustrates the change of $\gamma_e - \gamma_{\text{eff}}$ as a function of supercooling for different values of γ, the side surface free energy. A

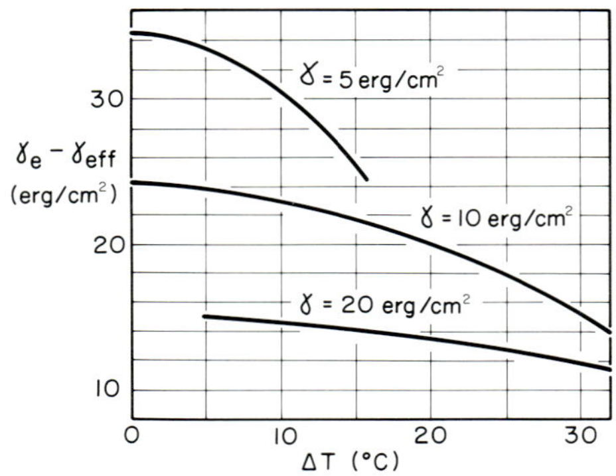

Fig. VI.16 *Change in $\gamma_e - \gamma_{\text{eff}}$ According to the Lauritzen and Passaglia Treatment of Chain Folding as a Function of Supercooling ΔT*
Data by Lauritzen and Passaglia (1967). The curves are not much dependent on γ_e. The apportionment of the free enthalpy is done such that $A_0{}^j$ is based on the total free enthalpy barrier and A^{ij} on the fold free energy and 50% of the other terms; γ is the side surface free energy in erg/cm^2.

significant portion of γ_{eff} is seen to depend upon γ. The value of ∂l is shown in Fig. VI.17. Quite analagous to the discussion of other fluctuation theories and to the Lauritzen and Hoffman (1960) nucleation theory, there is a temperature at which ∂l approaches infinity. This temperature is reached again where large

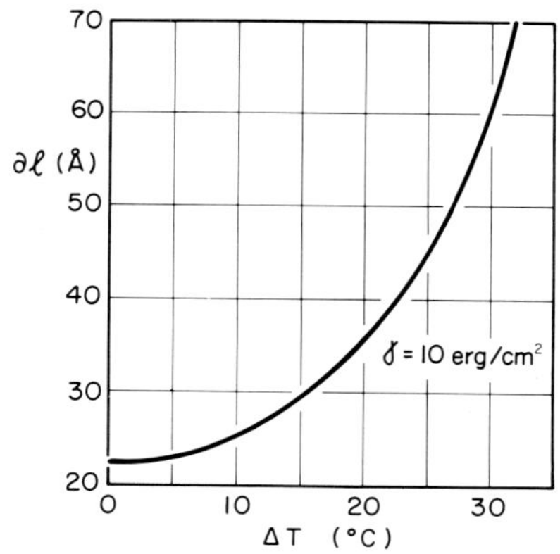

Fig. VI.17 *Change in ∂l According to the Lauritzen and Passaglia Treatment of Chain Folding as a Function of Supercooling ΔT.* Data by Lauritzen and Passaglia (1967). Conditions as in Fig. VI.16.

positive fluctuations are not restricted by the rate equations through large free enthalpy gains.†

6.1.4.4 Surface Roughness

The experimental data of Sect. 3.2.2.4 indicated a reasonable degree of fold surface roughness. The theoretical descriptions of crystallization in Sect. 6.1.4.3 should allow some insight into the possible fold irregularities. The simple nucleation theory does not account for surface roughness by definition. All other descriptions allow for a certain inherent roughness. However, only the descriptions which permit unlimited fluctuations are of interest for results on a distribution of fold-length. The treatment of Price (1961) which assumed minimum free enthalpy within each growing layer yields strongly peaked fold-length distributions, the peaks occurring at the substrate thickness. With typical values for polyethylene, Price (1961) calculated that at about 50°C supercooling a fold-length equal in length to the substrate length should occur ten times more often than one 13 Å shorter or longer, and a hundred times more often than one 20 Å shorter or longer than

† Assuming $a_0 = b_0$, the critical temperature occurs whenever $\Delta g_f = \gamma/a$. At that temperature Eq. (64) modified by Eq. (66) yields for $l_j > l_i$ and $l_j > l_s$ the equation $A^{ij}/B^{ij} = \exp[-2a^2\gamma_c + a\gamma(l_i + l_s)]$, which is an l_j-independent expression, indicating that even the largest fluctuations would be as likely as the smallest. Since growth is now not dependent on thickness, the average must also tend to infinity.

the substrate. The predicted roughness by this analysis is thus quite small, and the fold lengths should have a standard deviation of only a few angstroms, even at relatively large supercooling. The Lauritzen and Passaglia (1967) treatment (see Sect. 6.1.4.3) yields more realistic values for the surface roughness. At the melting or dissolution temperature the standard deviation of the fold-lengths approaches 8 Å at a specific side surface free energy of 10 erg/cm^2 (the surface roughness does not depend markedly on the fold surface free energy). Increasing the supercooling increases the roughness slowly. At a ΔT of about 35°C the roughness may have doubled. Figure VI.18 illustrates the

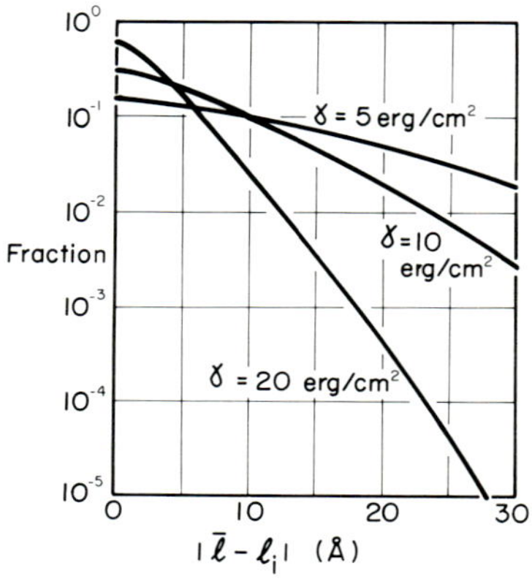

Fig. VI.18 *Distribution of the Fractions of Species $\bar{l}$ That Deviate from the Average Fold-Length $\bar{l}$.*
Data by Lauritzen and Passaglia (1967); $\Delta T = 3°C$; γ is the specific side surface free energy; other conditions as in Fig. VI.16. The specific end surface free energy would shift only the average fold length $\bar{l}$.

distribution curve of the fractions of species of given deviation from the average fold-length for several specific side surface free energies at a temperature close to the melting or dissolution temperature ($\Delta T = 3°C$). The standard deviations for the three curves are 3.7, 8.5, and 18.5 Å for decreasing specific side surface free energies. The order of magnitude of these fluctuations in fold-lengths is close to the experimental results described in Sect. 3.2.2.4.

Besides fluctuations in fold-length, surface roughness is also produced by noncrystallized chain ends, accidental long loops, and tie molecules (see Sects. 3.2.2.4 and 4.3.2). General statements about these effects are that chain ends are frequently excluded from the crystal (see Sect. 4.3.4) and can thus be estimated from the molecular weight distribution. Long loops and tie molecules are nonequilibrium effects, even the restricted equilibria assumed for the calculation of chain folding do not permit their formation. An estimate of their number may be accessible through the discussion of molecular nucleation (see Sect. 5.3). In order to make a long loop or a tie molecule, it is necessary to have a molecule long enough to yield at least two simultaneous molecular nuclei. Figure V.49 can thus be used to estimate the lower. limit for the appearance of tie molecules and long loops. Qualitatively it could be shown by Mehta and Wunderlich (1974) that, indeed, at constant crystallization temperature the number of tie molecules increases with increasing content of high molecular weight components, and that for a given molecular weight sample the number of tie molecules increases with decreasing crystallization temperature.

6.1.4.5 Growth Rates

The linear growth rate of a crystal is discussed in detail in Sect. 6.1.2. At this point it will be briefly shown how chain folding and molecular nucleation affect the growth rate. The discussion must naturally be in accord with the experimental observation that the crystal growth rate normal to the growth face is constant with time and increases with supercooling as expected for secondary or molecular nucleation (see 6.1.2.1).

To connect the theories of chain folded crystal growth permitting single or unlimited fluctuations in the fold-length (see Sect. 6.1.4.3) with the growth rate, one assumes proportionality of the growth rate and the total flux, as expressed for example in the derivation of Eq. (75), Chapter V. Frank and Tosi (1961) derived the total flux to be

$$S_T = S_T^0[\exp(a_0\, \Delta g_f/\gamma) - 1]\exp[-4b_0\, \gamma\gamma_e/(\Delta g_f\, kT)] \qquad (68)$$

Lauritzen and Passaglia (1967) use the same equation for the case of unlimited fluctuations, except for the replacement of γ_e by the effective parameter γ_{eff} [see Eq. (67)]. Figure VI-19 shows a plot of Eq. (68) in terms of $\Delta T \log S_T$ versus ΔT with typical data for the Lauritzen and Passaglia (1967) analysis. For a moderate range of supercooling the curve is practically linear, indicating that only the last exponent of Eq. (68) is significant for the variation of S_T with supercooling. A comparison of Eq. (68) with Eqs. (68) and (81), Chapter V, shows that this term is identical to the key term for the secondary or molecular nucleation rate dependence on supercooling. The maximum which develops at about 30°C supercooling is, however, not solely dependent on the slow decrease in S_T^0 with decreasing temperature as one would expect from pure nucleation theory, but depends also on the increased value of γ_{eff} with increasing supercooling (see Sect. 6.1.4.3 and Fig. VI.16).

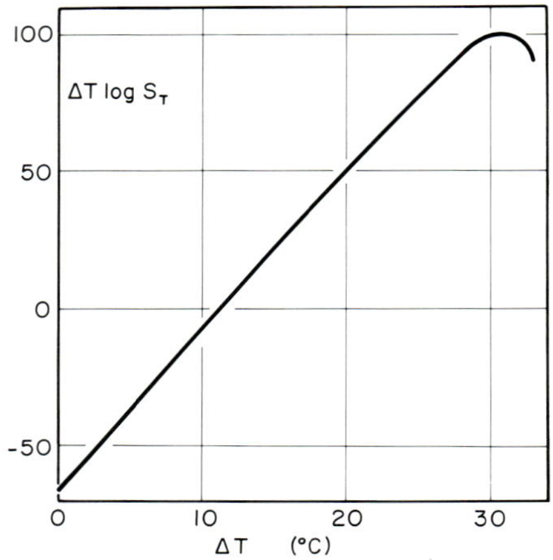

Fig. VI.19 *Plot of ΔT log S_T as a Function of Supercooling ΔT.*
Data by Lauritzen and Passaglia (1967); $\gamma = 10$ erg/cm^2; $\gamma_e = 60$ erg/cm^2; other conditions as in Fig. VI.16 [see Eq. (68) with $\gamma_e = \gamma_{\mathrm{eff}}$].

Finally, to arrive at an expression for the linear growth rate v of the crystal normal to the growth front, the expression S_T for the total flux per growth site must be coupled with the number of available nucleation sites N. Let us assume first that once started, crystallization is completed before a new nucleation event occurs. Under these conditions the following proportionalities must exist

$$v = \text{const.} \times S_\mathrm{T} N = \text{const.}' \times I^* \tag{69}$$

The physical picture under these conditions is that in the case of secondary nucleation, a crystal layer once started sweeps quickly across the surface of the crystal. After completion of the layer there is a pause in crystallization until the next secondary nucleus is produced and another layer can be completed. As a crystal grows in size, the number of nucleation sites should thus increase and the growth rate would also have to increase, in contrast to experimental observation. Sanchez and DiMarzio (1971a) could only suggest that perhaps the growth faces are sectorized to a constant size with constant nucleation sites. Instead of completing a whole layer, for unknown reasons, each nucleation event would then complete only one sector. In the case where molecular nucleation is accepted as the mechanism, such sectorization is a natural result of the termination of growth as soon as crystallization of one molecule is completed. As the crystal size increases, more nucleation sites are created, but simultaneously just the same fraction of increased molecular nuclei are needed to cover the complete crystal surface. A typical polyethylene crystal lamella of 127-Å fold-length and a 10-μm long (110) growth face needs, for example, about 225 molecules of 140 000 MW to complete one layer. Thus, molecular nucleation followed by completion of crystallization of a single molecule leads naturally to a size-independent linear crystal growth rate.

The general case of multiple nucleation before a single layer (or sector) is completed has been treated by Hillig (1966). It leads in the case of secondary nucleation to the proportionality†

$$v = \text{const.} \times (I^* v_\infty^2)^{1/3} \tag{70}$$

which is crystal size independent, even for secondary nucleation based crystal growth. Sanchez and DiMarzio (1971a) could show that the analogous expression for the one-dimensionally limited growth on a lamellar surface is (see also Lauritzen, 1973)

$$v = \text{const.} \times (I^* v_\infty)^{1/2} \tag{71}$$

This situation has been called "regime II crystallization" by Lauritzen and Hoffman (1973) in contrast to the "regime I crystallization" which permits completion of growth before new nucleation. Comparison of Eq. (71) with Eq. (69) shows that the exponent which determines the supercooling dependence of the linear growth rate derived from Eq. (68) is for regime II crystallization half of that for regime I crystallization. Transferring these ideas to molecular nucleation, analogous expressions are expected if regime II crystallization is equated with the multiple nucleation within the region of a single molecule. Such multiple nucleation should result in loops and tie molecules between the different independently grown segments of the molecules.

6.1.4.6 *Conclusions about Chain Folding*

To summarize the various discussions of chain folding one must recognize that a complete theory of chain folding should explain the following experimental facts:

1. Lamellar morphology (see Sects. 3.2.2 and 3.3.2).
2. Fold length/supercooling relationships (see Figs. III.12–19, V.41, and VI.13).
3. Surface roughness (see Sects. 3.2.2.4 and 4.3.2).

† $v_\infty = a_0$(A-B) in the nomenclature of the derivation of Eq. (75), Chapter V.

4. Crystal growth rates (see Sects. 5.2.1, 6.2 and 6.3).
5. Molecular weight fractionation (see Sect. 5.3).

At present the following picture emerges: After primary nucleation (see Sect. 5.1) further crystallization has its start in molecular nucleation (see Fig. V.40B), the step which governs reversible addition of the initial portion of a molecule. If the molecule is long enough, it will be included in the crystal, if not it will be rejected. The nucleus is of necessity chain folded at low supercooling, but may at large supercooling change to a single traverse (see Fig. V.49). Classical nucleation theory (see Sect. 5.3.4) can give an approximate description of the process. Further growth of the molecular nucleus may give rise to fluctuations in fold-lengths so that the final observed average fold-length is larger than the initial nucleus fold-length. At low to moderate supercooling, this increase in fold-length is not very large (10–40%). Classical, one-dimensional crystal growth theories treating fluctuations can be fitted to the experimental data at moderate to low supercooling to give a reasonable understanding of the data (see Sect. 6.1.4.3). The surface roughness is similarly understandable, although little more than qualitative experimental data are presently available for comparison. The crystal growth rates are shown in Sect. 6.1.4.5 to be easily understandable on the basis of a model of molecular nucleation followed by kinetically governed fluctuations in subsequently growing folds.

Difficulties exist still in the analysis of the reasons for the relative unimportance of secondary nucleation (see Sect. 5.3) which also seems to be coupled with the problem of the increase in fluctuations at high supercooling (see Sects. 6.1.4.2 and 6.1.4.3). Instead of a crystallization retarding effect solely dependent on the side surface free energy, an additional factor, proportional to the fold-length l, must be found to resolve these difficulties. Perhaps both problems can be treated by applying a two-dimensional growth theory to the problem of chain folding. Up to the present, two-dimensional theories have proven to be too complicated for solution. Another area which needs progress is the area of change from regime I to regime II of crystallization (see Sect. 6.1.4.5) to analyze tie molecules and loops in more detail. Finally, the area of transition from chain folded to fringed micellar crystallization has not been analyzed in detail (see Sect 5.1.2.1 for nucleation and Sect. 6.1.7 for crystallization).

6.1.5 Molecular Weight and Structure Dependence of Crystal Growth

In Sect. 6.1.4.4 the linear crystal growth rate v was directly linked to the secondary or molecular nucleation rate [Eq. (69) or (71)]. For the discussion of the influence of molecular weights and structure it is convenient to separate the growth rate expression into its three parts:

$$\ln v = \ln v_0 - \Delta G_\eta/(kT) - \Delta G^*/(kT) \tag{72}$$

The last two terms represent the effects of the free enthalpies of activation for the transport of the crystallizing unit across the phase boundary [see Eq. (6), Chapter V] and for formation of the critical size nucleus [see Eq. (68) or (82), Chapter V]. For regime II crystallization both terms have to be multiplied by 0.5 [see Eq. (71)]. The first term collects all other factors. The general shape of Eq. (72) is shown in Fig. V.5 for primary nucleation. The growth rate approaches zero in the vicinity of T_0, a temperature close to the glass transition temperature [see Eq. (6), Chapter V] and at T_m, the melting temperature [see Eqs. (68) and (82), Chapter V]. Between the two temperatures lies a maximum.

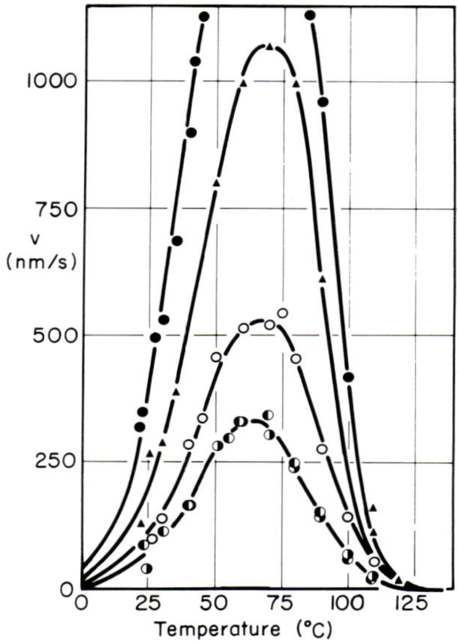

Fig. VI.20 *Linear Crystallization Rate v as a Function of Crystallization Temperature for Different Molecular Weights of Poly-(tetramethyl-p-silphenylene siloxane).* The parameters next to the curves represent the viscosity average molecular weights: (●) 8.7×10^3 (▲) 15.8×10^3 (○) 5.6×10^4 (◐) 3.7×10^5 (◑) 1.4×10^6. Data by Magill (1964, 1967).

Gandica and Magill (1972) showed that it is possible to find a reduced temperature $(T - T_0)/(T_m - T_0)$ in terms of which crystallization rate ratios relative to the maximum rates can be reproduced for all analyzed macromolecules by a single master curve. T_m is the melting temperature of the crystal and T_0 is a temperature about 50°C below the glass transition temperature. Low molecular weight compounds can be fitted to a similar curve (Magill *et al.*, 1973) but with a different position of the maximum crystallization rate. For macromolecules the maximum occurred at about 0.6 of the reduced temperature scale, and for small molecules, at about 0.8.

The absolute value of the crystallization rate varies greatly with molecular weight and chemical structure. Figure VI.20 illustrates a selection of the extensive set of crystal growth rate determinations on different molecular weight poly(tetramethyl-*p*-silphenylene siloxane)s of structural formula (I)

$$Si(CH_3)_3{-}O{-}[Si(CH_3)_2{-}C_6H_4{-}Si(CH_3)_2{-}O{-}]_xSi(CH_3)_3$$

(I)

by Magill (1964, 1967, 1969). At low molecular weight the maximum crystallization rate is manyfold faster. At high molecular weight, a common limit seems to be reached. The maximum crystal growth rates of a series of different high molecular weight macromolecules are listed in Table VI.3. The maximum linear crystal growth rates cover a range of more than five orders of magnitude.

Comparing data for polyethylene, polychlorotrifluoroethylene, and isotactic polystyrene, Boon *et al.* (1968) concluded that $4\gamma\gamma_e b_0/(k\Delta h_f \rho_c)$, the temperature-independent portion of ΔG^* [see Eqs. (68) and (82), Chapter V] is probably not much different for these three macromolecules, and neither, perhaps, for the others. An estimate for the transport term in Eq. (72) can be obtained using the WLF expression (see also the footnote on page 9)

$$\Delta G_\eta/(kT) = C_1/(C_2 + T - T_g) \tag{73}$$

with $C_1 = 2.07 \times 10^3$ and where T_g represents the glass transition temperature. Instead of using the universal value of 51.6 for C_2, a special fit of the growth rate data is necessary for some of the macromolecules.† The changes in glass transition and melting temperature alone, however, do not account for the variation in maximum growth rate. Instead, a variation in v_0 over several orders of magnitude is necessary for agreement between the data and Eq. (72). Boon *et al.* (1968) suggested that v_0 is largely determined by the mobility of the molecules and that for polystyrene the bulkiness of the phenyl groups reduces v_0 by a factor 2×10^{-5} relative to v_0 of polyethylene.

† That is, for isotactic polystyrene, $C_2 = 75$, for poly(tetramethyl-*p*-silphenylene siloxane) $C_2 = 90$, for nylon 6, $C_2 = 130$, but for polyethylene and polychlorotrifluorethylene, $C_2 = 51.6$.

TABLE VI.3

MAXIMUM CRYSTAL GROWTH RATES OF MACROMOLECULES[a]

Macromolecule	Growth rate v (nm/sec)	Ref.
Polyethylene, CH_2-	8.3×10^4	*b*
Nylon 6.6, $(CH_2-)_6NH-CO-(CH_2-)_4CO-NH-$	2.0×10^4	*b*
Polyoxymethylene, CH_2-O-	6.7×10^3	*b*
Nylon 6, $(CH_2-)_5NH-CO-$	2.5×10^3	*b*
Polychlorotrifluoroethylene, $CF_2-CClF-$	5×10^2	*b*
Poly(tetramethyl-p-silphenylene siloxane), $Si(CH_3)_2-C_6H_4-Si(CH_3)_2-O-$	3.5×10^2	*c*
Polypropylene, isotactic, CH_2-CHCH_3-	3.3×10^2	*b*
Poly(ethylene terephthalate), $(CH_2-)_2-O-CO-C_6H_4-CO-O-$	1.7×10^2	*b*
Polystyrene, isotactic $CH_2-CHC_6H_5-$	3.3	*b*
Poly(4,4′-isopropylidenediphenylene carbonate) $C_6H_4-C(CH_3)_2-C_6H_4-O-CO-O-$	0.2	*b*

[a] Molecular weights about 100 000.
[b] Boon *et al.* (1968).
[c] Magill (1969).

Godovskii (1969) found empirically by analyzing half-times of crystallization for twenty-three macromolecules that the maximum crystallization rate is found at $0.82T_m$ to $0.83T_m$, regardless of the absolute level of T_m. Furthermore, he suggests that the crystallization rate is higher for more symmetric and shorter repeating units. Privalko and Lipatov (1973) proposed that Eq. (73) represents only the enthalpy portion of the transport term and an entropy term must be added to get to the free enthalpy required for Eq. (72). They suggest the entropy term ΔS should be representative of the change in going from the melt to the nucleus. Assuming that ΔS is a function of the change in packing density [see Eq. (13), Chapter II] one can write

$$\Delta S = f(k_c - k_a) \tag{74}$$

An empirical relationship which fits several of the macromolecules of Table VI.3 was derived by Privalko and Lipatov (1973):

$$\log(\log v_0) = 7.55k_c - 5.0 \tag{75}$$

(v_0 in cm/sec; for values of k_c see Tables II.9–II.18). The amorphous packing fraction does not occur in Eq. (75) since it is approximately constant for the analyzed macromolecules.

The molecular weight dependence of the linear crystal growth rate can be seen from Fig. VI.21 for poly(tetramethyl-*p*-silphenylene siloxane). Over the range of temperatures shown in Fig. VI.21 the lower molecular weights are

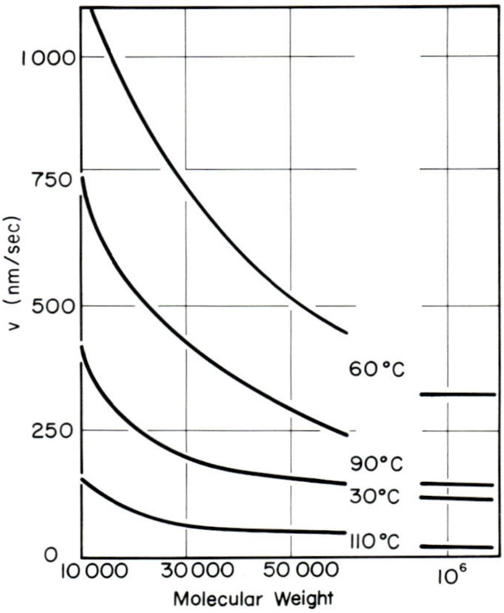

Fig. VI.21 *Linear Crystallization Rate v as a Function of Molecular Weight for Different Temperatures.*
Poly(tetramethyl-*p*-silphenylene siloxane). Temperatures in °C. The equilibrium melting temperature is estimated to be 152°C. Data by Magill (1964, 1967, 1969).

crystallizing faster. At sufficiently high temperature (low supercooling) the crystal growth rate trend reverses. Magill (1969) places this temperature at 130°C. For polyethylene, Lindenmeyer and Holland (1964) show little molecular weight dependence at about 113°C and increasing crystal growth rates with increasing molecular weight at 120°C and higher. This reversal of the crystal growth rate trend is connected with the free enthalpy of critical nucleus formation in Eq. (72). Figure V.4 shows the dominance of ΔG^* at low supercooling (drawn out curve). The melting point change of polyethylene with molecular weight can be seen from Fig. V.43 and is discussed in more detail in Chapter VIII. Being concerned with molecular weight changes in the macromolecule range, i.e., above about 10 000, one expects no more than perhaps 3–5°C change in equilibrium melting temperature, which is sufficient only at small supercooling to reverse the "normal" trend of crystal growth rates shown in Fig. VI.20.

Several approaches have been taken to pinpoint the major factor for the molecular weight dependence of the crystallization rate. Hoffman and Weeks (1962) using Eq. (73) for the transport term suggested that v_0 can be written

$$v_0 = b_0(kT/h)f(1/MW)^y \tag{76}$$

The factor $b_0 kT/h$ results from the assumption that every nucleation event leads to an increase in thickness of one layer thickness b_0 [see Eq. (5), Chapter V, and Eq. (69), this chapter]. The molecular weight (MW) function was suggested to have an exponent between zero and one. Additional complications in this model arise from changes to regime II crystallization [Eq. (70)] and from the fact that molecular nucleation requires decreasing numbers of nucleation events for larger molecular weights to cover one molecular layer. In the latter case v_0 should increase linearly with molecular weight, other factors being equal. Indeed, an increasing v_0 with molecular weights was observed by Magill (1964) when analyzing poly(tetramethyl-*p*-silphenylene siloxane) using the WLF expression [Eq. (73)] with $C_2 = 90°C$ and experimental values for the melting and glass transitions. The relationship Eq. (75) would call for a v_0 which decreases with molecular weights since the amorphous packing fraction increases with molecular weight while the crystalline packing fraction should stay constant.

A different approach to the analysis of the molecular weights dependence of crystal growth rate was taken by Lovering (1970), Devoy and Mandelkern (1969), and Magill (1969). These authors assume a molecular weight independent v_0 and also a molecular weight independent activation energy in the transport term for any one polymer. The proper fit of data and Eq. (72) can then only be achieved when one permits a variable $\gamma\gamma_e$ product in ΔG^*. The $\gamma\gamma_e$ value must increase with increasing molecular weight by about 50% to slow down the crystallization rate sufficiently.

Sanchez and DiMarzio (1971b) in a purely theoretical approach predict for dilute solution crystallization that there should be a broad maximum in a growth rate–molecular weight plot, a behavior which has not been observed to date.

An interesting observation was made by Magill *et al.* (1974) who found that the literature data on the molecular weight dependence of linear crystal growth rate of macromolecules can be fitted empirically at constant temperature to a function of the type

$$\log v = \text{const.} \times (\text{MW})^{-1/2} \tag{77}$$

Figure VI.22 illustrates the fit of the data to such an equation. The polythylene data are for relatively low supercooling and are thus plotted at constant supercooling rather than constant temperature.

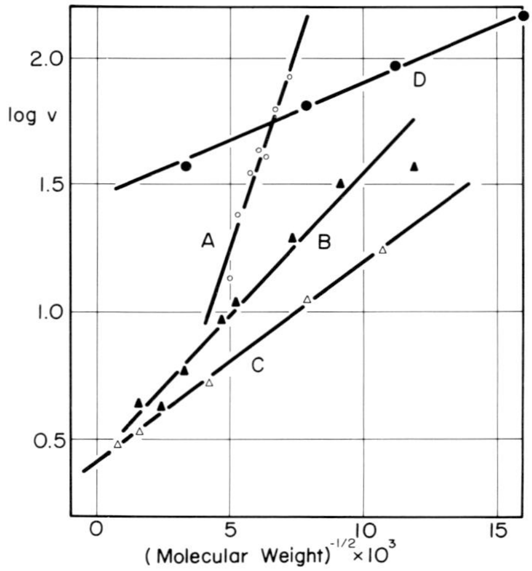

Fig. VI.22 *Logarithm of the Linear Crystallization Rate v as a Function of the Inverse Square Root of Molecular Weight.*
(A) Poly(ethylene terephthalate), data by van Antwerpen (1971). Thesis, Delft Technical University. Crystallization temperature 200°C. (B) *trans*-Poly-(2-methylbutadiene), high melting spherulites. Data by Lovering (1970). *J. Polym. Sci. Part C* **30**, 329, crystallization temperature 40°C. (C) Poly-(tetramethyl-*p*-silphenylene siloxane), data by Magill (1964, 1967), crystallization temperature 70°C, the ordinate is to be increased by 2.0 for these data. (D) Polyethylene, data by Lindenmeyer and Holland (1964) at constant supercooling $\Delta T = 25°C$; the ordinate is to be increased by 2.0 for these data. (Logarithm to the base 10, v in nm/sec).

A unique set of experiments leading to measurements of the linear crystal growth rate as a function of molecular weight and number of folds was done by Kovacs and Gonthier (1972) on poly(ethylene oxide) crystallized from the melt controlled by self-seeding. Poly(ethylene oxide) fractions of sufficient sharpness and low enough molecular weight crystallize depending on temperature as fully extended chains, singly, doubly, and higher folded lamellae (see Fig. III.17). Figure VI.23 shows the data for a 6000 MW fraction growing with zero, one and two chain folds plotted under the assumption that ΔG_η is 7.0 kcal/mole, a value derived for this temperature region from the temperature coefficient of viscous flow. Also given are the limiting curves for

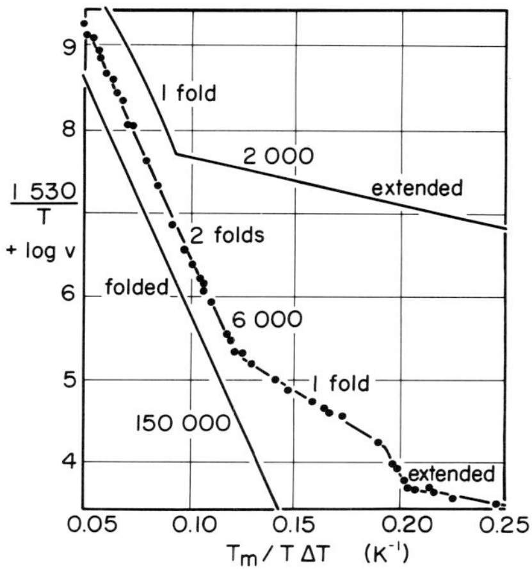

Fig. VI.23 *Logarithm of the Linear Crystallization Rate v of Poly(ethylene oxide) versus* $T_m/T\Delta T$.
Data by Kovacs and Gonthier (1972). Upper curve 2000 MW, middle curve 6000 MW, lower curve 150 000 MW. The different sections of the curve refer to molecules with the indicated numbers of folds in the crystal. The growth rate is given in nm/sec.

150 000 molecular weight, frequently folded chain crystals and 2000 molecular weight, extended, and once-folded chain crystals. The data are plotted versus $T_m/(T\Delta T)$ with T_m representing the extended chain melting temperature of the molecular weight in question, and ΔT giving the supercooling for the particular molecular weight fraction. The left portion of the figure shows again the decreasing rate of chain folded crystal growth with increasing molecular weight. The right portion reveals that the slope of extended chain crystal growth is much less than for chain folding with a considerably larger molecular weight dependence. The singly folded poly(ethylene oxide) of 6000 MW shows an intermediate slope. The simplest interpretation of these data would be that the specific fold surface free energy in the term for the critical nucleus size [see Eqs. (68) and (82), Chapter V] is much larger (slope about 55) than the specific surface free energy of the extended chain crystal (slope about 4.6). This does not agree with the observation from melting point studies that extended and folded surfaces behave as if they had similar surface free energies. The higher surface free energy may thus be an effective parameter in the growth rate expression. As was concluded in Sect. 6.1.4.6, it is likely that a factor other than surface free energy which is proportional to crystal thickness in the chain direction must be included in the description of crystallization of macromolecules.

TABLE VI.4

CRITICAL TEMPERATURES (°C) IN CRYSTALLIZATION OF POLY(ETHYLENE OXIDE)[a]

Molecular weight[b]	Extended to once folded	Once to twice folded	Twice to thrice folded
2000	42.0	—	—
3000	49.8	—	—
4000	54.6	50.6	—
6000	59.2	55.7	—
10000	(63.6)	(61.6)	57

[a] Data by Kovacs and Gonthier (1972).
[b] Fractions of M_n/M_w between 1.05 and 1.3.

Table VI.4 lists the critical temperatures where the changes in the number of folds of the molecules occur, i.e., where the crystal growth rates of the different crystals are the same. These temperatures increase with molecular weight for the same type of transition. The corresponding values of the logarithm of the crystal growth rates decrease approximately linearly with the increasing critical temperatures. The extended chain to folded chain transition temperature can be represented by

$$T_e = 69.5[1 - 168 \quad (\mathrm{MW})^{-1}] \tag{78}$$

The conclusions one must draw from this discussion are that chemical structure, molecular weight dependence, and also the number of fold dependences of the crystal growth rate are not understood sufficiently. The general equation (72) contains too many adjustable parameters to conclusively state which causes the effects observed.

6.1.6 Secondary Crystallization and Crystal Perfection

Secondary crystallization and crystal perfection during isothermal crystallization of macromolecules has been observed as soon as quantitative descriptions of the overall crystallization with the Avrami equation [Eq. (33)] was attempted. Hartley *et al.* (1954) observed a small, slow, linear density increase [10^{-5}–10^{-3} g/(cm^{3}hr)] when poly(ethylene terephthalate) was retained for longer times at the initial crystallizing temperature. Kovacs (1955) showed a logarithmic increase in the density of high pressure polyethylene at long times (up to 1000 hr) after primary crystallization. Gent (1955) reported similar results for 1,4-poly(2-methylbutadiene). Secondary crystallization of polyurethanes was reported by Dörffurt (1956) and Rohleder and Stuart (1960).

More detailed studies of secondary crystallization were made by Zachmann and Stuart (1960) for poly(ethylene terephthalate) and by Majer (1960a) for low pressure polyethylene. The general observations were a distinct drop in Avrami coefficient n [see Eq. (33)] after a major portion of the sample had crystallized. The final slow secondary increase in crystallinity was frequently observed to have a logarithmic time dependence for many decades before ceasing, and its overall rate was faster at higher crystallization temperatures. Experiments not carried over sufficiently long periods of time, or measurements of insufficient sensitivity may miss the secondary crystallization.†

The term secondary crystallization (German: *Nachkristallisation*) is frequently applied to all effects which increase the crystallinity after the primary crystallization (German: *Hauptkristallisation*), describable by an Avrami-type equation, is over. One can separate the delayed crystallinity increase into at least two processes. The first involves further crystallization of amorphous macromolecules or portions of molecules and will be called secondary crystallization in the further discussion. The second involves further perfection of initially poorly crystallized macromolecules and will be called crystal perfection.‡ As with any definition of related processes, it is possible to find cases which are in between the two categories, and even more frequently, one finds both mechanisms operating simultaneously, complicating the overall picture of crystallization. The discussion of secondary crystallization according to this new definition is an extension of Sect. 6.1.3.3 that dealt with complications of the Avrami analysis (see Table VI.2). The discussion of crystal perfection, in turn, is closely related to the processes to be analyzed in the discussion of annealing in Chapter VII.

Secondary crystallization as just defined must be discussed in connection with the rejection of solvent (diluent) molecules, of noncrystallizable portions of the macromolecules, and of too-short macromolecules at the crystal growth front. It is shown in Sect. 3.6.1 that such rejection of noncrystallizable species is frequently the cause of dendritic branching. The rejected material is in this case concentrated§ between the branches of the crystal. Its mixing by

† Note that in the case of Fig. VI.11 there may be almost no secondary crystallization because of the relatively low level of crystallinity reached at a very low temperature (cold crystallization, see Sect. 6.1.7). The crystals are probably small and grow under conditions of relatively little change in concentration of the surrounding melt which reduces the trend to secondary crystallization. Also, the low crystallization temperature will reduce crystal perfection (see text).

‡ For a discussion of such distinction of processes see Keith and Stuart (1969) *Kolloid Z. Z. Polym.* **231**, 428 and 429, and Peterlin (1964). *J. Appl. Phys.* **35**, 80.

§ The possibility of changes in crystallization rate caused by stresses in the sample due to blocked-off amorphous material between crystals has been considered by Peterlin and Roeckl (1963).

diffusion with the bulk of the remaining amorphous material is increasingly impeded as crystallization progresses. Schultz and Scott (1969) illustrated the possibility of a diffusion controlled one-dimensional secondary crystallization for polyethylene at 80 to 124°C which had undergone its primary crystallization at higher temperature (125°C, 30 min, 70–80% completion). Although the total amount of secondary crystallization after about 5000 min went through a maximum at approximately 110°C, which would be typical for new crystal growth, the long time increase in crystallinity (after 100 min) showed increasing rates with increasing temperature, indicative of crystal perfection. A similar rise in rates of crystallinity increase at long times with increasing temperature was reported for polyethylene (crystallized at one temperature only) by Majer (1960a) and Nakamura *et al.* (1973).

Taking into account the complications at the crystal growth front at the time secondary crystallization begins, it is not enough for the description of crystallization to know the exact geometry of the crystal, (as given for example by Fig. III.98). It is also necessary to know the changing degree of supercooling at the growth front in going from primary to secondary to tertiary branches and the possible mechanical hindrances to partially crystallized molecules. The more the secondary growth occurs in the interior of the dendrite or spherulite, the more will further crystallization be slowed down by the impeded diffusion of the noncrystallizable species and by mechanical hindrance. One can imagine that under such conditions a simple mathematical description of crystal growth is not possible.† Secondary crystallization will continue until defective, metastable interfaces of a structure described in Sect. 4.3.2 are produced (see also Keith, 1969).

Zachmann and Stuart (1960) treated the secondary crystallization hindered by noncrystallizable parts of molecules, loops and entanglements. They assumed that through thermal motion these physical hindrances to further crystallization could be partially removed in time. Each time such a hindering conformation is removed, a constant portion of uncrystallized material can add to the crystal. The change in N_i, the number of hindering conformations of type i, can be written most simply as a first-order process with the rate constant k_i

$$dN_i = -k_i N_i \, dt \tag{79}$$

This process is governed as usual by an enthalpy and entropy of activation. The total amount of remaining amorphous polymer should now be proportional to the remaining hindering conformations:

$$(1 - w^c) = \sum \beta_i N_i \tag{80}$$

† Some of the problems of changing concentrations in crystallization are discussed further in Sect. 6.3.4 dealing with copolymer crystallization and in Sect. 6.2.1 dealing with crystallization from solution.

where w^c is the crystallinity and the β's are the proportionality constants. The sum has to be extended over all i. The final expression for the secondary crystallization is thus

$$(1 - w^c) = \sum B_i \exp[-k_i t] \tag{81}$$

where the B_i are the respective uncrystallized portions of the sample at time $t = 0$, the beginning of secondary crystallization for the hindering conformations. Equation (81) proved useful for times larger than triple the half-time of the primary crystallization. Zachmann and Stuart (1960) used Eq. (81) for the description of secondary crystallization of poly(ethylene terephthalate) between 140 and 220°C from the glassy state. A large portion of intermediate times of secondary crystallization could be fitted with a single hindering conformation. The sum in Eq. (81) was necessary for a more than linear increase in the logarithm of crystallinity at shorter times, and a less than linear increase at very long times. A similar effort to describe secondary crystallization of polyethylene was made by Hay and Booth (1972) by combining a first-order secondary process with the primary Avrami expression (see Sect. 6.1.3.3 for other combinations of two Avrami expressions). These analyses still lead to fractional n in the primary Avrami terms [Eq. (61)] so that Hay and Booth (1972) concluded that perhaps the crystal perfection must be taken into account separately.

Crystal perfection discussions lead to the idea that during crystallization at relatively high temperatures the once-crystallized material may have enough time to reorganize† into a more stable structure. Evidence for a time-dependent increase in fold period, the largest imperfection of macromolecules, is shown in Fig. III.18. The similarity of the observed rate of increase in fold-length and the slow crystallization at long times was observed through dilatometry (see Sect. 4.1.1) and NMR (see Sect. 4.1.6) by Peterlin and Roeckl (1963). A detailed discussion was presented by Peterlin (1964). According to this concept, the change of density and crystallinity during isothermal crystallization and annealing of polymer samples is a consequence of the gradual reduction of the surface to volume ratio by fold-length growth. On the assumption that the same relationships between density and fold-length that applied to annealing applied to isothermal crystallization, an overall expression for the crystallization of polyethylene was derived by Peterlin (1964). Since on annealing the fold-length increases linearity with the logarithm of time (see Fig. III.18) and density ρ seems to increase on

† An attempt to link the increase in crystallinity after primary crystallization to a general relaxation process as suggested for glasses by Hirai and Eyring (1958) was made by Banks *et al.* (1963a) and Gordon and Hillier (1965).

annealing linearly with the reciprocal fold-length, the crystalline density as a function of time t can be written in this case as

$$\rho_c(t) = \rho_c\{1 - A/[1 + B \ln(t - \tau)]\} \tag{82}$$

where A and B are constants, ρ_c is the final crystalline density, and τ is the time of the initial creation of the crystal in question. The specific volume $1/\rho$ of the overall sample at anytime t can be written as

$$1/\rho = (1 - w^c)/\rho_a + \int_0^t [(dw^c/d\tau)/\rho_c(t)]\, d\tau \tag{83}$$

where w^c is the morphological weight fraction crystallinity to be determined from the Avrami equation† and ρ_a is the amorphous density. The integral takes care of the increasing density of each newly crystallized portion as time progresses, starting at its time of creation τ. The apparent, observed crystallinity is then from its definition [Eqs. (3) and (4), Chapter IV, and Eq. (83)]

$$w^c_{\text{app}} = w^c + \frac{1}{(1/\rho_a) - (1/\rho_c)} \left([w^c/\rho_c] - \int_0^t [(dw^c/d\tau)/\rho_c(t)]\, d\tau \right) \tag{84}$$

Equation (84) can now be coupled through w^c with Avrami-type expressions of various degrees of complexity to fit the experimental data. The advantage of Eq. (84) lies in the proposition that annealing and initial crystallization follow a similar mechanism so that A and B of Eq. (82) may be available from different experiments.

Comparison of secondary crystallization as described by Eqs. (81) and (61) and crystal perfection as described by Eq. (84) leads to the conclusion that by fitting with proper parameters probably all these equations can describe the experimental data. For full interpretation of the crystallization process, additional morphological data must be sought for every case under investigation.

Experimental evidence for crystal perfection during crystallization at short times was presented by Wunderlich and Cormier (1966) and Wunderlich *et al.* (1967) through fast calorimetry of a seeded polyethylene melt at different stages of crystallization. Figure VI.24 illustrates the data for polyethylene crystallizing at 125.5°C and 128°C. All samples were seeded with thin slices of extended chain crystals (which melt at higher temperature, see Chapter VIII). As time progresses, the initially grown materials seem to be perfect up to an initial metastable limit. Table VI.5 lists the times needed to initially reach this limit after the start of crystallization; the total amounts crystallized at the time the limit was reached on 1 mg of extended chain seeds (approximately 7 cm^2 of crystallizable surface), and the melting temperature

† Equation (33), for conversion of volume fraction to weight fraction see Eq. (9), Chapter IV.

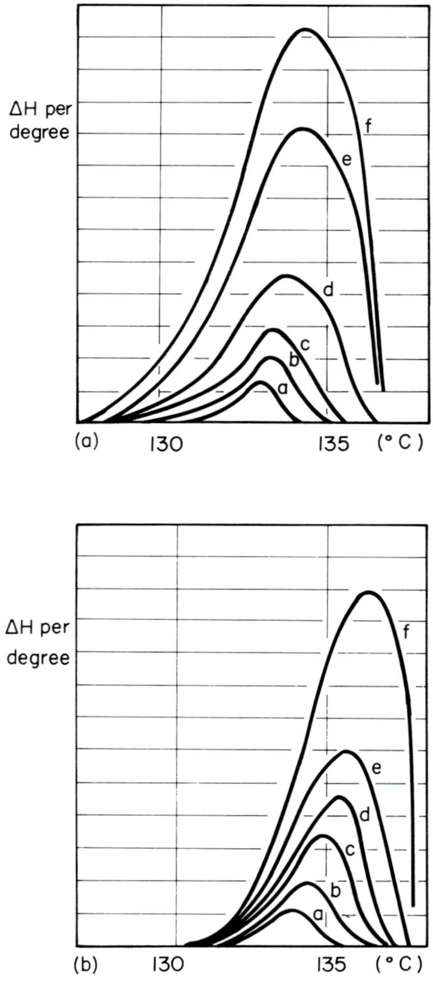

Fig. VI.24 *Thermal Analysis Curves of Polyethylene Crystallized for Various Lengths of Time in the Presence of Extended Chain Crystal Seeds.*
(**a**) Crystallized at 125.5°C for (a) 0.5 min; (b) 1 min; (c) 2 min; (d) 5 min; (e) 8 min; (f) 10 min. (**b**) Crystallized at 128°C for (a) 5 min; (b) 10 min; (c) 15 min; (d) 20 min; (e) 30 min; (f) 69 min. Approximately 7 mg of ployethylene melt was crystallized in the presence of 0.5 mg of 3-μm thick slices of extended chain crystals for the given time, followed by analysis at 5°C/min. See also Table VI.5 and text (Wunderlich *et al.*, 1967).

TABLE VI.5

CONDITIONS WHEN THE INITIAL METASTABLE CRYSTAL PERFECTION IS REACHED[a]

Cryst. temp. (°C)	Cryst. time (min)	Amount crystallized (mg/mg of seeds)	T_e (°C)
121.5	$\ll \frac{1}{2}$	—	134.5
122.5	$< \frac{1}{2}$	0.78	135.0
123.5	2	0.26	135.0
124.5	4	1.81	136.0
125.5	8	2.02	136.5
126.5	10	0.84	137.0
127.5	15	0.65	137.0
128.5	30	0.60	137.5
129.5	60	0.48	138.0
130.5	120	0.05	138.0
131.5	~600	0.03	138.5
132.5	—	—	139.0

[a] Data by Wunderlich *et al.* (1967), the total crystallizable melt was 10–15 mg per mg of seed extended chain crystals. Differential scanning calorimetry at 5°C/min.

(end of melting) of the crystals of the metastable limit. The times listed in Table VI.5 are much shorter than the previously discussed crystal perfection at longer times. At the lower temperatures of crystallization the times for perfection are at the limit of the experimental time scale; in addition, the crystals produced are metastable to such a degree that annealing occurs during thermal analysis. Figure VI.25 shows the melt begin and peak temperatures of the first observed crystals (curves a in Fig. VI.24), and the melt begin, peak, and end temperature of the crystals after the limiting perfection has been reached for the first time (curves e in Fig. VI.24). The perfection involves an increase in melting point by at least 2°C. Crystal perfection persists also for the later grown material as can be seen from Fig. VI.26 where thermal analysis was carried out after longer times of crystallization. Changes in crystal perfection in these experiments can be noticed on the low temperature side of the melting peak. As crystallization time proceeds, the initially increasingly larger amounts of poorer crystals improve, and the melting peak sharpens. Again, the order of magnitude of the crystal perfection seems to be 1–2°C. Also, from Fig. VI.26 it is possible to gain some indication of later growth of some considerably lower melting crystals at the foot of the melting peak (128–131°C melting temperature). These crystals may fit the

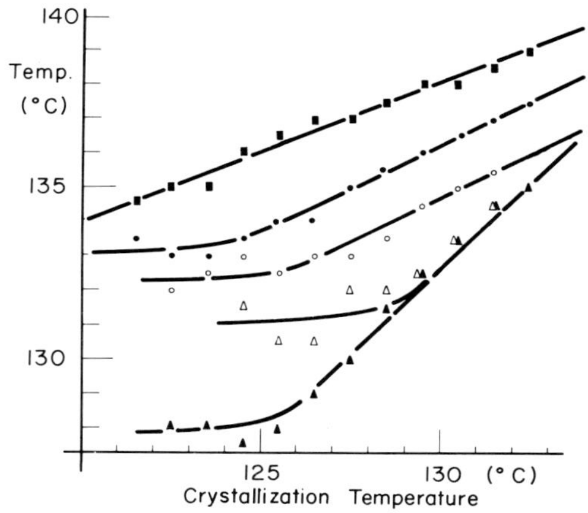

Fig. VI.25 *Changes in Characteristic Melting Temperatures of Polyethylene Crystallized at Various Temperatures.*
Samples and analysis as in Fig. VI.24. (■) Limiting melt end temperatures at 5°C/min heating rate; (curves (e) in Fig. VI.24). (▲, ●) melt begin and peak temperatures under the same conditions; (△, ○) initial temperature of beginning of melting and peak (curves (a) in Fig. VI.24). The horizontal portions of the curves may be caused by annealing during heating for analysis rather than be an indication of a change in mechanism (Wunderlich *et al.*, 1967).

description of secondary crystallization given above, indicating the large range of effects which may alter the classical Avrami assumptions.

The present discussion has led to the following sequence of events which should be considered in a description of crystallization:

1. Growth on seeds as assumed in the classical crystallization description.
2. Immediate crystal perfection of the initially poorly grown crystals.
3. Secondary crystallization of initially rejected or hindered material.
4. Slow crystal perfection through processes similar to those observed in long time annealing.

Which of the four steps needs to be treated in detail for a particular sample must be decided by detailed analysis which varies not only from macromolecule to macromolecule, but also with temperature and crystallization condition.

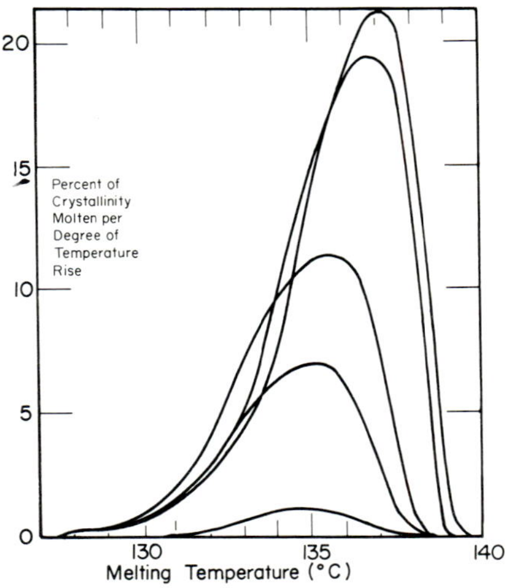

Fig. VI.26 *Thermal Analysis Curves of Polyethylene Crystallized for Various Lengths of Time in the Presence of Extended Chain Crystal Seeds.*
Crystallized at 126.4°C for 23, 30, 60, 300, and 3660 min, respectively. The first curve corresponds to the curves (f) in Fig. VI.24; note the change in ordinate. The shift of the melting peak beyond the initial limiting melt end is mostly caused by the increasing instrument lag at the melt end for the much larger amounts of melting compared to Fig. VI.24 (Wunderlich *et al.*, 1967).

The nature of the short-time crystal perfection is not well explored. One may speculate that it involves perfection on a nearest-neighbor level which also can lead to fold-length increase whenever there is a proper mechanism available. For example, the fold-length extension in the growth of extended chain crystals from the macromolecular melt seems to occur immediately after initial crystallization (see Sects. 3.3.1 and 6.3.2 and Figs. III.30, III.40, III.41 and VI.60). Later fold-length extension through annealing seems much less effective (see Sect. 7.3.3). The crystal width in which rearrangement occurs, as indicated by the region of rounding, is comparable to the thickness of polymer crystals grown on seeds before the initial metastable structure is reached in the experiments described by Figs. VI.24–VI.26 and Table VI.5 (0.03–3 μm).

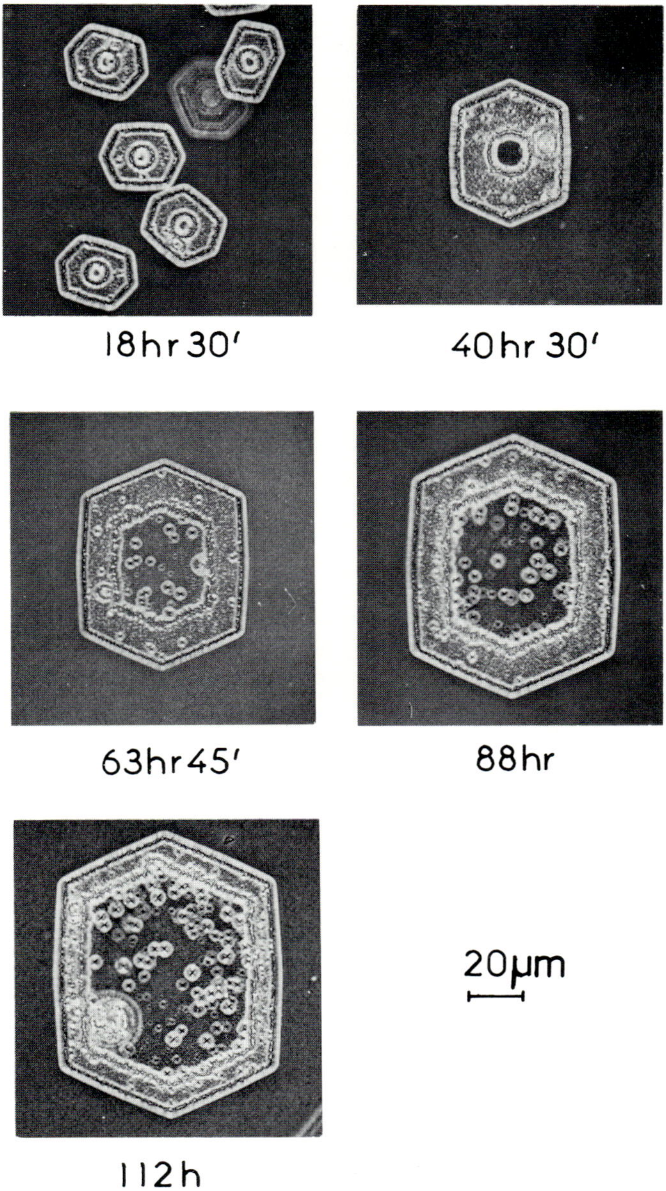

Fig. VI.27 *Isothermal Crystal Perfection of 6000 MW Poly(ethylene oxide) at Different Times of Crystallization from the Melt.*

Crystallization temperature $T_c = 59.16°C$. Optical micrograph decorated with additional crystal overgrowth caused by quenching in ice water after the indicated crystallization times (see also Fig. IV.36). The inner, little decorated portion represents fully extended chain crystal; the highly decorated borders and inner circle at 18.5 hr indicate large surface irregularity; the intermediately decorated area is made up of once folded molecules. Kovacs and Gonthier (1972).

A specially interesting observation of short time crystal perfection involving chain extension has been reported by Kovacs and Gonthier (1972) for poly(ethylene oxide) fractions of low molecular weight. A 6000 molecular weight fraction when crystallized from the melt above 59.25°C (see also Figs. VI.23 and III.17) grows to a fully extended chain crystal. Somewhat below that temperature a once-folded macroconformation is crystallized first, which quickly (in the time scale of this rather low supercooling crystallization) perfects itself. Starting in the center of the single crystals which grow under such conditions, a double thickness portion develops, as is shown in the series of optical micrographs of Fig. VI.27 taken at 59.16°C. The extended chain portion is distinguishable by its relative lack of surface decoration which is introduced by quenching the samples to low temperature for observation. The early central circular portion of the reorganizing crystal is highly decorated on cooling. This indicates great surface irregularity in contrast to the less decorated interior portions of crystals of later stages of crystallization. The rate of growth of the diameter of this early round area is only slightly less than the linear growth rate of the extended chain crystals when extrapolated to lower temperature (see Fig. VI.23). The final growth rate of the boundary of the extended chain interior portion is identical to the growth rate of the outer boundary so that a constant frame of folded material develops. Both areas of the crystal remain separated by a highly decorated narrow band. Kovacs and Gonthier (1972) suggest that the overall process involves both an extension of the originally folded chains and an added crystallization of new molecules to fill-in the holes developed in the crystal due to the chain extension. It is of interest to note that the final rate of growth of the interior extended chain portion of the crystal remains nucleation controlled, but is faster than the direct growth of extended chain crystals from the melt (see Fig. VI.23).

6.1.7 Crystal Growth without Chain Folding

Crystal growth of a macromolecule may be thought to occur along two basically different mechanisms, one leads to predominant chain folding, and the other to little folding as in a fringed micelle structure. The same two paths were described in the discussion of homogeneous nucleation in Sects. 5.1.2.1 and 5.1.2.2 (see also Figs. V.10 and V.40). For folding it was concluded that the molecule must have sufficient mobility in the time scale of crystallization to go from the melt or solution conformation to the chain folded macroconformation. This time requirement is set by the entropy reduction of the left-over uncrystallized molecular portions. The entropy reduction increases the free enthalpy barrier for crystallization beyond the free enthalpy barrier for chain-folded crystallization and thus slows down fringed micellar crystallization in favor of chain folding (see Sect. 5.1.2.1). Ideally the folded chain crystal should be lamellar with a relatively smooth surface which represents the phase boundary separating crystalline and amorphous molecules (Fig. III.51), while the fringed micelle crystal should be small with a relatively rough surface which marks the boundary where the crystalline and amorphous portions of molecules join (Fig. III.7). The discussion of the morphology of chain folded lamellae has revealed that these contain also some loose folds, free chain ends (cilia), and tie molecules (see Sects. 3.2.2.4 and 4.3.2), defects which are otherwise only expected in fringed micellar crystallization. Similarly, chain folds are expected as defects in strictly fringed micellar crystallization as is expressed by the general scheme of possible macroconformations in the solid state depicted by Fig. III.5. Thus, crystallization of macromolecules should be treated as a case somewhere inbetween the two limits: chain folded or not chain folded. Other things being equal, the two mechanisms of crystallization may point also to the differences in crystallization from dilute solution and the melt. In dilute solution one expects normally a more or less random coil structure of sufficient mobility which crystallizes in form of lamellar, regularly chain folded crystals. In the melt the amorphous macroconformation may already be somewhat more extended (see the following), and together with reduced mobility, a more irregular crystal is favored tending more to the fringed micellar structure. The discussion of morphology in Chapter III has shown, however, that at present the true limit of fringed micellar structure has not been analyzed in much detail.

In this section crystallization with little or no regular chain folding, the fringed micellar crystallization, will be explored. Since the fringed micelle was the dominating early model of macromolecular crystallization (see Sect. 3.2.1), it is natural that most early discussion of crystallization of macromolecules was in the framework of adding chain segments with two loose ends to a growing crystal (see for example Stuart, 1955). Figure VI.28 illustrates schematically nucleation and growth of fringed micellar crystals as proposed by Richards (1945). Growth may occur, both, along and at right angles to the molecular chain axes. More recently the structure of the melt has received intensive attention and the proposed models of increased chain extension in the melt discussed in Sects. 2.4.9 and 5.1 and shown in Fig. V.1 have led naturally to an increased interest in crystallization with little folding or with irregular chain folding leaving long loops (switchboard model, Flory, 1962; Fischer and Lorenz, 1963, see also Sect. 3.2.2.3).

The basic principles of secondary nucleation and growth of crystals developed in Sects. 5.2 and 6.1 must also hold for the case of no chain folding. The molecular nucleus (Sect. 5.3) is assumed to consist in this case

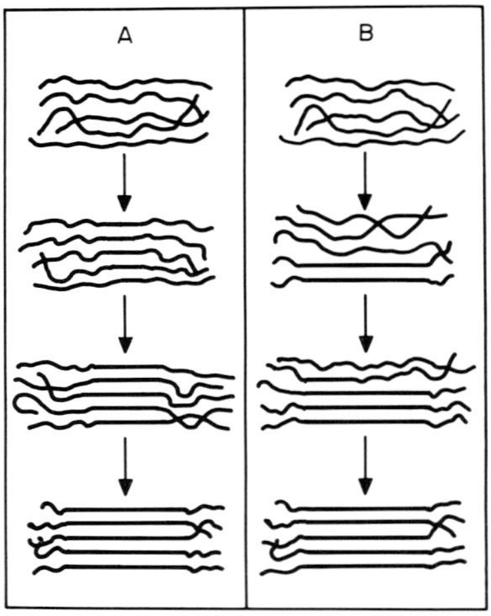

Fig. VI.28 *Possible Nucleation and Growth of Fringed Micelle Crystals.* Drawing after Richards (1945). (A) Growth mainly along the chain axis. (B) Growth mainly normal to the chain axis.

of only one or a portion of one traverse. Calvert and Uhlmann (1972) collected the necessary equations to describe nucleation and growth under no chain folding conditions. For the case of multiple nucleation on a growth face Eqs. (70) or (71) lead to the proper crystallization rate equation for the linear growth normal to the molecular chain direction. On using these equations, the major change from the prior discussed chain folded case is that γ_e, the fold surface free energy, must be replaced in the case of no chain folding by the proper term $(\gamma_e + \gamma_e')$. The extra quantity γ_e' accounts for most of the added entropy effects of the uncrystallized chain ends (see Sect. 5.1.2.1). Another distinction of crystallization without chain folding is that the crystal thickness in the chain direction is not limited as before by the folds, but at least some further growth of the crystals in the chain direction should be possible. Large amounts of growth should be hindered by low segmental diffusion rates, chain entanglement, occurrence of loops, noncrystallizable defects, and chain ends. Thus, fringed micelle-type crystallization leads to rather small crystals normally.

Calvert and Uhlmann (1972) discussed the case that crystal growth in the chain direction is limited only by the rate the randomly coiled molecule can convert by segmental diffusion to the linear, crystalline macroconformation. This they described in terms of the rate with which a single chain unit can diffuse over a distance L. The distance L, they assumed, limits growth in the chain direction and could furnish a reason for lamellar fringed micellar growth with reasonably broad thickness distribution.

When the diffusion rate is small relative to the crystallization rate, only crystallization of nearest-neighbor crystallizable segments can occur. Wunderlich (1958) has calculated the resulting crystal size distribution for a random distribution of noncrystallizable defects. This limit of "cold crystallization" (Dole, 1959) may be realized on crystallization close to the glass transition temperature (see also Sect. 6.3.4). Figure VI.29 illustrates the large decrease in final crystallinity on cold crystallization in comparison to equilibrium crystallization. Besides calculation of crystallinity, this simple model of crystallization with minimum motion permits also the computation of side surface area as shown in Fig. VI.30. Chain entanglements, loops, and chain ends are collected together with the inherent copolymerized noncrystallizable defects in the adjustable parameter $X_B = 1 - X_A$, the mole fraction of all causes for interruption of crystallization along neighboring chains in the melt.

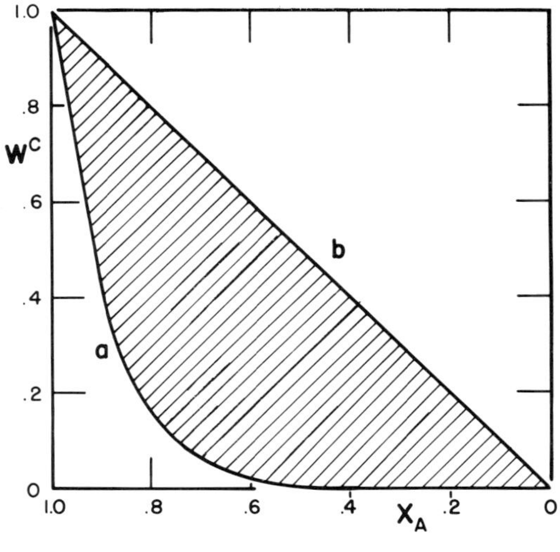

Fig. VI.29 *Crystallinity on Cold Crystallization and on Equilibrium Crystallization as a Function of Concentration.*

The cold crystallization curve (a) is calculated using the assumption that only crystals longer or equal to 3 repeating units are stable. $X_B = 1 - X_A$ is the mole fraction of all crystallization hindering defects. The equilibrium curve (b) represents crystallization of all crystallizable repeating units (for simplicity the mole fraction is assumed to be equal to the weight fraction) Wunderlich (1958).

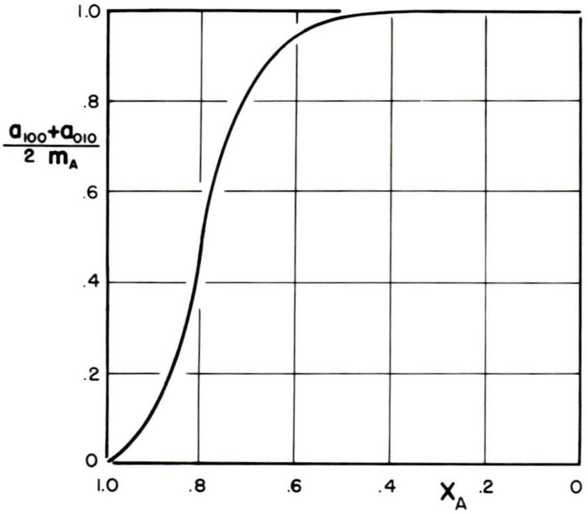

Fig.VI.30 *Side Surface Area of Cold Crystals.* The surface area is calculated as the fraction of repeating units which have three neighbors in the crystal. Corner areas caused by only two crystalline neighbors are not considered. The surface area is given in units $(a_{100} + a_{010})/2m_A$ which is the size of one average side surface per repeating unit in the assumed square lattice (CN = 4). Wunderlich (1958).

The two attempts to understand fringed micellar crystallization just described are limited in their use to the description of crystallization rate and crystal size and shape, respectively. Another line of enquiry involves attempts to illuminate the chain conformations on the crystal surfaces using statistical thermodynamics (Zachmann, 1969). These considerations are of value when one assumes that crystallization is slow relative to diffusion processes. Some details of the statistical thermodynamics of fringed micelle surfaces are given in the discussion of homogeneous nucleation (Sect. 5.1.2.1) and nonequilibrium melting (Chapter IX). Assuming that all of the crystal morphology is given, the major conclusion of such treatment is that in the early nucleation stage the fringed micelle crystal is much less stable than the folded chain crystal. Under slow crystal growing conditions one would thus not expect fringed micellar nucleation and growth, but rather chain folding (see Sects. 5.1.2.1 and 6.1.4).

The next step is to describe the changes in the structure of the surface layer as one goes from slow crystallization to increasingly faster crystallization. The description given by Ewers *et al.* (1972) is of interest to gain understanding of the distribution of the end-to-end distances of loops on the crystal surface.

The model is based on the assumptions that no large scale diffusion exists, i.e., molecule portions which are neighbors in the melt before crystallization will remain in close proximinity in the crystal. Furthermore, the probability of the formation of a surface loop with an end separation h is assumed to be proportional to the number of ways in which this loop can be realized. Figure VI.31 shows schematically the situation for crystallization on the

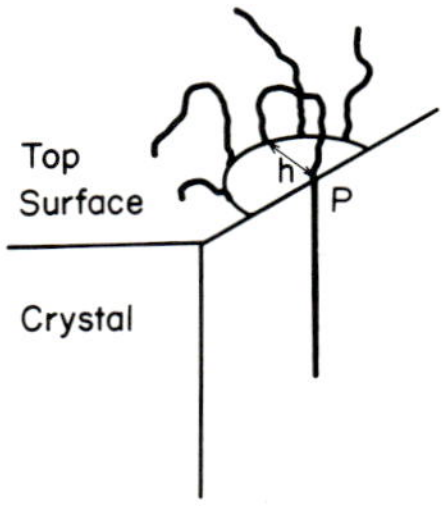

Fig. VI.31 *Schematic Drawing of the Crystallization of a Chain Segment at Point P.* Assumed process is the addition of a free end from the top crystal surface. The other two permitted possibilities are addition of a new, previously not crystallized molecule and the inclusion of a free end from the bottom surface.

growth face of a crystal. The probability that the new molecular chain segment of length l, equal to the thickness of the crystal, added at point P is from a new, previously uncrystallized molecule is called p_1. Two other possibilities are recognized. The newly crystallizing chain segment may be a part of a free chain end left from previous crystallization on the top (shown in Fig. VI.31) or on the bottom surface. The probabilities of these free ends crystallizing and forming surface loops are assigned the values $(1 - p_1)/2$ each.† After a certain distance of crystal growth a steady-state must be reached. On further crystallization just as many crystallizable free chain ends disappear due to loop formation and too great a distance from point P, as are formed due to loop formation at the opposite surface and addition of new molecules

† The case of inclusion of chain segments of molecules partially crystallized in other crystals (tie molecules) is also treated by Ewers *et al.* (1972), but brings little added information since the probabilities are treated as adjustable parameters.

with free ends on both surfaces. The available free ends on the surface should have a broad distribution. Instead of considering all possible lengths, Ewers *et al.* (1972) assume that all free ends have the average length† $\overline{N}c_0$.

$$\overline{N}c_0 = (N_0 c_0 - l)(1 + 3p_1)/[4(1 + p_1)] \tag{85}$$

where c_0 is the length in the chain direction per repeating unit, N_0 is the number of repeating units in the molecule, and l is the crystal thickness. The resulting probability densities for loops of end separation h, $\omega(h)$ as a function of p_1 for molecules of different lengths of N_0 are shown in Fig. VI.32. The

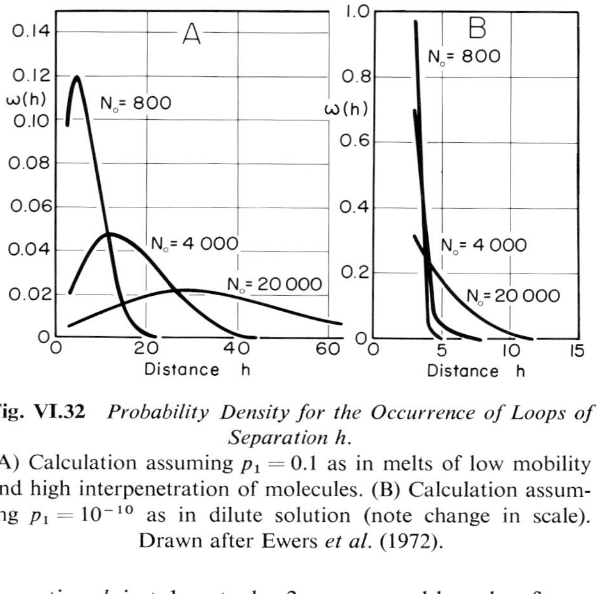

Fig. VI.32 *Probability Density for the Occurrence of Loops of Separation h.*
(A) Calculation assuming $p_1 = 0.1$ as in melts of low mobility and high interpenetration of molecules. (B) Calculation assuming $p_1 = 10^{-10}$ as in dilute solution (note change in scale). Drawn after Ewers *et al.* (1972).

minimum separation h is taken to be 3, a reasonable value for many crystal structures (see Tables II.9–II.18). As one would expect, larger molecules have larger separations of the termination points of the loops. Also, for small p_1 the minimum separation (adjacent re-entry) is most probable for all lengths. The latter case is most likely for dilute solution. On melt crystallization p_1 must be discussed in terms of the melt structure. The more the molecules are interpenetrating each other, the larger should p_1 be, and the more loops of larger separation should occur.

† A newly crystallized molecule leaves two free ends of average length $(N_0 c_0 - l)/2$, while on added inclusion in the crystal the length may decrease to zero, giving an average length of $(N_0 c_0 - l)/4$. Adding the two averages multiplied with their respective probabilities $2p_1/(1 + p_1)$ and $(1 - p_1)/(1 + p_1)$ leads to Eq. (85).

A detailed model of the melt was proposed by Pechhold *et al.* (1972) (see also Sect. 4.2.3 and Sect. 5.1). It is based on the description of the melt as a high defect concentration phase of largely parallel molecules. The free enthalpy of such a system is lowered by the cooperative collection of defects in planes which cause a change in direction of bundles of molecules as is shown in Fig. VI.33 (meander model). Such a model seems to be in accord with the

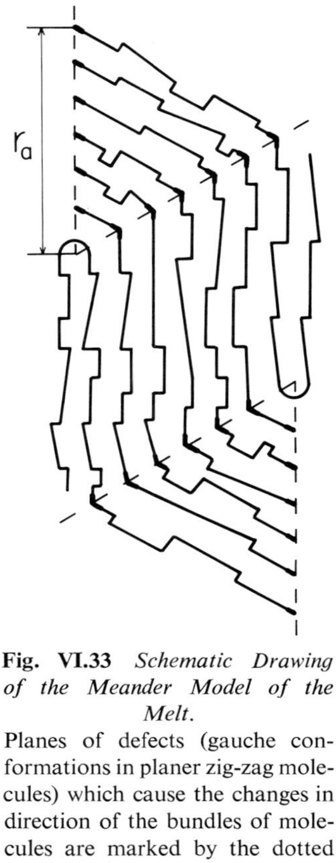

Fig. VI.33 *Schematic Drawing of the Meander Model of the Melt.*
Planes of defects (gauche conformations in planer zig-zag molecules) which cause the changes in direction of the bundles of molecules are marked by the dotted lines (see also Fig. V.1). Drawn after Pechhold *et al.* (1972).

observation of a nodular structure of the melt (see Sect. 2.4.9 and following) which indicates larger nonisotropic regions in the melt. Several thermodynamic and mechanical properties have been shown to be in accord with the predictions of the meander model. Crystallization of the molten meander is assumed to be simply a transport of matter by diffusion of defects along the chain, increasing the parallel portion of the chains, and decreasing their defect concentration. Typical values calculated on the basis of polyethylene are a meander radius (r_a, see Fig. VI.33) of about 50 Å. In this crystallization model much of the crystal structure is already formed in the melt. Most chains have nonadjacent reentry, and much of the molecular weight segregation must have occurred prior to crystallization.

A somewhat less detailed model of the melt with implications on the crystallization process was proposed by Yeh (1972). The major elements of this model are shown in Fig. VI.34. They are ordered domains (O.D.) of

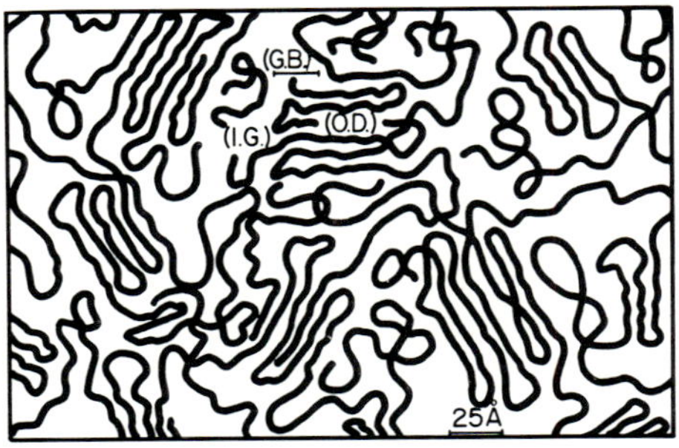

Fig. VI.34 *Schematic Drawing of the Folded-Chain Fringed-Micellar Grain Model of the Melt.* O.D. = ordered domains, G.B. = grain boundaries, I.G. = intergrain region. Yeh (1972).

20–40 Å in which the chain segments are parallel and an intergrain region I.G.) which consists of more random conformations. The ordered domains are limited in the chain direction by the grain boundaries (G.B.) which are 10–20 Å thick and contain folds (mostly adjacent reentry). The intergrain region should contain the low molecular weight molecules segregated in the melt and chain ends, loops and tie molecules. Again, in this case, crystallization is a relatively nondistinct process of perfection and fusion of grains.

The experimental basis of the assumption of the latter model is based on electron microscopic evidence provided by Yeh and Geil (1967) on the crystallization of macromolecules like poly(ethylene terephthalate) from the glassy state. Figure VI.35 shows that thin films of glassy poly(ethylene terephthalate) quenched from the melt (280°C) into ice water have a nodular structure of 45–100 Å diameter. On heating to 60°C small branched aggregates seem to form out of these nodules as is shown in Fig. VI.36. At even higher temperature (66°C) these aggregates take clearly the form of incipient spherulites as is illustrated in Fig. VI.37 (compare with Figs. III.117B and III.118). Crystalline order has been observed parallel with the aggregation of nodules by electron

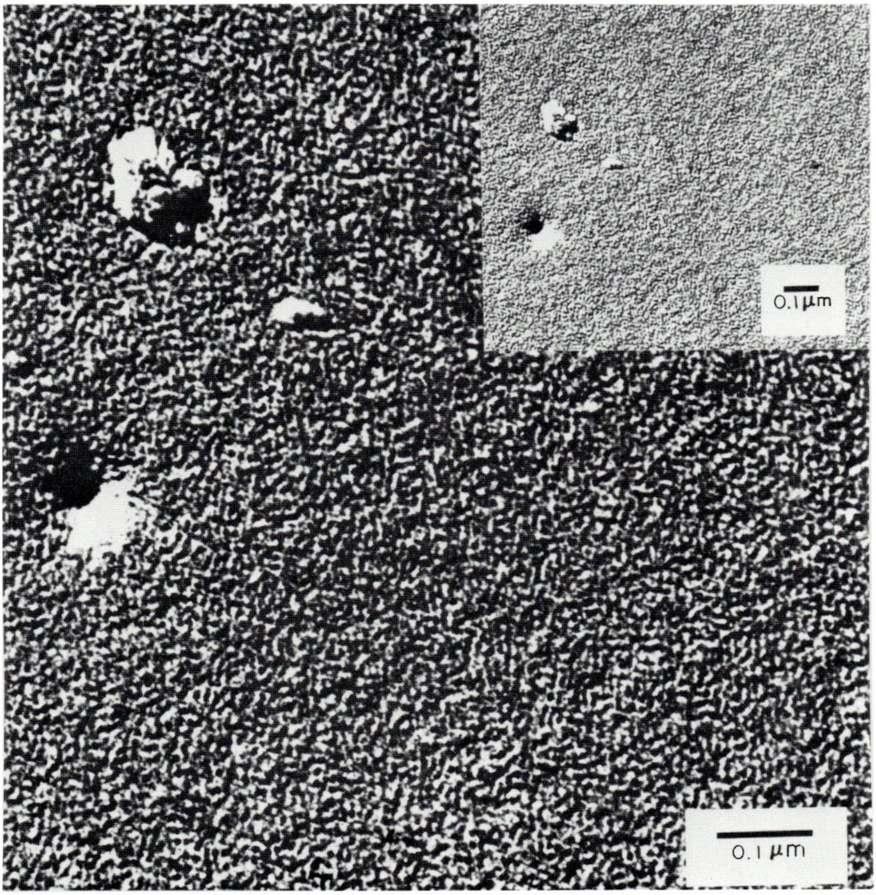

Fig. VI.35 *Nodular Structure of Glassy Poly(ethylene terephthalate).* Electron micrograph of a replica of the polymer quenched in ice water from the melt at 280°C. Yeh and Geil (1967).

diffraction. Diffraction patterns corresponding to the crystal structure of poly(ethylene terephthalate) have been recorded starting on structures as shown in Fig. VI.36. The crystallization seems clearly to be caused by alignment and fusion of the nodular structures at temperatures somewhat below the glass transition temperature (about 70°C).

To summarize, the knowledge on crystallization without chain folding has still many gaps. In particular, the delineation of its applicability is still in dispute, although it is clear that increasing immobility of the molecules and

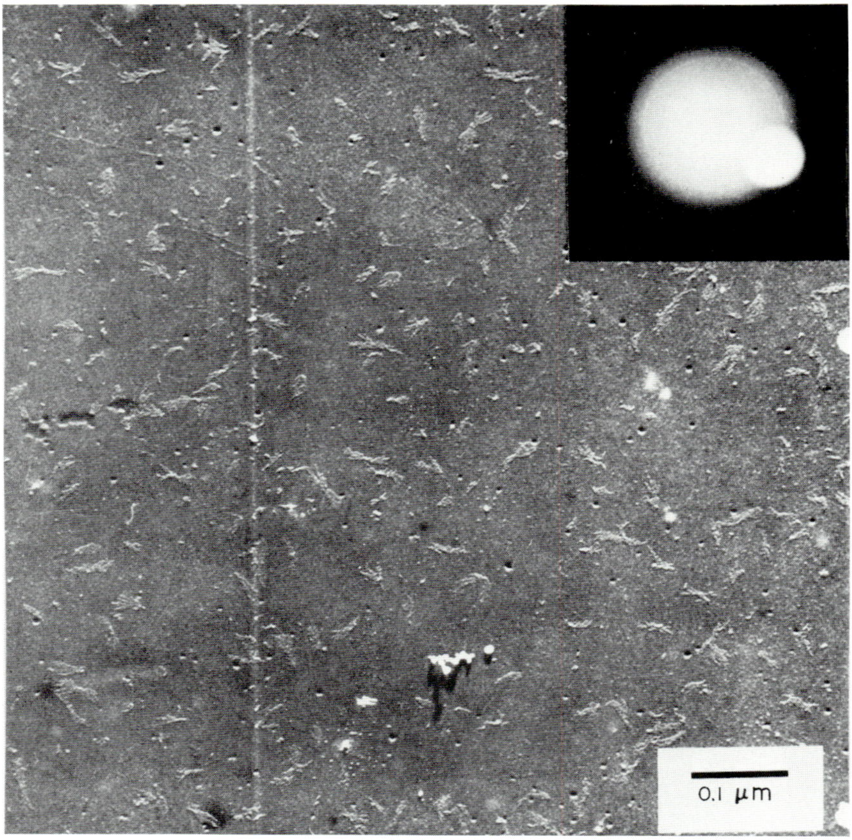

Fig. VI.36 *Beginning Ordering of Glassy Poly(ethylene terephthalate).* Electron micrograph of a replica as Fig. VI.35, sample heated for 24 hr at 60°C after quenching. Yeh and Geil (1967).

chain stiffness should favor such crystallization. The basic idea running through all the attempts to describe crystallization without chain folding is the crystallization of chains in close proximity of the crystal surface with a minimum of motion. One of the key experiments should thus be the observation of rejection of species from crystals. The basic information needed as a starting point for theories of crystallization without chain folding is the detailed structure and motion of the macromolecular melt and (concentrated) solution. This is a field of research which was neglected in the past 20 years, but has seen signs of progress more recently.

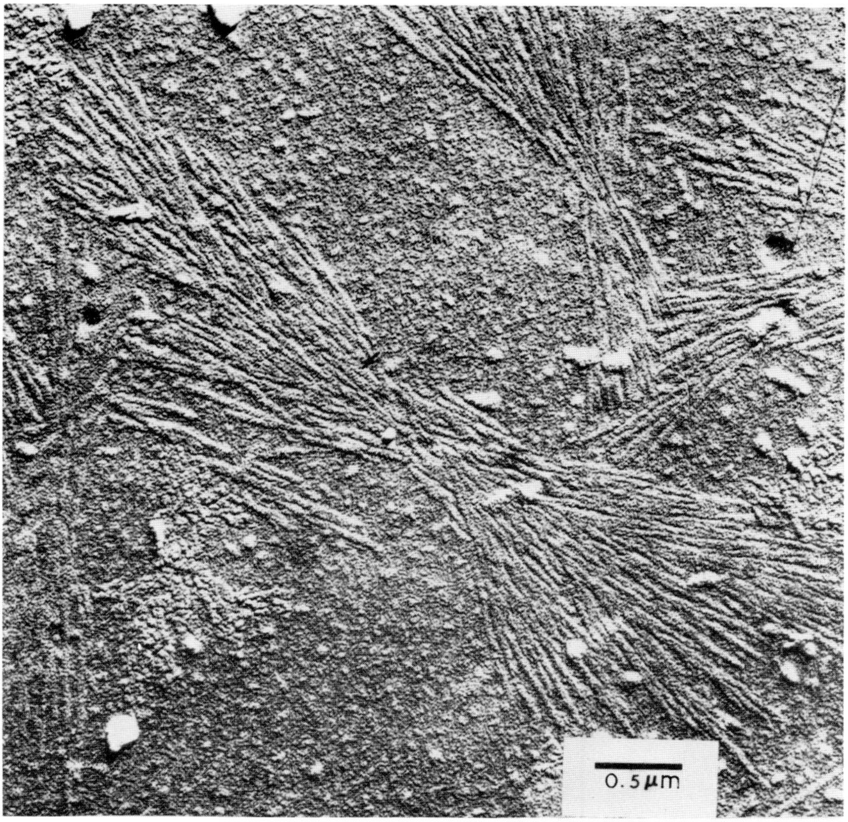

Fig. VI.37 *Incipient Spherulites of Poly(ethlylene terephthalate).*
Electron micrograph of a replica as Fig. VI.35, sample heated for 144 hr at 66°C after quenching. Yeh and Geil (1967).

6.1.8 Crystallization during Polymerization

Crystals of linear macromolecules can be grown not only from the macromolecular melt or solution, but may be formed directly from the monomer by crystallization during polymerization† as is shown in Fig. VI.38. Under conditions of coupled polymerization and crystallization special crystal morphologies have been observed (see Sects. 3.8.1, 3.8.3, and 3.9). Another point of difference is the macroconformation (see Sect. 3.2) of the molecules grown under conditions of crystallization during polymerization.

† Crystals grown during polymerization are also called nascent macromolecular or as-polymerized crystals.

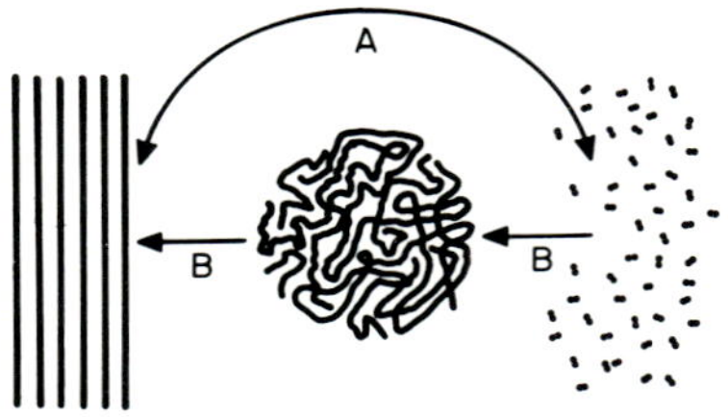

Fig. VI.38 *Schematic Representation of Crystallization of Macromolecules.* Path A: Crystallization during polymerization; Path B: Polymerization followed by crystallization (separate polymerization and crystallization). Wunderlich (1968).

The problems of chain folding (see Sect. 6.1.4) or fringed micellar crystallization (see Sect. 6.1.7) are now connected in addition with the polymerization mechanism. It is thus possible when growth occurs from the monomer to obtain chain folded crystals at as much as 100°C below the glass transition temperature (of poly-*p*-xylylenes, for example) (see Sect. 6.4.1). Crystallization from the melt, in contrast, would have led to chain folding only at temperatures above that of the glass transition. At lower temperature, crystal formation would have been more of the fringed micelle type and would cease completely at the glass transition temperature.

A first survey of the field of crystallization during polymerization was made by Wunderlich (1968) after observing by thermal analysis the increased crystal perfection of macromolecules transformed directly from the monomer to the macromolecular crystalline state (Wunderlich *et al.*, 1965). The possibility of simultaneous polymerization and crystallization had been recognized by Staudinger *et al.* (1927) (see Sect. 3.2.1, footnote p. 186). For many years this potentially different path to macromolecular crystals was neglected.† Even now the amount of information about crystallization during polymerization is much less than that available about crystallization from the macromolecule.

To assess the possible types of mechanisms, it is useful to distinguish between simultaneous polymerization and crystallization, and successive polymerization and crystallization. The first mechanism is easy to visualize. Primary and secondary bonds are set at the same time and the energetics governing polymerization and crystallization must be considered simultaneously. The successive polymerization and crystallization is more difficult

† For a summary of the limited information available by the early 1960 s see Geil (1963), Chapter VIII.

to elucidate. In this case the polymerization and crystallization sites are separated and the question of the conformation and length of the chain segment not yet crystallized becomes of importance. A typical example is schematically shown in Fig. VI.39. The dots denote the polymerization sites which have been located on a surface (as in heterogeneously catalyzed polymerization), but may also be randomly distributed in solution (as in homogeneously catalyzed polymerization). The resulting crystal morphology and final crystalline macroconformation of the completed chain will now depend upon the number of actively polymerizing chains contributing to the crystal, the crystal interface, the rates of polymerization and crystallization of the different growing chains, and the chain conformation before crystallization, which may be affected by temperature, solvents, adsorption on a substrate,

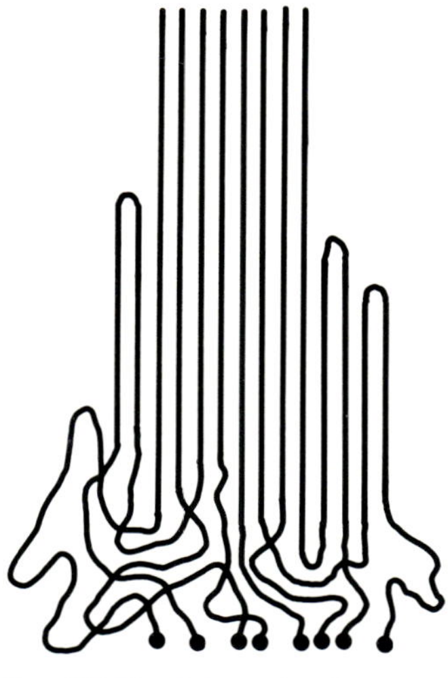

Fig. VI.39 *Schematic Representation of Successive Polymerization and Crystallization.*
In this drawing it is assumed that the polymerization sites (large dots) are aligned as, for example, in heterogeneous catalysis and the produced chains desorb readily. The indicated chain folding in the overall shish-kebob-like morphology (see Sect. 3.8.2) may come about because of faster polymerization on the outer polymerization sites due to better monomer supply (see also Keller and Willmouth, 1969).

and for short-site separation, the heat of polymerization (Wunderlich, 1968). Finally, there should be a continuous transition from successive polymerization and crystallization to separate polymerization and crystallization.

While crystallization of macromolecules is limited to the melt or solution, monomers may also crystallize from the gas phase. Thus, crystallization during polymerization must be discussed in terms of the monomer in the gaseous and condensed phases.

There is also a path of polymerization and crystallization which starts with crystallized monomer. Chemical reactions which involve a crystalline reactant (or product) and are influenced by the crystal structure are known as topochemical reactions (Kohlschütter, 1927). Topochemical polymerization reactions have been reviewed for example by Magat (1962), Morawetz (1963, 1966), Charlesby (1965), Tabata (1968), and Bamford and Eastmond (1969). Most of the reactions which start with a crystalline monomer lead to amorphous macromolecules and are thus of little interest to the present discussion.

When there is a direct path from the reactant crystal to the product crystal without intermediate loss of order, the reaction is called topotactic.† Topotactic polymerizations lead to special crystal morphologies and molecular macroconformations. In such reactions the main influence is the crystal structure and morphology of the monomer.

The further discussion of crystallization during polymerization in this section (general description) and in the description of examples (Sect. 6.4) will be divided into polymerization from the gaseous, liquid (melt and solution). and solid monomer. Successive and simultaneous polymerization and crystallization will be looked for whenever possible. The overall goal of this discussion is to connect as far as possible the special crystal morphologies (fibrillar and isometric) and molecular macroconformations (extended) described in Sects. 3.8.1, 3.8.3, and 3.9 and the conclusions gained from nucleation of actively polymerizing crystals in Sects. 5.1.2, 5.1.2.3, and 5.1.2.4).

Crystallization during polymerization from the gaseous or liquid phases may follow the mechanism outlined in Sect. 6.1.2.2 as long as polymerization and crystallization are simultaneous and sufficiently reversible so that surface diffusion is possible. As the monomer molecules get larger in size and acquire internal degrees of mobility, the description of surface diffusion, chemical

† "In topotaxy, a single crystal of a starting material is converted into a pseudomorph containing one or more products in a definite crystallographic orientation; the conversion takes place throughout the entire volume of the crystal. For true topotaxy there must be some three-dimensional correspondence between the structures of the product and its host (in contrast to epitaxy in which the correspondence need only be two-dimensional)." Dent Glasser *et al.* (1962).

reaction, and crystallization take on added complications which have not been analyzed in detail. A trend to fibrillar crystals has been observed and linked with the faster formation of the covalent bonds (Sect. 6.1.2.2) and limited nucleation of new molecules on the crystal surface (Sect. 5.1.2.3); no systematic analysis is, however, available. As soon as the crystallization occurs far below the ceiling temperature, it is much more difficult to reverse the once-made covalent bond of the molecular chain for surface diffusion to the proper crystallization site; cooperative setting of molecular and crystal structure becomes less likely. The crystallization mechanism tends to change under such conditions to successive polymerization and crystallization.

The ceiling temperature occurs at the equilibrium of polymerization and depolymerization reaction and can be expressed thermodynamically in the usual way as

$$T_c = \Delta H_p/\Delta S_p \tag{86}$$

Assuming the reaction to be an addition reaction (see Sect. 1.2) which occurs through the addition of monomer M to the activated chain end M_x^*, the propagation and depropagation steps can be written as

$$M_x^* + M \underset{\mathrm{d}}{\overset{\mathrm{p}}{\rightleftarrows}} M_{x+1}^* \tag{87}$$

The kinetics at the ceiling temperature where the rates are equal leads to

$$k_p[M_x^*][M] = k_d\,[M_{x+1}^*] \tag{88}$$

where the quantities in brackets represent the concentrations (or, for the gas phase, partial pressures). Inserting the standard expression for the temperature dependence of the rate constants $k = A\exp[-\Delta E^{\ddagger}/(RT)]$ in Eq. (88) with the assumption $[M_x^*] \approx [M_{x+1}^*]$ leads us to

$$[M]_c\,A_p\exp[-\Delta E_p^{\ddagger}/(RT_c)] = A_d\exp[-\Delta E_d^{\ddagger}/(RT_c)] \tag{89}$$

The difference between the activation energies of polymerization and depolymerization is the heat of reaction ΔH_p, so that Eqs. (89) and (86) can be combined into an expression for the entropy of polymerization

$$\Delta S_p = R\ln(A_p[M]_c/A_d) = \Delta S_p^0 + R\ln[M]_c \tag{90}$$

with ΔS_p^0 representing the entropy of polymerization at the standard state (partial pressure or concentration in the chosen units equal to one). The equilibrium concentration or vapor pressure can then, with the use of Eq. (86) be expressed as

$$\ln[M]_c = [\Delta H_p/(RT_c)] - (\Delta S_p^0/R) \tag{91}$$

TABLE VI.6

CEILING TEMPERATURES OF CRYSTALLINE MACROMOLECULES[a]

Monomer	T_c (°C)	$[M]^b$	Notes/solvent
Chloral	96	76	Gaseous monomer
Chloral	12.5	0.10	Solvent pyridine
Tetrafluoroethylene	560	96	Gaseous monomer; above T_m, liquid polymer
Formaldehyde	126	760	Gaseous monomer
Formaldehyde	30	0.06	Solvent methylene chloride
Sulfur trioxide	27.0	279	Gaseous monomer
Sulfur trioxide	30.4	1.0	Liquid monomer
Selenium	83	1.0	Liquid monomer and polymer (floor temperture, ring-chain equilibrium in the melt above, Se_8 rings only below 83°C)

[a] Data from Brandrup and Immergut (1966).

[b] For gaseous monomer partial pressure in mm Hg, for dissolved monomer mole fraction.

Table VI.6 is a collection of the limited data on ceiling temperatures of crystalline macromolecules. For most of the examples discussed in Sect. 6.4 the ceiling temperature is higher than the crystalline melting point.

Crystallization during polymerization starting from the crystalline monomer can be discussed in terms of the principle of least motion in solid state chemistry. It is assumed that the reaction of least change in atomic positions and electronic configurations will be favored (see for example Hine, 1966). Indeed, Baughman (1974) showed that the greatest reactivity of related monomer is observed for the smallest motion, and the actual mode of polymerization follows the path of least motion.

As outlined by Kaiser *et al.* (1972), the polymerization in the solid state may follow a heterogeneous or a homogeneous path. In the heterogeneous path the polymer–monomer interaction (see the following) as well as the crystal structures are of insufficient match to permit isomorphous replacement† of monomer by polymer. The reaction may be started at suitable reaction sites such as surface, grain boundaries, or defects (see for example Thomas and Williams, 1971). After nucleation of the new phase, growth occurs exclusively at the phase boundary. The morphology of the original

† For a discussion of isomorphism see Sect. 2.4.8.

crystal may be completely lost, or be preserved due to original orientation of the nucleus of the polymer crystal by epitaxy. X-ray analysis during the reaction reveals the immediate appearance of the new macromolecular crystal phase and the persistence of the unchanged monomer crystal reflexes until practically all monomer is converted.

The path of homogeneous conversion of monomer to polymer is illustrated by a group of diacetylenes reacting by 1,4-addition to macromolecules (see Sect. 6.4.3). The conversion occurs statistically within the monomer crystal. No new phase is nucleated. Experimentally the reaction can be followed by X-ray diffraction. No new reflections are observed during the reaction and none disappear. As the reaction progresses, some of the diffraction angles change continuously to larger or smaller values while others remain constant (Kaiser *et al.*, 1972). The macromolecule is thus isomorphously replacing the monomer. The direct conclusion of such mechanism is that on complete conversion, a monomer single crystal has changed to a macromolecular crystal of identical morphology and a macroconformation of the chains which is most likely fully extended has resulted. Figures III.142 and III.143 are two examples of such single crystals.

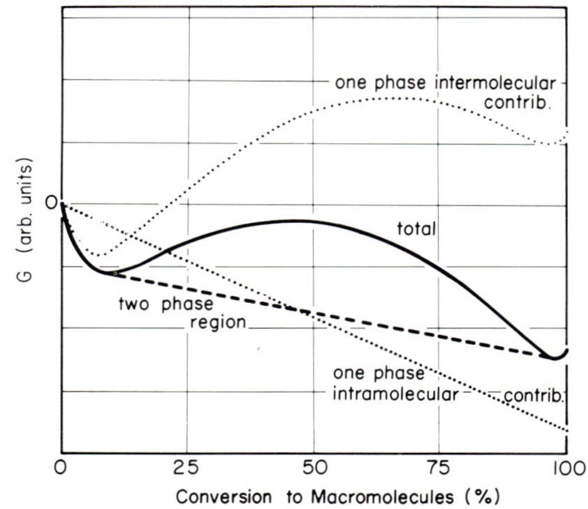

Fig. VI.40 *Free Enthalpy Diagram of Solid State Polymerization with Seperation into Two Phases.*
The intramolecular contribution is due to chemical reaction, the intermolecular contribution from the entropy of mixing and polymer repeating unit-monomer pair interaction (regular solution). The pair interaction is assumed unfavorable (positive free enthalpy). Drawn after Baughman (1974).

Criteria of the phase stability during the homogeneous conversion of monomer crystals to macromolecular crystals have been developed by Baughman (1974). These criteria involve the free enthalpy changes on reaction (intramolecular effects) as well as the changes in packing (intermolecular effects). Figure VI.40 illustrates schematically the case where the reaction would lead to separation into two phases. The intramolecular effect decreases the free enthalpy linearly with increasing conversion. The intermolecular effect has been calculated assuming regular solutions. The initial decrease in free enthalpy arises from the favorable entropy of the mixing and the change to a positive free enthalpy is caused by the assumed unfavorable pair interaction between polymer repeating unit and monomer. Figure VI.41, in

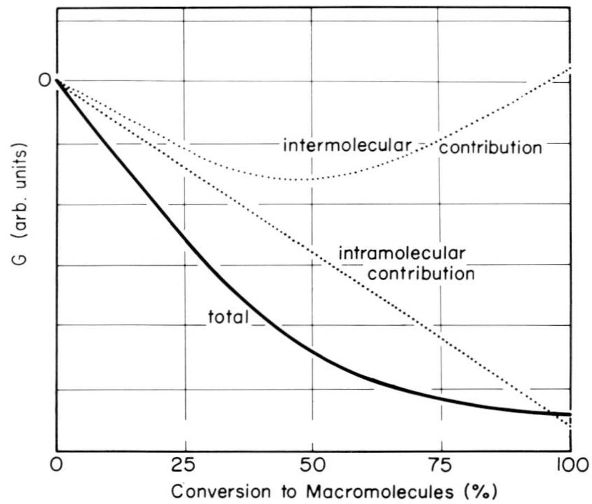

Fig. VI.41 *Free Enthalpy Diagram of Solid State Polymerization in One Phase.*
As Fig. VI.40. The pair interaction is assumed favorable (negative free enthalpy). Drawn after Baughman (1974).

contrast, illustrates the condition for successful one-phase polymerization. The polymer repeating unit-monomer pair interaction is negative in this case, adding to the entropy term and producing a continuously decreasing free enthalpy as conversion increases. Thermodynamic arguments alone, however, are not sufficient to predict homogeneous conversion. If symmetry and dimensional criteria are not fulfilled, the as-polymerized crystals are likely to be twinned, fibrillated, or otherwise disordered. Baughman (1974) postulates that the site symmetry of the monomer in its crystal structure must also be a possible site symmetry of the repeating unit in the polymer crystal, and that the symmetry elements that relate reacting monomer molecules must be possible symmetry elements of the macromolecular chain.

6.2 Solution Crystallization of Macromolecules

The subject of crystallization of the preformed macromolecule from solution is usually separated into two limits, crystallization from dilute solution and crystallization from concentrated solution (the latter is also called the effect of diluent on crystallization). In dilute solution crystallization one can imagine that the molecules assume macroconformations close to the random coil (see Sect. 1.4). Different molecules are separated from each other and arrive on slow crystallization independently, one after the other, at the crystal surface. These are conditions one expects to be suitable for growing large, perfect, equilibrium crystals. Carefully grown crystals from dilute solution are indeed large, but only in two dimensions (see Figs. III.68–III.70). The chain folding principle (Sect. 3.2.2.1) is overriding for the crystal dimension in the molecular chain direction. Slow crystallization from quiescent, dilute solution usually yields metastable 50–200-Å thin, laterally large lamellae. A stirred or otherwise stressed solution is shown in Sect. 6.2.2 to yield increasingly fibrillar crystals (shish-kebobs, see Sect. 3.8.2 for the morphology). Larger hydrostatic pressure is shown in Sect. 6.2.3 to have little effect on the crystal morphology beyond the normal effect of an increase in degree of supercooling caused by increasing pressure.

The major added variable in crystallization from solution is concentration. The change of the dissolution temperature with concentration is discussed in Chapter VIII. Crystallization from solution with similar thermodynamic driving force than from the melt can start only at a lower temperature. In addition, the concentrations change as crystallization proceeds, and with it changes the degree of supersaturation.† Since the supersaturation is the driving force for both molecular nucleation and linear crystal growth (see Sects. 5.3.4 and 6.1.2), the overall crystallization should slow down relative to the simple Avrami expression Eq. (33). For crystallization from concentrated solution this behavior has been observed (see Fig. VI.44, see also Figs. VI.66–67 for the related copolymer behavior).

For crystallization from dilute solution, the observed linear growth rate, in contrast to concentrated solutions, is almost constant until most of the macromolecules are crystallized (see Fig. VI.3). This different behavior on

† Baur (1967) could show that within the framework of linear phenomenological nonequilibrium thermodynamics, the crystallization does not necessarily occur with a monotonous decrease in concentration of the separate components. At given temperature and pressure, crystallization of a given component may see earlier or later partial melting.

crystallization from dilute solution is unique for macromolecules and is linked to the unique structure of dilute solutions. Macromolecules in dilute solution can be thought of as random coils invading only a volume fixed by the macroconformations (see Sect. 1.4). Outside of the random coil the repeating unit concentration drops to zero; inside, it does not change with macroscopic concentration as long as the macromolecules are sufficiently separated from each other (see, for example, Flory, 1953, Chapter XII). The supersaturation governing nucleation and growth is thus changing only as long as the overall concentration is larger than the average concentration within the random coil of a single molecule. The value of ΔT, the supercooling, in nucleation and growth expressions should thus be best determined experimentally from dissolution experiments of extended chain crystals into a dilute solution. The effect of factors of crystal growth not governed by ΔT is described in Sect. 6.2.1.

The need of molecular nucleation causes a temperature region of practically no crystallization below the dissolution temperature, as can be seen from Fig. III.99 (see also Fig. V.49). This temperature region is followed for many macromolecules by a small region of about 10°C width where single crystal lamellae and growth spirals of well defined morphology grow. At even lower temperature, dendrites develop (see Sect. 3.6.1). The molecular weight fractionation of samples with broad distributions of chain lengths is also based on molecular nucleation and is described in Sects. 5.3.2 and 5.3.4.

At increasing macromolecular concentration the conditions of bulk crystallizations are approached; with it, the usual problem of more fringed-micellar-like crystallization arises (see Sect. 6.1.7). The question of crystallization in the presence of a diluent has not received much attention. Some experiments are summarized at the end of Sect. 6.2.1. The presence of even a small amount of diluent in the crystallization of polyethylene under high hydrostatic pressure slows down the chain extension process immediately after crystallization (see Sect. 6.2.3). One may conclude that reorganization of once-formed folded chain lamellae after crystallization from solution may under these circumstances of better initial crystallization be at a minimum (see also Sect. 6.1.6). Frequently the finally observed fold-length of crystals grown from solution is close to the value expected from molecular nucleus sizes.

6.2.1 Normal Crystal Growth

Most work on crystallization from solution has been carried out on polyethylene. The first kinetic data were taken by Holland and Lindenmeyer (1962) through observation of the increasing size of single crystals with time. All crystals were nucleated (accidentally?) athermally at the start of crystallization. Solutions in xylene of 0.01 to 0.001 wt. % of a 120 000 MW fraction of polyethylene were crystallized at temperatures from 80 to 92.2°C, the temperature region where single crystals can be grown (see Fig. III.99). The growth rate stayed constant up to the last stages of crystallization as is shown in Fig. VI.3. The temperature dependence of the logarithm of the linear growth rate was shown to conform to a $1/(T\Delta T)$ dependence as expected for molecular nucleation controlled growth [see Eq. (82) Sect. 5.3.4, and Fig. VI.42].

With the development of the self-nucleation technique (Sect. 5.1.4) more precise nucleation control become possible. Blundell and Keller (1968) and Pedemonte (1973) seeded polyethylene fractions by dissolution in xylene and

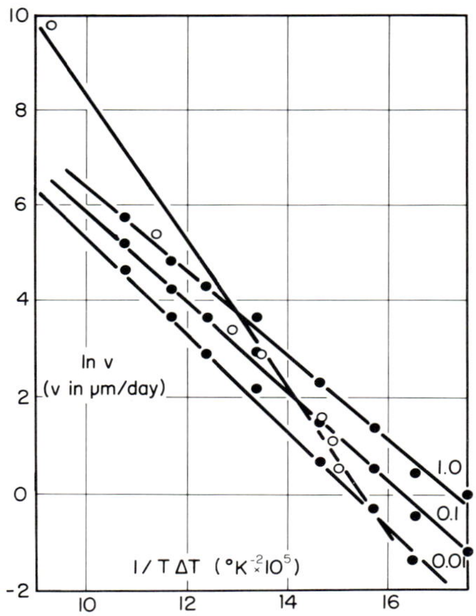

Fig. VI.42 *Change in Linear Crystal Growth Rate as a Function of* $1/(T\Delta T)$ *Normal to the {110} Surface of Polyethylene Crystallized from Dilute Solution in Xylene.*

The natural logarithm of the growth rate v (μm/day) is plotted versus $1/(T\Delta T)$, with $\Delta T = T_0 - T$, where T_0 is the experimental dissolution temperature of extended chain crystals under similar conditions (110°C). Data by Holland and Lindenmeyer (1962): (○) 0.01–0.001 wt.%, 120000 MW fraction; data by Keller and Pedemonte (1973); (●) 1.0–0.01 wt.%, as indicated, typical broad molecular weight sample $\overline{M}_n = 6700$, $\overline{M}_w = 100000$.

octane to a temperature somewhat below the equilibrium dissolution temperature of polyethylene to set the proper number of nuclei. Crystallization was then carried out at chosen temperatures. The linear growth rate was evaluated by electron microscopy as shown in Fig. VI.2. Again, the temperature dependence of the logarithm of the linear growth rate (normal to the {110} faces) was shown to be well represented by a $1/(T\Delta T)$ proportionality. Figure VI.42 illustrates some data by Keller and Pedemonte (1973) and Holland and Lindenmeyer (1962) for xylene as a solvent. The slopes of the shown data vary between 0.9×10^5 and 1.5×10^5. With the standard values for polyethylene (see pp. 25 and 55) this leads, using Eq. (82), Chapter V, to a $\gamma\gamma_e$ product of 560 to 980 erg^2/cm^4; the same order of magnitude as was derived from primary homogeneous nucleation using the droplet technique.†

The next observation one can make from Fig. VI.42 is the small concentration dependence. For polyethylene sufficient information is available to discuss the concentration dependence in more detail. First, it is of interest to visualize the actual mass distribution in solution. For concentrations of 1.0 and 0.01 wt. % polyethylene in xylene the centers of typical 280 000 MW molecules are on the average 375 Å and 1750 Å apart. The root mean square end-to-end distance at the theta temperature (see Sect. 1.4) is about 500 Å, meaning that below 1 wt. % concentration the molecules of less than 280 000 MW may be considered increasingly isolated and the degree of supercooling should not be concentration dependent. At a given molecular weight and temperature the linear crystal growth rate can then only be dependent on concentration through the number of available molecules N for crystallization [see Eqs. (69)–(71), and Eq. (5), Chapter V, for I^*, and Eq. (82), Chapter V, for ΔG^*]. In the temperature region where well developed growth faces are observed, crystal growth is not diffusion controlled (see Sect. 3.6.1 and also Blundell and Keller, 1968). The number of molecules of given size arriving at the growth face for crystallization must thus be proportional to the bulk concentration c. Such linear growth rate dependence with concentration has been observed by Seto and Mori (1969), but only at concentrations in the range of 10^{-3}–10^{-4} wt. %. As the concentration increases, the crystal growth rate dependence on concentration becomes less than linear as first observed by Blundell and Keller (1968). Figure VI.43 illustrates the exponent α of concentration dependence of the linear crystal growth rate v in the concentration range 0.01–1.0 wt % polyethylene in xylene and octane as a function of crystallization temperature (Keller and Pedemonte, 1973). The exponent α is low (0.2–0.5) for relatively low crystallization temperatures and high molecular weights, and increases sharply to as much as 2 at low supercooling of

† $\gamma_e\gamma^2$ is shown in Sect. 5.1.2.2 to be 11 000 erg^3/cm^6 with $\gamma \approx 10$ erg/cm^2, this lead to 110 erg/cm^2 versus the present 56 to 98 erg/cm^2 for γ_e.

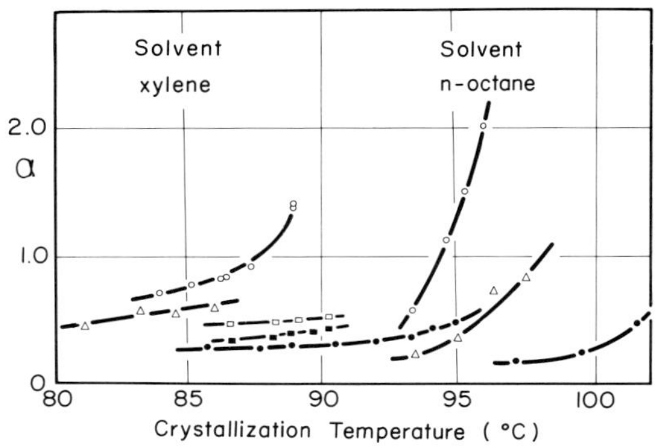

Fig. VI.43 *Slopes α of Logarithm of Crystal Growth Rate versus Logarithm of Concentration Plots for Polyethylene Crystallized from Dilute Solution in Xylene and Octane.*

Different symbols refer to different molecular weights: (○) $M_n = 10\,000$, $M_w = 16\,000$; (△) $M_n = 3250$, $M_w = 10\,000$; (●) $M_n = 6700$, $M_w = 100\,000$. Data by Keller and Pedemonte (1973). (□) fraction $M_w = 25\,200$; (■) fraction $M_w = 83\,900$. Data by Cooper and Manley (1973).

lower molecular weights. Seto and Mori (1969) report that the low concentration increase in α occurs at lower concentration as the crystallization temperature increases.

The upturn of α with crystallization temperature occurs within a few degrees of the crystallization temperature which requires the full molecule for molecular nucleation (see Fig. V.49), a reasonable result when one considers that molecular nucleation using the full molecule should lead to $\alpha = 1$. Nucleation of lower molecular weights which may need side-by-side nucleation of two molecules would, in turn, require an α with a value 2. In the region where the molecular nucleus is only part of a molecule, complications may arise by multiple nucleation in the domain otherwise covered by a single molecule [see Eqs. (70) and (71)] or by blocking of adjacent domains by the temporarily left uncrystallized chain ends after molecular nucleation (cilia) (see also Sanchez and DiMarzio 1971a,b). As the crystallization temperatures increase, the nucleation and growth rates decrease relative to the rate of arrival of new molecules, and in accord with this, the low concentration change to $\alpha = 1$ [Eq. (69)] moves to lower concentration.

The overall crystallization rates (see Sect. 6.1.3) of about 0.5 wt % solutions of polyethylene fractions ($M_n \approx 10^5$) in *n*-hexadecane, *p*-xylene, and decalin were measured by Devoy *et al.* (1970) by dilatometry. The results were qualitatively similar to bulk polymer data. They fitted an Avrami expression with an exponent $n = 4$ [see Eq. (33)]. No secondary crystallization or crystal perfection typical for bulk crystallization was observed (see Sect. 6.1.6). Unfortunately, no morphological evidence for the type of nucleation and crystal morphology was gathered for complete interpretation of these data. The temperature dependence of the time of 10% conversion agreed with the assumption of molecular nucleation control of crystal growth.

The crystallization of poly[3,3-bis(chloromethyl)oxacyclobutane]† of $M_n = 5 \times 10^6$–10^7 was studied by Johnsen and Lehmann (1969) from *m*-xylene solution of 0.1–0.001 wt. % polymer concentration. The solutions were analyzed by high resolution NMR (see Sect. 4.1.6) for their content of dissolved polymer, and samples were taken during the crystallization to check the process of nucleation and growth by electron microscopy. Self-nucleation (see Sect. 5.1.4) was achieved by initial dissolution at the boiling temperature of *m*-xylene (139°C) which is below the equilibrium dissolution temperature of the macromolecule (140–143°C). Athermal nucleations at levels from 4×10^{11}/g to 3×10^{12}/g were reached at crystallization temperatures of 45–112°C. Some data of N, the number of nuclei per gram of solute, and

TABLE VI.7

CRYSTALLIZATION OF POLY[3,3-BIS(CHLOROMETHYL)OXACYCLOBUTANE][a]

Initial conc. (g/cm³)	Cryst. temp. T_c (°C)	Growth rate v (μm/min)	Nuclei N(g⁻¹)
2×10^{-4}	45	0.42	2.5×10^{12}
	60	0.23	1.5×10^{12}
	78	0.13	0.7×10^{12}
	86	0.13	0.6×10^{12}
	95	0.07	0.9×10^{12}
	105	0.02	—
5×10^{-4}	60	0.27	0.9×10^{12}
	78	0.21	0.7×10^{12}
	86	0.12	0.9×10^{12}
	95	0.09	2.2×10^{12}
	105	0.02	0.4×10^{12}
2×10^{-3}	86	0.22	1.7×10^{12}
	95	0.17	1.3×10^{12}
	105	0.04	1.4×10^{12}
	112	0.02	2.4×10^{12}

[a] Selection of data by Johnsen and Lehmann (1969).

† Also called Penton, for morphology see Table III.3 and Fig. III.46.

crystallization rates v normal to the crystal growth faces are given in Table VI.7 for different initial concentrations and crystallization temperatures. From these data γ and γ_e were estimated to be 15 and 20 erg/cm^2, respectively. The overall crystallization rate was observed from 10^{-3} g/cm^3 to 10^{-1} g/cm^3 by NMR and compared at the low concentration end with calculations from the electron microscopic data. The crystallized mass should increase proportional to N, the number of nuclei and the square of time t for the observed morphology of constant thickness single lamellae (see Fig. III.46). The proportionality constant can be evaluated from the exact geometry. Qualitative agreement was achieved in the concentration region of overlap. For the whole region of initial concentrations analyzed by NMR the t^2 dependence remained. Comparison with the Avrami equation [Eq. (33) and Table VI.2] would indicate the expected athermal nucleation, followed by two-dimensional growth without impingement [Eqs. (50) and (28)].† The exponent α of the concentration dependence of the initial crystal growth rate was found in the chosen concentration range to vary almost linearly from 0.24 at 75°C to 0.44 at 100°C, a result similar to those for polyethylene above (see Fig. VI.43). In the final crystallization period (after about 60% had crystallized) an exponential decrease of the remaining crystallizable macromolecules was observed with a time constant decreasing from about 3/min to 0.02/min with increasing temperature of crystallization and decreasing initial concentration. This slowing of the cystallization cannot be caused by diffusion control since this would call for an increase in the time constant with temperature and not a decrease. Johnsen and Lehmann (1969) interpret the change as a change to regime I crystallization as expressed by Eq. (69). They calculate approximately the same activation free enthalpy from the final portion of crystallization as from the beginning of crystallization. Thus, the observation seems to be in accord with the polyethylene data of Seto and Mori (1969) at very low concentration.

The crystallization of poly(ethylene oxide) fractions (7×10^4–2×10^6 M_n) from about 0.3 g/100 cm^3 solution in ethanol was followed dilatometrically by Beech and Booth (1972) between 27 and 33°C. As for the polyethylene case, abrupt termination of the crystallization was observed, indicative of little or no secondary crystallization and crystal perfection (see Sect. 6.1.6). Also, the degree of crystallization was independent of the crystallization

† Because of the high dilution of the system, crystallization is completed before impingement becomes significant, Eq. (28) can thus be approximated by the unimpeded crystal growth expression $\exp(-E) = 1 - E$ [see discussion of Eq. (33)]; E gives directly the crystallized volume fraction [Eq. (50) for circular geometry] [note that both v^c and N refer to the whole solution, but because of their position on different sides of Eq. (50) the dilution effects of the solvent cancel].

temperatures used, in contrast to crystallization from the melt. Analysis of the data gave an Avrami coefficient† 4 as in polyethylene, but again no nucleation or morphology data are available for a more detailed interpretation (see Table VI.2).

Extensive kinetic data of crystallization of poly(ethylene oxide) copolymerized with styrene to a two-block molecule were collected by A. J. Kovacs, *et al.* (1966). The morphology of such crystals is quite similar to poly(ethylene oxide) homopolymer (see Figs. III.20 and III.45). The crystallizable poly(ethylene oxide) block was of 10.9×10^3 MW and the noncrystallizable, atactic polystyrene block was of 7.3×10^3 MW. Measurements by dilatometry were carried out by crystallization of a 1.15 g/100 cm^3 solution in ethylbenzene between 17.5°C and 30°C after setting the number of seeds at 35°C and 50°C. At 50°C the number of remaining self-nuclei is close to zero since the dissolution temperature of copolymers in ethylbenzene is about 45°C. The nucleation is in this case thermal (homogeneous or/and heterogeneous). Analysis of the crystallization isotherms shows that they were not superimposable by shift along the logarithm of time axis. A maximum amount of crystallization occurred at 26°C. The morphology of the crystals changed from single crystal lamellae at high temperatures (as in Fig. III.45) to complex aggregates at low temperatures (see for example Fig. III.72). At a temperature of 35°C, a fixed number of seeds is set (self-nucleation see Sect. 5.1.4) and crystallization after heating to 35°C led to a uniform crystallization of single crystal lamellae. In this case, the analysis of the isotherms with an Avrami expression [Eq. (33)] gives good agreement for over 90% of the crystallization with a value of $n = 0.9$ at 30°C crystallization temperature. For lower crystallization temperatures n increases slightly to reach about 1.2 at 20°C. This low exponent n is supported by direct microscopic measurements. It indicates a diameter of the single lamellae proportional to $t^{1/2}$ instead of a constant linear growth rate. The morphology which is shown in Figs. III.20 and III.45 reveals, however, that standard diffusion control, which could account for a $t^{1/2}$ dependent crystal size (see Sect. 6.1.2.3), is not possible because of the extended crystallographically perfect growth faces. A. J. Kovacs, *et al.* (1966) suggest that the rejection of copolymer species with particularly long polystyrene sequences must be at the root of the abnormal linear growth rate‡ (see also Sect. 6.3.4). The small increase of the exponent n with decreasing temperature can be linked to the slowly increasing number of

† Again, the analysis is possible using the simple expression $v^c = Kt^n$ since little impingement is expected.

‡ The sample had an overall weight fraction of 0.40 in polystyrene. After crystallization at 25°C to a fraction 0.86 only 0.34 weight fraction polystyrene was found connected to the poly(ethylene oxide) crystal surfaces.

growth spirals observed in the samples. These would change the two-dimensional growth into a three-dimensional growth with correspondingly larger n. The crystallization isotherms after heating to 50°C (thermal nucleation) were much more complicated. The simple Avrami expression [Eq. (33)] applied only over a much smaller region of crystallization ($0.4 < v^c < 0.8$) and the exponent needed to fit the equation varied from about 2 at 28°C to 6 at 20–21°C, to decrease again at lower temperatures. The complicated nucleation, morphology, and rejection processes could not be sorted out for a detailed explanation.

Isotactic polypropylene ($2\text{–}4 \times 10^5$ MW) solutions in decalin and tetralin (about 0.4 wt. % polymer) were crystallized by Shah and Lahti (1973). Dilatometrically determined crystallization isotherms gave a good fit to an Avrami expression [Eq. (33)] with an exponent n of 4.

Some data on the crystallization of poly(decamethylene terephthalate) (40 000 MW) from 1 wt% to 2 wt.% solution in 1,2-dichloroethane were gathered by Lanceley and Sharples (1966) using optical microscopy and light scattering. Again, over a long range of crystal growth the dimensions of the crystals increased linearly, indicating a very small concentration dependence. The growth rate of the bulk polymer was found to be only about three times larger than the dilute solution at the same degree of supersaturation.

The crystallization of isotactic polystyrene ($M_n = 185\,000$) from 3–5 wt. % solution in toluene/ether mixtures was followed by Helms and Challa (1972) by observation of the crystalline and the helix infrared absorption. From the observation that on initial occurrence of the crystalline absorption (565.5 cm^{-1}) no significant change in the helix content (585.5 cm^{-1}) occurred, it was concluded that perhaps preformed helices in solution aggregate to the initial crystals (see Sect. 6.1.1). Extracting the Avtami exponent (Eq. 33) from the absorption data showed, as before, good fit over the whole crystallization range. The values of n were 2.3–2.4 for the 565.5 cm^{-1} band and close to 3.0 for the 585.5 cm^{-1} band.

The effect of changing viscosity of 2 wt. % poly(ethylene sebacate) ($M_n = 40\,000$) in rosin-xylene solution was analyzed through optical microscopy by Asaubekov *et al.* (1972). The viscosity was varied by adjusting the rosin-xylene ratio (overall from 4 cP to 10^8 cP at various crystallization temperatures). Within the accuracy of the measurement there was no effect of the viscosity on the growth rate of the spherulite and dendrite diameters. This result is surprising since for dendrites and spherulites one usually assumes a diffusion controlled mechanism of growth (Sect. 3.6.1). The explanation may be, however, that the observed growth at the boundaries of the superstructure is not diffusion controlled (see point O in Fig. III.101). A more detailed analysis of the change in morphology would have been of interest in connection with these experiments.

The kinetics of crystallization of higher concentration was analyzed by Mandelkern (1964a). The effect of diphenyl ether and tetralin on the crystallization of poly(ethylene oxide) was investigated in the concentration range of 0.1 to 0.7 volume fraction diluent, that of dimethyl formamide on poly-(decamethylene adipate) from 0.2 to 0.6 (Mandelkern, 1955). In addition, the complete range of concentration of polyethylene in α-chloronaphthalene was discussed by Mandelkern (1964b). Above about 0.05 volume fraction in polymer, the crystallization isotherms changed from the simple Avrami type described above. The curves of a given initial concentration are still superimposable by shift along the log t axis; the growth slows down, however, as crystallization proceeds. Figure VI.44 shows a typical sample of polymer

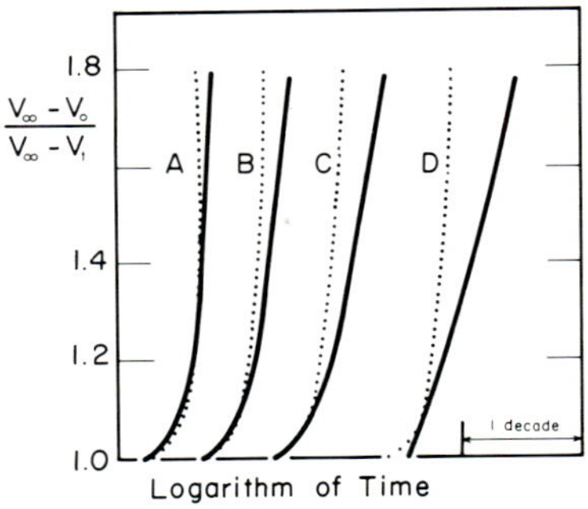

Fig. VI.44 *Crystallization Isotherms of Poly(ethylene oxide) Diluted with Diphenyl Ether.*
Curve A, pure polymer; curve B, 0.90 dilutent; curve C, 0.30 diluent; curve D, 0.69 diluent (concentration in volume fraction). The dotted lines present attempts to fit the experimental data with an Avrami expression [Eq. (33)] with $n = 4$. The position along the time axis is arbitrary. (Note that the ordinate is different from the normally plotted crystallinity, compare with Fig. VI.10) Data by Mandelkern (1955).

diluent crystallization isotherms. Qualitatively this can be accounted for by the change in supercooling ΔT (see Sect. 6.2) and in N, the number of molecules available for crystallization and growth (see the discussion of dilute solution crystallization above and also Mandelkern, 1964a). Detailed discussion is, however, only possible after more of nucleation and crystal growth of diluent mixtures has been analyzed to separate the different effects entering into the Avrami expression (see also the discussion of copolymer crystallization, Sect. 6.3.4 and melting, Chapter X).

Crystallization from concentrated solution of macromolecular solvents has been analyzed in the systems isotactic polystyrene–atactic polystyrene and poly-ε-caprolactone–poly(vinyl chloride). In the first case, the glass transition temperature of both components is about the same; in the latter, increasing amounts of poly(vinyl chloride) increase the glass transition temperature continuously from the value of pure (amorphous) poly-ε-caprolactone (−71°C) to that of poly(vinyl chloride) (+82°C). In the polystyrene case Yeh and Lambert (1972)† observed on a 550 000 MW isotactic sample with various ratios of atactic polymer between 900 and 1 800 000 MW that the linear growth rate observed by optical microscopy decreased almost linearly with increasing noncrystallizable atactic polystyrene (see also Table VI.8). The maximum in the linear growth rate versus crystallization temperature curves occur all in the same temperature region (178–183°C) as in undiluted isotactic polystyrene. The molecular weight of the atactic solvent had a smaller and somewhat irregular effect on the growth rate, despite the fact that melting experiments showed considerable diluent effects of the lower molecular weights (below 19 000), indicating changes in the degree of supercooling of perhaps up to 10–30°C. Changes in the morphology may, however, obscure the various effects.

The system poly-ε-caprolactone–poly(vinyl chloride) is one of the rare polymer systems miscible over the whole concentration range. The changes in overall crystallization rate were measured by Robeson (1973) on a 40 000 MW poly-ε-caprolactone mixed in various proportions with the polymeric solvent. Measurements of the tensile modulus were correlated to the Avrami crystallization expressions [Eq. (33)]. Induction times deduced from modulus–log time plots increased approximately logarithmically with increasing solvent concentration (4.3 min for a 70/30 and 4 900 min for a 40/60 solution). The experimental data could be represented approximately assuming a linear crystallization rate decrease with increasing solvent concentration as observed for the polystyrene case and a change in the transport term ΔG_η which accounts for the change in the glass transition temperature of the initial solution (see Eq. (6), Chapter V, and footnote on p. 9).

6.2.2 Effect of Stirring

In the previous section crystallization from quiescent solutions was discussed. Stirring changes the morphology of the obtained crystals from lamellar

† For earlier work on this system see Keith and Padden (1964). *J. Appl Phys.* **35**, 1286; and Boon and Azcue (1968). *J. Polym. Sci. Part A-2*, **6**, 885.

single crystalline or dendritic (see Sects. 3.3.2 and 3.6.1) to fibrillar (see Sect. 3.8.2). The nucleation of crystallization under such conditions is discussed in Sect. 5.1.4.5.

The kinetics of crystallization of polyethylene from stirred solution has been studied by Pennings and Pijpers (1970). Figure VI.45 illustrates the

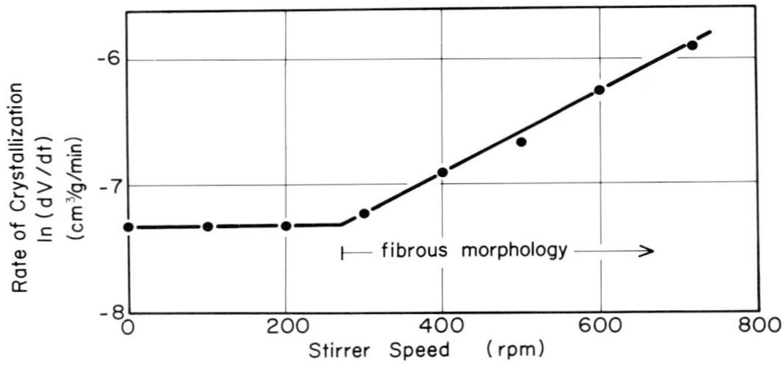

Fig. VI.45 *Logarithm of Crystallization Rate of Polyethylene Grown with Stirring.*
Fraction of 7.2×10^5 MW, 1% solution in xylene, rate at the 50% crystallization time at rates of stirring indicated. Couette-type dilatometer with magnetic internal stirring. Crystallization temperatures 91.0°C. Data by Pennings and Pijpers (1970).

dilatometric data on a fraction of 7.2×10^5 MW at various stirring speeds. Below 270 rpm no fibers were observed and crystal growth is independent of stirring. Similarly, a too low molecular weight will show no effect of stirring at any stirring rate. Above 270 rpm at 91°C the logarithm of the crystallization rate increases linearly. At the same time the fibrillar crystals are observed. The appearance of the shish-kebobs was linked with the onset of formation of Taylor vortices necessary to induce extensional flow needed for nucleation (see Sect. 5.1.4.5).

The separation of nucleation and crystal growth of shish-kebob was accomplished by Pennings *et al.* (1973) through the study of seeding. While for the initiation of crystallization at higher temperatures than possible under quiescent conditions it was necessary to have extensional flow, seeds inserted in flowing solution would continue to grow under conditions of simple shear flow. Only a small minimum velocity gradient is necessary for temperatures above the limit of growth under quiescent conditions. Tips of several fibers are shown in Fig. VI.46. The taper observed is related to the growth mechanism (see also Sect. 3.8.2, Fig. III.137). Figure VI.47 shows schematically the proposed mechanism of their growth (Pennings *et al.*, 1970). After nucleation of a fibril (or seeding with a preformed fibril) longitudinal growth will take

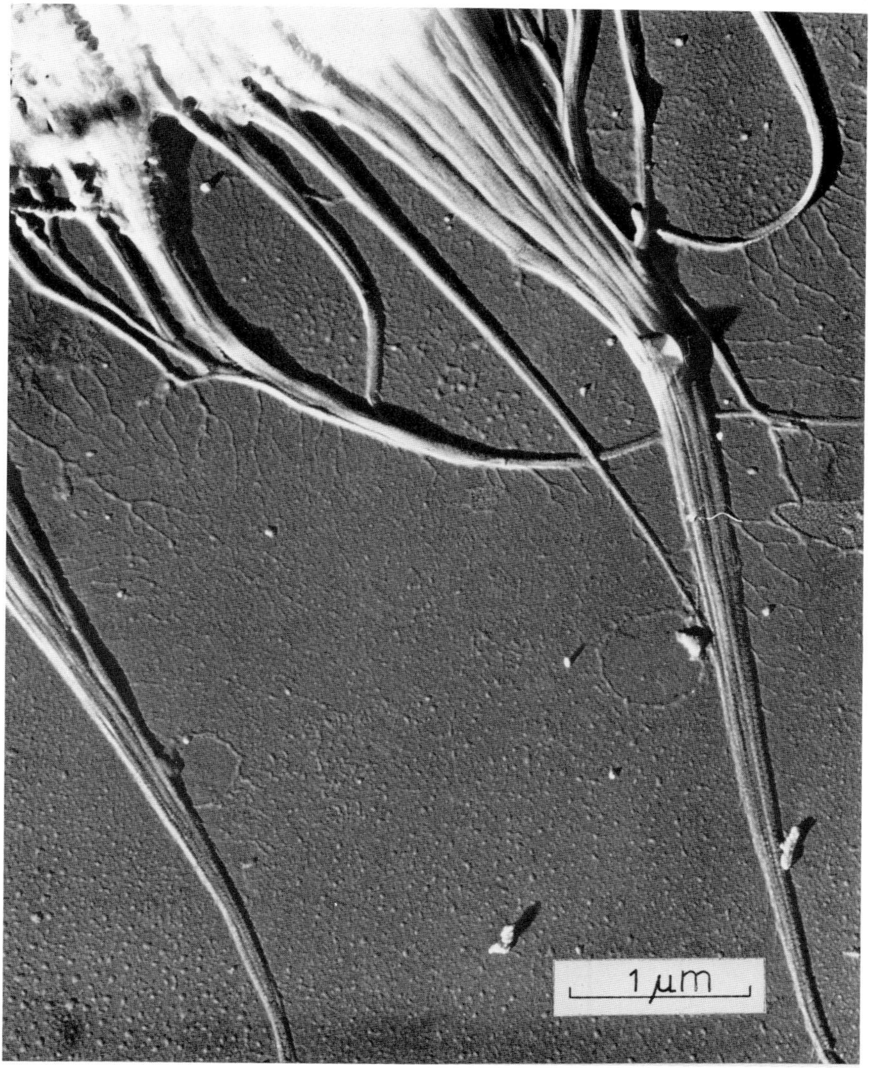

Fig. VI.46 *Electron Micrograph of Fiber Ends of Stirrer Crystallized Polyethylene.* Couette-type instrument. Micrograph by Pennings *et al.* (1970).

place by trapping of new molecules on the (110) growth faces. The free ends of the molecules reach into the solution where they are oriented and deformed. The loss in entropy associated with this process is the principle factor for the

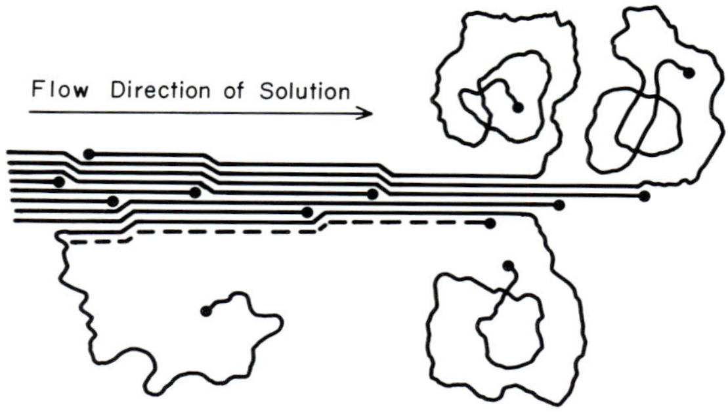

Fig. VI.47 *Schematic Drawing of the Process of Growth of Fibrous Crystals under Shear.*
The circles indicate chain ends. Drawn after Pennings *et al.* (1970).

accelerating action of the flowing solution. Measured longitudinal growth rates (along the chain direction) fitted a growth rate with a critical nucleus free enthalpy

$$\Delta G^* = 4c_0 \gamma^2/\Delta g_f \tag{92}$$

[compare for example with Eq. (68), Chapter V]. A typical growth rate of 8 cm/min was observed at a velocity gradient of 0.7/sec from xylene solution at 104°C. An expression like Eq. (92) would be expected for crystallization governed by the formation of new secondary nuclei of thickness c_0, the unit cell length in the chain direction. Such growth should be classified as fringed micellar growth (see Sect. 6.1.7). The geometry of the taper of Fig. VI.46 permits the computation of the growth rate at right angles to the chain axis which is governed by the rates of attachment of new molecules.

An alternate mechanism of formation of the shish-kebob structure, which is based on deformation of multilayer crystals (see Sect. 3.3.3) after initial growth, has been proposed by Nagasawa and Shimomura (1974). McHugh and Schultz (1973) in another mechanism suggests that fluctuating axial–radial fluid strains are the cause for shish-keboblike morphology.

6.2.3 Effect of Pressure

Crystallization of polyethylene from the melt under sufficiently elevated pressure (see Sect. 6.3.3) leads to extended chain crystals (Sect. 3.3.1) with a broad distribution of thickness in the molecular chain direction. Similar experiments using 0.05 wt. % solutions in toluene were carried out by Wunderlich (1963) up to 6.0 kb. Contrary to the observation of increasing lamellar thickness in melt crystallization, practically no change of the morphology was observed at constant supercooling on crystallization from solution at various pressures (see Sect. 3.2.2.2, Fig. III.14). These results have been linked to the observation of practically no crystal perfection on crystallization from dilute solution (Sect. 6.2.1) and the mechanism of thickening on crystallization from the melt under elevated pressure (Sect. 6.3.3). The reason that thin crystals grow from solution should thus be the relatively higher initial perfection of the folded chain crystals which prohibits solid state perfection.

The concentration dependence of the formation of extended chain crystals of polyethylene was studied by Treiber *et al.* (1973a). Figure VI.48

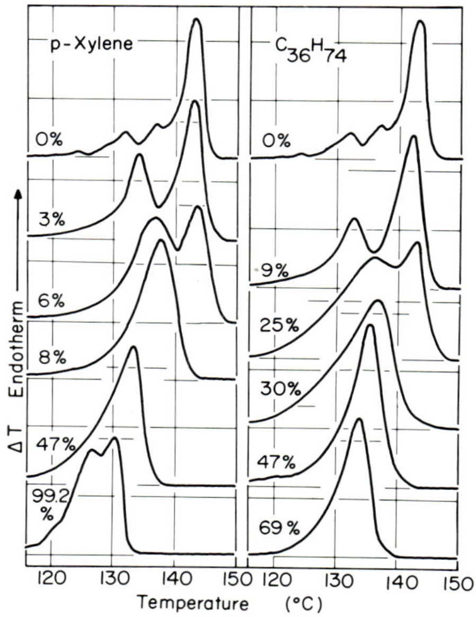

Fig. VI.48 *Differential Thermal Analysis Curves of Polyethylene Crystals Grown with Various Amounts of Solvent at 4.2 kb Pressure.* Concentrations in wt.%. The highest melting peak corresponds to extended chain crystals, low molecular weight crystals and folded chain lamellae of various degrees of perfection melt at lower temperatures. Crystallization on cooling at 4°C/hr from 220°C at 4.2 kb pressure. Analysis by DTA at 10°C/min heating rate. Drawn after Treiber *et al.* (1973).

illustrates the changes in the resulting crystals through their melting characteristics as determined by differential thermal analysis. The presence of even small amounts of a solvent causes a portion of the crystals to be less perfect.

For xylene–polyethylene crystallized at 4.2 kb it is sufficient to add 8 wt. % solvent to grow no high melting extended chain crystals. Electron microscopy confirmed the thermal analysis. Using hexatriacontane as a solvent leads to similar results, but the concentration for no extended chain crystal growth is now at about 30 wt. %. Converting the concentrations of both solvents into mole ratios (solvent molecules to repeating units) shows in both cases a limit of about 0.01 for the growth of extended chain crystals. The high solvent concentration case for the toluene-polyethylene system reaches into the dilute solution region (see above) and exhibits again a double melting peak which is explained by the formation of relatively perfect, thin lamellae which reorganize less during the heating process than the initially poorer crystals melting finally at the higher temperature (see Chapter IX).

6.3 Melt Crystallization of Macromolecules

Crystallization from the molten state is discussed in this section based on experimental data. In Sect. 6.3.1 crystal growth data from the undisturbed melt are analyzed. The effect of orientation in the melt on crystallization is treated in Sect. 6.3.2. A large body of literature has been generated in order to explain the growth of extended chain crystals of polyethylene under elevated pressure. This work is reviewed in Sect. 6.3.3 and broadened to other macromolecules. Finally, in Sect. 6.3.4 some information on the crystallization of copolymers is collected.

There is still little known about the overall process of transport of crystallizable molecules or segments to the crystallization site, their adsorption and possible mobility on the crystal surface before final crystallization, the sequence of steps suggested for crystal growth of small molecules from the vapor phase (see Sect. 6.1.2.2, Fig. VI.4). Also, the overall questions of how much of the macroconformation of the molecules in the final crystal is already preformed in the melt and how molecules are entangled or possibly segregated according to molecular weight or chemical composition remains speculative. All these problems are under active investigation (see, for example, Sect. 6.1.7).

6.3.1 Normal Growth

The crystallization kinetics of macromolecules from the melt has found much attention in the past. It was discovered early that crystal growth from the melt had to be nucleation controlled because of its strong negative temperature coefficient (see Fig. V.37, Mandelkern *et al.*, 1954; Mandelkern, 1958). The derivation of evidence for the detailed mathematical form to decide between primary or secondary nucleation control was more difficult, as was pointed out by Frank (1958). For experiments on most polymers both expressions, that for primary nucleation [Eq. (21), Chapter V] and that for secondary nucleation [Eq. (68), Chapter V] can equally well be fitted to the experimental data (see for example Mandelkern *et al.*, 1968). A wide enough range of crystallization temperatures was finally measured on poly(tetramethyl-*p*-silphenylene siloxane) by Magill (1969) to decide clearly in favor of a secondary nucleation (but note also earlier data on nylon 11, Table VI.8). Furthermore, the discussion of molecular nucleation (see Sect. 5.3) makes it likely that molecular nucleation rather than secondary crystal nucleation is the rate determining step in crystal growth from the melt [Eq. (82), Chapter V].

The description of melt crystallization appears initially simpler than crystallization from solution (Sect. 6.2) since the thermodynamic driving force should stay constant throughout the crystallization. Even in mixtures of different molecular weights, changes in melting temperature and chemical potential of the melt with changing composition due to segregation are small as long as the sample contains no really low molecular weight fractions (below 10 000 MW; but see the larger effects of molecular weight on growth rate, Sect. 6.1.5). A remaining problem to the determination of the thermodynamic driving force is the knowledge of the equilibrium melting temperature T_m^0 (see Chapter VIII). Since normally grown crystals are metastable, equilibrium melting temperatures are frequently not available experimentally (see Chapter IX). Too low temperatures are often reported due to small crystal size, and too high values are possible due to disregard of superheating. The uncertainty in T_m^0 may reach as much as 5–20°C. Naturally, this greatly influences other adjustable parameters in the overall equation for crystal growth [see, for example, Eq. (72)]. Often, values for T_m^0 have been derived by trial and error on the assumption that certain nucleation expressions must be correct (see, for example, Mandelkern *et al.*, 1968).

A series of data on spherulitic growth rates on polyethylene, isotactic polypropylene, poly(ethylene oxide), and poly(ethylene sebacate) by Asaubekov *et al.* (1972) from the melt and various diluent concentrations indicate that at constant supercooling the investigated polymers crystallize faster from the melt than from solution. For poly(ethylene sebacate), for example, a ratio of almost 10^5 is observed for the ratio of melt to solution crystallization (2 wt. % in xylene) at constant supercooling (20–30°C). Faster crystallization from the melt can also be derived from data on polyethylene presented by Hoffman *et al.* (1969, Fig. 7). At 15–25°C, supercooling the linear crystallization rate from the melt is about ten times as large as from solution (0.001 wt. % in xylene). The linking of these differences to specific factors in the growth rate equation [Eq. (72), for example], however, is not quite certain. Concentration changes alone are not sufficient to explain the differences. Lower values of $\gamma\gamma_e$

have been reported for polypropylene and poly(ethylene oxide) crystallizing from solution (see Asaubekov *et al.*, 1972), but Jain (1969) reports no change for similar poly(ethylene oxide) experiments. The polyethylene data discussed above show also only little difference in $\gamma\gamma_e$. Similarly, different activation energies for the transport term have been assumed in some cases, while in others no change has been proposed for the same macromolecule. The main reasons for these different interpretations may be sought in the frequently too narrow temperature range of analysis to separate all the factors influencing the growth rate, the uncertainty in the equilibrium melting and dissolution temperatures, and major imperfections in the idealized theory applied to as largely different phase changes as solution and melt crystallization (see also Sect. 6.1.7).

The first detailed study of isothermal crystallization of a linear macromolecule was done by Bekkedahl (1934) on *cis*-1,4-poly(2-methylbutadiene). A series of additional experimental data on melt crystallization experiments on some thirty different macromolecules is summarized in Table VI.8, and some of the linear growth rate data are plotted in Fig. V.37. A more detailed description of results on polyethylene, polypropylene, poly(ethylene oxide), poly(ethylene terephthalate) and nylon 6 is given below.

6.3.1.1 Polyethylene

The long series of experimental data on the kinetics for the melt crystallization of polyethylene was started with the work of Kovacs (1955).† For low density polyethylene (MW = 15 000) he observed by dilatometry an Avrami exponent of 1.0 [Eq. (33)] after an induction time which varied with crystallization temperature (100–112°C). In addition to the fast exponential crystal growth, he observed a slow logarithmic decrease in volume of 4.0–4.65 $\times 10^{-3}$ cm^3/g for each decade of time which he followed up to 1000 hr (see Sect. 6.1.6). The most extensive dilatometric work was carried out by Ergoz *et al.* (1972) on seventeen fractions of linear polyethylene covering the molecular weight range from 4200 to 8×10^6 and the temperature range of 125–132°C. All samples were heated to 160°C to destroy all self-nuclei (see

† For other measurements see also: Buchdahl *et al.* (1959). *J. Polym. Sci.* **36**, 215; Buckser and Tung (1959). *J. Phys. Chem.* **63**, 763; Matsuoka (1960). *J. Polym. Sci.* **42**, 511; Price (1960). *J. Phys. Chem.* **64**, 169; Rybnikar (1960, 1963) *J. Polym. Sci.* **44**, 517 and *Part A***1**, 2031; Banks *et al.* (1963). *Polymer* **4**, 61, 281; Heber (1964). *J. Polym. Sci. Part A* **2**, 1291; Keith and Padden (1964). *J. Appl. Phys.* **35**, 1286; Lindenmeyer and Holland (1964), *J. Appl. Phys.* **35**, 55; Banks *et al.* (1964). *Polymer* **5**, 163; Gordon and Hillier (1964, 1965). *Trans. Farad. Soc.* **60**, 763, *Phil. Mag.* **11**, 31; Godovskii and Barskii (1966), *Vysokomol. Soed.* **8**, 395, *Polym. Sci. USSR* **8**, 431; Iwanami *et al.* (1972). *Kobunshi Kagaku* **29**, 139; Stafford (1973). *J. Polym. Sci. Symposium* **42**, 837; Era and Venäläinen (1973). *J. Polym. Sci. Sympos.* **42**, 879.

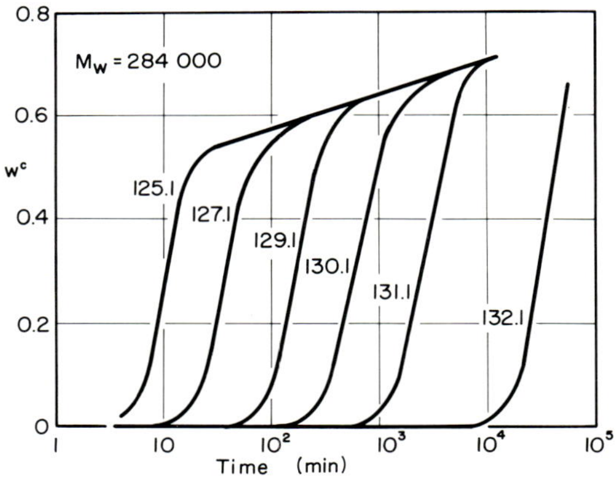

Fig. VI.49 *Plot of the Weight Fraction Crystallinity of 284 000 MW Polyethylene as a Function of Crystallization Time at the Indicated Temperatures in °C.* Dilatometric data after heating for 80 min at 160°C. Drawn after data by Ergoz *et al.* (1972).

Fig. V.27). Figure VI.49 illustrates the change in crystallinity with time for the 284 000 MW fraction for the different crystallization temperatures. All curves seem to reach a common linear asymptote, a time-dependent pseudo-equilibrium value which is independent of the crystallization temperature. All curves for any of the fractions can be brought to superposition at least for the early part of crystallization† by shifting along the log t axis as shown in Fig. VI.50. Also shown in the figure are curves calculated with the Avrami expression [Eq. (33); see also Fig. VI.10].‡ Up to 6000 MW the curves are well represented with a coefficient $n = 4$; this is followed up to 1.2×10^6 MW by a fit with a coefficient $n = 3$; the highest molecular weights, finally, are best represented by a coefficient $n = 2$. The deviations from these curves start much earlier for the higher molecular weights. Ergoz *et al.* (1972) noted that this point of deviation from the Avrami expression occurred at about the same level of crystallinity where the free growth approximation which disregards impingements of crystals (see Sect. 6.2.1) could no longer represent the data. One could conclude from these results that by the time impingement becomes of importance, polyethylene crystallization does not fit the overall

† Complete superposition is usually possible for crystallization temperatures below 129°C only.

‡ Note that the data are not fitted to a maximum degree of crystallinity (as, for example, Fig. VI.11). Thus, deviations from the Avrami expression occur much earlier, but seemingly better agreement with whole numbered exponents n is reached.

TABLE VI.8

CRYSTALLIZATION OF MACROMOLECULAR MELTS[a]

Macromolecule	Avrami exponents	Description	References
Polyethylene CH_2—	1–4 and fractional	See Sect. 6.3.1 for details; nucleation data Sects. 5.1.2.2, 5.1.3.3 and 5.1.4.2 (Tables V.2 and V.4) (Figs. V.13, V.14, V.27 and V.29); dilatometric overall growth rate data, Figs. VI.49–VI. 52; linear crystal growth rate data, Figs. VI. 53 and V.37; for segration on crystallization, see Sect. 5.3.3 (Table V.6 and Fig. V.49).	*See Sect. 6.3.1.1.*
Polytetrafluoroethylene CF_2—		Optical microscopy on cooling at 0.5°C/sec from above T_m; linear crystal growth rates 12.4 μm/sec at 316.5–317°C.	*Katz (1970). J. Polym. Sci.* ***B8***, *789.*
Selenium Se—	1.7–4.0	Overall rates and linear growth rates; heterogeneous athermal nucleation; for T_c 70–160°C the Avrami coefficient changes from 2.4 to 1.7 (100°C) to 4.0; two successive processes above 120°C, second with an Avrami coefficient 0.6–1.5; linear growth rates 80°C: 0.0103 μm/min, 130°C: 0.350 μm/min (max), 160°C: 0.091 μm/min; secondary crystallization may involve chain rearrangement (see Sect. 6.4.2.4 for details).	*Crystal (1970). J. Polym. Sci. A2,* ***8***, *2153.*

Polypropylene, isotactic CH_2-CHCH_3-	3–4	See Sect. 6.3.1.2 for details; nucleation data Sects 5.1.2.2, 5.1.3.3 and 5.1.4 (Tables V.2 and V.3, Figs. V.21 and V.22); for dilatometric and X-ray data on overall crystallization see Figs. VI.54 and VI.55; linear crystal growth rate data Fig. V.37.	*See Sect. 6.3.1.2.*
Poly-1-butene, isotactic $CH_2-CHC_2H_5-$	3–4	For nucleation data see Sects. 5.1.2.2 and 5.1.4.2 (Tables V.2 and V.4, Figs. V.28 and V.29); dilatometric data for melt crystal form II are reported from 95–110°C with fractional n within the limits shown; transition rates of form II to I have a $t_{1/2}$ of 3900 min at 40°C and 1475 min at 10°C; linear crystal growth rates were constant throughout the measurement, data from 90–107°C.	*Powers et al. (1965). J. Res. Nat. Bur. Stand.* ***69A****, 335, Sastry and Patel (1972). Eur. Polym. J.* ***8****, 63.*
Polystyrene, isotactic $CH_2-CHC_6H_5-$	~3	For nucleation data see Tables V.2 and V.4, Sect. 5.1.4.3 and Fig. V.30; from 180–210°C n was practically constant; optical microscopy shows initially increasing numbers of nuclei with time ($<$40 min), then constant (145–204°C); the initial crystallization followed in accord with this observation an initial crystallinity increase with $n = 4$; a slow log t proportional process seems to occur at the end of crystallization; for linear crystallization rate data see also Fig. V.37.	*Hay (1965). J. Polym. Sci. Part A,* ***3****, 433.*

Continued

TABLE VI.8 *(continued)*

Macromolecule	Avrami exponents	Description	References
		Mixtures of isotactic and atactic polymers show an almost linear decrease of growth rate with atactic polymer concentration without shift in the maximum crystallization rate (see also Sect. 6.2.1).	*Yeh and Lambert (1972). J. Polym. Sci. Part A–2* ***10***, *1183.*
Poly(vinyl chloride), partially syndiotactic $CH_2-CHCl-$	—	Standard poly(vinyl chloride) may develop 10–20% crystallinity which shows no clear spherulites for nucleation and crystallization rate measurement; also, an Avrami type analysis was not carried out; kinetics are reported for crystallization from 90°C to 170°C; the maximum crystallization rate seems to be at about 130°C.	*Kovacs (1956). C. R. Acad. Sci. Paris,* ***243***, *50; Rybnikar (1970). Makromol. Chem.* ***140***, *91.*
Polychlorotrifluoroethylene $CF_2-CFCl-$	3	For nucleation see Sects. 5.1.2.2, 5.1.3.3 and 5.1.4 (Table V.4), heterogeneous athermal above 180°C, 166–180°C additional thermal nucleation; the overall crystallization data can generally be fitted with a simple Avrami expression up to 80% of the total crystallization which is commonly 0.5–0.7; a slow secondary crystallization and crystal perfection process occurs at long times; for linear crystal growth data see also Fig. V.37.	*Hoffman and Weeks (1962). J. Chem. Phys.* ***37***, *1723; see also Price (1952). J. Amer. Chem. Soc.* ***74***, *311 and Rybnikar (1962). Coll. Czech. Chem. Commun.* ***27***, *2307.*

trans-1,4-Polybutadiene $CH_2{-}CH{=}CH{-}CH_2{-}$	2.2±0.4	The molecular weight dependence was measured by dilatometry (2×10^4–3×10^5).	*Fernandez-Bermudez et al. (1970). An. Quim.* ***66****, 779.*
cis-1,4-Poly(2-methylbutadiene) $CH_2{-}CCH_3{=}CH{-}CH_2{-}$	3	Dilatometry and microscopy show similar variation of overall and linear crystal growth rate with temperature; a maximum rate is found at −20 to −26°C (measurements were carried out between 0 and −40°C); athermal heterogeneous nucleation is assumed; a slow secondary crystallization or crystal perfection has been observed; for the effect of strain which affects mainly nucleation see Sects. 5.1.4.5 and 6.3.2.	*Andrews et al. (1971). Proc. Roy. Soc. London,* ***A324****, 79; Gent (1955) J. Polym. Sci.* ***18****, 321.*
trans-1,4-Poly(2-methylbutadiene) $CH_2{-}CCH_3{=}CH{-}CH_2{-}$	3	Crystallization occurs in two crystal forms, the low melting β form (see Table II. 13) and the high melting γ form (see Sect. 2.4.3); below 40°C only β can be found, above, increasing γ is observed; the Avrami overall rate constant K [see Eq. (33)] changes its temperature dependence at 40°C; primary nucleation and crystal growth are different for the polymorphs, primary nucleation of β is always faster, but linear crystal growth of γ is faster at the temperatures of observation of both polymorphs (when referred to equal supercooling, however, β linear crystal growth rates are	*Fischer and Henderson (1967). J. Polym. Sci. Part A–2,* ***5****, 377.*

Continued

TABLE VI.8 *(continued)*

Macromolecule	Avrami exponents	Description	References
		higher); overall and linear crystal growth rate data could be made to agree.	
		Increasing molecular weight decreases the linear crystal growth rate of the β form from 10^4 to 10^5, after which little change was observed up to 10^6 (see Sect. 6.1.5).	*Lovering (1970). J. Polym. Sci. Part C*, ***30***, *329.*
trans-1,4-Poly(2-chlorobutadiene) $CH_2{-}CCl{=}CH{-}CH_2{-}$	4	The dielectric constant was measured from -17°C to $+21$°C for a sample of finally 0.25 crystallinity; crystallization isotherms could be superimposed when shifted along log t; only small deviations from the Avrami exponent occur at the end of crystallization	*Simek and Müller (1969). Kolloid Z. Z. Polym.* ***234***, *1092 (see there for earlier references).*
Polyoxymethylene $CH_2{-}O{-}$	2	Usually heterogeneous, athermal nucleation is reported, for homogeneous nucleation see Table V.2; dilatometry is complicated due to decomposition; superposition was possible for data from 140°C to 160°C; the linear crystal growth rates reported by different authors vary, but recent, more precise data show a break when plotted versus $1/\Delta T$ at 158–160°C;	*Inoue and Takayanagi (1960). J. Polym. Sci.* ***47***, *498; Rybnikar (1966) Collect. Czech. Chem. Commun.* ***31***, *4080; Pelzbauer and Galeski (1972). J. Polym. Sci. Part C* ***38***, *23, see also Mihajlov and Nedkov (1972). J. Polym. Sci. Part C* ***38***, *33 for*

		below, the slope indicates about half the surface free energy product as is also shown for polyethylene (Fig. VI.53) and suggested by Eqs. (69) and (71); spherulites grow below 158°C, hedrites above.	*comparison of linear crystal growth rate data.*
Poly(ethylene oxide) $CH_2{-}CH_2{-}O{-}$	2 ± 0.6	See Sect. 6.3.1.3 for details; for nucleation data, see Sect. 5.1.4.2 (Table V.2 and V.4, Fig. V.29); linear crystal growth rate data depend on macroconformation in the crystal (see Fig. VI.23 and Table VI.4); crystal perfection is described in Sect. 6.1.6 (Fig. VI.27); the molecular weight dependence of the crystal growth rate is discussed in Sect. 6.1.5.	*See Sect. 6.3.1.3*
Poly(propylene oxide), isotactic $CH_2{-}CHCH_3{-}O{-}$	2.0–2.6	Samples of a weight fraction tacticity 0.93 to 0.98 with number average isotatic sequences of 15–55 repeating units and 5000 average degree of polymerization were studied dilatometrically and by optical microscopy at 40–55°C; an Avrami exponent of 3 fits usually the first 60% of crystalization, the deviations from this value are linked to 17–30% of secondary crystalization inside the spherulite boundaries; increasing tacticity yields faster linear crystal growth rates; for data see Fig. V.37.	*Booth et al. (1970). Polymer* ***11****, 11.*

Continued

TABLE VI.8 *(continued)*

Macromolecule	Avrami exponents	Description	References
Poly[3,3-bis(chloromethyl) oxacylobutane] $CH_2-C(CH_2Cl)_2-CH_2-O-$	1.7–3.3	The crystallization of the α crystal form was followed from 140–160°C by dilatometry, the Avrami exponents were fractional and varied with crystallization temperature, but described the first half of crystallization well; some linear growth rates are available; the maximum in crystallization rate accurs between 40°C and 140°C, too fast to be measured by these authors.	*Hatano and Kambara (1961). Polym.* **2**, *1; Inoue (1962) J. Polym. Sci.* ***61***, *343.*
		Crystallization from the glassy state leads to the β-polymorph; for nucleation see Table V.2.	*Collier, Thesis, Case Inst. Tech. (1966). See also Sifleet et al. (1973). Polym. Eng. Sci.* ***13***, *10.*
Polycaprolactone $(CH_2-)_5-CO-O-$	3	Measurements of dilatometry, optical microscopy, and X-ray diffraction are available in the temperature range 40–57.8°C for molecular weights varying between 700 and 22 000; at 45, 48, and 54°C the above listed Avrami exponent was observed; agreement between calculation from nucleation and linear growth rate data by optical microscopy and measured dilatometry were found up to the late stages of crystallization where secondary crystallization increases the crystal-	*Goldfarb et al. (1974). Makromol. Chem.* ***175***, *2483; Remi and Skoulious (1972). Makromol. Chem.* ***156***, *157.*

		linity to higher values than expected from microscopy; the slow increase in crystallinity was followed up to 1000 h; extended chain crystals observed for low molecular weight samples changed smoothly into folded chain crystals at higher molecular weights (120–165Å fold-length); only a small change in fold-length was observed on changing the crystallization temperature from 20°C to 55°C.	
Poly(ethylene succinate) $CH_2-CH_2-O-CO-$ $CH_2-CH_2-CO-O-$	3	Various samples of 820–4400 MW were analyzed by dilatometry and microscopy; the nucleation was shown by droplet and melt observation to be mainly heterogeneous between 15°C and 50°C; initially the number of nuclei increased rapidly with time to reach than a final level determined by temperature; the nucleation rates increased with molecular weight; the linear growth rates increased from −10°C to 30–35°C (maximum) to decrease again to 80°C, the maximum temperature used; increasing molecular weight decreased the rates but kept the maximum rate at the same temperature; for 1450 MW, Avrami exponents could be determined for crystallization below 10°C and above 60°C; below 10°C, $n = 3.0 \pm 0.2$; above 60°C after about 60% crystallization, a change from $n = 2.5$–3.0 to $n = 1.4$.	*Ueberreiter and Steiner (1966). Macromol. Chem.* ***91****, 175; Steiner et al. (1966). Kolloid Z. Z. Polym.* ***214****, 23; Ueberreiter et. al. (1955). J. Polym. Sci.* ***16****, 53; see also Takayanagi and Kusomoto (1959). Kogyo Kagaku Zasshi* ***62****, 587.*

Continued

TABLE VI.8 *(continued)*

Macromolecule	Avrami exponents	Description	References
Poly(ethylene adipate) $CH_2-CH_2-O-CO-(CH_2)_4-CO-O-$	—	Fractions of 1400–9900 MW have been crystallized between 15°C and 55°C and their linear crystalline growth rates measured by optical microscopy (see Fig. V.37 for data); the fractions show discontinuities at the temperatures (molecular weight dependent) where the spherulites change from negative, simple to alternating, ringed and finally to positive, simple; maximum growth rate at about 20–30°C.	*Takayanagi and Yamashita (1956). J. Polym. Sci.* ***22****, 552, see also (1957) Memoirs Fac. Eng. Kyushu Univ.* ***16****, 111.*
Poly(ethylene terephthalate) $CH_2-CH_2-O-CO-C_6H_4-CO-O-$	2–4	See Sect. 6.3.1.4 for details; nucleation has been observed to be thermal and athermal, for self-nucleation see Sect. 5.1.4; the linear crystal growth rate as function of temperature and molecular weight is described by Eq. (93) (Fig. V.37); some data are shown in Table VI.11; Avrami coefficients as low as $n \leq 1$ have been reported.	*See Sect. 6.3.1.4.*
Poly(tetramethylene terephthalate) $(CH_2-)_4O-CO-C_6H_4-CO-O-$	2.6–2.8	Half-times of crystallization, overall crystallization rates, and linear crystal growth rates of a 0.32 crystalline polymer of 13 400 MW have been determined from 208.8 to 214.9°C ($T_m \approx 260$°C); at equal supercooling the melt crystallizes faster than the	*Gilbert and Hybart (1972), Polymer* ***13****, 327.*

		ethylene glycol ester.	
Poly(tetramethylene isoterephthalate) $(CH_2-)_4-O-CO-C_6H_4-CO-O-$	2.5	Half-times of crystallization, overall crystallization rates, and linear crystal growth rates of a 0.13 crystalline polymer of 12 800 MW have been determined from 78°C to 109.7°C ($T_m \approx 185$°C); the maximum rate of crystallization occurs at about 60°C supercooling ($t_{1/2} = 200$ min); the crystallization is the slowest of all terephthalates analyzed.	*Gilbert and Hybart (1972), Polymer* ***13****, 327.*
Poly(pentamethylene terephthalate) $(CH_2-)_5O-CO-C_6H_4-CO-O-$		Half-times of crystallization and linear crystal growth rates have been determined (10 600 MW); no constant Avrami exponent was obtained; the maximum crystallization rate occurs at 70°C supercooling and is lower in temperature and magnitude than that of poly(ethylene terephthalate).	*Gilbert and Hybart (1973), Polymer* ***13****, 327.*
Poly(hexamethylene adipate) $(CH_2-)_6-O-CO-(CH_2-)_4CO-O-$	~5	At 51°C, this polymer has been reported to crystallize for the first 10% with the Avrami exponent given; a branching mechanism has been suggested.	*Booth and Hay (1972). Brit. Polym. J.* ***4****, 9.*
Poly(hexamethylene terephthalate) $(CH_2-)_6-O-CO-C_6H_4-CO-O-$	2.9	Half-times of crystallization and overall crystallization growth rates have been determined for a 0.38 crystalline polymer of 11 800 molecular weight ($T_m \approx 186$°C); the overall crystallization is faster than found for poly(ethylene terephthalate).	*Gilber and Hybart (1972). Polymer* ***13****, 327.*

Continued

TABLE VI.8 *(continued)*

Macromolecule	Avrami exponents	Description	References
Poly(decamethylene adipate) $(CH_2-)_{10}O-CO-(CH_2)_4CO-O-$	3, 4	Melts of 11 500 MW were analyzed by dilatometry between 60°C and 74°C; the data were analyzed assuming constant nucleation rate and two- or three-dimensional crystal growth.	*Mandelkern et al. (1954). J. Appl. Phys.* **25**, *830.*
Poly(decamethylene sebacate) $(CH_2-)_{10}O-CO-(CH_2)_8-CO-O-$	4	Dilatometric and microscopic data on 10^4 MW samples are available from 67°C to 73°C; thermal nucleation; for linear crystal growth rates see Fig. V.37; calculated Avrami constants K (Eq. 33) from microscopy and measured constants from dilatometry agreed within experimental error.	*Flory and McIntyre (1955). J. Polym. Sci.* **18**, *592; see also Mandelkern (1958).*
Poly(decamethylene terephthalate) $(CH_2-)_{10}O-CO-C_6H_4-CO-O-$	2.6–3.8	Dilatometry and optical microscopy were carried out on a sample of above 10 000 molecular weight from 118–126°C; no extended secondary crystallization processes were observed; a discontinuity in the volume–time curve interrupts the good fit to an Avrami exponent of fractional value between 50 and 98% of crystallization, a plot of the rate constant K [Eq. (33)] or the total volume change on crystallization versus temperature reveal a transition at about 122.8°C; nucleation was found to be thermal (constant rate) which leveled off to zero in the later stages of crystallization (see Sect.	*Sharples and Swinton (1963). Polymer* **4**, *119.*

		5.1.2.2); combination of linear crystal growth rate and nucleation rates led to approximate agreement with dilatometry; two-step Avrami processes (see Sect. 6.1.3.3) do not improve the data treatment.	
Polypivalolactone $CH_2-C(CH_3)_2-CO-O-$	~2.5	Four samples of molecular weight 71–241 × 10^3 (polydispersity 2.3–3) were analyzed by thermal analysis between 175 and 200°C; the overall crystallization rate decreased with increasing molecular weight; the fractional Avrami exponent fitted the data well.	*Borri et al. (1971). Eur. Polym. J.* ***7****, 1515.*
Poly(4,4′-isopropylidenediphenylene carbonate) $C_6H_4-C(CH_3)_2-C_6H_4-O-CO-O-$	1–3.7	Crystallization of five samples between 3 × 10^4 MW and 7 × 10^4 MW was measured at 190°C; the total degree of crystallinity was between 18% and 28%; the samples exhibited long induction times; Avrami expressions could be fitted to all data but varied greatly for the different samples and for different pretreatment (dissolution and reprecipitation); optical microscopy revealed feathery spherulites, nucleation was athermal, at 90°C the linear crystal growth was about 0.005 μm/min; ploy(4,4′-methylenediphenylene carbonate) and poly(4,4′-thiodiphenylene carbonate) were also analyzed by optical microscopy and were found to crystallize considerably faster.	*Turska et al. (1968). J. Polym. Sci. Part C* ***16****, 3373; B. von Falkai and Rellensmann (1964). Makromol. Chem.* ***75****, 112.*

Continued

TABLE VI.8 *(continued)*

Macromolecule	Avrami exponents	Description	References
Nylon 6 $(CH_2-)_5CO-NH-$	2–4	For details see Sect. 6.3.1.5; homogeneous nucleation occurs only below 128°C (see Table V.2); self-seeding and heterogeneous nucleation is likely in most cases (athermal and thermal) (see Table V.4, Sect. 5.1.3.3); for linear crystal growth rates see Fig. V.37; overall growth rates show Avrami coefficients which vary with temperature and molecular weight (see Table VI.12).	*See Sect. 6.3.1.5.*
Nylon 8 $(CH_2-)_7CO-NH-$	5–6	Samples of average from 9 400 MW to 44 500 MW were analyzed by dilatometry, differential thermal analysis and microscopy between 182°C and 191°C; kaolin and lead phosphate nucleation had little effect on the Avrami exponent, but increased the rate somewhat; higher molecular weights crystallize slower, but have similar Avrami exponents.	*Ceccorulli and Manescalchi (1973). Makromol. Chem.* ***168****, 303.*
Nylon 11 $(CH_2-)_{10}CO-NH$	3, 4	The overall crystallization kinetics was measured from 173°C to 184°C by dilatometry; the change in Avrami exponent *n* [Eq. (33)] from 3 to 4 occurred at about 175°C; the linear crystal growth rate (see Fig. V.37) in	*Kahle (1957) Z. Electrochem.* ***61****, 1318.*

		cm/min can be expressed by log $v = -1.925\text{–}1.45 \times 10^4/T\,\Delta T$, (data from 150° to 184°C, T in °K), a $1/\Delta T^2$ expression does *not* fit the data; heterogeneous (a thermal and thermal) nucleation was assumed (see Sect. 5.1.3.3) and spherulitic growth observed.	
Nylon 12 $(CH_2-)_{11}CO-NH-$	4–6	Polydisperse samples and fractions of average molecular weight between 9 000 and 41 000 were crystallized between 161°C and 169°C and studied by thermal analysis and dilatometry; the high Avrami exponent was attributed to sheaf like crystal growth.	*Manescalchi et al. (1973). Eur. Polym. J.* ***9****, 601.*
Nylon 5.6 $(CH_2-)_5NH-CO-(CH_2-)_4CO-NH-$	—	Linear crystal growth rates only were measured from 220° to 242°C by optical microscopy; spherulites of different appearance (see Sect. 3.7.4) fit on the same growth rate–temperature curve; extrapolated maximum growth rate at about 185°C.	*Magill (1965). J. Polym. Sci. Part A* ***3****, 1195.*
Nylon 6.6 $(CH_2-)_6NH-CO-(CH_2-)_4CO-NH-$	2, 3(?)	Some data on overall crystallization rate were derived by density balance and depolarization of light (see Sect. 6.1.3.2), both sets of data are difficult to interpret;	*Allen (1952). Trans. Faraday Soc.* ***48****, 1178; Magill (1961). Polymer* ***2****, 221.*
		Optical microscopy on samples of 2500 to 20 000 MW showed thermal nucleation with dependence on thermal history (see also Sect. 5.1.4); moisture	*McLaren (1963). Polymer* ***4****, 175.*

Continued

TABLE VI.8 *(continued)*

Macromolecule	Avrami exponents	Description	References
		in the starting material speeded up crystallization; films of less than 50 μm thickness showed increasingly faster nucleation and crystallization; for linear crystal growth rates see also Fig. V.37;	
		Different types of spherulites observed (see Sect. 3.7.4) showed different growth rates.	*Lindegren (1961). **J. Polym. Sci.** **50**, 181.*
Nylon 9.6 $(CH_2-)_9NH-CO-(CH_2-)_4CO-NH-$	—	Linear crystals growth rates only were measured from 200°C to 234°C by optical microscopy; spherulites of different appearance (see Sec. 3.7.4) fit on the same growth rate–temperature curve; extrapolated maximum growth rate at about 180°C.	*Magill (1965). **J. Polym. Sci. Part A** **3**, 1195.*
Nylon 6.10 $(CH_2-)_6NH-CO-(CH_2-)_6CO-NH-$	—	Linear crystal growth rates of two samples (11 600 and 26 300 MW) were measured by optical microscopy from 110°C to 215°C, the maximum growth rate is at 120°C, the lower molecular weight has always the higher growth rate, above 205°C negative spherulites which grow faster than the normal positive spherulites were also observed.	*Lindegren (1961). **J. Polym. Sci.** **50**, 181.*

Poly(N,N′-sebacoylpiperazine) $-N\langle CHCH_3-CO, CO-CHCH_3\rangle N-CO-(CH_2-)_8CO-$	4	Three samples of widely differing molecular weight were analyzed by dilatometry between 152°C and 171°C; the data analysis was carried out assuming thermal nucleation, spherical crystal growth seems to fit the data better.	*Mandelkern et al. (1954). J. Appl. Phys.* ***25****, 830.*
Polyurethanes $(CH_2-)_4O-CO-NH-$ $(CH_2-)_6NH-CO-O-$	2.3	A 9000 MW sample was analyzed by density weighing between 163°C and 176°C; the main crystallization follows an about constant Avrami coefficient for the whole temperature region; slow crystallization at the end of the fast Avrami portion;	*Rohleder and Stuart (1960). Makromol. Chem.* ***41****, 110.*
		For the trimethylene glycol urethane an Avrami exponent of $n = 2$ has been recently proposed.	*Kercha et al. (1972). Polim. Mashinostr.* ***6****, 16.*
Poly(tetramethyl-p-silphenylene siloxane) $Si(CH_3)_2-C_6H_4-$ $Si(CH_3)_2-O-$		A wide range of molecular weight samples was analyzed mainly by optical microscopy; details are given in Sect. 6.1.5 (Fig. VI.21); the nucleation is heterogeneous after destruction of self-nuclei at 220°C for 5 min; at lower temperatures increasing thermal nucleation; for linear crystal growth rates see Figs. V.37 and VI.20; only a a $1/(T\,\Delta T)$ dependence fits the data.	*Magill (1969). J. Polym. Sci. Part A-2* ***7****, 1187; (1968). Polym. Lett.* ***6****, 853; (1964). J. Appl. Phys.* ***35****, 3249.*

[a] Arrangement of polymers as in Sect. 2.4 in the following sequence: class 1 polymers, vinyl polymers, diene polymers, polyoxides, polyesters, polyamides (and urethanes), other polymers.

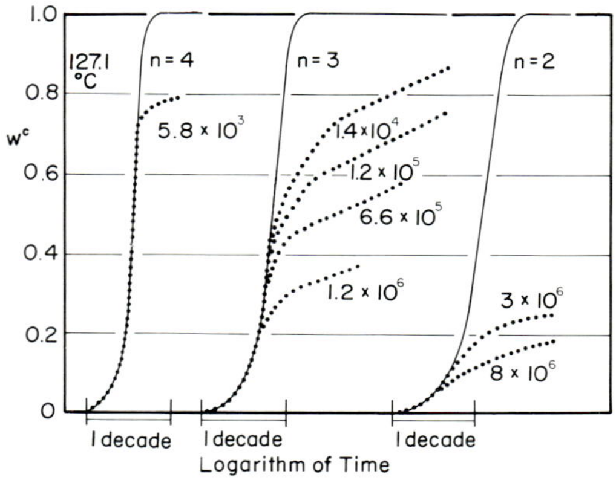

Fig. VI.50 *Shifted Weight Fraction Crystallinity Curves at 127.1°C of Polyethylene as a Function of Logarithm of Time.*

Drawn-out curves: Avrami curves [Eq. (33)] with $n = 2$, 3, and 4 (see Fig. V.10). Note that the experimental curves are not matched at a maximum crystallinity as in Fig V.11, for example. Different molecular weights are indicated in the graph. Drawn after data by Ergoz *et al.* (1972).

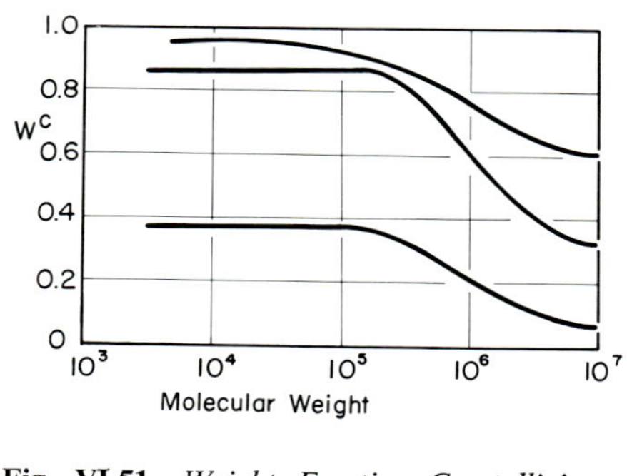

Fig. VI.51 *Weight Fraction Crystallinity of Polyethylene as a Function of Molecular Weight.*

Bottom curve: at the first deviation from the Avrami curve (see Fig. VI.50). Middle curve: at the end of crystallization at the chosen crystallization temperatures (long time limit). Upper curve: after cooling to room temperature. Ergoz *et al.* (1972).

model assumed. Figure VI.51 illustrates the level of crystallinity at which the first deviation from the Avrami expression occurs as a function of molecular weight. Also shown are the final attained levels of crystallinity at the crystallization temperature and after cooling to room temperature. All three curves follow a similar molecular weight dependence, pointing to similar causes for deviation from ideal crystallization behavior and final level of crystallization. The overall rate of crystallization as a function of crystallization temperature and molecular weight can be judged from Fig. VI.52 where the reciprocals of the times for 1% crystallization are plotted. Up to 10^5 MW the crystal growth rate increases with molecular weight at low supercooling as expected from the discussion of the linear growth rates (see Sect. 6.1.5); at higher molecular weights the overall rate decreases again. A detailed interpretation of these data must await information on nucleation and linear crystal growth rates on the chosen samples as well as morphological analysis. As expected, at 130°C, all molecular weights led to basically lamellar crystals (see Sect. III.3.3.2).

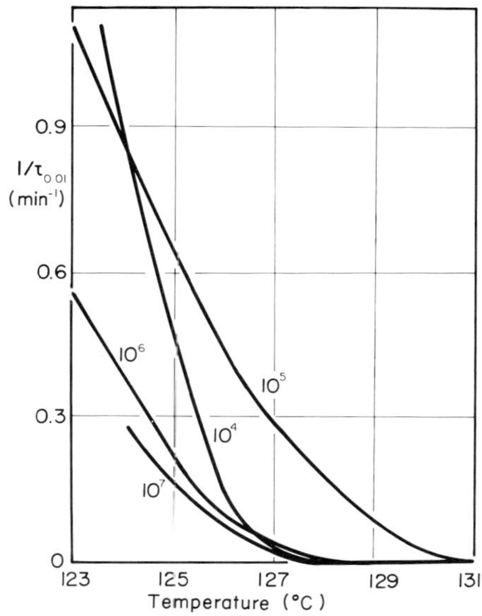

Fig. VI.52 *Reciprocal of the Time to Reach 1% Crystallinity of Polyethylene as a Function of Crystallization Temperature.*
Overall crystallization rate, compare with linear growth rates of a typical macromolecule Fig. VI.20. Constructed using data by Ergoz *et al.* (1972) and Fatou and Barrales–Rienda (1969).

A 115 000 MW sample analyzed by Vidotto *et al.* (1969) yielded, as in the previous samples, an Avrami exponent of $n = 3$, but showed values tending towards $n = 1$ as the number of self-nuclei increased. This increase in self-nuclei was accomplished by heating to temperatures of 138°C and below before crystallization (see Fig. V.27 and also Banks *et al.* 1963b).

More details on the later stages of crystallization were derived by Booth and Hay (1971) on three linear polyethylene fractions (weight fraction molecular weight 41 000–165 000, polydispersity ~2) and a broad distribution sample ($\overline{M}_w = 142\,000$, $\overline{M}_n = 19\,000$). By the use of Eq. (61), which postulates two successive crystallization processes, a fit of the crystallization over the whole crystallization interval was attempted (see Sect. 6.1.3.3). With Avrami exponents of $n = 3$ or $n = 4$ and a linear expression for the secondary crystallization, only small errors of 3–4% remained. A detailed error analysis showed systematic trends in the error, indicative of a wrong functional relationship. Better agreement was possible using fractional exponents ($n = 2.85$, 2.70, 3.20, 3.55 for the above samples, respectively) (see also Booth and Hay, 1972).

Direct microscopic determinations of linear growth rates v of a polyethylene fraction of 30 600 MW were carried out by Hoffman *et al.* (1973) and are shown in Fig. VI.53. The two different sections of the curve have been

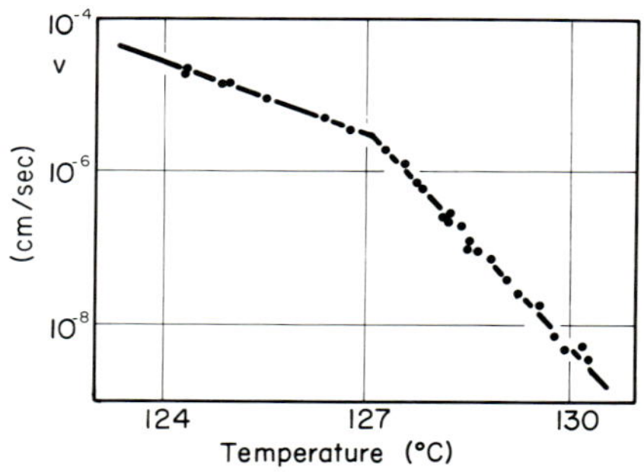

Fig. VI.53 *Change of Linear Crystal Growth Rate of Polyethylene with Temperature.* Fraction of 30 600 MW, the low temperature crystals show spherulitic morphology (see Sect. 3.7, Figs. III.8 and III.116), the high temperature crystals were identified as axialites (see Sect. 3.3.3). Data by Hoffman *et al.* (1973).

taken as an indication of change from regime I to regime II crystallization (Sect. 6.1.4.5). Below 127°C the growth rate is assumed to be represented by Eq. (71), while the growth rate above 127°C is assumed to be represented by

Eq. (69). The ratio of the two slopes of the data when plotted versus $1/(T\Delta T)$ are indeed close to 1:2 as expected. The data of Fig. VI.53 can be used to evaluate the product $\gamma\gamma_e$ through Eq. (68), Chapter V. Hoffman *et al.* (1973) report a product of about 1250 erg^2/cm^4 when using an equilibrium melting temperature of 417.8 K. With a change in the analysis to the experimental melting point value 414.6°K (see Chapter VIII), a shift of only 3.2°, a $\gamma\gamma_e$ product of 825 erg^2/cm^4 results, illustrating the extreme sensitivity of the low supercooling data on the choice of the melting temperature. Overall and within the rather wide margin of error, both sets of parameters can be brought into agreement with nucleation and solution crystallization data (see Sects. 5.1.2.2 and 6.2.1).

Thus the experimental data on polyethylene cover many aspects of nucleation and growth, but are not yet sufficient to resolve all the questions raised in the general discussion of Sect. 6.1. Remaining key problems are the fact that not all data are generated on the same sample, the uncertainty of the parameters used in the analysis of the data, and the questions of the detailed geometry of the growing crystals.

6.3.1.2 *Polypropylene*

Polypropylene in its isotactic configuration (see Sect. 1.5) became commercially available in 1958 and with it began an intensive study of its crystallization. The basic problem was presented and resolved in first approximation by von Falkai and Stuart (1959), Padden and Keith (1959), Marker *et al.* (1959), Griffith and Rånby (1959), von Falkai (1960), and Majer (1960b). Dilatometric data could be fitted up to high values (from 95% to 99%) of the maximum crystallinity (usually 0.50–0.65) with an Avrami expression [Eq. (33)] with an exponent $n = 3$. Table VI.9 illustrates that the dilatometric data could be perfectly matched with the microscopic data. The nucleation was in these experiments heterogeneous and athermal.† Equation (31) could be used to calculate the Avrami coefficient K [Eq. (33)]. Thus, polypropylene when crystallized after heating to 180°C seemed to be a clear-cut example of heterogeneous athermal nucleation followed by a three-dimensional sherulitic crystal growth where at the growth front most of the crystallization takes place to the level of the final crystallinity. A sizeable decrease in the number of nuclei and a small change in the growth rate was observed when the melt was initially heated to 200°C and above. Majer (1960b) observed that this thermal history finally increased the Avrami exponent to $n = 4$ (on heating the melt to 220°C), and he interpreted this as a change from athermal to thermal nucleation.

† For a detailed discussion of nucleation of polypropylene melts see Sect. 5.1.3.3, Tables V.2 and V.3 and Figs. V.21 and V.22.

TABLE VI.9

CRYSTALLIZATION PARAMETERS OF ISOTACTIC POLYPROPYLENE[a]

Cryst. temp. T_c (°C)	Linear rate v (cm/min)	Nuclei N(cm^{-3})	$(4/3)N\pi v^3$ K(calc.)	Dilatometry K(meas.)
122.0	1.8×10^{-3}	2.36×10^{6}	5.76×10^{-2}	—
125.0	1.2×10^{-3}	1.56×10^{6}	1.13×10^{-2}	—
127.5	7.0×10^{-4}	1.02×10^{6}	1.47×10^{-3}	1.40×10^{-3}
130.0	4.3×10^{-4}	0.85×10^{6}	2.83×10^{-4}	2.88×10^{-4}
132.5	2.6×10^{-4}	0.73×10^{6}	5.37×10^{-5}	5.37×10^{-5}
135.0	1.6×10^{-4}	0.65×10^{6}	1.12×10^{-5}	1.07×10^{-5}
138.0	8.6×10^{-5}	0.58×10^{6}	1.54×10^{-6}	1.60×10^{-6}
140.0	5.9×10^{-5}	0.53×10^{6}	4.56×10^{-7}	4.69×10^{-7}
145.0	2.7×10^{-5}	0.47×10^{6}	3.88×10^{-8}	3.85×10^{-8}

[a] Approx. 51 200 MW; n-heptane extracted; before crystallization heating for 15 min to 180°C; Avrami coefficient $n = 3$ for all measurements; melting temperature 173°C after crystallization at 150°C. Data: von Falkai and Stuart (1959), see also von Falkai (1960).

A study of crystallization of isotactic polypropylene (melting temperature 172°C) under conditions of thermal nucleation was made by Hoshino *et al.* (1965).† The number of heterogeneous nuclei was set by heating for 60 min to 200°C. Under such conditions optical microscopy revealed an increasing number of nuclei with time for the chosen sample. Initially the nucleation rate was constant, then the rate decreased. The linear crystal growth rate was constant throughout the crystallization, as usual. Dilatometry gave typical isotherms which could be superimposed for the different temperatures by shifting along the log t abscissa. Figure VI.54 illustrates the 140°C crystallization data converted from the reported graphs to volume fraction crystallinity. The drawn-out curve shows the deviation of the experimental data from a best fit of a $\log[-\ln(\{v_m^c - v^c\}/v_m^c)]$ versus $\log t$ plot with $n = 3$ [see Eq. (34) and Fig. VI.11]. The experimental data increase initially (up to 60 min) somewhat slower, then considerably faster than the calculated curve, indicative of higher Avrami exponents (see Fig. VI.10). A better fit can be achieved with an

† Additional, often repetitious data on polypropylene crystallization can also be found in the papers by Magill (1961, 1962). *Nature* **191**, 1092, *Polymer* **3**, 35; Limbert and Baer (1963). *J. Polym. Sci. Part A* **1**, *3317*; Parrini and Correri (1963). *Makromol. Chem.* **62**, 83; Chiu (1964). *Analyt. Chem.* **36**, 2058; Donald *et al.* (1964). *J. Polym. Sci. Part C* **6**, 93; Heyns, and Heyez (1972). *Thermal Analysis*, **3**, 341; Ikushima *et al.* (1972). *Kobunshi Kagaku* **29**, 186; Johnsen and Spilgies (1972). *Kolloid Z. Z. Polym.* **250**, 1174; Iwanami *et al.* (1972). *Kobunshi Kagaku* **29**, 139; Malinskii *et al.* (1973). *Dokl. Akad. Nauk. SSSR* **208**, 1142.

exponent $n = 3.25$. Hoshino *et al.* (1965) derived for the weight fraction crystallinity [Eq. (47)] by point-by-point analysis a variable exponent changing from $n = 4$ at low conversion to $n < 2$ at high conversion. Comparison with measured nucleation rates and linear growth rates using optical microscopy did not lead to agreement with the dilatometry results. The Avrami exponent calculated using microscopy showed a constant $n = 3.9$ for the whole crystallization range. Unfortunately the sample used for microscopy was not the same as that used for dilatometry. Hoshino *et al.* (1965) suggested that the crystallinity may change through secondary crystallization within the boundaries of the spherulites (see Sect. 6.1.6). To prove changing crystallinity within the spherulites the crystallinity was measured separately by X-ray diffraction (see Sect. 4.1.2). Figure VI.55 summarizes the change of weight fraction of spherulites with time as determined by optical microscopy and the weight fraction crystallinity as determined by X-ray diffraction for the crystallization temperature 140°C. The average crystallinity of the spherulites can be calculated as the ratio of weight fraction crystallinity to weight fraction spherulites. A definite increase in the average crystallinity of the spherulites with crystallization time is shown in Fig. VI.55. Applying a two-stage crystallization model (see Sect. 6.1.3.3) Hoshiono *et al.* (1965) found a reasonable description of the experiments with a primary crystallization using an Avrami exponent $n_1 = 3.9$ and a secondary crystallization with $n_2 = 1.8$.

Without the separate determination of X-ray crystallization shown in Fig. VI.55, the dilatometric data of Fig. VI.54 would have hardly permitted the suggestion of a secondary crystallization. Thus one must assume that the

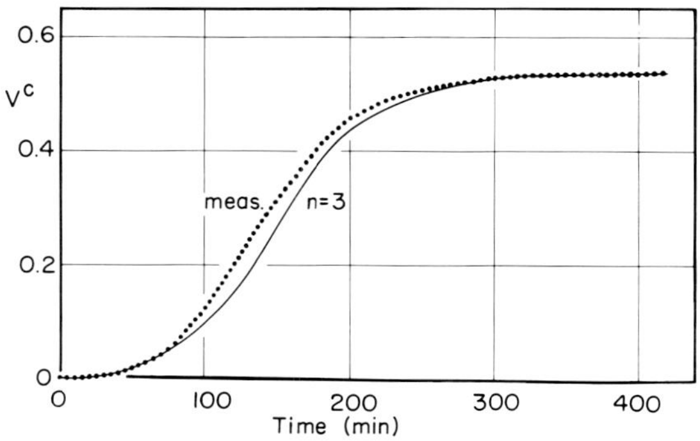

Fig. VI.54 *Change in Volume Fraction of Crystallinity of Isotactic Polypropylene on Crystallization at 140°C.* Experimental points taken from graphs of Hoshino *et al.* (1965), recalculated to volume fraction crystallinity v^c (see Eq. (9), Chapter IV. The drawn-out curve represents the best fit using the Avrami analysis of Eq. (34).

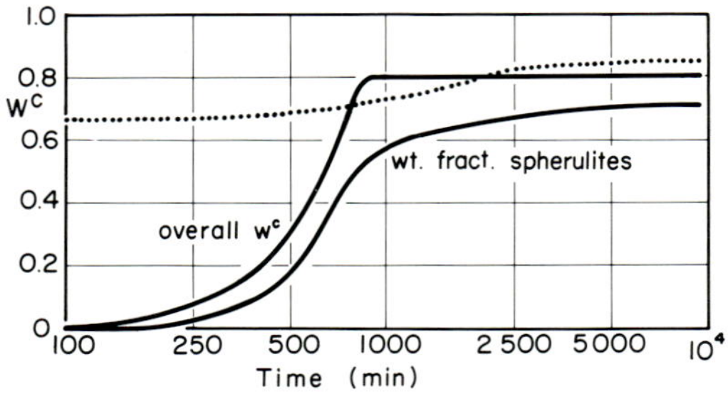

Fig. VI.55 *Change in Weight Fraction of Spherulites and of Crystallinity for Propylene Crystallized at 140°C.* Unextracted whole polymer of melting temperatures 176°C. Dotted curve: average crystallinity calculated as ratio of the weight fraction of crystallinity (by optical microscopy) and the weight fraction of spherulites (by X-ray diffraction). Data by Hoshino *et al.* (1965).

simple interpretation of the data of Table VI.9 is also not fully correct; most likely there is a similar change in spherulite crystallinity, which cannot, however, be detected with sufficient accuracy by dilatometry. Another discrepancy, but this time towards an Avrami coefficient $n = 2$ instead of $n = 3$, was found by calorimetry and analyzed by Godovskii and Slonimskii (1966) and Godovskii (1969) in terms of a possibly different nucleation mechanism.

Usually, the different morphologies of the polypropylene spherulites described in Sect. 3.7.4 are not considered separately despite the fact that Padden and Keith (1959) suggested a 20% faster linear crystal growth rate of the weakly birefringent monoclinic spherulites (see Fig. III.129). A further complication of the crystallization of polypropylene is represented by the remaining atactic sequences or molecules in the sample. As the tacticity decreases it is no longer possible to superimpose the different crystallization isotherms and the later stages of crystallization slow down markedly (Hoshino *et al.* 1965). The slowing down of the isothermal crystallization of the isotactic–atactic mixtures of polypropylene is caused in part by the decreasing crystallization rate as the atactic, noncrystallizable fraction increases. Table VI.10 illustrates the decrease of the linear crystallization rate with changing tacticity. An even larger portion of the retardation may come from the slow diffusion of the rejected atactic polymer away from the growth front which increases the effective concentration of the atactic polymer in the interior of the spherulite beyond the average melt concentration. Keith and Padden (1964) observed that for reasonably high molecular weight of atactic polypropylene (above 3000) and crystallization below 125°C to 130°C there was no change in the initial linear crystallization rate observable by optical

TABLE VI.10

LINEAR CRYSTALLIZATION RATE OF POLYPROPYLENE OF VARYING TACTICITY[a]

T crys. (°C)	100%[b] isotact.	90%[b] isotact.	80%[b] isotact.	60%[b] isotact.	40%[b] isotact.
120	29.4	29.4	26.9	22.8	21.2
125	13.0	12.0	11.0	8.90	8.57
131	3.88	3.60	3.03	2.37	2.40
135	1.63	1.57	1.35	1.18	1.12
T_m [c]	171°C	169°C	167°C	165°C	162°C

[a] Data in μm/min, by Keith and Padden (1964).

[b] Mixtures refer to non extracted isotactic polypropylene (which itself has a 20% atactic fraction) viscosity average 178 000 MW and an atactic polypropylene of a viscosity average 87 000 MW.

[c] Experimental melting temperatures.

microscopy of the spherulites. Under such conditions, diffusion was not fast enough to mix the melt effectively and the growth tips of the spherulite always encountered the same initial average atactic concentration throughout growth. While the perimeter of the spherulites grows faster than predicted from the average remaining isotactic polypropylene under such conditions, the secondary and tertiary interior branches must grow considerably slower (see Sect. 3.6.1 for a discussion of the branching mechanism). This behavior changed whenever diffusion was fast enough for mixing; then the spherulite growth rate slowed down and, in addition, asymmetries in the spherulites developed when growing close to other crystallizing spherulites due to interaction of the concentration gradients of the neighboring spherulites (see also the discussion of Copolymers, Sect. 6.3.4).

6.3.1.3 *Poly(ethylene oxide)*

The first investigation of the crystallization kinetics of poly(ethylene oxide) was made by Mandelkern *et al.* (1954) using dilatometry. A sample of 9–10 000 MW was shown to crystallize in a fashion approximated by an Avrami expression [Eq. (33)] with an exponent $n = 3$. Subsequent measurements which included microscopy and calorimetry revealed a more complicated picture.† The initial stages of crystallization were fitted sometimes with

† For data and discussions see also Barnes *et al.* (1961). *J. Phys. Chem.* **65**, 1742; Banks and Sharples (1963). *Makromol. Chem.* **59**, 233, **67**, 42; Keith and Padden (1964). *J. Appl. Phys.* **35**, 1286; *Hillier* (1965). *J. Polym. Sci. Part A* **3**, 3067; Jain and Swinton (1967). *Europ. Polym. J.* **3**, 371; Jain (1969). *Ind. J. Technol.* **7**, 123; Hay *et al.* (1969). *Polymer* **10**, 187, 203.

a value of $n = 4$ or $n = 3$, but for the later stages a decrease to a constant $n = 3$ or $n = 2$ was reported. Others could fit their data over the whole range with $n = 2.0$ or $n = 2.5$. No extended slow secondary crystallization or crystal perfection as in polyethylene was observed, and consequently, a fit with a two-process Avrami expression as given by Eq. (61) with $n_1 = 3.0$ gave little improvement of the data fitting over fractional or varying Avrami exponents. In some cases it was found by optical microscopy that nucleation in the initial phases of the crystallization occurring at a constant rate was to be followed by athermal heterogeneous nucleation. This nucleation behavior could account for a changing Avrami exponent (see Table VI.2). For samples with low molecular weight contents, the linear crystal growth rates were observed to slow down with progressing crystallization, a possible indication of molecular weight rejection. Frequently, the observed spherulites were of such a size that in the later growth stages they would exceed the sample dimensions and stop growing, thus reducing the overall crystallization rate. These effects alone could explain many of the observed discrepancies.

Dilatometric, calorimetric, and microscopic analysis of the crystallization of a series of fractions from 10^3 to 6×10^5 MW of poly(ethylene oxide)s were carried out by Beech *et al.* (1972a,b) and Godovskii *et al.* (1972). These newer measurements confirmed much of the older work of fractional and varying Avrami exponents of values which are usually between 2 ± 0.6. In addition, they provided strong evidence for a large effect from crystal perfection. Beech *et al.* (1972a) showed that partially or completely crystallized samples, when heated to a temperature slightly below the melting point, where otherwise no crystallization occurs (63.3°C), continued to increase in crystallinity proportional to the logarithm of time. Godovskii *et al.* (1972) illustrated a molecular weight dependent break in the linear crystal growth rate versus $1/(T\Delta T)$ plot and linked it with extended (high temperature, low molecular weight) and folded chain (low temperature, high molecular weight) crystallization.

The most detailed work on the crystallization of poly(ethylene oxide) was carried out by Vidotto *et al.* (1969) and Kovacs and Gonthier (1972) through a study of nucleation and single crystal growth rate, morphology, and thickening. The nucleation experiments on a sample of 780 000 MW ($M_w/M_n = 2$) revealed a critical melting temperature at about 69°C (see Fig. V.29). Samples heated between 68° and 69°C had a practically amorphous density, but retained enough self-nuclei to lead to an Avrami exponent of 3 for the first half of isothermal crystallization after cooling. The second half of the crystallization process was considerably slower with an exponent close to 1. Samples heated first to temperatures between 69°C and 100°C had much fewer nuclei (see Fig. V.29, Sect. 5.1.4.2) and the second process was superimposed on the first, yielding an average Avrami exponent of $n = 2$ as reported earlier. The second process was shown by Kovacs and Gonthier (1972) to involve for fractions of given molecular weight mainly crystal perfection to larger fold-length and not secondary crystallization. Details of the crystal perfection are given in Sect. 6.1.6 (see Fig. VI.27). In Sect. 6.1.5 and Fig. VI.23 it is shown that for a sharp molecular weight fraction distinct temperature regions of extended, once and multiply folded initial crystal growth exist. Each type of macroconformation has a different crystal growth rate. Table VI.4 shows that the temperatures of change from one macroconformation to the other are molecular weight dependent.

A summary of the complicated crystallization behavior of poly(ethylene oxide) is as follows: One may observe athermal self-nucleation by melting a crystalline sample between 68°C and 69°C before crystallization. By heating to between 69°C and 100°C the number of nuclei is increasingly reduced, but nucleation is still largely athermal. Heating the melt first to higher temperatures and crystallizing at low temperatures may give initially increasing contributions of thermal nucleation. The major determining factor for the linear crystal growth rate, besides temperature, is the particular macroconformation of the growing crystal. Once grown, folded macroconformations may perfect by thickening. For broad molecular weight distributions, segregation and simultaneous, separate crystallization of different macroconformations must be expected. Crystal morphology may vary from lamellar, single crystalline to branched, spherulitic. The crystal aggregates are often so large that sample geometry may also influence the results. All these effects account most likely for all the observed, often contradictory, data. Poly(ethylene oxide) represents thus a typical example for the case where, for the interpretation of the overall crystallization rate, much microscopic detail must be determined independently.

6.3.1.4 *Poly(ethylene terephthalate)*

Poly(ethylene terephthalate) crystallization, despite a large list of reports,† is still not as well understood as that of the macromolecules described above. A detailed microscopic analysis of the processes occurring during the crystallization is still not available. The first more detailed study of the crystallization

† For additional work not described in detail see Kolb and Izard (1949). *J. Appl. Phys.* **20**, 571; Cobbs and Burton (1953). *J. Polym. Sci.* **10**, 275; Turska and Przygocki (1964). *Faserforsch. Textiltech.* **11**, 732; Mayhan *et al.* (1965). *J. Appl. Polym. Sci.* **9**, 3605; Mitsuishi and Ikeda (1966). *J. Polym. Sci. Part A2* **4**, 283; Anokhin *et al.* (1969). *Tepl. Svoistva Veshchestv.* 14; Ozawa (1971). *Polymer* **12**, 150; Fielding-Russell and Pillai (1970). *Makromol. Chem.* **135**, 263 (1972), *Kolloid Z. Z. Polym.* **250**, 2; Przygocki (1972). *Polimery* **17**, 197 and (1973). *Przegl. Wlok.* **27**, 138; Misra and Stein (1972). *J. Polym. Sci. Polym. Lett.* **10**, 473 and (1973). *J. Polym. Sci. Polym. Phys. Ed.* **11**, 109.

kinetics of poly(ethylene terephthalate) was carried out by Keller *et al.* (1954), Morgan (1954), and Hartley *et al.* (1954). They observed that, besides the dependence of crystallization on temperature, there was a sizeable influence on crystallization by the following factors:

1. Molecular weight (decreasing rate with increasing molecular weight).
2. Chemical purity (larger percentage of diethylene glycol content through dehydration increases crystallization rate).
3. Sample diameter (filaments of smaller diameter crystallized faster).
4. Mechanical stress (bending or drawing increased the rate of crystallization).
5. Melt condition (higher melt temperature decreases the crystallization rate, particularly at higher temperature, and can change the Avrami exponent n).†
6. Added nuclei or impurities (Rybnikar, 1960, found a decrease in n when adding 0.5% TiO_2).

Thus to obtain useful data it was found necessary to use a reproducibly made large sample (1 mm cubes) of given molecular weight (16 800) which was fused first by heating to 294 C. Typical data are listed in Table VI.11. The overall crystallization kinetics changes from slow crystallization with an Avrami exponent $n = 2$ at 90–160°C to $n = 3$ in the region of fastest crystallization, to finally $n = 4$ above 230 C where the crystallization rate is slow again. Between 210°C and 240°C nucleation and linear spherulite growth rate information was gathered and, as is shown in Table VI.11, reasonable agreement could be achieved between the two sets of data. Rybnikar (1960) found also spherulitic growth in the region where $n = 2$ which does not fit the possibilities listed in Table VI.2. The detailed interpretation must await clearer morphological evidence. Further support of these kinetic data was published by Zachmann and Stuart (1960), who also described the secondary crystallization in more detail (see Sect. 6.1.6).

The linear growth rate of the spherulites as a function of 19 000–39 000 MW was measured by van Antwerpen and van Krevelen (1972) over the complete temperature range by light scattering. The linear crystal growth rate v decreases with increasing molecular weight over most of the range of

† Note that Hartley *et al.* (1954) observe a change of the temperature of shift of the exponent n from 3 to 4 at a higher crystallization temperature (250°C instead of 230°C) when the melt is heated higher before crystallization (294°C instead of 268°C) as one would expect from increasing removal of self-nuclei (see Sect. 5.1.4), but Rybnikar (1960) finds the opposite effect (crystallization at 235°C shows an Avrami exponent 4 on heating the melt 15 min to 265°C before crystallization, but 3 on heating 15 min to 290°C).

TABLE VI.11

CRYSTALLIZATION OF POLY(ETHYLENE TEREPHTHALATE)

T_c (°C)	Exp. n	Avrami const. K (min^{-n}) from density	Avrami const. K (min^{-n}) from microscopy	Nuclei (or rate)	Crystal growth[c] rate v (cm/min)	Ref.
110	2	3.5×10^{-4}	—	—	—	*a*
160	2	3.2×10^{-1}	—	—	—	*a*
180	3	1.4	—	—	—	*a*
200	3	3.3×10^{-1}	—	—	—	*a*
210	3	1.1×10^{-1}	2.1×10^{-1}	Not reported	Not reported	*a*
220	3	2.6×10^{-2}	2.1×10^{-2}[d]	3.2×10^{8}/cm^3	2.5×10^{-4}	*a*
230	3	1.3×10^{-4}	2.3×10^{-4}[d]	1.7×10^{6}/cm^3	3.2×10^{-4}	*b*
240	4	5.1×10^{-8}	3.5×10^{-8}[e]	4.9×10^{6}/cm^3/min	1.9×10^{-5}	*a*

[a] Hartley *et al.* (1954).
[b] Rybnikar (1960).
[c] For other values see Figs. V.37 and Eq. (93).
[d] Use Eq. (32A) for evaluation.
[e] Use Eq. (32B) for evaluation.

crystallization temperatures. Their data could be well represented by

$$\ln v = \ln\left(-5 \times 10^9 + \frac{2.93 \times 10^{14}}{M_n}\right) - \frac{777}{T - 316} - \frac{3093 \times 10^5}{T^2(573 - T)} \tag{93}$$

(v in nm/sec, T in °K). The equation is based on Eq. (72) with reasonable parameters. Data on crystallization from the melt and from the glass fitted on the same curves.†

Some of the morphological complications in the crystallization of poly(ethylene terephthalate) are described by Konrad and Zachmann (1971). By determining the overall density and comparing it with absolute density measurements from X-ray small angle determination of crystal and amorphous regions as outlined in Sect. 3.2.2 (p. 211) it seems likely that both amorphous and crystalline area densities increase with increasing crystallization temperature by as much as 5%, this makes overall crystallinity determinations unreliable. In particular, it is surprising that the density between crystal

† Baranov *et al.* (1970) presented more limited, but similar data which seem to illustrate a higher maximum crystallization rate temperature and lower maximum rate for poly(ethylene terephthalate) crystallized from the glass. However, their linear growth rate data from the melt as well as from the glass do not agree well with any of the discussed measurements.

lamellae seems to be lower than extrapolated from the melt. The strong change in crystalline area density may point to the inclusion of more amorphous defects (see Sect. 4.3.1) at lower crystallization temperatures. The study of annealing of poly(ethylene terephthalate) (see Sect. 7.3.4) has shown that the macroconformation of the once-grown crystals can at higher temperature be changed by chemical reaction through transesterification. The initially chain-folded macroconformation changes slowly to a more extended, fringed micellar macroconformation.

6.3.1.5 Nylon 6

Nylon 6 crystallization has been studied for many years with somewhat contradictory results.† As with the poly(ethylene terephthalate), there exists no comprehensive analysis of all aspects of the problem. Droplet experiments suggest that homogeneous nucleation of nylon 6 occurs only at 128°C or lower (see Table V.2). Thus practically all reported analyses have been carried out under conditions of thermal or athermal heterogeneous nucleation (see Sect. 5.1.5). Like all other macromolecules, nylon 6 has the possibility of starting crystallization by self-seeding (see Sect. 5.1.4 and Table V.4). A 30-min heating to 260°C is the condition usually adopted to eliminate self-seeding.

Detailed optical microscopy over a wide temperature range was carried out by Burnett and McDevit (1957). The linear crystal growth rates of a 27 400 MW nylon 6 are plotted in Fig. V.37 (curve 1).

The overall crystallization was studied through dilatometry and with a density balance. The crystallization increases steeply from about 215°C to reach a maximum at about 185°C. The main portion of the crystallization is frequently described by an Avrami exponent $n = 4$ [Eq. (33)], which would agree formally with thermal nucleation and spherical crystallization (see Table VI.2). The early stage of crystallization usually shows much higher exponents. Rybnikar (1962) reports values as high as $n = 10$. However, his data on four different nylon 6 samples could be expressed in the temperature range measured (195–215°C) by an exponent $n = 4$‡ from the beginning of crystallization if the zero of time was set at the first appearance of crystals, i.e., the time scale was corrected for an induction period (see Sect. 5.1.5).

† Besides the references given in the text the following papers are of interest; Kanamaru *et al.* (1950). *Kobunshi Kagaku* 7, 1, 4, 8; Hartley *et al.* (1954). *Ricerca Sci.* (*Suppl.*) **25**, 577; Cefelin *et al.* (1960). *Coll. Czech. Chem. Comm.* **25**, 1267; Inoue (1961). *J. Polym. Sci.* **55**, 753; Rybnikar (1961). *Chem. průmysl* **11**, 157; Magill (1962). *Polymer* **3**, 43; Ueda and Nukushina (1966). *Kobunshi Kagaku* **23**, 769; Mitomi and Tonami (1971). *Kobunshi Kagaku* **28**, 630.

‡ One sample, which was possibly nucleated, showed an Avrami coefficient of $n = 3$.

More recent data by Turska and Gogolewski (1971) from 201–213°C on four commercial samples led to similar results, but showed a change from $n = 2.3$–3 to $n = 4$ at 209–210°C. Fractions of 17 000 to 49 000 MW with polydispersities of 1.3 to 1.4 lead to the exponents n listed in Table VI.12. The overall

TABLE VI.12

AVRAMI EXPONENTS (n) OF NYLON 6[a]

MW	Crystallization temperatures (°C) 201	205	206	207	209	210	213
17 000	5.0	5.0	5.0	5.0	5.0	6.0	6.0
23 000	5.0	5.0	5.0	5.0	5.0	6.0	6.0
32 500	5.0	5.0	5.0	5.0	5.0	6.0	6.0
37 000	3.9	3.9	3.9	3.9	3.9	5.0	5.0
49 000	2.8	2.8	2.8	2.8	2.8	4.0	4.0

[a] Data by Turska and Gogolewski (1971).

rate decreased with increasing molecular weight as usual (see Sect. 6.1.5). A clear increase in exponent between 209°C and 210°C is seen as in the broad molecular weight distribution samples. The high values of n were suggested by Turska and Gogolewski (1971) to indicate sheaflike morphology (see Table VI.2, Fig. III.118 and Sect. 6.1.3.3). Crystallization data via measurement of the intensity of transmitted light through nylon 6 of different molecular weights could be described with $n = 4$ for all molecular weights (2.6×10^3 to 1.9×10^4) and temperatures except for the early stage of crystallization where n increased to 6 (Ishibashi and Tani, 1969). Again, the explanation offered was sheaflike crystallization (for possible complications of light intensity measurements see Sect. 6.1.3.2).

Addition of nucleating agents was shown to speed up the crystallization of nylon 6 (see Sect. 5.1.3.3). Adding LiCl and KCl, DeCandia *et al.* (1973) could demonstrate a considerable slowing down of the overall crystallization rate at 195–205°C. The Avrami coefficient also decreased by about one unit with 1–2 mole% LiCl, or more than 5 mole% KCl. The pure nylon 6 was, as in several other instances, reported to need already a low Avrami exponent of 2 for its description. After the addition of the salts, exponents as low as 0.9 were found.

A low Avrami exponent of 2.2–2.3 was also found on isothermal crystallization from the glassy state at 48–54°C (Kozlowski, 1972). An attempt to analyze the faster crystallization in the temperature range up to 63°C by nonisothermal crystallization (16°C/min heating rate) decreased the computed n to as low as 0.5.

Finally, Tomka *et al.* (1967) analyzed the crystallization of nylon 6 polymerized from the monomer melt at the crystallization temperature under conditions which did not permit crystallization during polymerization. The polymerization was complete before crystallization started. In this way, the melt should show no thermal history (but a possible chemical history). The analysis of these data was complicated by the somewhat arbitrary choice of t_0 the time-zero for crystallization. It may serve as an illustration of the sensitivity of the Avrami analysis of dilatometric data that the exponent $n = 2$ whenever t_0 was taken at the beginning of volume change, but $n = 3.2$–3.6 when t_0 was chosen at the somewhat earlier times when crystallization initially could be detected by light scattering. Separating the slower secondary crystallization gave best fits with values between $n = 3.5$ and $n = 4.0$, in reasonable agreement with other measurements in the chosen temperature interval of 195–202°C.

Thus, at present it is not possible to clearly link different mechanisms of crystallization with the observed overall rates of nylon 6. A number of complications have been suggested which make it imperative that morphological and overall measurements should be carried out on the same, well-defined material which is nucleated reproducibly. As with poly(ethylene terephthalate), it should also not be overlooked that polyamides can undergo chemical reactions which change the backbone bonds and thus alter macroconformations after crystallization (see Sect. 7.3.4).

6.3.1.6 Conclusions

The experimental evidence reviewed in Table VI.8 and Sects. 6.3.1.1–6.3.1.5 have shown that the overall crystallization of many linear macromolecules from the melt can be described only as a first approximation with an Avrami expression [Eq. (33)]. The average value of the Avrami exponents listed in Table VI.8 is $n = 3.05$ with a standard deviation of ± 0.7. For many macromolecules n is close to three and a picture of athermal heterogeneous nucleation followed by spherulitic growth is acceptable; others require $n = 4$, indicative of thermal nucleation, which is most often thermal heterogeneous nucleation followed by spherulitic growth (see for example Tables VI.9 and VI.11). The also frequently observed exponent $n = 2$ could well be related to fibrillar or lamellar crystal growth following thermal or athermal nucleation, respectively (see Table VI.2).

However, a more detailed look at the crystallization reveals a much more complicated picture. The nucleation usually is not identifiable as athermal or thermal, but rather a mixture of both, showing initially increasing numbers of nuclei which level off with time. The growth rates are constant with time, their description in terms of surface free energies is hampered, however, because of uncertainties in the equilibrium melting temperatures.† The actual geometry of the crystalline superstructure is much more complicated than the assumed spherical, lamellar, or fibrillar structures. The linear crystal growth rates, in addition, are different for different molecular weights, different macroconformations, and different polymorphs. Finally, secondary crystallization and crystal perfection occur almost always. With these effects proven for most of the discussed cases in Sects. 6.3.1.1 to 6.3.1.5, one must conclude that the main reason for the remaining usefulness of the simple Avrami expression is based on a certain amount of cancelation of opposing secondary effects and the strong correlation between the different parameters in the Avrami equation (K, n, and t_0, the starting time of crystallization) (see Sect. 6.1.3.2).

Thus, a good description of the crystallization of macromolecules needs measurement of the overall crystallization rate and a detailed analysis of the nucleation process, the crystal morphology, and secondary crystallization as well as crystal perfection and changes in the amorphous structure remaining on partial crystallization. If the sample is not crystallized at low rates of growth where heat effects are negligible, the actual temperature at the growth front needs to be discussed. Furthermore, if the sample is not chemically uniform (i.e., consists of a reasonably sharp fraction of a homopolymer with negligible imperfections in form of noncrystallizable repeating units), a detailed study of segregation and diffusion processes and their effects on chemical potentials and crystallization rates is necessary. Even for the chosen examples described in Sect. 6.3.1.1 to 6.3.1.5 these conditions are not met for any one sample at the same time.

6.3.2 Effect of Deformation

An effect of deformation on macromolecular melts is shown in Sect. 5.1.4.5 to be the occurrence of fibrillar self-nucleation. In particular, higher molecular weight molecules are extended in the melt and initiate fibrillar crystal nuclei in direction of the stress with causes the deformation. Figure V.33 shows the typical appearance of a sheared polymer when observed between crossed polarizers under an optical microscope.‡ More detailed structure is revealed by the thin film electron micrographs of Fig. III.80 (see also Sect. 3.8.2). This enhanced nucleation leads necessarily to a faster overall crystallization on deformation. Measurements of the rate of the subsequent growth of lamellar crystals of *cis*-1,4-poly(2-methylbutadiene) on the fibrous self-nuclei were made by Andrews *et al.* (1971). The effect of deformation on

† The various surface free energies suggested in the literature are not further discussed for that reason.

‡ See also Haas and Maxwell (1969) and Wereta and Gogos (1971) for deformation studies on polyethylene and poly-1-butene melts.

the linear crystal growth rate, measured on single lamellae growing epitaxially on the fibrous nuclei, was surprisingly small. Increasing the linear deformation of a thin film from 50% to 250% increases the growth rate by only 25% and leaves the maximum of the curve of crystal growth rate versus temperature almost unchanged (see Fig. VI.57). Thus, the molecules that in the growth step add epitaxially to the fibrous nuclei are most likely already almost completely relaxed by the time of crystallization and behave accordingly not much different from normal melt (Sect. 6.3.1).

To maintain the state of deformation of the molecules up to the time of crystallization, Mackley and Keller (1973) used a dynamic crystallization technique. Continuous partial extension of the molecules in the melt was maintained by flow out of opposing jets as sketched in Fig. VI.56. At the

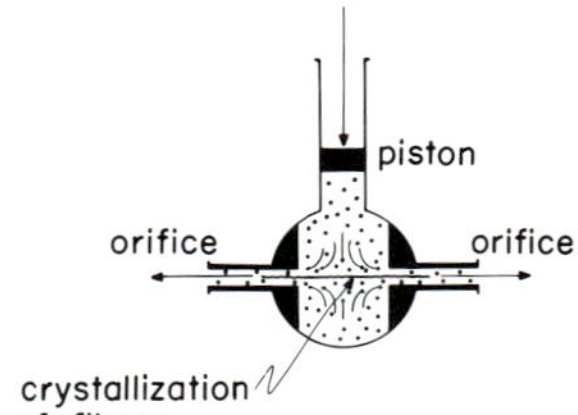

Fig. VI.56 *Schematic Drawing of the Experimental Condition to Produce Pure Shear on Crystallization from the Melt.*

point of greatest longitudinal velocity gradient (estimated to be 50 sec) fibrillar crystallization was observed at temperatures as high as 140°C. This compares with crystallization without shear at about 15°C lower temperatures (see Fig. VI.49). At lower temperatures multiple fibrils grew close to the first fiber, finally to plug the capillary jets by crystal growth at 134°C. Analysis of the extruded crystallized melt still showed two different portions, one highly oriented, the other somewhat less so. One may conclude that the structure is still similar to the skish-kebobs described in Sect. 3.8.2.

The classical approach to maintain melt deformation throughout crystallization is to use cross-linked melts. Figure VI.57 shows the linear crystal growth rate for 50% and 200% elongation of *cis*-1,4-poly(2-methylbutadiene) measured on single lamellae. The maximum for the crystal growth rate curve is depressed when compared with that of the uncross-linked macromolecules, as one would expect from the introduction of noncrystallizable segments into the macromolecules (see Sect. 6.3.4). The enhanced low temperature drop of crystal growth rate of the cross-linked sample to zero has been linked to the

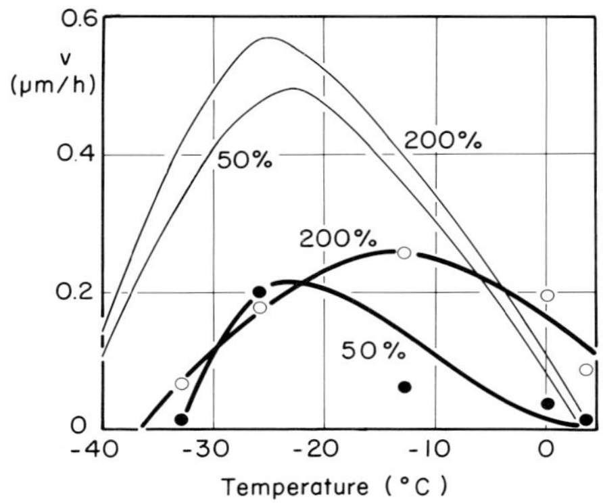

Fig. VI.57 *Linear Crystal Growth Rate for Cross-Linked cis-1,4-Poly(2-methylbutadiene).* Molecular weight between cross-links is about 2500. The drawn-out thin curves refer to the same melt, but not cross-linked. Extension given in the figure. Data by Andrews *et al.* (1971).

higher glass transition temperature of the cross-linked polymer. Deformation by 200% speeds up crystallization at the higher temperatures. This faster crystallization of the crystalline lamellae would be expected solely from the increase in melting temperature for strained molecules (see Chapter VIII). Correcting the degree of supercooling ΔT in the free enthalpy of molecular nucleation [Eq. (81), Chapter V], or secondary nucleation [Eq. (68), Chapter V], does not lead to quantitative agreement of the data in Fig. VI.57 for different deformation. Andrews *et al.* (1971) pointed out that correcting the melting temperature for deformation leads to too high a temperature for the maximum rate of crystal growth.

The change in ΔT due to deformation would qualitatively also account for the increase in nucleation, and should lead, according to the description of chain folding, at constant crystallization temperature to decreasing fold lengths with increasing deformation. The latter has been reported by Kobayashi and Nagasawa (1970) on polyethylene melts crystallized at constant shear strain.

The quantitative evaluation of all these effects is at present hampered by the difficulty in description of the crystal morphology and molecular macroconformation in a particular crystal grown from a deformed melt. In rough outline it is known that the morphology changes as the deformation of the melt increases from spherulitic to deformed spherulitic, to the various forms of shish-kebob, to fibrillar. There are, however, not enough details to discuss the changes in growth kinetics caused by the changes in morphology and macroconformation. Recent work by electron microscopy on strained thin films revealed that the fibrils seem to be discontinuous along their axis, indicating a possible nodular structure (see Sect. 6.1.7 and Figs. VI.34–36 for similar ideas on unstrained crystallization) [Hardin and Yeh (1973) on *trans*-1,4-poly(2-methylbutadiene), Luch and Yeh (1972) on *cis*-1,4-poly(2-methylbutadiene), Klement and Geil (1972) on a polycarbonate, isotactic poly(methyl methacrylate), and polystyrene, and Yeh and Lambert (1971) on isotactic polystyrene]. Presently these observations are not well understood and contribute to the uncertainty in discussing the effect of deformation in crystallization.

A much larger effort was spent over the years on the elucidation of the effect of deformation on the overall crystallization rate of cross-linked macromolecules.† In the light of the above discussion of the underlying processes,

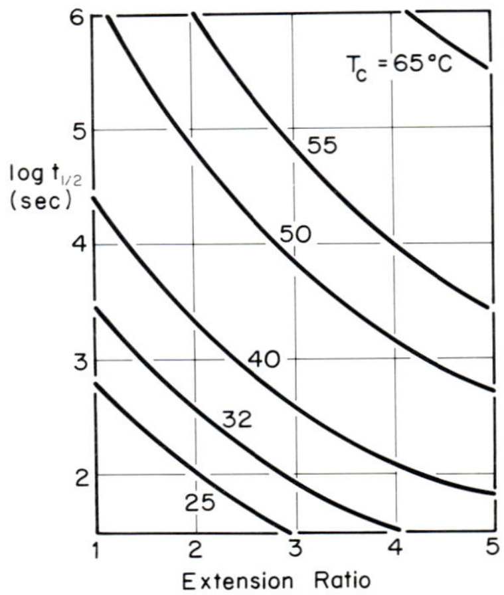

Fig. VI.58 *Half-time for Crystallization of trans-1,4-Poly(2-chlorobutadiene) as a Function of Temperature and Extension Ratio α.*
Molecular weight between cross-link points 20000, crystallization temperatures given in the figure. The test piece was 5 × 1 × 0.07 cm and stretched at 95°C before quenching to the crystallization temperature. Data by Gent (1965).

† For general reviews of overall crystallization see Mandelkern (1964a) and Treolar (1958). For corrections of the Avrami expressions (Sect. 6.1.3) and for orientation dependence of nucleation and growth see Ziabicki (1974).

the observed results on the overall process must be treated as a first approximation. The seemingly good fit with equilibrium considerations (see Chapter VIII) must rest on considerable cancelation of errors and inherent experimental uncertainties. Gent (1954, 1955, 1965, 1966) studied the crystallization of cross-linked *cis*- and *trans*-1,4-poly(2-methylbutadiene), and *trans*-1,4-poly(2-chlorobutadiene). The crystallization was followed by the stress decay and linked at zero extension to dilatometric data. Figure VI.58 illustrates the half-time for crystallization of *trans*-1,4-poly(2-chlorobutadiene) derived from stress decay and dilatometric measurements. Increasing the extension ratio α decreases the half-time for crystallization by several orders of magnitude. The other analyzed samples behaved similarly; overall, the increase in crystallization could be accounted for approximately by the corresponding increase in the melting temperature (see Chapter VIII). Figure VI.59 shows crystallization isotherms of *cis*-1,4-poly(2-methylbutadiene) at different deformation. The isotherms for the extension ratio

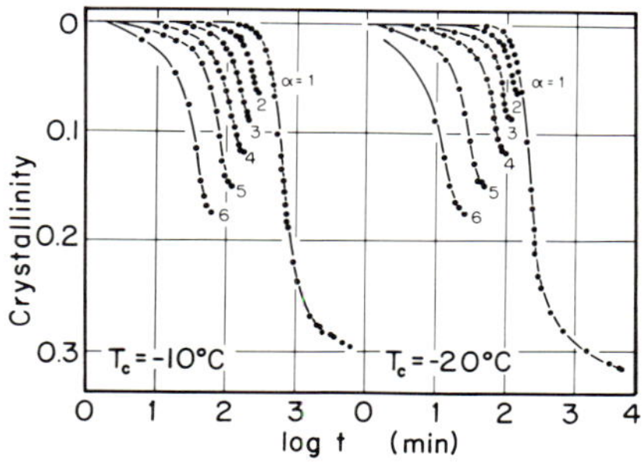

Fig. VI.59 *Crystallization Isotherms of cis-1,4-Poly(2-methylbutadiene) at Different Extension Ratios* α.
Molecular weight between cross-linked points about 25 000, extension ratios α given in the figure. The test piece was $3.5 \times 0.3 \times 0.04$ cm and heated well above its melting point for at least 1 hr to remove all prior crystals before stretching. Data by Kim and Mandelkern (1968).

$\alpha = 1$ were obtained by dilatometry, the others by measurement of stress decay. Again the marked increase in overall rate of crystallization with elongation is evident. Fitting the isotherms to Avrami expressions leads us to the exponent listed in Table VI.13. For higher elongation the Avrami exponent decreases rapidly towards $n = 1$, a value in formal agreement with one-dimensional crystal growth from athermal nuclei which would be expected

TABLE VI.13

VALUES OF THE AVRAMI EXPONENT n FOR DIFFERENT ELONGATION RATIOS α OF *cis*-1,4-POLY(2-METHYLBUTADIENE)[a]

T_c (°C)	$\alpha = 1$	$\alpha = 2$	$\alpha = 3$	$\alpha = 4$	$\alpha = 5$	$\alpha = 6$
−20	3.5	2.85	2.1	1.55	1.4	1.13
−15	3.5	2.95	2.15	1.57	1.37	1.35
−10	3.6	2.9	2.15	1.60	1.55	1.30
0	3.6	2.45	2.1	1.65	1.4	1.35
Average	3.55	2.79	2.13	1.59	1.43	1.28

[a] Determined from the free growth approximation, see Eq. (33) and Sect. 6.2.1.

from fibrillar crystallization. The usual deviations from the Avrami expression occur at $\alpha = 1$ at a crystallinity of about 0.1, while for the deformed samples, practically the whole crystallization range could be fitted ($w^c = 0.07$ to 0.17 depending on α).

Poly(*trans*-cyclopentene) crystallization was analyzed in the slightly cross-linked state as a function of time and strain by Kraus and Gruver (1972). Within experimental error Avrami expressions with exponents [see Eq. (33)] between $n = 1$ and $n = 2$ modified for changes of ΔT for strain could represent the data.

A study of the crystallization of isotactic polypropylene from oriented melts by Leugering and Kirsch (1973) led to the observation of increased amounts of trigonal crystals (type II, see Sect. 2.4.2). The spherulites were row nucleated and could be easily distinguished from the more stable monoclinic spherulites by their higher birefringence and lower melting temperature (see also Sect. 3.7.4).

Biaxial deformation of cross-linked *cis*-1,4-poly(2-methylbutadiene) was studied by Oono *et al.* (1973). The formation of crystallites was followed by X-ray diffraction. The Avrami exponent n was shown to decrease linearly from ~3 to 0.7 at a 3 × 3 extension ratio. The molecular weight of the sample between cross-link points was about 5000 and the crystals were oriented in the stretched samples with their (010) planes parallel to the film plane.

6.3.3 EFFECT OF PRESSURE

The effect of high pressure on the crystallization of linear macromolecules from the melt gained interest with the first observation of polyethylene extended chain crystals (3–7 kb crystallization pressure, see Sect. 3.3.1). A comparison of the morphology of crystals of polyethylene and polychlorotrifluoroethylene grown under elevated pressure (Figs. III.32 and III.33) with those of polytetrafluoroethylene grown from the melt without added pressure (Figure III.11 and III.125) and those of selenium and polyphosphates grown under conditions where crystallization is possible during polymerization (Figures III.31 and III.141) establishes that extended chain crystal growth is not coupled exclusively to the effect of pressure. Furthermore, a study of the molecular nucleation of polyethylene on crystallization at elevated pressure under conditions of chain extension revealed a similar rejected molecular size as is observed at atmospheric pressure (Sect. 5.3.4, Fig. V.49, curve 2). Specifically, a broad molecular weight distribution polyethylene crystallized for 8 hr at 238°C and 5 kb pressure, followed by slow cooling, showed a critical molecular weight for segregation of 3×10^4. When compared with Fig. V.49, this corresponds to a critical molecular weight expected for crystallization without added pressure at 129.6°C (Mehta, 1973, see also Table V.6).

The lamellar thickness of the pressure crystallized polyethylene is larger for less supercooling and longer crystallization times (Sect. 3.2.2.2). The lamellar thickness–temperature curve (Fig. III.15) cannot be fitted to a critical nucleus dimension with reasonable values for the surface free energies. From observation of the rounding of the ends of the extended chain lamellae (Fig. III.40), narrowing of the lamellae at points of crystal intersections (Fig. III.41), and the increase in lamellar thickness at larger distances from the spherulite centers (Figs. III.122 and III.123), it was concluded by Wunderlich and Melillo (1968) that the extended chain crystals grown under such conditions chain fold initially. They extend in a second, solid state process. The carrots shown in Fig. VI.60A may serve as an illustration of the intersection of two initially slender objects which lead on subsequent growth to morphological features such as illustrated in Fig. III.41. Similar morphological observations and conclusions of crystal growth were drawn by Rees and Bassett (1971).

The detailed growth process is shown schematically in Fig. VI.60B. New polymer is added on the growth face with fold-lengths and molecular nucleus dimensions not unlike those for atmospheric pressure crystallization. The immediately following crystal perfection step is, however, much enhanced at the elevated temperature and pressure. The rounded crystal edges were directly observed during the crystallization by Jackson *et al.* (1972) by optical microscopy through a diamond anvil high pressure cell.

The isothermal crystallization kinetics of polyethylene at 5 kb pressure was measured by Hatakeyama (1974). Figure VI.61 reveals that the sharp fractions seem to crystallize in two stages, while broader molecular weight distributions show more normal behavior. Fitting an Avrami expression [Eq. (33)] to the

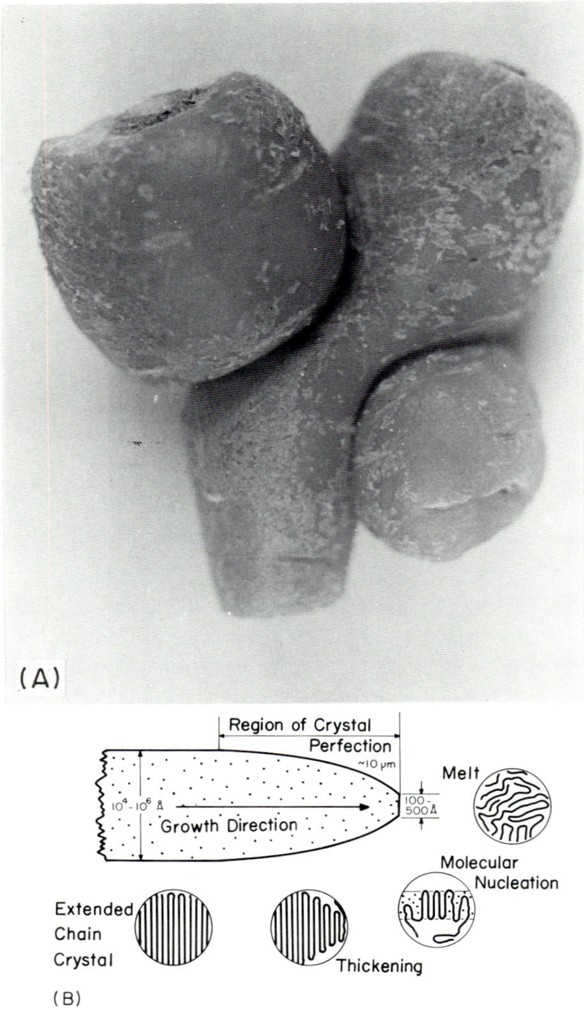

Fig. VI.60 *Extended Chain Crystal Growth.*
A. Interpenetration of carrots growing in a restricted space. On thickening, the structure produced can serve as a model for extended chain crystal intersection as shown in Fig. III.41 (observed, photographed, and contributed by A. Keller). B. Schematic drawing of the changes in a polyethylene lamella during crystallization at elevated pressure. Estimates are for 230°C, 5 kb crystallization conditions. It is assumed here that new polymer is added only to the growth face; compare with poly(ethylene oxide), where it is possible to add new chains from the lamellar surface (Sect. 6.1.6).

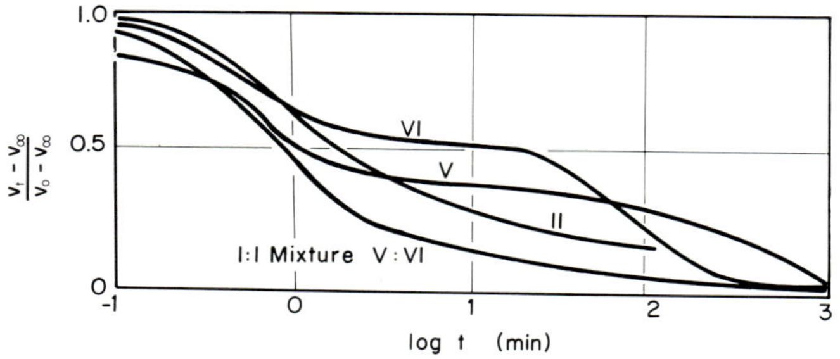

Fig. VI.61 *Crystallization Isotherms at 5 kb for Polyethylene.* VI: fraction of 5×10^4 MW; V: fraction of 1.7×10^4 MW; II: broader molecular weight distribution sample of average 5.4×10^4 MW; V + VI: mixture of fractions V and VI in equal weight ratio. Data by Hatakeyama *et al.* (1974).

data yields an exponent n close to one for the one-step crystallization. From the data of Fig. VI.50 an exponent of $n = 3$ would be expected on crystallization of these polymers at atmospheric pressure. At intermediate pressures of 1–2 kb where still folded chain lamellae are observed, the exponent was found to be 2–3; only at higher pressures, accompanied with extended chain crystallization, did n drop below 2, to reach 1 at about 5 kb.

The reason for the distinct two-stage crystallization kinetics is at present not fully explained. Bassett and Turner (1974) report similar observations through thermal analysis at elevated pressure. Analysis of the crystallized fractions after release of pressure shows normal melting behavior with a single endothermic peak, excluding the possibility of two-stage crystallization to different crystal perfection or size. Thus, Hatakeyama *et al.* (1974) suggested that the first-grown network of large lamellae, as illustrated to some degree by Fig. III.40, hinders the trapped melt temporarily from reaching suitable crystallization sites. Indeed, on inspection of larger surface areas, one frequently gains the impression that the exceptionally large crystals are grown first, to be followed by somewhat smaller ones, usually nucleated densely on the chain end and fold surfaces of the earlier lamellae as shown in Fig. VI.62 (see also Jackson *et al.* 1972). It is not known at present, however, whether the samples showing two-stage crystallization have a particularly well "stacked" morphology, in contrast to the branched morphology which should lead to more spherulitic crystals with less chance of trapping melt. A different explanation was offered by Bassett and Turner (1974) who found that although melting at atmospheric pressure showed no bimodal distribution of crystal perfection or size, melting at elevated pressure developed a double peak, seemingly the reverse of the two-stage crystallization. Similar

Fig. VI.62 *Fracture Surface of Extended Chain Crystals of Polyethylene, Illustrating Secondary Lamellar Growth.*

Broad molecular weight sample, crystallized for 8 hr at 227°C and 4.5 kb. The fraction surface is decorated with folded chain lamellae by heating to approximately 130°C (Sect. 3.4.1). Scale 1 μm (Wunderlich and Melillo, 1968).

observations were also made by Yasuniva *et al.* (1973). Based on the double endotherms on melting and also the two-step crystallization described above, Bassett and Turner (1974) postulated a new crystalline phase of polyethylene at elevated pressure. This new phase could be identified by high pressure X-ray techniques and shown to have a 4.24 Å chain separation with hexagonal symmetry (Bassett, 1974). From volume measurements one would expect for this new crystal structure a shortened *c* dimension, interpreted by Bassett (1974) as an indication of an irregular helix. Such a high pressure phase could also be the cause of the ease of chain extension. Polytetrafluoroethylene is known to have a similar irregular helix of increased mobility in the temperature range where extended chain crystallization is possible (see Sect. 2. 4.1). Yasuniva *et al.* (1973) emphasize that extended chain crystallization can be a fast process (see also Wunderlich and Arakawa, 1964) and is certainly distinct from slow annealing which can also lead to an increase in the chain extension as discussed in Sect. 7.3. The low Avrami exponent has not found any explanation yet.

A series of other macromolecules have been crystallized from the melt under elevated pressure. Polychlorotrifluoroethylene forms extended chain crystals similar to polyethylene (see Fig. III.33). Miyamoto *et al.* (1972) showed that pressure less than 1 kb is sufficient for chain extension.

Thin films of *cis*-1,4-poly(2-methylbutadiene) were crystallized at 4 kb pressure by Phillips and Andrews (1972). Chain extension by a factor of up to 3 was observed relative to atmospheric pressure crystallized samples; the latter have normally only a 70–80 Å fold-length. The crystals were distinctly elliptical when viewed normal to the chain axis, in agreement with considerable chain extension after initial crystallization (see Fig. VI. 60B). The shift of the maximum crystallization rate of *cis*-1,4-poly(2-methylbutadiene) with pressure to higher temperature was explained by Martin and Mandelkern (1963) through the increase in melting temperature with pressure. The rate of crystallization at the maximum decreases with pressure.†

The crystallization rate and number of nuclei of poly(hexamethylene succinate) were measured by Asmussen *et al.* (1972) at pressures of 0.1–5.0 kb and temperatures between 50°C and 100°C. At constant temperature both quantities increased, but at constant degree of supercooling, a decrease was noted.

Nylon 6 was crystallized from the melt at pressures up to 8 kb by Gogolewski and Pennings (1973). After purification, decomposition could be reduced and it seems that extended chain crystals similar to those grown by crystallization during polymerization described in Sect. 6.4.2.3 were grown (slow cooling to room temperature from 240°C to 310°C).

Siegmann and Harget (1973) reported the possible growth of extended chain crystals of poly(ethylene terephthalate) from melts under 2 kb pressure.

Isotactic polypropylene was shown by Kardos *et al.* (1966) to crystallize at elevated pressures to increasing amounts in a different high pressure crystal form (γ). Strained samples (500–800%) were shown by Hasegawa *et al.* (1970) not to change from the normal crystal structure (see Table II.10), but reach much higher order.

† For a discussion of early data on crystallization of *cis*-1,4-poly(2-methylbutadiene) see also Jenckel (1955). Paragraph 42 in H. A. Stuart, Ed. "Die Physik der Hochpolymeren." Vol. 3, Springer Verlag, Berlin.

Different high pressure crystal structures were also reported for poly-(vinylidene fluoride) (Doll and Lando, 1968), polytetrafluoroethylene (Weir, 1953), syndiotactic polypropylene (Hasegawa *et al.*, 1970), and isotactic poly(4-methyl-1-pentene) (Hasegawa *et al.*, 1970).

The combined effects of pressure and orientation on the crystallization of polyethylene have been described by Southern *et al.* (1972).† The samples were produced by melt extrusion at 134–139°C through a capillary under conditions of crystallization. The polyethylene begins to crystallize at a pressure in the vicinity of 600 b in the conical capillary entrance region. This partially crystalline sample is then forced through the narrow capillary, causing melting and recrystallization due to shear (final barrel pressure 1.9 kb, shear gradient 185/sec). Transparent strands of polyethylene are produced that are distinctly higher melting (139°C by thermal analysis at 5°C/min) and more crystalline (0.82) than similarly crystallized nonsheared samples (131°C, 0.64). In addition, the strands are of the highest degree of uniaxial orientation thus far recorded (*c* axis along the strand direction). The transparent filaments are composed of a large number of microfibers of 200–250 Å diameter in its core and 3000-Å diameter fibers in its outer sheath. Continuous crystalline order seems to persist along the axis of the microfibers for approximately 5000 Å. The outer fibers are significantly less ordered than the inner micro-fibers and show an X-ray low angle spacing of 290 Å. The core microfibers consist thus of extended chain crystals, while the outer fibers are composed of folded chain lamellae, both morphologies having the same *c* axis orientation. It was found by measurement of the infrared dichroism (Miller *et al.*, 1973) that the remaining amorphous chain portions are much less oriented than in highly drawn polyethylene (see Sects. 6.3.2 and 7.3.2).

Southern and Wilkes (1973) could show that films extended under conditions similar to the fibers discussed above, have a similar structure, but with additional orientation. Besides orientation of the *c* axis in the extrusion direction, the *a* axis is preferentially aligned in the thickness direction and the *b* axis in the traverse direction.

By compression of lightly cross-linked polyethylene at 160–180°C, Kitamaru and Hyon (1974) produced melt films of $\frac{1}{10}$ their initial thickness. This was followed by crystallization through fast cooling. The orientation of

† See also Sect 6.3.2 for the effect of orientation on crystallization and Sect. 7.3.3 for the effect of pressure on the annealing of oriented samples. For earlier studies see Van der Vegt and Smit (1967). "Advances in Polymer Science." Soc. Chem. Ind. (London), Monograph 26, 313; Sieglaff and O'Leary (1969). *Am. Chem. Soc. Preprints, Div. Polym. Chem.* **10**, 57; Southern and Porter (1970). *J. Macromol. Sci.* **B4**, 541, and *J. Appl. Polym. Sci.* **14**, 2305; Desper *et al.* (1970). *J. Appl. Phys.* **41**, 4284; Crystal and Southern (1971). *J. Polym. Sci. Part A-2* **9**, 1614.

the (110) plane was observed parallel to the film surface. The crystallites were found to be up to 1500 Å in the *c* direction and 170–350 Å in the other directions. At lower compression the (200) plane orientated parallel to the film surface.

6.3.4 Effect of Copolymerization

A copolymer is described in Sect. 1.5 as a molecule consisting of different repeating units. The various repeating units may differ in chemical composition or only in configuration, structure, or position. Their sequence is fixed during the polymerization. The order of the different repeating units along the molecular chain is usually characterized as random, blocky, or regular. Since regularly repeated sequences can have any length, a practically unlimited variety of different regular copolymers can be made.

For the discussion of crystallization of copolymers it is of basic importance to find the degree of isomorphism possible between the various repeating units (see Sect. 2.4.8). In solution crystallization (Sect. 6.2) noncrystallizable solvent molecules can simply be rejected at the crystal interface and thus affect crystallization through the lowering of the chemical potential of the crystallizable units in the dissolved phase and through diffusion effects. Noncrystallizable copolymer units, in contrast, interrupt or terminate crystal growth in the molecular chain direction because their position along the chain is fixed. A copolymer unit within a crystal represents a defect with various possible disruptive effects. The defect can range from a simple point defect (Sect. 4.3.4) to an amorphous defect (Sect. 4.3.1). Frequently the severity of the crystal disruption is not only a fixed property of the crystal structure and the nature of the copolymer unit, but depends also on the crystallization conditions. Adding to these problems the complications of chain folding (Sect. 6.1.4), the effects of copolymer units rejected to crystal surfaces, and the segregation of molecules of varying composition, one can realize that copolymer crystallization represents a most difficult problem which is far from being solved. In this section attempts at the theory of copolymer crystallization are summarized first, to be followed by a discussion of some experimental results.

The basic two-phase equilibrium theory of crystallization of copolymers was developed by Flory (1955). It contains the stringent conditions that all crystallizable repeating units, called A, are freely available to add to crystals. The A units occur in sequences of various lengths ζ along the chain. Every time a non-crystallizable B unit terminates an A sequence, the crystal must also be terminated. Ultimate equilibrium is reached when all sequences of crystallizable A units are in crystals of lengths matching the sequence lengths ζ. Such morphology could thus be characterized as ideal fringed micellar (see also Sect. 6.1.7). Noncrystallizable B units after completion of crystallization must make up the surface layers of the crystals. No chain folds, crystal defects, or crystal-size effects other than the limitation to ζ repeating units in the molecular chain direction are taken into account. Although unrealistic when compared with actual copolymer crystallization, one expects such equilibrium theory to be useful in providing an expression for the calculation of the thermodynamic driving force towards crystallization.

The quantitative description of a copolymer melt is based on the probability P_ζ that a given A unit is located in a sequence at least ζ A units in length. Next we represent by w_j the probability that a randomly chosen repeating unit is an A unit and is also a member of a sequence of j A units terminated on both ends by B units. Finally $P_{\zeta,j}$ represents the probability that a specific A unit is a member of a sequence j in length and is followed in a given direction by at least $\zeta - 1$ additional A units. With these definitions the following relationships become obvious:

$$P_{\zeta,j} = w_j(j - \zeta + 1)/j, \qquad j \geq \zeta \tag{94}$$

$$P_{\zeta,j} = 0, \qquad j < \zeta \tag{95}$$

$$P_\zeta = \sum_{j=\zeta}^{\infty} P_{\zeta,j} \tag{96}$$

Writing out Eq. (96) for successive values of ζ reveals to us the validity of the following difference equation

$$w_\zeta = \zeta(P_\zeta - 2P_{\zeta+1} + P_{\zeta+2}) \tag{97}$$

The quantities P_ζ and w_ζ must now be related to the structure of the two phase system. In a copolymer before crystallization, the mole fraction of A units is n_A, the number of A units is N_A, the number of sequences of length ζ is N_ζ^0. Knowledge of these quantities permits the calculation of P_ζ^0 and w_ζ^0, the initial probabilities. Choosing a random copolymer, we find that the probability p of the occurrence of an A unit next to a chosen repeating unit of the polymer is equal to n_A, so that we can easily see that†

$$P_\zeta^0 = p^\zeta \tag{98}$$

and

$$w_\zeta^0 = \zeta(1 - p)^2 p^\zeta \tag{99}$$

† For nonrandom copolymers

$$P_\zeta^0 = n_A p^{\zeta-1} \tag{98a}$$

and

$$w_\zeta^0 = n_A \zeta N_\zeta^0 / N_A \tag{99a}$$

Equilibrium between a crystal consisting of sequences of A units ζ in length [which has ΔG_ζ^0 as its free enthalpy of fusion into a melt of sequences of length ζ only ($P_\zeta = 1$)] is reached under ideal solution conditions whenever melting can occur without change in free enthalpy or

$$\Delta G_\zeta = \Delta G_\zeta^0 + kT \ln P_\zeta^e = 0 \tag{100}$$

where P_ζ^e is the temperature dependent equilibrium probability (note that all sequences of A-units ζ or longer contribute to P_ζ). For any ΔG_ζ^0 one can write in analogy to the prior used expression for nucleation of crystals (see Eqs. V.2 and V.11)

$$\Delta G_\zeta^0 = a_0 \, b_0 \, c_0 \, \zeta \, \Delta g_\mathrm{f} - 2a_0 \, b_0 \, \gamma_\mathrm{e} \tag{101}$$

where Δg_f is the free enthalpy of fusion/cm^3 and a_0, b_0, and c_0 are the appropriately chosen crystallographic dimensions of a repeating unit; $c_0 \zeta$ is the crystal length in the chain direction, previously labeled l. The parameter γ_e represents a surface free energy of a fringed-micelle-type crystal (see Sect. 6.1.7) and all limitations discussed for fringed micellar nuclei in Sect. 5.1.2.1 apply naturally also to the copolymer crystals. In particular, when the sequences of A units are long enough, chain folding must intervene. This drive towards chain folding might also permit inclusion of occasional B units within the crystal to extend the limit of chain-folded crystallization.

A typical equilibrium distribution of w_ζ^e for different temperatures is shown in Fig. VI.63. The plotted data are calculated using Eqs. (97), (100), and (101). Where A and B units are of equal weight, w_ζ^e would also represent the weight fraction of A sequences ζ in length remaining in the melt at equilibrium. Also shown in Fig. VI.63 is the initial distribution for a random copolymer melt computed from Eq. (99). Given both the initial and the equilibrium distributions of the different sequences ζ, the driving force towards crystallization can be calculated. From Eq. (100) one concludes that a necessary and sufficient condition for crystallization is that

$$P_\zeta^0 > P_\zeta^e \tag{102}$$

for one or more values of ζ (Flory, 1955). Comparing the initial values of P_ζ^0 (Eq. 98) with the equilibrium values P_ζ^e fixed by Eqs. (100) and (101) one can see that at any one temperature a critical length ζ^* exists below which no crystal growth can start and above which there is a driving force towards crystallization. At the length ζ^* P_ζ^e and P_ζ^0 are equal so that [from Eqs. (98) and (100)]

$$kT \ln P_{\zeta^*}^e = kT\zeta^* \ln p = -\Delta G_{\zeta^*}^0 \tag{103}$$

The temperature for which ζ^* becomes very large is the equilibrium melting temperature T_m of the copolymer in question. When large crystals are

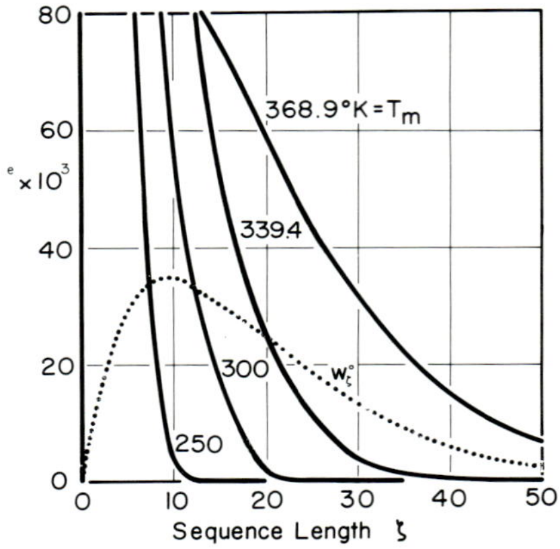

Fig. VI.63 *Equilibrium Weight Fractions of A Sequences in the Melt.*
Calculated assuming $\Delta H_f/R = 500°$K, $T_m{}^\circ = 400°$K, $2a_0b_0\gamma_e/k = 340°$K, $n_A = p = 0.9$. Dotted curve: initial distribution $w_\zeta{}^\circ$. The temperatures of equilibrium in °K are listed next to the curves. T_m, the melting temperature of a large crystal in contact with the initial melt is 368.9°K [Eq. (104)].

contained in a crystallized copolymer this temperature would be realized as the melt end. It can be calculated from Eqs. (103) and (101)† letting ζ^* increase to large values (see also Chapter X).

$$(1/T_m) - (1/T_m^0) = -(R/\Delta H_f) \ln p \tag{104}$$

where T_m^0 is the equilibrium melting temperature of the pure homopolymer of A units only, and ΔH_f is the heat of fusion per mole of repeating units of A units.

Below T_m, crystallization is thermodynamically possible. It was shown by Baur (1965) that even when one neglects all nucleation difficulties, the change from w_ζ^0 to the equilibrium value w_ζ^e illustrated in Fig. VI.63 is not necessarily monotonous. Crystallization can occur such that the time-dependent $w_\zeta(t)$ increases or decreases. This conclusion was reached by assuming that the rate of change in P_ζ on crystallization is always proportional to the thermodynamic driving force $\Delta G_\zeta/T$, as required by the linear laws of the phenomenological nonequilibrium thermodynamics. The change of P_ζ with time is expressed by

$$dP_\zeta/dt = -L_\zeta \, \Delta G_\zeta/T \tag{105}$$

† Δg_f is approximated as before by $\Delta h_f'[1-(T_m/T_m{}^\circ)]$, see Eqs. (16)–(18), Chapter V.

where L_ζ is the phenomenological coefficient. A first approximation to the differential equation is the first term of the Taylor expansion

$$P_\zeta(t) = P_\zeta^e + [P_\zeta^0 - P_\zeta^e] \exp(-t/\tau_\zeta) \tag{106}$$

and

$$1/\tau_\zeta = L_\zeta[\partial\, \Delta G_\zeta/\partial P_\zeta]_e/T$$

where τ_ζ is the isothermal and isobaric time constant for crystallization of sequences ζ in length. Using Eq. (100)

$$\tau_\zeta = P_\zeta^e/kL_\zeta \tag{107}$$

Figure VI.64 illustrates the change of τ_ζ for three lengths ζ as a function of temperature. At any given temperature a certain sequence fractionation is

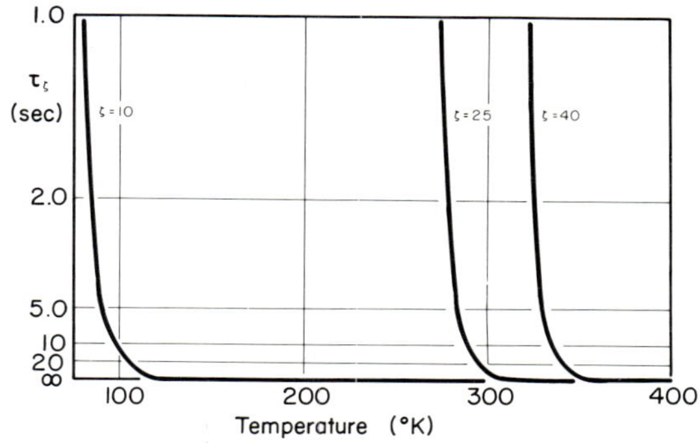

Fig. VI.64 *Change of τ for Different ζ as a Function of Temperature.*
Calculated from Eqs. (107), (100), and (101) assuming $\Delta H_f/R = 500°\text{K}$, $2a_0 b_0 \gamma_e/k = 4000°\text{K}$ (about 150 erg/cm^2), $T_m^\circ = 400°\text{K}$ and $kL_\zeta = 1/\text{sec}$. Drawn after Baur (1965).

possible because of the different time scales for crystallization. Using Eqs. (97) and (106), the time dependence of w_ζ can be calculated. Figure VI.65 shows the changes which should occur under the chosen conditions during the course of crystallization leading from w_ζ^0 to finally w_ζ^e. In the initial time interval $w_\zeta(t)$ decreases for any value of ζ above ζ^*. Then, however, it increases for the smaller ζ steeply above w_ζ^0 since a sequence of A units may crystallize initially on to any crystal of its own length or shorter. Any crystallization onto a crystal of shorter length produces one or two amorphous ends which increase the concentration of shorter amorphous A sequences in the melt. The overall crystallization shown in Fig. VI.65 proceeds along the following sequence of steps: Initially the very long sequences reach the

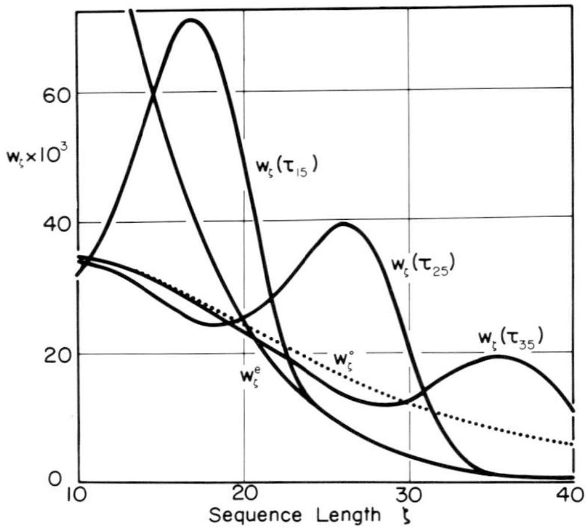

Fig. VI.65 *Change in $w_\zeta(t)$ as a Function of ζ for Several Successive Times.*
Calculated for $p = n_A = 0.9$ so that for the conditions of Fig. VI.64, $T_m = 368.9°K$ (Eq. 104);

$$\exp\left[\frac{\Delta H_f}{R}\left(\frac{1}{T_m^\circ} - \frac{1}{T}\right)\right] = 0.8$$

which corresponds to a supercooling of $T = T_m - T$ of about 29°C; the surface free energy has been chosen such that $2a_0 b_0 \gamma_e / kT = 1.0$. The three distributions $w_\zeta(t)$ correspond to the times τ_{35}, τ_{25}, and τ_{15}, respectively. Drawn after Baur (1965).

equilibrium value w_ζ^e. Through their crystallization an enrichment (relatively small) of long sequences (shorter than the initially crystallized ones) results. As crystallization continues, this "wave" of increased concentration in the distribution curve grows in amplitude and runs towards smaller values of ζ, to finally settle into the equilibrium distribution. To stay all the time as close to equilibrium as required by the treatment, all crystals must continuously readjust their composition to comply with the changing forces [Eqs. (105) and (100)]. Experimentally we know that this condition cannot be met. The system will thus freeze long before equilibrium is reached.

The connection of the above calculations to the experimentally observed overall crystallization rate (Sect. 6.1.3) is given through the time dependent crystallinity $w^c(t)$ (see also Sect. 4.1.1). Since all quantities w_ζ above refer to the molten part of the two-phase system only,† $w^c(t)$ is given for the case of A

† The equation $w^c = \sum_{\zeta=\zeta^*}^{\infty}[w_\zeta^0 - w_\zeta^e]$ given by Flory (1955) for the final crystallinity at a chosen temperature is only approximately valid for small degrees of crystallization as can be seen from Eq. (111), Baur, (1965). Its use in crystallization of copolymers of large crystallinity should be discontinued.

and B units of equal weight by

$$w^{c}(t) = \frac{N_{A} - N_{A}^{a}(t)}{N_{A} + N_{B}} \tag{108}$$

where $N_{A}^{a}(t)$ is the number of A units remaining in the melt at time t and N_{A} and N_{B} are the total numbers of A and B units in the crystallizing system. Since

$$N_{A}/(N_{A} + N_{B}) = \sum_{\zeta=1}^{\infty} w_{\zeta}^{0} \tag{109}$$

and

$$N_{A}^{a}(t)/(N_{A} + N_{B}) = [1 - w^{c}(t)] \sum_{\zeta=1}^{\infty} w_{\zeta}(t) \tag{110}$$

where $[1 - w^{c}(t)]$ is the total fraction of repeating units of A and B which remain molten, one can easily see that

$$w^{c}(t) = \frac{\sum_{\zeta=1}^{\infty} [w_{\zeta}^{0} - w_{\zeta}(t)]}{1 - \sum_{\zeta=1}^{\infty} w_{\zeta}(t)} \tag{111}$$

Equation (111) can be used to calculate through Eqs. (105)–(107) the development of crystallinity at constant temperature, and it can also be used to compute the change of equilibrium crystallinity with temperature by setting $w_{\zeta}(t) = w_{\zeta}^{e}$, accessible through Eqs. (97) and (100).

A number of efforts exist to analyze the experimental copolymer crystallinity in terms of Eq. (111) or its approximation (see footnote p. 264). Since crystallization studies are governed by nucleation, usually melting experiments after careful (slow) crystallization are chosen to stay closer to the equilibrium predicted by Eq. (111). The results provide clear evidence that actual copolymer crystal distributions are far from equilibrium. Adjusting the one not fully fixed parameter γ_{e} (Eq. 101) to get a best fit, poor correspondence is found at low temperature (the observed crystallinity is much less than predicted) and at high temperature [the last crystal melts much earlier than predicted by Eq. (104)] (see Richardson *et al.* 1963 and Chapter X). Qualitatively these deviations are easily understood from the general knowledge of crystallization of macromolecules (see Sect. 6.1). The low temperature crystallinity is reduced because small size crystals cannot form since the longer crystallizable sequences are frozen in crystals, prohibiting the shorter sequences to diffuse to suitable crystallization sites. In addition, a general reduction in crystallinity is expected from the crystallization of A sequences on crystals shorter than their length in the early stages of crystallization which later cannot be changed as required by the ultimate equilibrium (see Fig. VI.65). The poor fit at high temperatures, finally, is caused by nucleation difficulties of the largest crystals which results in chain folding (see Sect. 5.1.2.1) with the accompanying reduction in crystallinity (see Chapter IV) and melting point (see Chapter IX).

Efforts by Kilian (1965) to take into account the hindered equilibrium by reducing ζ^* of Eq. (103) by assuming that each sequence of A units crystallizes on a crystal which is a given amount (ζ_k) shorter in the chain direction, permits better fitting of the experimental data. Baur (1966) showed that the introduction of this additional fitting parameter corrects not only for the actual crystal sizes, but in an indirect way, must account for the mixing effects which cause crystallization on slow cooling not to go strictly in order of sequence length.

Crystallization under even more restrictive conditions such that only nearest neighbors in the melt can crystallize, leads to the limit of lowest crystallinity. Wunderlich (1958) has analyzed this simple case of "cold crystallization" (see Sect. 6.1.7 and Fig. VI.29). To compare with experimental data, it is necessary again for us to have two parameters for fittings, namely the shortest possible crystal in the molecular chain direction, and an end surface free energy γ_e.

Observations of isomorphism (see Sect. 2.4.8) leaves the possibility that in some copolymers all or a limited amount of the foreign B units can be included in the crystal lattice. This case of cocrystallization was treated by Sanchez and Eby (1973) with the rather simple assumption that only the heat of fusion is affected (linearly) by copolymerization, and by Helfand and Lauritzen (1973) who based their description on the general fluctuation theory of chain folding by Lauritzen and Passaglia (1967) (see Sect. 6.1.4.3).

While the equilibrium considerations are up to now of limited use for the actual description of copolymer crystallization, they form the basis for the understanding of the kinetics by providing the necessary input for the free enthalpy difference needed for the formalism of nucleation. Gornick and Mandelkern (1962) generalized the nucleation expression [Eq. (5), Chapter V], to take account of the copolymer effect. For the nucleation rate they wrote

$$I^*(\theta) = I^*(0)\frac{n_A(\theta)}{n_A(0)}\exp\left(-\frac{\Delta G^*(\theta) - \Delta G^*(0)}{RT}\right) \tag{112}$$

The variable θ characterizes the extent of crystallization at time t for the copolymer. It is defined by

$$\theta = w^c(t)/w^e \tag{113}$$

and represents a variable changing from 0 to 1 during complete crystallization.

The necessary free enthalpy expressions at the various θ can be calculated from Eq. (104) which takes the form

$$(1/T_\theta) - (1/T_m) = -(R/\Delta H_f)\ln[n_A(\theta)/n_A(0)] \tag{114}$$

so that $\Delta g_f(\theta) = \Delta h_m(T_\theta - T)/T_\theta$ in analogy to Eqs. (16)–(18), Chapter V, just that the heat of fusion is replaced by the appropriate heat of dissolution Δh_m. Equations (112)–(114) can be applied to primary as well as secondary and molecular nucleation. The main effect is the attenuation of nucleation rates due to a lowering in equilibrium melting temperature T_θ because of increasing copolymer units in the melt. Figure VI.66 illustrates the expected

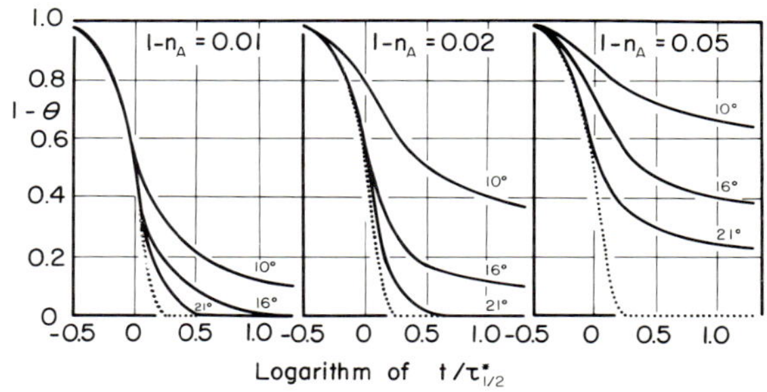

Fig. VI.66 *Change in Relative Crystallinity θ as a Function of Logarithm of Time for Copolymers.*
Calculated using an Avrami-type calculation based on nucleation as described by Eqs. (112)–(114). Time in multiples of the half-crystallization time τ^* which holds if Eq. (33) with $n = 4$ were obeyed (constant nucleation rate dotted curve). Random copolymers of $n_A = 0.01$, 0.02, and 0.05. Numbers next to curves show the undercooling $T_m - T$. Drawn after Gornick and Mandelkern (1962).

overall effect of the changing copolymer concentration on crystallization of three different initial values of n_A as calculated by Gornick and Mandelkern (1962). The time scale is expressed in multiples of the half-time expected for constant concentration. A characteristic retardation at long times is observed even at relatively low copolymer content. Since defects such as chain ends and branch points in homopolymers may also often be treated as noncrystallizable copolymer units, there is a possibility that at least part of the secondary crystallization discussed in Sect. 6.1.6 is caused by the attenuation due to the "copolymer" units. It is interesting to note that in the case of "cold crystallization" (see Sect. 6.1.7 and above) insufficient mobility exists to change the concentration in the amorphous phase and, as a result, the overall crystallization rate shows none of the above predicted retardation as is illustrated by Fig. VI.11.

Experimental evidence which qualitatively supports the attenuation of crystallization through the presence of copolymer units was presented by Mandelkern (1964a, Section 8-6) from literature data on crystallization of *trans*-1,4-polybutadiene with 20% cis configurations, branched polyethylene, cross-linked polymers, and polypropylene of limited stereo-regularity.

A more detailed analysis of a copolymer system was carried out by Nachtrab and Zachmann (1970) and Johnsen *et al.* (1970) on the ethylene–vinyl acetate system (0–24 wt. % vinyl acetate). For all copolymers a spherulitic morphology was observed which became increasingly feathery at higher crystallization temperatures. For higher copolymer content the spherulites were strongly nonspherical. Crystallization slowed down when two spherulites approached each other. Figure VI.67 shows the change of the linear spherulite

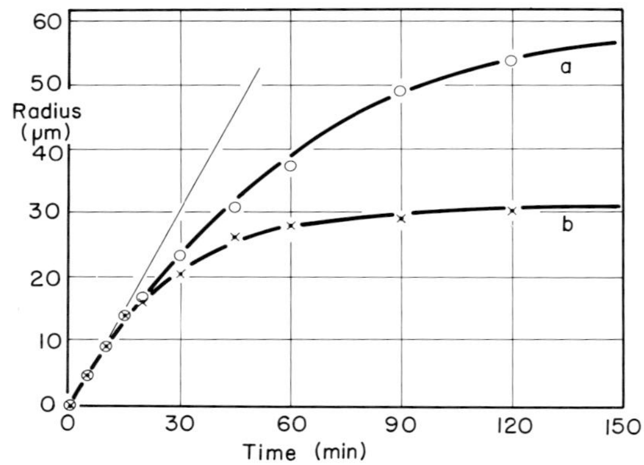

Fig. VI.67 *Spherulite Dimensions of a Copolymer of Ethylene and Vinyl Acetate as a Function of Time.* Crystallization temperatures 85°C; 17 wt.% vinyl acetate. The spherulite dimension was measured along a radius pointing towards a distant neighboring spherulite (curve a) and a closer neighboring spherulite (curve b). Drawn after Nachtrab and Zachmann (1970).

dimensions of a typical copolymer spherulite in two directions. One direction was chosen so that it would point to another spherulite, about 160 μm away, the other was chosen in line with a much closer neighboring spherulite, 60 μm away. In contrast to homopolymers (see Fig. VI.3) the linear crystal growth rates are not constant, and in addition, growth slows down more whenever the neighboring spherulites are approached within $\sim$20 μm. The slowing down of crystallization with time is expected from the previous discussion, but for the spherulite boundary a slow-down in growth rate should occur only if different sequences of crystallizable units are freely mobile and can readjust the concentration in the bulk of the remaining amorphous polymer. Obviously a copolymer molecule cannot move beyond its own length after a single segment has crystallized. One can thus explain the changing crystal growth rates only if segregation of whole molecules is possible. The deviation from spherical shape would then be an indication of beginning diffusion control. In agreement with possible diffusion of whole molecules less suitable to crystallization, it was observed by Nachtrab and Zachmann (1970) and Johnsen *et al.* (1970) that the nonlinearity of the crystal growth rate decreases at lower crystallization temperature. At the lower temperatures the fraction of molecules without crystallizable sequences decreases rapidly (but the mobility of the molecules relative to the crystal growth rate decreases also as long as one carries out experiments on the high temperature side of the growth rate versus temperature curve).

An approximate calculation of the decrease in growth rate was possible assuming simplified concentration gradients. The initial growth rate given in Fig. VI.67 by the tangent at time $t = 0$ should be describable by nucleation expressions like Eq. (5), Chapter V, with free enthalpy and concentration adjusted to the copolymer case. Plotting the logarithms of the initial growth rates versus $T_m/(T\Delta T)$ as in Figs. VI. 23 and VI. 43 gives us a series of parallel lines for the various copolymers. The same slopes must mean that the $\gamma\gamma_e$ product and Δh_f, the heat of fusion of Eqs. (68) or (81), Chapter V, cannot be much different for all the copolymers (300 erg^2/cm^4 and 70 cal/cm^3, respectively). The preexponential factor however, seems to have more than a simple n_A dependence of crystal growth; increasing copolymer content slows down the growth rate more than suggested by Eq. (112). Attempts to fit Avrami expressions [Eq. (33)] to the data for overall crystallization measured by calorimetry failed as expected from Fig. VI.66 after a relatively short initial period (74–10 % of total crystallization for the analyzed homopolymer to copolymer series). The slowing of crystallization must be a major reason for the earlier deviation. The final crystallinity was found to increase for any given copolymer with increasing supercooling (up to more than twice the low supercooling values). This indicates that the change in activity of copolymer units expressed by Eq. (100) must be the major influence in setting the final crystallinity level (see Fig. VI.66) although the maximum level permitted by Eq. (111) is never reached. If crystallization were limited by freezing of the residual melt after growth of initial crystals alone, higher temperatures should lead to higher crystallinity due to higher mobility of the molecules.

Ethylene terephthalate copolymerized with up to 30 % ethylene sebacate or ethylene azelate was studied by measurement of transmitted light between crossed polarizers by Jackson and Longman (1969). The curves of the half-times for crystallization when plotted versus crystallization temperature are moved to lower temperatures for higher concentration of noncrystallizable units. No quantitative analysis was attempted. The ethylene terephthalate–ethylene sebacate copolymer crystallization kinetics shown in Fig. VI.11 was only studied under cold crystallization conditions (see above).

Nylon 11 copolymers with up to 20 mole % ω-aminocaproic acid were studied by Hybert and Pepper (1969) by dilatometry. Increasing amounts of comonomer increased the half-time of crystallization when plotted versus experimental supercooling (taken as the difference between the observed melting temperature and the crystallization temperature). The total crystallinity decreased with increasing concentration of B units and increased with increasing supercooling for any one polymer. Avrami plots with exponents of $n = 4$ were fitted to the early stages of crystallization of copolymers and homopolymer. Details of possible cocrystallization were not investigated.

The whole concentration range of copolyamides from nylon 6 with nylon 6.6 have been analyzed by Harvey and Hybart (1970) using light transmission techniques. The plots of the temperatures for constant half-times of crystallization (60 min and 10 min) as a function of composition followed closely the shape of the experimental melting temperature versus composition. The 60 min and 10 min half-time curves are displaced to lower temperatures by about 20°C and 30°C, respectively. All curves show a minimum at 40 mole % of hexamethylene adipate ($T_m \approx 180$°C).

Gogolewski and Turska (1972) carried out dilatometry of crystallization of copolymers of ω-caprolactam with piperazine adipate and piperazine terephthalate. The experimental melting temperatures decreased from 225°C to ~190°C for up to 8 mole % comonomer. The observed heats of fusion by scanning calorimetry decreased from 13.9 cal/g for the homopolymer to 10.2–11.6 cal/g for the copolymers with between 2.5 and 8 mole % comonomer units. All kinetic data fitted Avrami expressions with exponents between 2 and 4 depending on crystallization temperature (see also Sect. 6.3.1.5). The piperazine terephthalate copolymers crystallized slower than the piperazine adipate copolymers when compared at equal concentration and (experimental) supercooling (factor $\frac{1}{4}$ at 4 mole %). ω-Caprolactam copolymerized with various C-methyl substituted caprolactams was studied by Tomka and Sebenda (1972) by measurement of induction times by optical microscopy. Changes of the time for observation of first crystals seem to go parallel to the depression of the melting temperature with copolymer content.

All measurements on the polyamids are unsupported by structure data. The discussion of isomorphism (Sect. 2.4.8) suggests, in line with the relatively good fit of an uncorrected Avrami expression and high crystallinity of the copolymers, that some or all of the polyamids analyzed form mixed crystals. More precise measurements are necessary to study the changes in concentration and their effects on crystallization for the case of complete or partial isomorphism.

The crystallization of blockcopolymers of sequences of poly-ε-caprolactone/poly(ethylene oxide)/poly-ε-caprolactone of various ratios were studied by Perret and Skoulious (1972). The center-section of all tri-block molecules had a length of 540 Å, the two poly-ε-caprolactone sections were varied from 1200 Å to 313 Å to 86 Å. The homopolymers have similar melting temperatures (65°C), but such different crystal structures (see Tables II.14 and II.15) that isomorphism is not possible. Even in the molten state the triblock copolymers are most likely a two-phase system (see also Sect. 2.4.9 and Fig. II.65 for amorphous–amorphous triblock copolymers, and Sect. 3.2.2.2 and Fig. III.20 for crystalline–amorphous two-block copolymers). On crystallization the basic structure is always a multilayered lamella. The overall structure is determined by the thickness of the first growing lamella. The caprolactone-rich copolymer shows a constant 275 Å lamellar repeat thickness. Such constant thickness is characteristic of the homopolymer. The folded chain poly-ε-caprolactone lamellae have a thickness of about 113 Å, covered on either side with an about 52 Å thick layer of amorphous poly-ε-caprolactone. This agrees with the observed crystallinity of about 0.5. Between the poly-ε-caprolactone lamellae are poly(ethylene oxide) lamellae of 59 Å thickness which crystallize for this small layer thickness only on annealing. The poly-(ethylene oxide)-rich copolymer, in contrast, shows a varying overall lamellar thickness with crystallization temperature as is characteristic for the homopolymer (150–415 Å total lamellar thickness for crystallization from 20 to 55°C, 121–334 Å for the poly(ethylene oxide) portion, compare with Fig. III.17). The copolymer of about equal proportions shows at crystallization temperatures below 51°C initial poly(ethylene oxide) crystallization followed by poly-ε-caprolactone crystallization in a two-step process (dilatometry). Above 51°C the polycaprolactone crystallizes first. The morphology changes parallel with the kinetics. From 20°C to 53°C crystallization temperature the lamellar repeat changes from 240 Å to 385 Å, to develop a second repeat starting from 48°C crystallization temperature (initially 270 Å). Finally, this second lamellar repeat length becomes the only one at a value of 330 Å above 53°C crystallization temperature. The kinetics shows little of the slow-down expected for random copolymers (see Fig. IV.66), although secondary crystallization or crystal perfection does persist over longer times. The block-copolymer components influence each other thus only through the geometrical restriction of the first component crystallizing to its most suitable folded chain lamellar structure.

6.4 Crystallization during Polymerization

As with the crystallization of the already polymerized macromolecule described in Sects. 6.2 and 6.3, nucleation is also of basic importance for the crystallization during polymerization. The two nucleation types, intermolecular and intramolecular nucleation, are discussed in Sects. 5.1.2.3 and 5.1.2.4 (see Fig. V.10). The basic mode of nucleation applies to primary and secondary nucleation as well as to molecular nucleation. Since in most cases of crystallization during polymerization the nucleation steps are not fully investigated, the description of subsequent crystallization is complicated.

Particularly severe are the restrictions when there is need of a homogeneous reaction for the production of oligomers or polymer portions capable of crystal and molecular nucleation. This reaction must change later to heterogeneous crystallization and polymerization (see Sect. 5.1.2.3). Also important is the possibility of occlusion of chain ends within the crystal as an additional special physical termination reaction of the polymerization to the crystalline state (see Sect. 6.4.2).

Rather than trying to present an extensive list of the many possible cases of crystallization which would have to expand on an earlier review (Wunderlich, 1968) and involves reinterpretation of data usually collected without attention to the possibility of crystallization during polymerization, this section collects the better studied examples of crystallization during polymerization of gaseous, liquid (dissolved or molten), and solid monomers (Sect. 6.4.1–6.4.3, respectively). General aspects of crystallization during polymerization are given in Sect. 6.1.8; others are discussed with the examples on which they were explored. The special morphologies of selected crystals grown during polymerization are discussed in Sects. 3.8.1, 3.8.3, and 3.9.

6.4.1 Gaseous Monomer Crystallization

Growth of crystals from vapor has been studied less frequently than crystallization from solution or melt. This neglect is particularly surprising in view of the relatively advanced description of crystal growth from the vapor phase (Sect. 6.1.2.2) The general techniques available for growing organic crystals have been reviewed by Reynolds (1963). Vapor pressures, in the range of 10^{-3} atm to 1 atm at a temperature up to the melting point, are most suitable for the growth of large crystals. Lower vapor pressures cause inconveniently slow crystallization, and the chances of dendritic growth are increased. The best crystals are expected close to the ceiling temperature [Eq. (91), Table VI.6], with relatively small differences in temperature between the vapor and the crystal. The polymerization of selenium, formaldehyde, tetrafluoroethylene, ε-aminocaproic acid, and *p*-xylylene are discussed next as examples of macromolecular crystals grown from the gaseous monomer phase.

Selenium boils at 684.5°C. In the vapor phase at low temperature, it consists of a number of cyclic forms. Up to 200°C, one finds Se_5 and Se_6

rings mainly, with only a few Se_7 and Se_8 rings. At increasing temperatures, Se_7 and Se_8 disappears. At higher temperatures, Se_2 molecules appear, and dominate above 900°C (Fujisaki *et al.* 1966).

Single crystals in the form of long needles with the molecular helix axis oriented along the needle axis were grown by Straumanis (1940) at 0.01–0.11 mmHg by evaporation at 180–230°C. The melting temperature of trigonal selenium is 220.5°C. Best crystals were grown from vapors of only slightly higher temperature. The end surfaces of the needles, which one would expect to be (001), were exclusively $(1\bar{1}1)$ (see Sect. 6.1.2.2; for the crystal structure, see Table II.9 and Fig. II.39). At the lower temperatures, the needles are increasingly hollow. The needlelike morphology and the hollow ends are caused by faster crystal growth in the helix direction which does not allow long distance diffusion on the $(1\bar{1}1)$ surface and by limited nucleation of new molecules on the side surfaces.

At higher supersaturation ($p/p_0 = 25$–100) Furuta (1964) grew dendritic selenium crystals at 100°C which maintained the basic fibrillar growth habit and showed branching through twinning. Quenching of gaseous selenium produces amorphous (red) selenium (Straumanis, 1940).

Ehinger (1961) found that early during crystallization, needles grow in the surface of the support. This observation is in line with the need of initial oligomer molecules to be present before nucleation which must be adsorbed in the plane of the support (Sect. 3.4.3). Later, the needles grow mainly at an angle to the support. At crystallization temperatures close to the melting temperature, lamellae were observed in addition to needles. The lamellae grew apparently by addition of chains through successive polymerization and crystallization which leads to chain folding (see also Sect, 6.4.2), while the needles grew as a result of simultaneous polymerization and crystallization.

From comparison of thermal analysis of needlelike selenium crystals with bulk crystallized lamellae, one can deduce that needles are more perfect than normal melt grown crystals (Jaffe and Wunderlich, 1969).

A probable mechanism of selenium crystallization based on these observations is as follows. Initial nucleation occurs on the substrate by liquid oligomer molecules so that the nuclei have their [001] direction parallel to the substrate. Further growth in the temperature region of fast crystallization (from 80 to 160°C, Table VI.8) is by simultaneous polymerization and crystallization on the $(1\bar{1}1)$ surface. Diffusion occurs in the side surfaces over longer distances because of molecular nucleation difficulties. At faster rates of crystallization, this longer range diffusion can lead to hollow needles because of faster supply of crystallizable molecules to the needle edge from the side surface. At temperatures close to the melting temperature, polymerization may be faster than crystallization, changing the crystal habit to lamellar and the mechanism to successive polymerization and crystallization. Similarly.

successive polymerization and crystallization is possible at much lower temperatures; but at these temperatures an amorphous polymer results because of the proximity of the glass transition temperature, which is at about 45–55°C (Wolpert *et al.* 1971). Detailed kinetic data on crystallization of selenium from its vapor are not available.

Gaseous formaldehyde polymerizes readily below its ceiling temperature of about 120°C (Tables VI.6 and VI.14). Crystalline fibers grown on sublimation) were shown by Staudinger and co-workers to be macromolecular and to have the chain axis oriented along the fiber axis (see Sects. 3.2.1 and 3.8.1, Figures III.6 and III.127 for fibrous morphology, and Table II.14 and Fig. II.49 for crystal structure). The gas phase above the macromolecules contains an equilibrium mixture of formaldehyde and trioxane [$(CH_2{-}O{-})_3$-rings]. Formaldehyde equilibrates rapidly with the crystal and is thought to be the initial decomposition product; the trioxane equilibrium is reached more slowly.

Equilibrium data for the different monomer phases in contact with crystalline and liquid polymer have been derived by Berlin *et al.* (1970) from vapor

TABLE VI.14

THERMODYNAMIC DATA[a] PER MOLE OF MONOMER

Monomer	Phases[b] Monomer	⇌Polymer	ΔH^0_{298} (kcal/mole)	ΔS^0_{298} (cal/deg mole)	ΔG^0_{298} (kcal/mole)	T_c (°C)
Formaldehyde[c] $CH_2{=}O$						
	Gas	Crystal	−16.7	−42.6	−4.0	119
	Liquid	Crystal	−11.1	−20.7	−4.9	263
	Gas	Liquid	−14.9	−38.7	−3.4	112
Trioxane[e] $(CH_2{-}O{-})_3$						
	Gas	Crystal	−15.2	−37.0	−4.2	138
	Gas	Liquid	−9.9	−25.3	−1.45	118
	Liquid	Liquid	−0.1	~0	−0.1	—
	Liquid	Crystal	−5.4	−11.7	−1.9	187
	Crystal	Crystal	−2.0	−1.6	−1.5	~1000

[a] Table of Berlin *et al.* (1970), the changes of ΔH and ΔS with T are negligible in the temperature region of interest (25–200°C) relative to the experimental error.

[b] For the reaction $CH_2O \rightleftharpoons \frac{1}{3}C_3H_6O_3$, one can calculate $K_p = 2.63 \times 10^{-7} \exp(5840/T)\,\text{atm}^{-2/3}$.

[c] The small difference to Table VI.6 may arise because Berlin *et al.* (1970) considered formaldyhyde and trioxane separately; this was not done for the data in Table VI.6.

[d] $T_m = -118°C$; $T_b = -19°C$.

[e] $T_m = 62°C$; $T_b = 115°C$.

pressure and infrared measurements, and are listed in Table VI.14. According to these data formaldehyde as well as trioxane in the vapor phase below 120°C are thermodynamically metastable relative to crystalline polyoxymethylene. Jaacks and Kern (1962) show that trioxane can be sublimed without polymerization when all traces of formaldehyde are removed (with Ag_2O). The active polymerizing species in vapor phase polymerization is thus formaldehyde. Small amounts of polyoxymethylene found on condensation of trioxane are polymerized in the solid state during the phase transition which can be initiated by traces of formaldehyde (Sect. 6.4.3).

The kinetics of the condensation of pure formaldehyde to polyoxymethylene in a glass vessel was followed by Boyles and Toby (1967) by measurement of the disappearance of the monomer in the temperature range 0–60°C, and the pressure range 16–216 mm Hg. Writing the polymerization–depolymerization equilibrium as

$$CH_2O + I{-}(CH_2{-}O{-})_x \underset{k_d}{\overset{k_p}{\rightleftarrows}} I{-}(CH_2{-}O{-})_{x+1} \tag{115}$$

leads to a change in monomer concentration given by

$$\ln\left\{\frac{[CH_2O] - [CH_2O]_\infty}{[CH_2O]_0 - [CH_2O]_\infty}\right\} = \int_0^t -k_p[I]\,dt \tag{116}$$

where $[CH_2O]_0$ is the initial concentration (partial pressure) and $[CH_2O]_\infty$ is the equilibrium concentration k_p/k_d (assuming no formation of trioxane). The time-dependent concentration of once initiated, actively growing chains is [I]. Boyles and Toby (1967) found the right hand side of Eq. (116) proportional to S/V, the surface to volume ratio of the vessel. This indicates the polymerization is initiated at a fixed number of sites on the surface. On filling the clean vessel (bare walls), we can describe the reaction with a second-order termination reaction†

$$I{-}(CH_2{-}O{-})_x + I{-}(CH_2{-}O{-})_y \xrightarrow{k_t} I{-}(CH_2{-}O{-})_{x+y}I \tag{117}$$

which for Eq. (116) leads to the expression

$$\ln\left\{\frac{[CH_2O] - [CH_2O]_\infty}{[CH_2O]_0 - [CH_2O]_\infty}\right\} = -\frac{k_p S}{2k_t V}[\ln(2k_t[I]_0 t + 1)] \tag{118}$$

Subsequent fillings of the vessel without removal of the prior made polymer leads to increasingly slower reaction rates and approaches a rate described by

$$\ln\left\{\frac{[CH_2O] - [CH_2O]_\infty}{[CH_2O]_0 - [CH_2O]_\infty}\right\} = -\frac{k_p S[I]_0}{V}t \tag{119}$$

† No detailed reaction mechanism was assumed by Boyles and Toby. It is generally known that the reaction does not follow a free radical mechanism, but is ionic.

which suggests an approximately constant number of chain ends throughout the reaction. Boyles and Toby (1967) did not consider effects of nucleation and crystallization. One may suggest that the initial expression of Eq. (118) applies mainly to the homogeneous reaction phase during which monomer is consumed making oligomer suitable for nucleation, and that Eq. (119) applies to the growth phase during which termination occurs mainly by occlusion (see Sect. 6.4.2) and is counterbalanced by new molecular nucleation. The effect of added catalyst, the change in molecular weight with reaction, and morphological details are not known.

Tetrafluoroethylene ($T_m = -142.5°C$, $T_b = -76.3°C$) has a ceiling temperature of 560°C, sufficiently far above the melting temperature of polytetrafluoroethylene ($T_m = 327°C$) to permit only crystallization during polymerization at a high degree of supersaturation or low vapor pressure. On sublimation of the polymer, a small amount of monomer deposits on the colder walls in the form of fibrillar macromolecular crystals (Sect. 3.8.1, Fig. III.131). The normal reaction mechanism is a free radical process, suggesting the low yield results from slow initiation of the reaction, followed by a complicated nucleation and growth process. Denser films have been produced on fluorine-activated metal surfaces by Toy (1971), but were not analyzed further (100°C, 4 atm monomer pressure, 1–18 hr deposition time). Changes in the morphology can be observed when vapor and substrate temperature are changed (Melillo and Wunderlich, 1972). Further discussion must await additional experiments.

The first study of crystallization during polymerization of a more complicated monomer from the gas phase was carried out on ε-aminocaproic acid. The original observation of oriented fiberlike nylon 6 crystals grown on heating of crystals of ε-aminocaproic acid by Morosoff *et al.* (1964) (Sect. 3.8.1) could be identified as epitaxial crystallization during polymerization of gaseous monomer on crystalline monomer (Morawetz, 1966, Sect. 3.4.1). The details of the reaction and resulting morphologies were studied by Macchi *et al.* (1968) and Macchi (1972, 1973). The reaction kinetics on KCl surfaces is shown in Fig. VI.68. The temperature chosen for the study was 170°C, at which the vapor pressure is 0.12 mm Hg. The reaction can be subdivided into three stages. No polymer appears in the nucleation stage which lasts several hours. It is considerably shortened on the surface of monomer crystals. A homogeneous condensation reaction must produce oligomers long enough to cause crystal nucleation without loss of activity towards further reaction. Monomer, dimer, and trimer crystals contain unreactive zwitterions. Only the tetramer or higher oligomer crystals contain neutral amino acid molecules and are capable of further polymerization. The reaction is known to occur by electrophilic attack of the carbonyl carbon on the amine. The initial nuclei must thus be at least 35 Å long in order to be

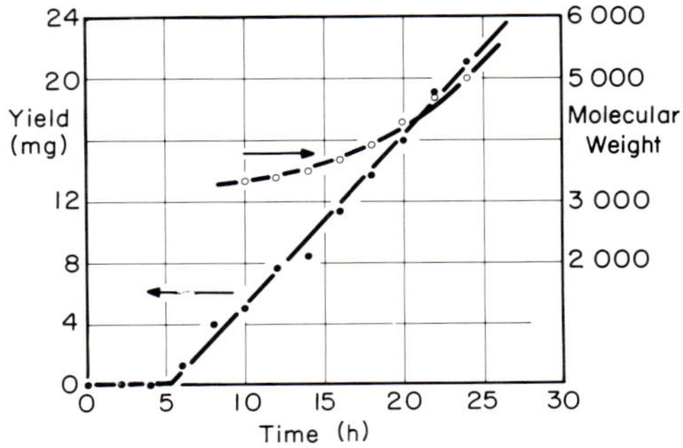

Fig. VI.68 *Yield and Molecular Weight of Nylon 6 as a Function of Time, Crystallized during Polymerization.*
Reaction temperature 170°C, 0.12 mmHg ε-aminocaproic acid vapor pressure (equilibrium pressure at 170°C). The reaction is carried out on approximately 10^3 cm^2 KCl (100) surfaces. Drawn from data by Macchi (1972).

reactive. Thin layer chromatography did not reveal any detectable concentration of oligomers at any stage of the crystallization, so that crystallization had to occur rather quickly after nucleation.

The growth rate of the second stage of the reaction is shown in Fig. VI.68 to be constant at constant monomer vapor pressure. The change in molecular weight with time indicates many new chains are created initially, while at a later stage, most monomer goes into extension of chains. While the nylon 6 grown on a substrate of monomer crystals was oriented and is of a more extended chain nature, most crystals on the KCl surfaces had lamellar morphology (90 Å fold-length) with spherulitic superstructure (Macchi, 1972). In addition, a smaller number of single crystals of 180 Å fold-length or chain extension and up to 1 μm width were observed by Macchi (1973). The spherulites were of the normal α-crystal structure (Table II.16), while the single crystals show a structure related to the γ-crystal (see Fig. II.58). It was suggested by Macchi (1973) that in these particular γ-crystals the molecular chains are not only parallel within the hydrogen bonded sheets, but all chains of a crystal point in the same direction. It may thus be possible that the crystallization of the spherulitic structures occurs mainly by successive polymerization and crystallization, while the polymer single crystals show growth closer to simultaneous polymerization and crystallization. An indication that the formation of the spherulites cannot be crystallization after polymerization is, in addition to the absence of oligomers, the missing epitaxy which nylon 6 shows when crystallized on KCl surfaces (see Sect. 3.4.2, Table III.4).

The third stage of the reaction occurs after exhaustion of the monomer. It produces an increase in degree of polymerization only. This last reaction stage seems to be related to the normal increase in molecular weight of polyamides on annealing (see Sects. 6.4.2.3 and 7.3.4). This would indicate that some or all of the produced crystals are not completely crystalline and have disordered regions which contain chain ends capable of further reaction.

Paraxylylene (I) gas can be produced in high yield from [2.2]-*p*-cylophane (II) and crystallized during polymerization on a solid substrate to poly-*p*-xylylene (III) (Gorham and Niegisch, 1971). The reaction is of the free radical

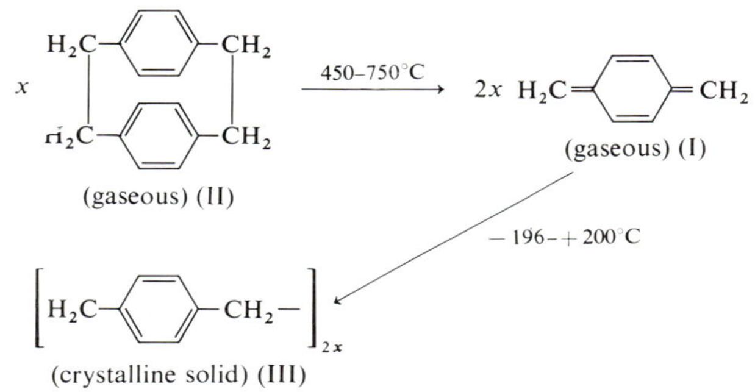

type (Szwarc, 1947), verified by ESR spectroscopy of the final polymer which contains up to 10^{-3} radicals/repeating unit.

The monomer is diamagnetic, which is in accord with the quinoid singlet ground state, illustrated as structure (I). The high reactivity can be linked to the high free valence (0.92) calculated for the methylene carbons of the monomer. The monomer is metastable in the gaseous state. This metastability was traced by Errede and Szwarc (1958) to the instability of the linear dimer (IV), which has lost the conjugation of the double bonds and contains two

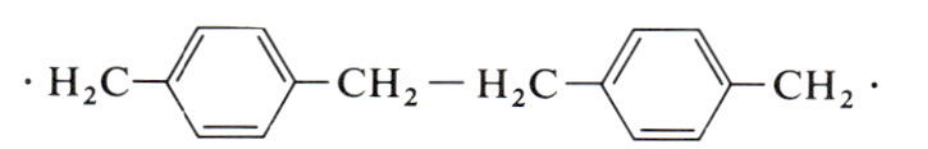

(IV)

radical ends. An estimate of the heats of formation (Kubo, 1971) from bond and resonance energies yields for the monomer, dimer, trimer, and tetramer the values −1731, −3446, −5196, and −6946 kcal/mole, respectively. The heats of reaction from the monomer are accordingly +8.0, −1.0, and −5.5 kcal/mole of monomer at 298°K for the dimer, trimer, and tetramer. A reasonable value of the entropy of reaction is −5 e.u., so that at 298°K dimer formation should be strongly hindered and only tetramer and higher polymers are thermodynamically favored. The free enthalpy barrier to dimer formation is a special type of hindrance to molecular nucleation and accounts for the absence of polymerization in the gas phase where multiple collisions of proper sequence are rare.

Polymerization of paraxylylene from the gas phase has been observed on surfaces. The polymer production was shown to follow a second-order rate law. The rate constant is shown in Fig. VI.69. It has the appearance of a

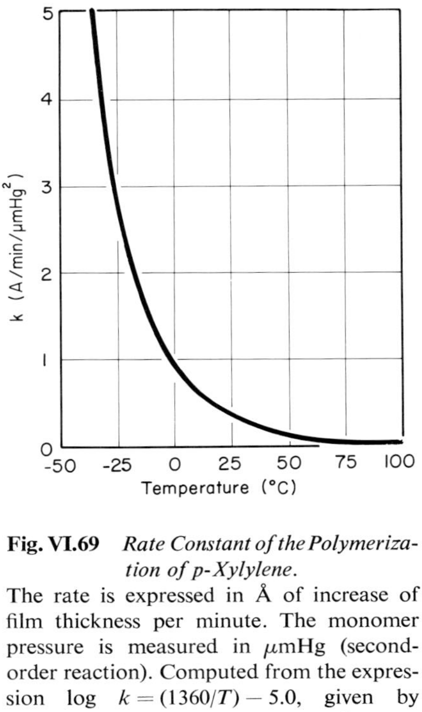

Fig. VI.69 *Rate Constant of the Polymerization of p-Xylylene.*
The rate is expressed in Å of increase of film thickness per minute. The monomer pressure is measured in μmHg (second-order reaction). Computed from the expression $\log\ k = (1360/T) - 5.0$, given by Carion *et al.* (1965). *Proc. IEEE Electronic Components Conf.* Washington, May 5–7.

reaction between weakly adsorbed reactants and shows a positive exponent of the Boltzmann factor, linked to an exothermal heat of adsorption. Above 30°C, the rate becomes very slow. The crystalline macromolecule is stable up to its melting point, usually observed at about 420°C. Fast decomposition occurs at about 500°C. Thus, polymerization occurs far from equilibrium.

The glass transition temperature of poly-*p*-xylene is between 50°C and 80°C, so that most polymerizations occur below the glass transition temperature. Since crystallization of macromolecules below the glass transition temperature is impossible, crystallization must occur during polymerization.

The morphology of poly-*p*-xylylene crystallized during polymerization between -17°C and 30°C was studied by Kubo and Wunderlich (1971, 1972). The samples were of the α-crystal form, had a crystallinity of about 0.6 and were chain folded (about 80 Å fold-length) with the polymer chain oriented in the support plane as required for epitaxial crystallization of a macromolecule (Sect. 3.4.3 and Fig. III.83). The crystallinity and melting points decreased with decreasing polymerization temperature. These observations are in accord with successive polymerization and crystallization (Sect. 6.1.8). The polymerization and crystallization sites must be sufficiently removed from each other to permit regular chain folding. From the negative temperature dependence of the reaction rate (Fig. VI.69) and the positive temperature dependence of crystallization on the low temperature side of the rate-temperature curve (see for example Fig. VI.20), one must conclude that polymerization and crystallization sites become more separated as the temperature is lowered. This agrees with the observed decreasing crystal perfection when the polymerization temperature decreases. At -78°C almost all of the chain folded α-crystals have disappeared. Instead of producing a completely glassy deposit, however, the reaction mode changes. At such low temperatures, it becomes possible to crystallize the monomer at least temporarily without reaction (Treiber *et al.*, 1973). Polymerization can then proceed later in the solid state.

Lowering the reaction temperature to -196°C, we find the crystallinity to be greater, but the crystal form is the metastable β-form and contains some oligomers (trimer). The molecular chains are no longer oriented parallel to the substrate, and chain folding occurs irregularly as a result of two ways of monomer addition (Kubo and Wunderlich, 1972). Figure VI.70 proves that at the lowest temperatures the polymerization reaction is intermittent. The heat of reaction at these lower temperatures is sufficient to raise the sample temperature substantially and cause evaporation of some monomer (pressure increase). At -20°C, the reaction parallels the adsorption of monomer, as expected from successive polymerization and crystallization of the α-form.

If the reaction temperature is increased above the region of fast reaction (see Fig. VI.69), one would expect that polymerization and crystallization sites would approach each other and possibly merge to show simultaneous polymerization and crystallization. First experiments by Iwamoto *et al.* (1974) have shown the crystallization mode to change at about 80°C. Again, the metastable β-polymorph is found instead of α; chain folding and orientation is not observed, and the produced crystals are rather poor.

In summary, the discussion of crystallization during polymerization from the gas phase for several systems has led to reasonable initial mechanisms. The main problem is to understand the interplay between polymerization and crystallization. Under suitable conditions all examples seem to indicate

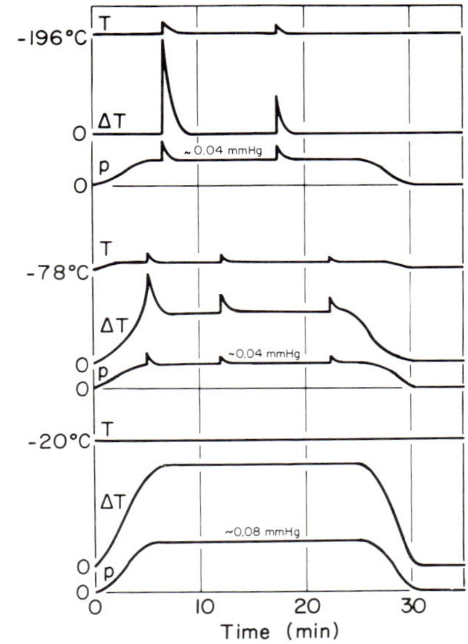

Fig. VI.70 *Schematic Representation of the Differential Temperature, Sample Temperature, and Monomer Pressure during Polymerization of p-Xylylene at Different Temperatures.*
Temperature difference measured relative to a shielded inert reference. Monomer pressure set by the rate of evaporation of [2.2]-*p*-cyclophane. The temperature is kept constant by a liquid bath surrounding the reaction vessel. Amplitudes are not to scale (Treiber *et al.*, 1973).

simultaneous polymerization and crystallization, and under changed conditions, successive polymerization and crystallization. The former is more likely to yield extended chain fibrils or isometric crystals, while the latter tends to give folded chain lamellar crystals. Refined mechanisms and information on kinetics are necessary for further progress. For the *p*-xylylene polymerization it was observed that differences in monomer temperature may cause differences in the resulting product. Fibrillar shaped macromolecular substances produced by polymerization from the gas phase have also been observed for amorphous macromolecules. Poly(methyl-2-cyanoacrylate) when produced from the monomer vapor under proper conditions may show fiberlike growth with the major chemical activity restricted to the ends (Smith-Johannsen, 1971).

6.4.2 Liquid or Dissolved Monomer Crystallization

Most polymerization reactions have been studied in solution or in the melt, and in many cases it has been observed that the product was a solid, crystalline polymer. However, only few of the observed cases have been analyzed in detail (Iguchi *et al.*, 1970). Rather than our listing monomers suspected of crystallization during polymerization, crystallization during polymerization of polyolefins, polyoxymethylene, some polyamides, and selenium are described in more detail in Sects. 6.4.2.1–6.4.2.4. Several other polymers are described as examples in Sect. 6.4.2.5. The details of nucleation of polyphosphates, polyoxymethylene, nylon 6, and polyethylene are discussed in Sect. 5.1.2.3. A possible nucleation and growth process for polytetrafluoroethylene is given in Sect. 5.1.2.4. The recurring problems for these crystallizations during polymerization involve the possible changes in the reaction details on crystallization, the question of simultaneous or successive polymerization and crystallization (see Sect. 6.18), the possibility of termination of the reaction by occlusion of chain ends within the crystal, the detailed final macroconformation of the macromolecule, and the interpretation of the kinetics of the process.

6.4.2.1 Polyolefins

The polymerization reaction of olefins usually must be initiated. It then proceeds as an addition reaction (see Sect. 1.2). The catalysis may either be homogeneous through a dissolved initiator, or heterogeneous, occurring on the surface of a solid catalyst particle. The cases of interest for the present discussion are the ones which lead to a crystalline macromolecule. In the homogeneous reaction, oligomer or folded chain primary nucleation of actively polymerizing molecules must occur as described in Sects. 5.1.2.3 and 5.1.2.4. In the heterogeneous reaction, the catalyst particle itself may serve as a primary heterogeneous nucleus for crystallization.

Polytetrafluoroethylene provides an example of a free radical polymerization leading to highly crystalline polymers at reaction temperatures far below the ceiling temperature (see Table VI.6). A possible mechanism of the reaction is described in Sect. 5.1.2.4 and involves successive polymerization and crystallization on the once-nucleated polymer particles (path 4 in Fig. V.10). The morphology of the aspolymerized (nascent) polytetrafluoroethylene was shown by Rahl *et al.* (1972) to be much closer to that of an extended chain crystal (see Sect. 3.3.1) than that expected on crystallization from solution (see Table III.3).

A similar polymerization and crystallization process for poly(vinylidene chloride) has been proposed by Wessling *et al.* (1973). The reaction is also a free radical initiated process and has been carried out in a wide variety of media ranging from nonswelling to strongly swelling solvents and temperatures of 20–70°C. The morphology of the produced crystalline polymer varied with increasing solvent power from small core particles with twisted branches (which may have been formed by initial liquid–liquid phase separation followed by crystallization) to dense curved lamellar aggregates (similar to solution grown material). The suggested mechanism is that of successive polymerization and crystallization by a free radical separated from the crystallization site by an amorphous chain segment.

Of the large group of olefins polymerized by an ionic, homogeneously catalyzed addition reaction, only ethylene has been studied with respect to the produced crystal morphology, mechanism, and kinetics. Fibrous crystals as well as chain folded lamellae and dendrites were observed in the low temperature polymerization of ethylene with Ziegler–Natta type catalysts and γ-rays (see Sect. 3.8.1 and Fig. III.134). Georgiadis and Manley (1972) studied the polymerization of ethylene between −25°C and +95°C in toluene using a solution (3 mmole/liter) of dicyclopentadienyltitanium dichloride dialkylaluminum chloride as catalyst $[(C_5H_5)_2TiCl_2{-}R_2AlCl]$. At the lowest temperatures (−25°C to +15°C) a granular, epitaxially interlaced network of 100 Å thick lamellae was observed. The structures were not regular enough to give a defined low angle X-ray diffraction pattern. At 25–70°C increasingly better lamellae resulted. Between 65°C and 80°C typical solution-grown crystals as shown in Fig. III.51 resulted. The mechanism proposed involves separate polymerization and crystallization for the high temperature crystals and successive polymerization and crystallization for the lower temperatures.

From the details of the low temperature epitaxial interlacing of the granular crystal aggregates it was suggested by Georgiadis and Manley (1972) that the catalyst molecule remains dissolved during the crystallization process and continues polymerizing ethylene. The initial portion of the growing molecule adds to the (110) growth front of a lamellar crystal as soon as its solubility limit is reached. Portions of the same molecule which are polymerized at a later time must add epitaxially to the fold surface since the crystal growth front has advanced in the meantime by addition of other molecules.

The observation of fibrillar crystals under seemingly similar conditions (see Sect. 3.8.1) was linked by Georgiadis and Manley (1972) to the possible formation of colloidal catalyst particles or to shear in the crystallizing solution. Both effects are known to yield fibrillar structures (see below for the effect of heterogeneous catalysts, and Sect. 6.2.2 for the effect of stirring). Arai *et al.* (1973) suggested that γ-ray induced polymerization may follow a similar crystallization during polymerization path.

The cationic polymerization of diazomethane using boron trifluoride etherate as a catalyst was studied by Mucha and Wunderlich (1974). The reaction was carried out in diethyl ether at temperatures between 0°C and 25°C. The first stage of the reaction is homogeneous. An insertion mechanism at the boron-end of the molecule is a likely description of the reaction†

$$CH_2{-}N{\equiv}N + BF_3 \xrightarrow{\text{slow}} F_3\overset{\ominus}{B}{-}CH_2{-}\overset{\oplus}{N}{\equiv}N \qquad (120)$$

$$F_3\overset{\ominus}{B}{-}CH_2{-}\overset{\oplus}{N}{\equiv}N \xrightarrow{\text{spontaneous}} F_3\overset{\ominus}{B}{-}\overset{\oplus}{C}H_2 + N_2 \qquad (121)$$

$$F_3\overset{\ominus}{B}{-}\overset{\oplus}{C}H_2 \xrightarrow{\text{very fast}} F_2B{-}CH_2F \qquad (122)$$

$$BF_2{-}CH_2F + CH_2{-}N{\equiv}N \xrightarrow{\text{slow}} F_2\underset{\underset{CH_2F}{\|}}{\overset{\ominus}{B}}{-}CH_2{-}\overset{\oplus}{N}{\equiv}N \qquad (123)$$

$$\text{Rate of decomposition of } CH_2N_2 = k[CH_2N_2][\text{catalyst}] \qquad (124)$$

Step (123) of the reaction is followed by spontaneous loss of N_2 and migration of CH_2F. The migration of a second fluoride would lead to a second chain growing on the same boron atom. This is presumed to be less likely since alkylboron compounds show reduced reactivity. If there are no termination reactions, polymerization can continue until the diazomethane in solution is exhausted. The resulting $F_2B{-}(CH_2{-})_xF$ should still be an active catalyst for further reaction, a fact which has been substantiated by Bawn *et al.* (1959). The macromolecule is thus a living polymer and the molecular weight distribution should be described by a Poisson distribution as long as no crystallization occurs [Eq. (11), Chapter I; see also curve **B** in Fig. I.1].

Figure VI.71 illustrates the decrease in monomer concentration as a function of time for different catalyst concentrations. While the dependence of the reaction on monomer is always first order, as expected from Eq. (124); the dependence upon catalyst concentration is first order only in the initial, homogeneous phase of the reaction (two minutes or less in the examples of Fig. VI.71). As soon as the first solid polymer occurs, the overall reaction slows down. For very low catalyst concentration there is a third reaction stage which begins when all oligomer in solution is exhausted. A once-made oligomer molecule can grow larger only by reacting with additional monomer. It is not possible (as in the step reaction of Sect. 5.1.2.3) for the oligomer molecule to add itself to another growing chain.

The final molecular weights in all the reactions of Fig. VI.71 were as high as 10^7 instead of the 1700 to 7000 predicted from the monomer to catalyst

† For a review of the reaction mechanism from the point of view of organic chemistry, see Bawn and Ledwith (1969). *Enc. Polym. Sci. Technol.* **10**, 337; Cowell and Ledwith (1970). *Quart. Revs.* **24**, 119; Imoto and Nakaya (1972). *J. Macromol. Sci. Revs. Macromol. Chem.* **C7**, 1. In addition, Davies *et al.* (1963). *J. Chem. Soc.* 4461 discusses the possibility of an additional cationic chain reaction in which the polar chain ends are strongly solvated by diazomethane molecules.

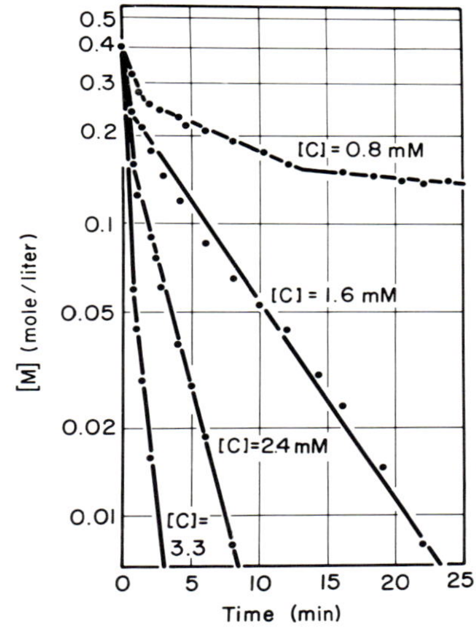

Fig. VI.71 *Logarithm of Concentration of Monomer as a Function of Time for Diazomethane Polymerizing in Dimethyl Ether with Boron Trifluoride as Catalyst.* Temperature 23°C, initial concentration 0.4 moles/liter. Catalyst concentrations in the figure as mmole/liter. The three straight-line regions correspond from left to right to the homogeneous, oligomer producing phase, the crystallization phase in the presence of dissolved oligomer, and the slow, final phase which shows only polymerization during crystallization at the solid surface. Mucha and Wunderlich (1974).

ratios for living polymers. This indicates that the reaction in the solid state is probably much faster than the solution reaction. To account for the overall slower reaction and the higher molecular weight, the majority of the oligomer molecules must become deactivated during the crystallization step.† In agreement with this reduction in the number of growing molecules following the homogeneous reaction phase, the yield of macromolecules decreases drastically with increasing catalyst concentration. For the example in Fig. VI.71, the high catalyst concentration reaction yields only 10% of the

† For a specific model of occlusion, see Iguchi (1970).

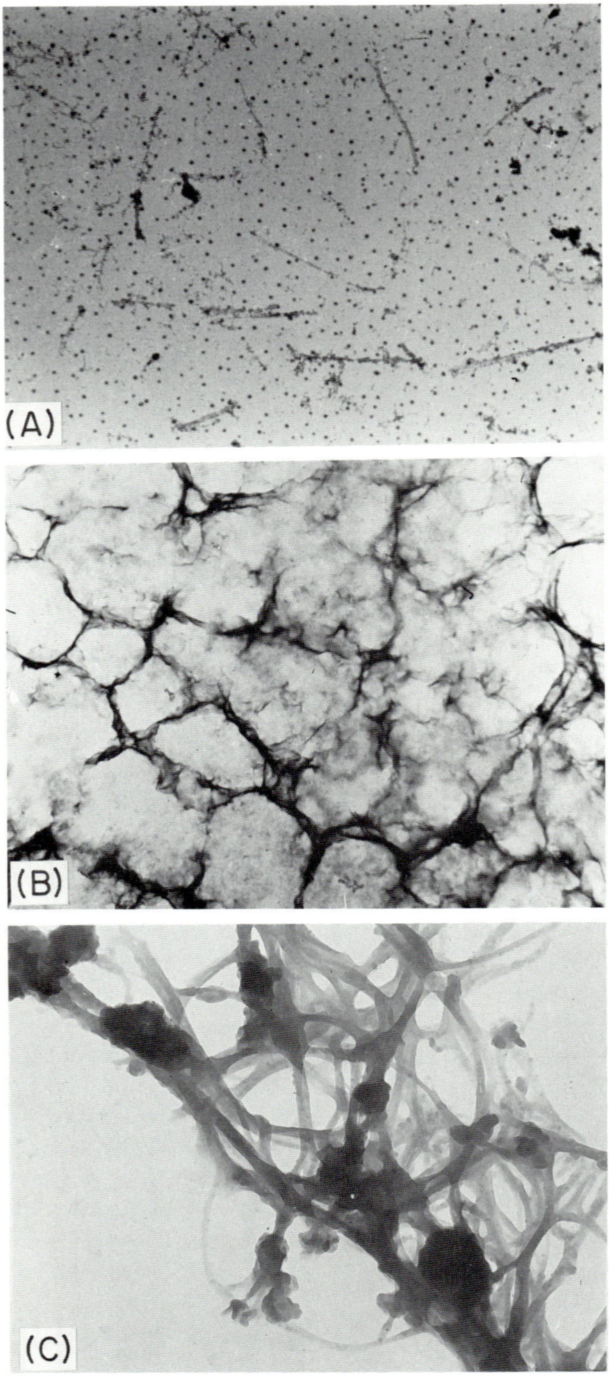
(A)
(B)
(C)

polymer of the low catalyst concentration reaction. Most of the monomer not recovered as macromolecules remains as low molecular weight oligomer.

The morphology of the polymer crystals from diazomethane is shown in Fig. VI.72. The initial crystals are fibrillar. Particularly at high oligomer (catalyst) concentration, well formed and high melting fibers are observed. Later stages of crystallization show increasing coalescence of fibers and formation of irregular connecting materials. Again, it is expected that polymerization and crystallization are successive. Crystallinity and melting points are higher for the as-polymerized samples than for crystals grown by cooling the polymer melt. Reducing the polymerization temperature to as low as $-50°C$ produces less crystalline samples (Magill *et al.*, 1965).

Heterogeneous catalysis of crystallization and polymerization has been studied intensively on the example of Ziegler catalysis.† In the case of such heterogeneous catalysis, the reaction site is fixed. The reaction occurs by an insertion mechanism at the catalyst site, illustrated by the following sequence of reactions of propylene to isotactic polypropylene (Rodriguez *et al.*, 1966, see also Buls and Higgins, 1970, 1973, and Keii, 1972)

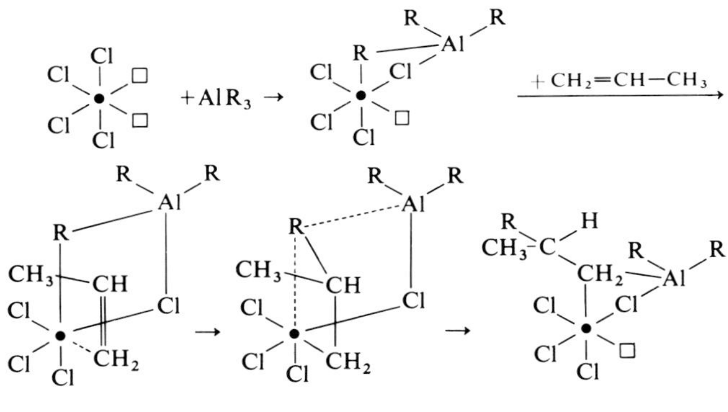

(125)

(•) represents Ti and two more coordinating Cl above and below the plane of projection; (□) represents a vacancy in the octahedral Ti coordination.

† For a review of "the nature of the active site in the Ziegler-type catalyst", see Boor (1967). *Macromol. Rev.* **2**, 115. A Ziegler-type catalyst is generally formed by combining a metal alkyl or hydride with a transition metal salt. Metal alkyls or hydrides are of groups I–III, the transition metal salts from groups IV–VIII. Most studied catalysts are formed from $TiCl_3(\alpha)$ and $Al(C_2H_5)_3$.

Fig. VI.72 *Morphology of Polymethylene Crystals Grown on Crystallization during Polymerization of Dizaomethane.* (A) Fibrillar structures grown in the initial stage of the reaction. (B) Final, packed morphology. (C) Fibrillar and lamellar components observed after etching a sample as shown in (B). Mucha and Wunderlich (1974). Scale 1 μm.

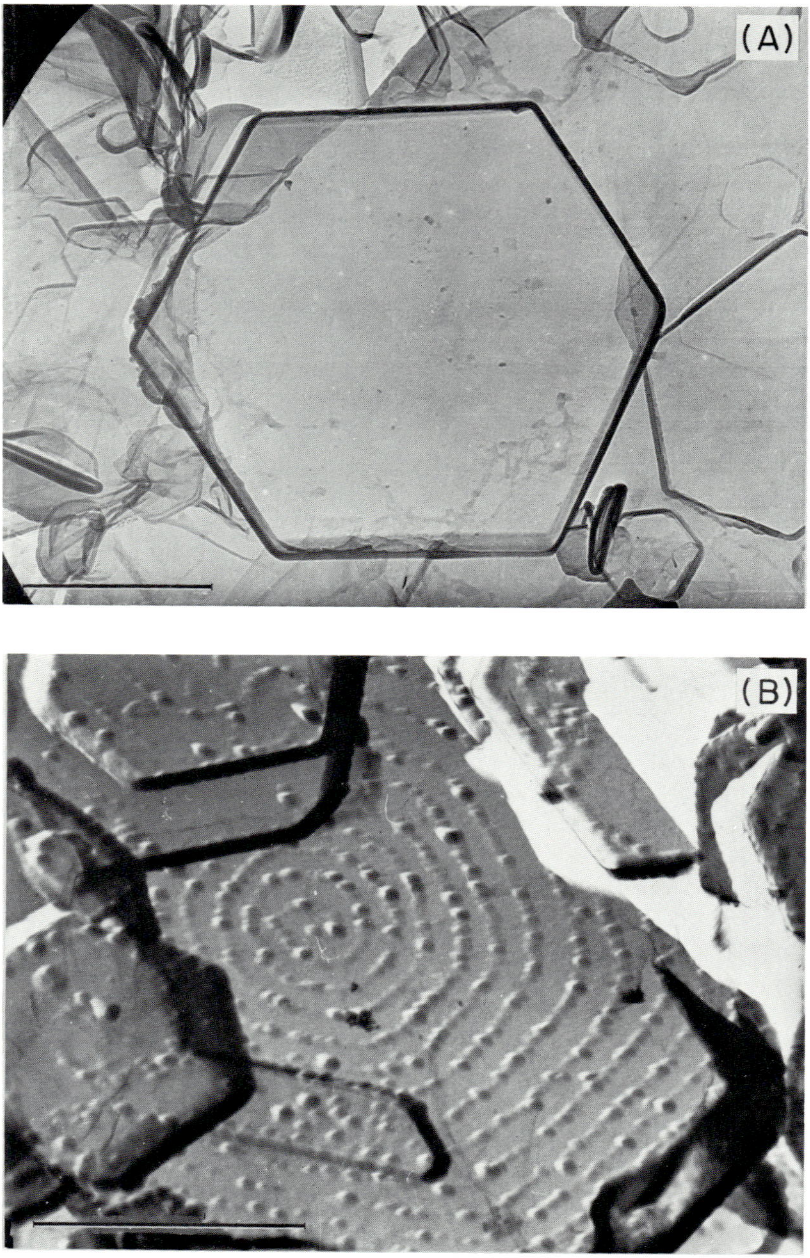
(A)
(B)

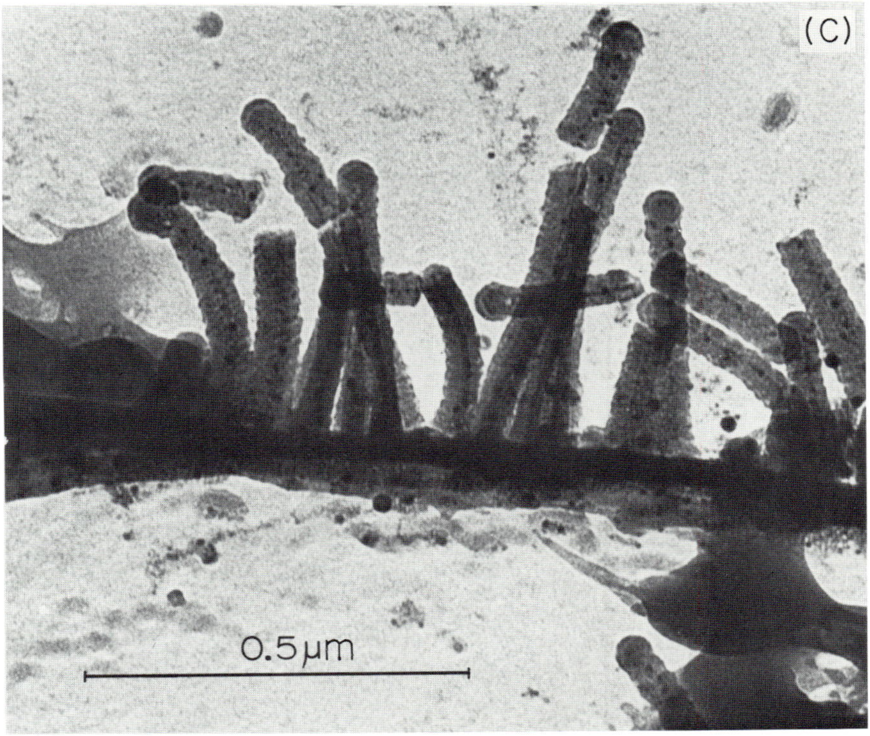

Fig. VI.73 *Electron Micrographs of Ziegler-Catalyst and Initial Polymerization Products Grown from the Gaseous Phase at Room Temperature.*
(A) α-$TiCl_3$ crystal prepared by reduction of $TiCl_4$ with hydrogen. Electron micrograph of a surface replica (Rodriquez *et al.*, 1966, p. 1973). (B) Crystal as in (A), activated with $Al(CH_3)_3$ after short time exposure to propylene, illustrating initial polymer development (Rodriquez *et al.*, 1966, p. 1976). (C) Later stage of growth of polypropylene on $Al(CH_3)_3$ activated α-$TiCl_3$ (Guttman and Guillet, 1970).

The active site has been proposed to be linked to a surface titanium atom with two chlorine vacancies on the otherwise octahedrally coordinated titanium. On reaction with the aluminum alkyl, the catalyst is activated. The monomer enters the vacancy left in the activated catalyst site, forming a π-complex. A stereospecific orientation is dictated by the requirement of minimal steric interaction with the external groups carried by the aluminum atom. In the subsequent reactions, the π-complex is inserted in the Ti—C bond. In addition, the Al—C bond breaks, so that the CH_2—CHR—CH_3 group is only linked to the titanium atom at the position of the original

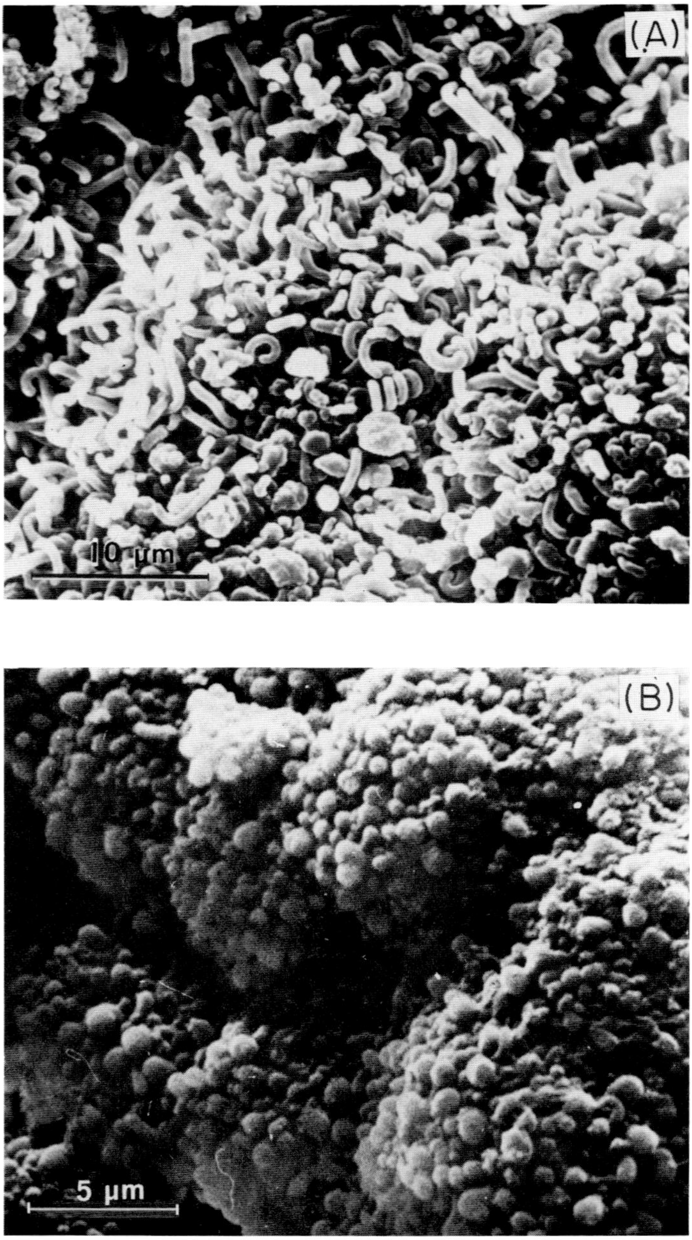
(A)
10 μm
(B)
5 μm

Fig. VI.74 *Scanning Electron Micrographs of Typical Texture of Polypropylene and Polyethylene Crystallized during Polymerization.*
(A) Worm-like morphology of polypropylene grown at high catalyst efficiency (Wristers, 1973a). (B) Globules of polypropylene, grown at low catalyst efficiency (Graff *et al.*, 1970; see also Wristers, 1973a). (C) "Cobweb morphology" of polyethylene (Graff *et al.*, 1970).

vacancy. A shift of the CH_2 group (attached to the titanium atom) to the previous coordination site of the R group restores the original vacancy configuration and allows the next step of the polymerization to proceed.† The polymer chain is thus growing out of the surface as an isolated molecule (root growth). Depending upon temperature, environment (type of solvent), density of active sites, and rate of polymerization, different crystal morphologies are expected (Wunderlich, 1968). The morphology of polyolefins produced by crystallization during polymerization in such systems has been reviewed by Marchessault *et al.* (1972) (see also Sect. 3.8.1).

Historically Bort *et al.* (1962) and Kargin *et al.* (1964) studied the texture‡ of heterogeneously polymerized and crystallized polyethylene for the first time. Their catalyst system consisted of partially chlorinated aluminum mixed with α-$TiCl_3$ dispersed in benzene. Polymer protuberances were observed to grow perpendicular to the aluminum surface. The appearance of the polymer was described as "bushlike" with filaments which were thought to be composed of smaller parallel ribbons with a transverse structure.

Considerably more detail is known today. A typical large catalyst single crystal is shown in Fig. VI.73A. After brief treatment with propylene gas [after activation with $Al(CH_3)_3$], decoration of the growth spiral edge and other defects occurs, suggesting the nature of the active site (Fig. VI.73B). The initial polymer has a globular texture. Guttman and Guillet (1970) suggested that the rate of polymerization is at the beginning of the reaction high enough to produce a molten (or swollen) hemispherical mass of amorphous polymer (heat of polymerization ~ -25 kcal/mole; heat of crystallization ~ -2.1 kcal/mole). Upon continuation of the reaction, the adsorbed monomer must diffuse through the polymer.

Recent schemes of the reaction kinetics which have been based on an adsorption governed mechanism by Tait *et al.* (1972) and additional diffusion control by Crabtree *et al.* (1973) are able to describe the reaction. The further development of the Polymer texture must involve solidification at the outer regions of the initial mass. Nucleation and crystallization can now occur through chain folding with various amounts of defects of the fringed micelle type (see Fig. III.5). The sample texture at a somewhat later stage of the reaction has been described as protuberances, as is shown in Fig. VI.73C (diameter ~ 0.05 μm).

† For "molecular orbital calculations on the propagation step in Ziegler-Natta polymerization," see Cossee *et al.* (1971). *Proc. 4th Int. Congr. Catal.* 1968 **1**, 207 (publ. by Akad. Kiado, Budapest, Hungary).

‡ The term texture rather than morphology is applied since the detailed arrangement of crystalline and amorphous areas, folding of chains, etc. is often not investigated.

At later stages of the polymerization "globular, wormlike,† and cobweb textures" were observed as shown in Fig. VI.74. Wristers (1973a) illustrated on a series of catalysts and monomers (ethylene, propylene, 1-butene, and 4-methyl-1-butene) that the occurrence of globular or wormlike morphology is dependent on catalyst activity. The diameter of the wormlike structures ($\sim 0.5\ \mu m$) seems related to the size of the actual catalyst particles which may, if aggregated, separate on polymerization. A special texture which can arise on the separation of clusters of catalyst particles during polymerization is the "cobweb structure", shown in Fig. VI.74C (Graff *et al.*, 1970). The fibrils are clearly secondary, being created by deformation due to the increase in polymer volume inside the aggregate of particles, schematically shown in Fig. VI.75. The influence of strain created by catalyst fracture and movement of

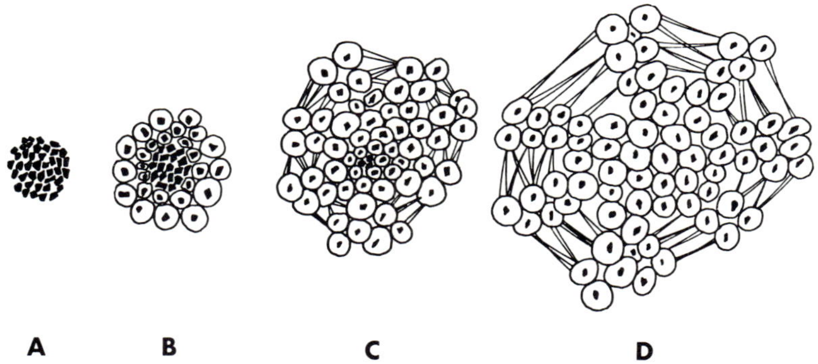

Fig. VI.75 *Schematic Representation of the Production of the Cobweb Texture.* Drawn by Graff *et al.* (1970), depicting subsequent stages of polymerization.

growing polymer particles has a major influence on the texture of the produced polymer as shown by Chanzy and Marchessault (1969), Marchessault and Chanzy (1970), and Chanzy *et al.* (1973, 1974). Chanzy *et al.* (1967) suggested that many of the fibrillar structures from Ziegler catalysis discussed in Sect. 3.8.1 may also be due to such strain effects or the effects of flow of the solvent.

The observations of the crystallization and polymerization on Ziegler catalysts has presented an example for the large breadth of textures which may result. In all cases it is reasonable to assume that polymerization and crystallization are successive. The crystallization is usually enhanced over melt or solution crystallization under similar conditions of supercooling. In the presently analyzed kinetics, there seems to have been little or no influence of the crystallization on the polymerization.

† The distinction of "wormlike" from "fibrillar" is somewhat arbitrary, but obvious when comparing, for example, Fig. VI.74B and III.134.

6.4.2.2 Polyoxymethylene

Polyoxymethylene is the first member of the series of polyaldehydes produced by reaction to polyacetals. The polymerization was first described more than one hundred years ago.† In a review of the polymerization of formaldehyde, Brown (1967) pointed out that, depending on the reaction path, either successive polymerization and crystallization or simultaneous polymerization and crystallization is possible. The two paths lead to largely different materials. Successive crystallization and polymerization occurs on cationic or anionic addition reaction. The mechanisms of anionic chain reactions with tertiary aliphatic amine and aluminum isopropoxide have been proposed as follows.

Initiation:

$$HO^{\ominus\oplus}HNR_3 + CH_2{=}O \longrightarrow HO{-}CH_2{-}O^{\ominus\oplus}HNR_3$$

Propagation:

$$\begin{aligned} &HO{-}(CH_2{-}O{-})_xCH_2{-}O^{\ominus\oplus}HNR_3 + CH_2{=}O \longrightarrow \\ &\quad HO{-}(CH_2{-}O{-})_{x+1}CH_2{-}O^{\ominus\oplus}HNR_3 \end{aligned} \tag{126}$$

Transfer:

$$\begin{aligned} &HO{-}(CH_2{-}O{-})_xCH_2{-}O^{\ominus\oplus}HNR_3 + R'OH \longrightarrow \\ &\quad HO{-}(CH_2{-}O{-})_xCH_2{-}OH + R'O^{\ominus\oplus}HNR_3 \end{aligned}$$

Initiation:

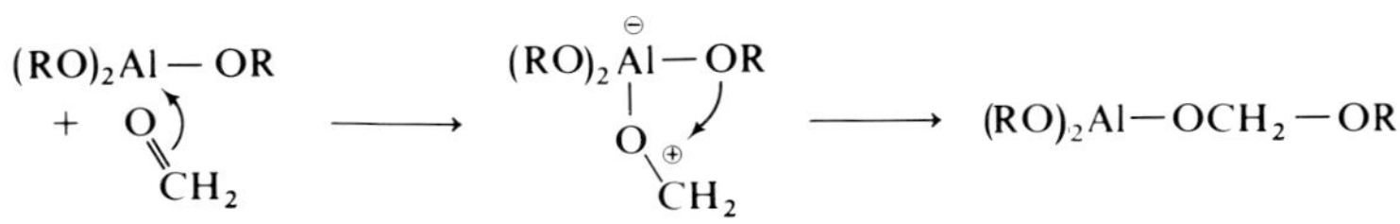

Propagation:

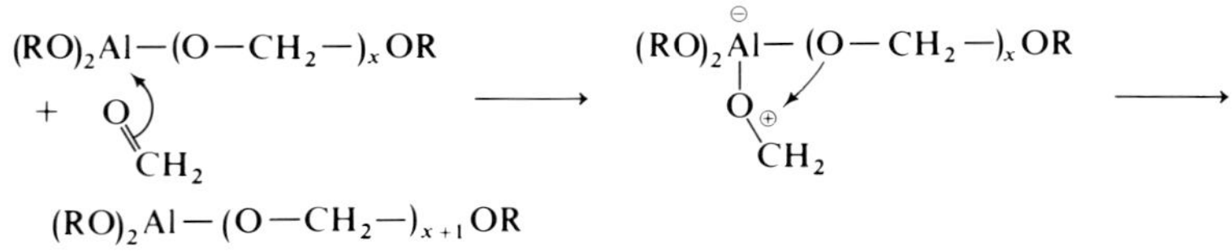

(127)

† Butlerov (1859). *Ann.* **111**, 242; for comprehensive reviews on the chemistry of formaldehyde and its polymerization, see Walker (1964) and Furukawa and Saegusa (1963); the polymerization from gaseous formaldehyde is discussed in Sect. 6.4.1, solid state reactions in Sect. 6.3.3.

Transfer:

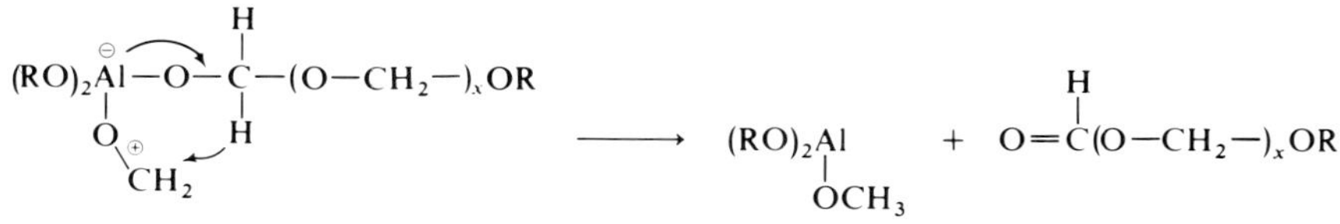

The transfer reactions reduce the molecular weight without influence on the rate since the active species does not get lost and reinitiates a new chain. Equation (126) indicates a simple addition mechanism while Eq. (127) involves a π-complex. Equation (127) also explains the tendency of metal containing catalysts to produce polymers containing a high concentration of methoxyl and formate end groups. Cationic initiators such as protonic and Lewis acids are also effective catalysts, but only fairly strongly acidic species are effective in producing high polymers at fast rates. In all cases, the produced polymer is insoluble under reaction conditions and forms dense particles or feathery aggregates. The crystallinity of the polymer is about 90%, readily distinguished from the 100% crystalline polymer grown under close to equilibrium conditions to be described below. The polymerization is thought to occur in the liquid phase on a chain end emerging a few monomer units out of the solid surface; thus crystallization follows successively, a few units behind the growing chain end (Brown, 1967). The detailed morphology of the as-polymerized crystals is not known. Radiation initiated polymerization of formaldehyde is also ionic and leads to crystalline polyoxymethylene of less than 90% crystallinity (Yamaoka *et al.*, 1964).

Simultaneous polymerization and crystallization is possible from aqueous or alcohol solution close to the ceiling temperature (Table VI.6). Nucleation occurs as soon as oligomers of an approximate length of ten repeating units have been formed in a homogeneous step reaction [see Sect. 5.1.2.3, Eq. (42)]. Further growth under close to equilibrium conditions can yield practically 100% crystalline polymer. An interesting interplay of nucleation and crystallization is presented by the epitaxy from aqueous formaldehyde solution on already polymerized, drawn polyoxymethylene. The molecular weight is limited in this case by the separation of the initial nuclei on the substrate (see Sect. 5.1.3.2, Fig. V.20).

Careful control of nucleation conditions is also necessary for the polymerization of trioxane to high crystallinity and high molecular weight polyoxymethylene from solution (Baccaredda, *et al.*, 1964). This reaction is cationic and may be started by a large number of initiators (Furukawa and Saegusa, 1963). For the polymerization step of the reaction in solution, catalyzed by the etherate of boron trifluoride, the following mechanism has been proposed (Kern and Jaacks, 1960):

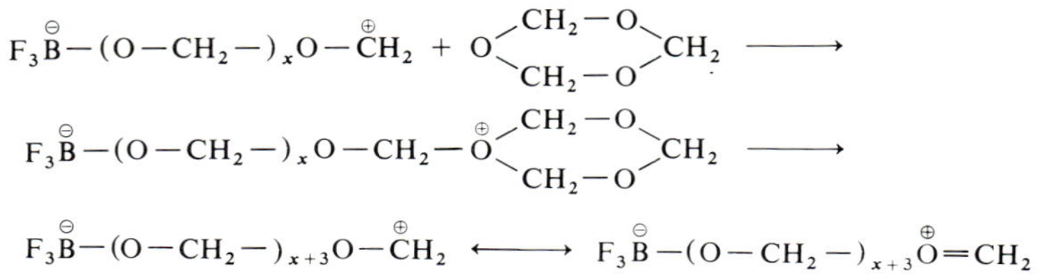

(128)

An oxonium ion at the active chain end alternates with a resonance stabilized carbonium ion. There seems to be no need for a cocatalyst when strong Lewis acids like $TiCl_4$ and $AlCl_3$ are used as initiators, the propagating species may be a macrozwitterion (Jaacks and Kern, 1963).

An induction period and an acceleration of the reaction as time goes on characterize the trioxane polymerization in solution (Kern and Jaacks, 1960). The sequence of reactions leading to actively growing polymer crystals can be described by the following steps (Price and McAndrew, 1967):

1. The etherate exchanges quickly with the trioxane in a reaction of low heat of reaction (Leese and Baumber, 1965):

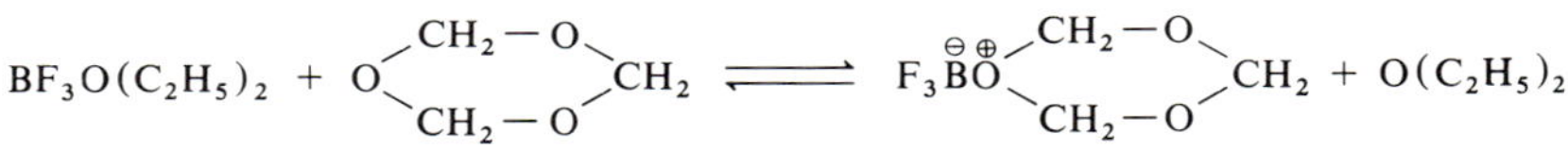

(129)

2. Ring opening similar to Eq. (128) sets up equilibria between oligomers and formaldehyde

$$F_3\overset{\ominus}{B}-(O-CH_2-)_x\overset{\oplus}{O}=CH_2 \rightleftharpoons F_3\overset{\ominus}{B}-(O-CH_2-)_{x-1}\overset{\oplus}{O}=CH_2+CH_2O \quad (130)$$

which has an endothermic heat of reaction of about 14 kcal/mole of formaldehyde produced. Initially formaldehyde is produced almost exclusively. This stage of the induction period can be eliminated by supplying the equilibrium concentration of formaldehyde (60 mmole/lites at 30°C in $C_2H_2Cl_2$) (Kern and Jaacks, 1960).

3. An almost athermal period follows where oligomerization involving linear oligomers and formaldehyde continues. As long as the oligomers remain dissolved, the heat of reaction is only about -0.3 kcal/mole (Leese and Baumber, 1965). The ring opening reaction is greatly reduced in this stage.

This intermediate phase of reactions seems to be dependent upon reaction conditions and disappears at higher trioxane concentrations (Kern and Jaacks, 1960).

4. Finally, the oligomers form a crystalline nucleus and the reaction becomes now exothermic due to the added heat of crystallization during simultaneous or successive polymerization and crystallization. The crystals of polyoxymethylene observed are hexagonal with the polymer chain parallel to the *c* axis (Leese and Baumber, 1965). They increase with time in lateral dimensions as well as in thickness, taking a pyramidal shape. The addition of new oligomers on the lateral surface would explain an autocatalytic acceleration of the reaction with time (Leese and Baumber, 1965).

The details of the crystal growth of trioxane to polyoxymethylene with boron trifluoride etherate catalyst have been studied by Iguchi *et al.* (1969b), Iguchi (1973) and Mateva *et al.* (1973). Iguchi *et al.* carried out the reaction at 30°C and 50°C in a poor solvent (cyclohexane). The initial, fast reaction (about 10% conversion) produced a precipitate of needle-like crystal nuclei which were proposed to be of very low molecular weight, perhaps consisting of catalyst and monomer. Subsequent growth occurred at a constant rate. The crystals grew at a 51° angle to the initial needle axis and changed to a lamellar morphology. The chain axis was in the direction of growth (in the plane of the lamella). The crystallinity of the lamellae was almost 100%, although the molecular weight was insufficient to match the crystal dimensions. The overall kinetics of the polymerization resembles that of the diazomethane in Fig. VI.71. The initial, fast homogeneous phase of the reaction varies in time and conversion depending upon catalyst and solvent (Jaacks and Kern, 1963).

Mateva *et al.* (1973) observed as initial crystals hexagonal lamellae of 50–100 Å thickness and 1000–2000 Å diameter when polymerizing in the better solvent nitrobenzene. The crystals grew quickly in width without initially changing much in thickness; quite analogous to normal, solution growth of single crystals (Fig. III.44). As conversion increased, the crystals grew in thickness through growth spirals with a Burgers vector of only a few angstrom in length. Figure VI.76A illustrates this thickening which leads finally to typical extended chain crystals (Fig. VI.76B). This thickening continued even after the equilibrium monomer concentration was reached† and could also be initiated on folded chain single crystals grown from solution by addition of catalyst and trioxane. It represents a special form of crystal

† For data on equilibrium concentration in various solvents, see Miki *et al.* (1970). *J. Polym. Sci. Part A-1*, **8**, *157*; and Bogdanova *et al.* (1972). *Vysokomol. Soedin.* **A14**, 1976.

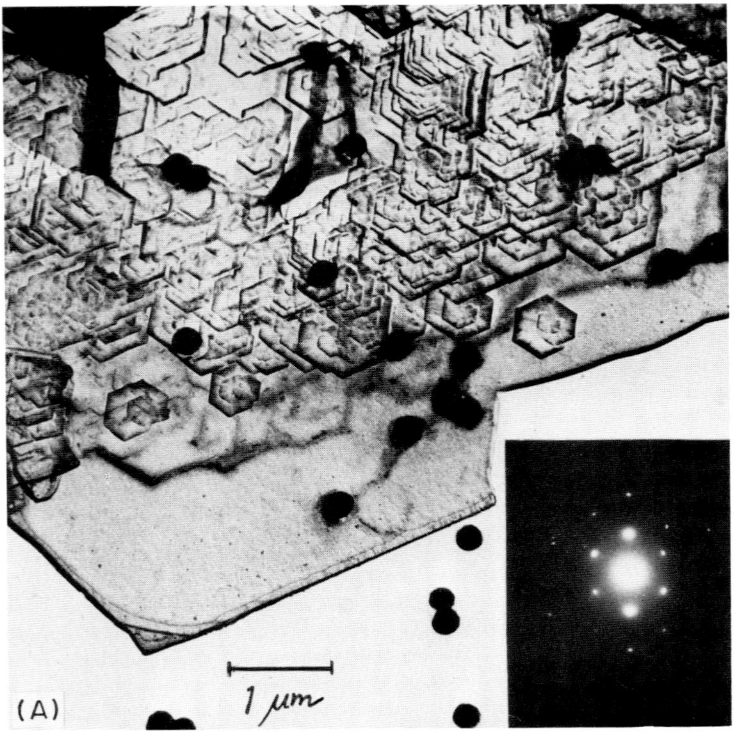

Fig. VI.76 *Thickening of Polyoxymethylene Crystals by Fold Extension.* (A) Growth spirals on the (001) surface of the folded chain crystal. Electron Micrograph by Mateva *et al.* (1973). (B) Electron micrograph of a fracture surface through a crystal as shown in (A) showing typical extended chain crystal morphology (compare with Figs. III.30–III.33) (Mateva *et al.*, 1973).

perfection. The proposed chemical reaction involves ring expansion at the chain fold:

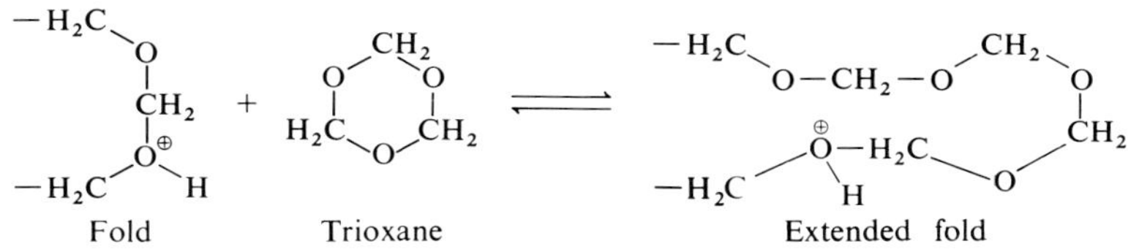

(131)

Thus the earlier observations of Leese and Baumber (1965) of the pyramid-shaped crystals has a surprising reason. Comparison with the different crystals observed by Iguchi, *et al.* (1969b) (see above) suggests that the changed habit results from a different nucleation mechanism. In the poor solvent it may not be possible to produce sufficiently long molecules for initial chain folded crystallization.

6.4.2.3 *Polyamides*

High melting polyamides are easily polymerized to the crystalline state (see also Sect. 6.4.1 for polymerization from the gaseous monomer). Most attention has been given to the polymerization of cyclic aliphatic lactams [Eq. (132)] and N-carboxyanhydrides of α-aminoacids [Eq. (133)].

$$x\ \overline{\mathrm{CO}-(\mathrm{CH_2}-)_m\mathrm{NH}} \longrightarrow [(\mathrm{CH_2}-)_m\mathrm{CO}-\mathrm{NH}-]_x \tag{132}$$

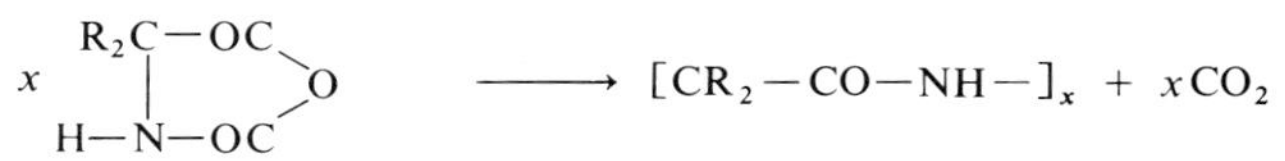

(133)

Most work on the chemistry has been carried out for the anionic polymerization of nylon 6 from the seven membered ring, the caprolactam (Wichterle *et al.*, 1961). The molecular chain is started by an *N*-acylimide which may be produced from two monomer molecules in a slow reaction during an induction period, or an *N*-acylamide produced by action of a cocatalyst, like carboxylic acid chlorides or anhydrides, on caprolactam. The cocatalyst action speeds up the reaction such, that fast polymerization below the melting point of the polymer becomes possible. A strong base, such as alkali metal, metal hydride, metal amid, or organometallic compound, activates the monomer by lactam anion formation. In this way the anion is not the

$$\overline{\mathrm{CO}-(\mathrm{CH_2}-)_5\mathrm{NH}} + \mathrm{B}^{\ominus}\mathrm{M}^{\oplus} \rightleftharpoons \overline{\mathrm{CO}-(\mathrm{CH_2}-)_5\mathrm{N}}^{\ominus}\mathrm{M}^{\oplus} + \mathrm{BH} \tag{134}$$

chain propagating species, but represents an activated monomer. The polymerization reaction proceeds through nucleophilic attack of the ion on the carbonyl group of the imid. Because of the greater basicity of the polymeric

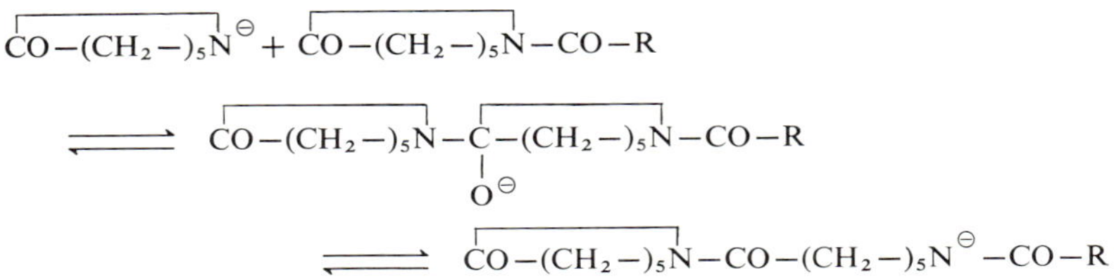

(135)

amid ion produced, a lactam anion is quickly reformed in a reaction identical to Eq. (134). Overall, the chain reaction is sustained by the continued reproduction of an imide and started by any imide producing reaction.

The maximum viscosity on melt polymerization is reached at the equilibrium concentration of caprolactam in the reaction mixture (about 5–10%, depending on temperature) Longer reaction times decrease the molecular weight due to side reactions. A different situation arises when the polymer is produced at temperatures low enough for crystallization. The crystalline phase does not contribute to the equilibrium and a much lower than equilibrium monomer content can be achieved (Wichterle *et al.*, 1961). Since it was also shown experimentally that polymerization stops when the imid is incorporated in the precipitate (crystal), but can be restarted on redissolution, it is likely that nylon 6 polymerization and crystallization is successive and not simultaneous. Polymerization temperatures of 75° and 90°C should correspond to precipitating oligomers of degrees of polymerization of 10 and 20 respectively.

On very fast caprolactam polymerization at low temperature, an amorphous "virgin polymer melt without thermal history" can be produced isothermally below the melting point (see Sect. 6.3.1.5 for its subsequent crystallization). When the reaction is sufficiently slowed down by zone polymerization (180°C and 0.02 in/hr zone travel rate) crystalline nylon 6 is produced directly from the monomer (Wunderlich and Liberti, 1966). The zone polymerized crystals have a higher density (1.17 versus 1.14 g/cm^3), a higher melting temperature (225.6 versus 215.5°C), and a higher heat of fusion (24.8 versus 14.3 cal/g) then melt crystallized of nylon 6 of similar molecular weight (40–80 000) (Liberti and Wunderlich, 1968). In addition, the crystals show a moderate degree of superheating on fast melting. Even on careful tempering of the melt crystallized nylon 6, the improved properties of the zone polymerized material could not be duplicated. X-ray analysis indicates that the molecular chains are arranged as in highly stretched polyamids with random or possibly even parallel directiveness of the crystallized polyamid chains rather than the antiparallel arrangement that is found in solution and melt crystallized nylon 6 (see Table II.16, Sect. 2.4.6). Crystallization of nylon 6 under pressure may yield similar perfection (see Sect. 6.3.3).

More details of the reaction and crystallization kinetics and on the obtained crystal morphologies were derived for nylon 4 and nylon 6 by Iguchi *et al.* (1969a) and Komoto *et al.* (1970). The monomer melts became turbid a few minutes after the reaction began. The observed morphological structures at the early stages of the reaction were spherulitic with a lamellar, α-crystalline substructure. The lamellae had a rough surface, in contrast to the melt- or solution-grown lamellae described in Sect. 3.3 and Fig. III.48, The molecular weight at the time of formation of crystals was about 400–600 and the crystals were 50–70 Å thick. The initial lamellae were thus fully extended chain oligomer lamellae with antiparallel chain placement (Sect. 2.4.6). Further polymerization and crystallization occurred on the surfaces of the initial lamellae. Figure VI.77 shows that the molecular weight increased

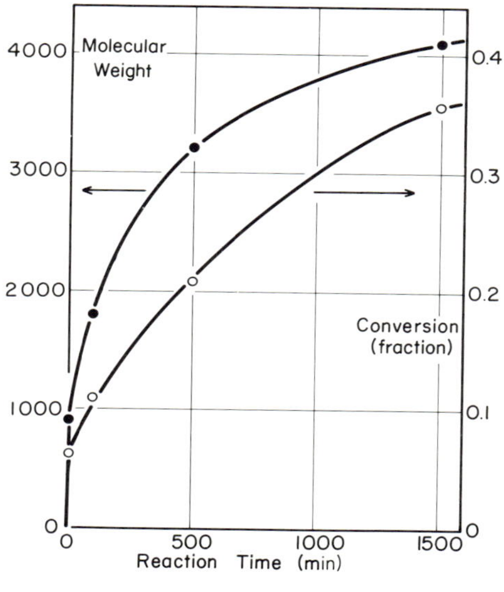

Fig. VI.77 *Increase in Conversion and Molecular Weight on Crystallization during Polymerization of Nylon 4.*
The reaction was carried out at 30°C. The reaction mixture was sealed under vacuum and contained a 1000:1:10 molar mixture of 2-pyrrolidone, *N*-benzoyl-2-pyrrolidone, and potassium. The cloud point was reached after 2 min. (Drawn after data by Komoto *et al.* 1970).

almost parallel to the conversion. Since there is no chemical reason for the increasing slow-down of the reaction, Komoto *et al.* (1970) postulated increasing loss of actively growing molecules due to occlusion, enhanced because only every second chain has the right directiveness for further polymerization. The final macroconformation of the spherulitic aggregates was assumed to resemble a fringed micellar structure (Fig. III.5) due to irregular growth between the lamellae which are correlated due to the spherulitic structure (compare with Fig. V.20 where correlation is caused by epitaxy).

The polymerization of *N*-carboxy-α-amino acid anhydrides [Leuchs' anhydrides, Eq. (133)] has been achieved by water and by a variety of bases like amines, alkali metal hydroxides, alkali metal oxides, and organometallic compounds in bulk, inert solvents, or starting with the solid monomer (Katchalski and Sela, 1958). Despite a great interest in the polypeptides produced in these reactions, little attention was initially paid to conditions which yield solid polymer. In case the solid was produced, no physical chemical analysis of the polymer was carried out. Polymerization of the crystalline monomer was already observed by Miller *et al.* (1940) (see also G. Kovacs *et al.* 1966).

In solution, polypeptides may still have a helical conformation depending on solvent and temperature. Inasmuch as the formation of a helix can be looked upon as a one-dimensional intramolecular crystallization,† crystallization during polymerization may be possible. Szwarc (1965) has reviewed the evidence for the influence of helix formation on the rate of reaction of Leuchs' anhydrides. One would expect a change in crystallization (and polymerization) rate after the critical length for stabilization of the helix in solution has been reached (7-13 repeating units). Although not fully conclusive, such evidence has been found. Even more striking is a strong retardation of the reaction in stereo-selective helical growth by a small amount of the other isomer. Addition of only 5% of D-isomer to an L-monomer reduces the polymerization rate by a factor of one-third. The 50:50 mixture polymerizes much more slowly than either isomer. This phenomenon of helix formation during polymerization may also play an important role in the polymerization of other systems (Ham, 1959, 1960, 1962). The helix formation in solution involves an ordering as in a three-dimensional crystal, but avoids the crowding of the reaction intermediate.

The crystallization mechanism of poly-L-alanine and poly-D,L-alanine from L- and D,L-alanine-*N*-carboxy anhydride was elucidated by Kawai

† The coil–helix transition has attracted much attention for biological polymers. A large specialized literature exists discussing the mechanistic details. For a monograph of this topic, see, for example, Poland and Scheraga (1970).

(1973) and Kawai and Komoto (1973). (See also Komoto *et al.*, 1972; Fujie *et al.*, 1973; Komoto and Kawai, 1973). From solution in acetonitril at 30°C, a solvent which favors helix formation, oligomer crystals (nuclei) of a degree of polymerization of 3 precipitated in form of ribbons with the chain axis normal to the ribbon surface. Trimers are still too short to form helices in solution; the crystal structure observed was accordingly the pleated sheet or β-structure (see Sect. 2.4.6). In this structure the polymers crystallize with antiparallel chains. Further polymerization was shown to be linear with time in conversion and molecular weight, indicating no occlusion (compare to Fig. VI.77). However, an analysis of the crystal structure after further polymerization revealed that α helices predominated. The original ribbons thickened and grew α helices epitaxially on the β chain ends. This surprising change in crystal structure from a β-type nucleus to an α-type growth seems to be possible since the helix cross section is more than twice that of the zig-zag chain (see Table II.16). Thus, helix-type crystallization would still lead to space filling when only half the chains can continue to grow in any one direction. Observation of the α crystals is made first when the degree of polymerization reaches 10–13. At this point in the polymerization, the β crystals seem not to have grown; rather, the additional 7–10 repeating units were attached without crystallization until (at just this length, see above) intramolecular helix formation is possible. The macroformation of these special crystals is shown schematically in Fig. VI.78. It represents a unique conformation not available by any other route.

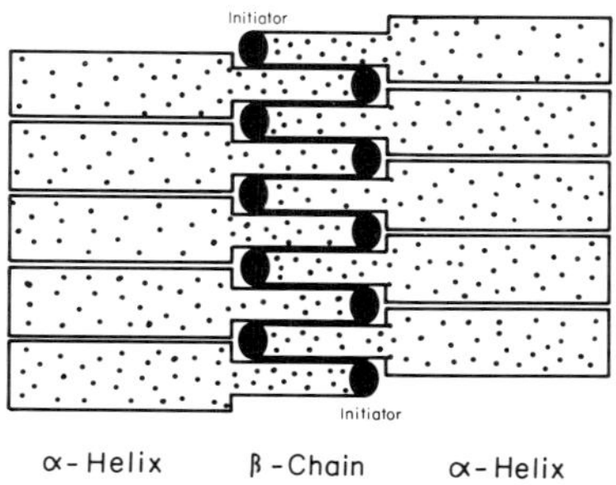

Fig. VI.78 *Schematic Drawing of the Chain Conformation of Poly-*L-*Alanine Crystallized during Polymerization from Its Leuchs' Anhydride*
(Drawn after Komoto *et al.*, 1972). For detailed structure of the α-helix and the β-chain see Fig. II.32.

Nylon 2 and nylon 3 (polyglycine and poly-β-alanine, respectively) when polymerized and crystallized from their *N*-carboxy anhydrides under similar conditions to the just described poly-L-alanine do not form helical crystal structures. The initially grown ribbons of antiparallel chain crystals show irregular further growth due to occlusion of many of the chains as was shown above for the nylon 4 and nylon 6 cases (Komoto *et al.*, 1973).

6.4.2.4 Selenium

The crystallization of selenium from the melt has in the past been treated as normal crystallization of a linear macromolecule (see Table VI.8). However, it has been known for a long time (Briegleb, 1929) that the selenium melt consists of a mixture of two allotropes, Se_8 rings and linear macromolecules. The ring–chain equilibrium has been described by Eisenberg and Tobolsky (1960). Above 83°C, the floor temperature, rings polymerize to chains. The equilibrium concentration of rings in the melt decreases gradually to about 20% at 400°C, while the molecular weight which is about 10^7 at low temperature decreases to about 10^5 at 400°C.

On crystallization, trigonal selenium of extended chain macroconformation results which contains only the macromolecular allotrope (see Table II.9 and Fig. II.39 for the crystal structure and Sects. 3.3.1 and 3.7.3 and Figs. III.31, III.124, and III.126 as well as Fig. IV.39 for crystal morphology). Thus, crystallization must occur at least partially similar to the type of polymerization which was observed in crystallization from the gas phase (see Sect. 6.4.1). The overall crystallization rate is shown in Fig. VI.79. The increase in crystallinity follows a fractional Avrami exponent ($n = 2.41$) and shows some slow crystal perfection towards the end of the crystallization, typical of the crystallization of macromolecules (see Sect. 6.1.3 and 6.1.6). From morphological observations, Coughlin and Wunderlich (1972) concluded that the growth involves initial chain folding, followed by chain extension as observed also for other extended chain crystals (see Sect. 6.3.3 and Fig. VI.60).

Figure VI.80 is an electron micrograph of the growth front of a selenium crystal growing from the melt at 126°C. The growth tips (Fig. VI.80B) are only 100–350 Å wide in the direction of the molecular chain, indicating chain folding. Further toward the interior, the crystal thickens, which can only occur by rearrangement of the once-crystallized chains. Finally, from the fusion of the fold surfaces of adjacent lamellae (see for example Fig. IV.39), it was concluded that the folds would open and connect in a third, final step of crystallization.

The concentration of chains and rings during crystallization was followed by quenching the crystallization at various stages and analysis of the frozen system (Coughlin and Wunderlich, 1973). Figure VI.81 illustrates the changes in overall concentration (weight per total weight of melt and crystal) of chains and rings at 90°C after heating from the glassy state. The ring concentration was determined by extraction of the quenched sample with carbon disulfide,

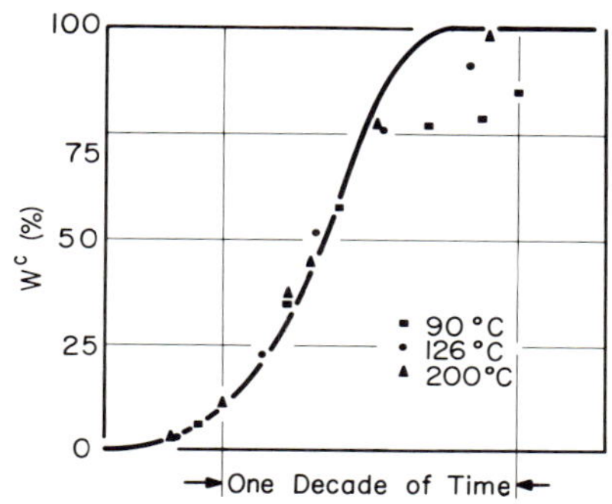

Fig. VI.79 *Percentage Crystallinity of Selenium as a Function of the Logarithm of Time.*
Data for the three crystallization temperatures are shifted along the time axis to coincide. The drawn out curve results from Eq. (33) with $n = 2.41$ (Coughlin and Wunderlich, 1973).

the crystallinity by density measurement. The initial concentration of rings (31.1%) was set by quenching to the glassy state from an equilibrated melt at 250°C. At this temperature, the chain molecules have an approximate average molecular weight of 10^5–10^6. Initially, practically no rings are involved in the crystallization. The ring concentration in the remaining melt must thus increase as is shown in Fig. VI.82. The dashed lines in Fig. VI.82 illustrate the expected equilibrium concentrations at the given temperatures. Under the proper conditions this concentration is even exceeded before fast reduction to zero occurs on further crystallization. Since the chemical approach to equilibrium is rather slow for all the chosen crystallization temperatures, the initial increase in ring concentration is mainly caused by the reduction in chains due to crystallization (compare with Fig. VI.81 which is plotted in terms of overall crystallization). The final rapid decrease in ring concentration

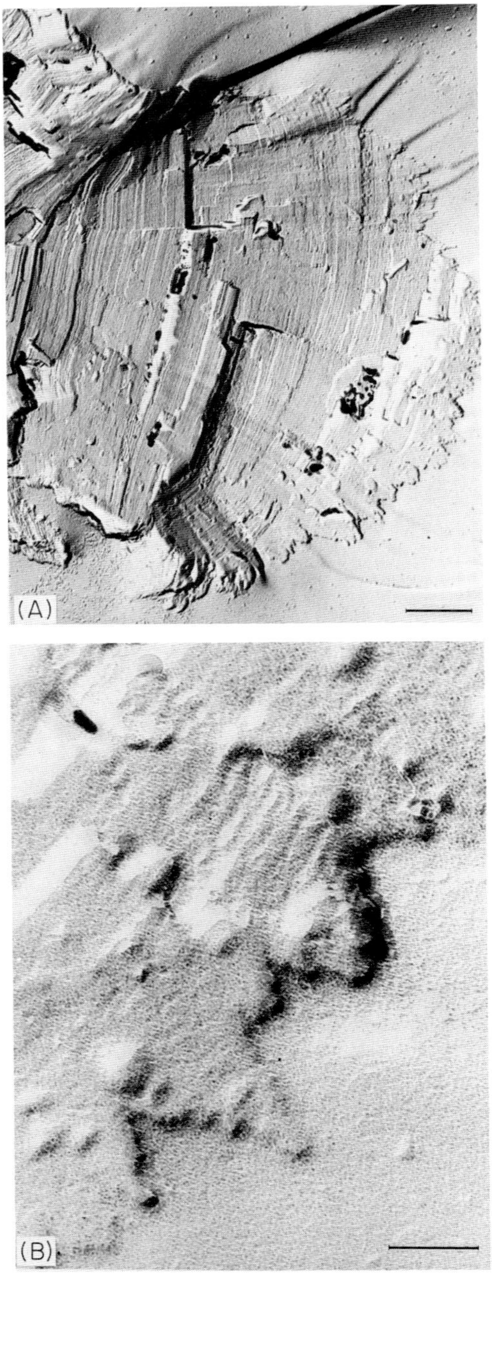
(A)
(B)

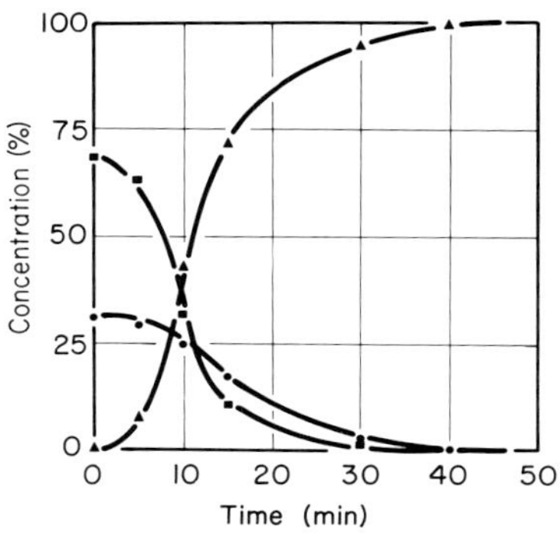

Fig. VI.81 *Concentration Changes during Crystallization of Selenium at 90°C from the Melt.*
All concentrations given as wt.% of the total sample; the symbols refer to rings (●), chains (■), and crystals (▲). The wt.% crystals is corrected for incomplete crystallization (Coughlin and Wunderlich, 1973).

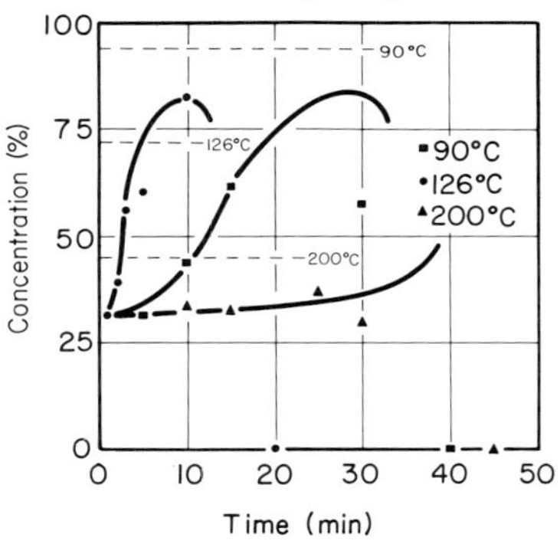

Fig. VI.82 *Concentration Changes of Selenium Rings in the Molten Phase during Crystallization.*
Concentrations given as wt.% of the remaining melt. Crystallization temperatures 90°C (■), 126°C (●), and 200°C (▲). Dashed lines represent the equilibrium concentrations in the melt (Coughlin and Wunderlich, 1973).

Fig. VI.80 *Electron Micrographs of a Fracture Surface of Quenched, Partially Crystallized Selenium.*
Crystallization was for 3 min at 126°C. The crystallographic c-axis is tangential to the incipient spherulite. Each separate lamella is narrow at its tip showing folded chain crystallization, followed by thickening. (A) Scale bar 0.5 μm; (B) Scale bar 0.1 μm (Coughlin and Wunderlich, 1972).

must then result from crystallization of rings during polymerization. The reaction site has been placed close to the crystal surface where a high concentration of free radical chain ends and of rings should be present. The overall process of crystallization during polymerization can be described by three rate constants using the following reaction scheme:

$$x\,\mathrm{Se}_8 \rightleftharpoons {}^*\mathrm{Se}_{8x}{}^* \longrightarrow \mathrm{Se}_{\mathrm{cryst}} \tag{136}$$

6.4.2.5 Other Macromolecules

The polymerization of tetrahydrofuran in bulk, initiated by triethyloxonium hexachloroantimonate, has been analyzed by Stejny (1973) from $-5°C$, where the produced polymer is soluble in its monomer, to $-45°C$, where early crystallization of the oligomer occurs. Figure VI.83 illustrates the kinetics data for the following reaction:

Initiation

$$(\mathrm{C_2H_5})_3\mathrm{O}^+\ \mathrm{SbCl_6}^- + \mathrm{OC_4H_8} \longrightarrow \mathrm{C_2H_5}-\overset{+}{\mathrm{O}}\mathrm{C_4H_8}\ \mathrm{SbCl_6}^- + (\mathrm{C_2H_5})_2\mathrm{O}$$

Propagation (137)

$$[\mathrm{O}-(\mathrm{CH_2}-)_4]_x\overset{+}{\mathrm{O}}\mathrm{C_4H_8SbCl_6}^- + \mathrm{OC_4H_8} \rightleftharpoons [\mathrm{O}-(\mathrm{CH_2}-)_4]_{x+1}\overset{+}{\mathrm{O}}\mathrm{C_4H_8SbCl_6}^-$$

The points of intersection of the curves for the lower temperature polymerization coincide with the first visual observation of the crystalline phase. Occlusion seems not to be the dominant factor since the rate increases after crystallization. The molecular weights reached when the first crystals grew was 9000 for the $-25°C$ reaction and 3900 for the $-45°C$ reaction. These particular length molecules crystallize as folded chain crystals with a fold-length of 110–122 Å at the reaction temperatures. The active chain end would be left dissolved and able to polymerize further. The reactions under the conditions illustrated in Fig. VI.83 are thus an example for the change from homogeneous, separate polymerization to heterogeneous, successive polymerization and crystallization. Extended chain oligomer nuclei would only be expected at even lower temperatures. With the chosen conditions reaction rates are at even lower temperatures inconveniently slow and stop almost completely at $-78°C$.

The crystallization of poly(1,4-epoxycyclohexane) leads to morphologies illustrated in Fig. VI.84. The reaction was carried out at $-15°C$ in methylene

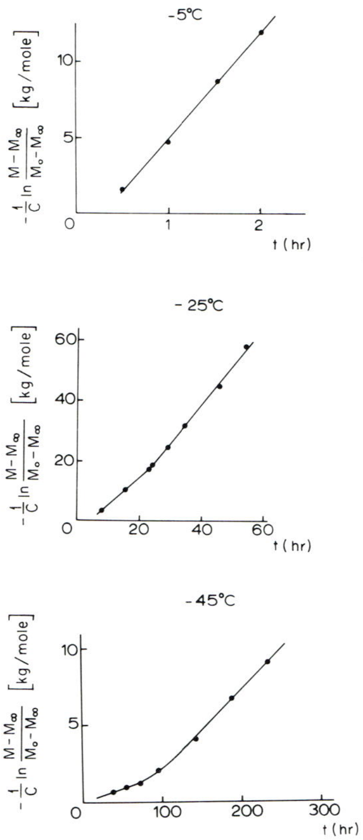

Fig. VII.83 *Kinetics of Polymerization of Tetrahydrofurane at Different Temperatures.*

Homogeneous, separate polymerization at −5°C; the polymer is soluble. At −25°C and −45°C, the homogeneous phase is followed by crystallization and heterogeneous polymerization. For the reaction see Eq. (137). Concentrations in mole/kg, c = catalyst concentration (living-type reaction); M = monomer concentration (M_0, initial; M_∞, equilibrium). Curves by Stejny (1973).

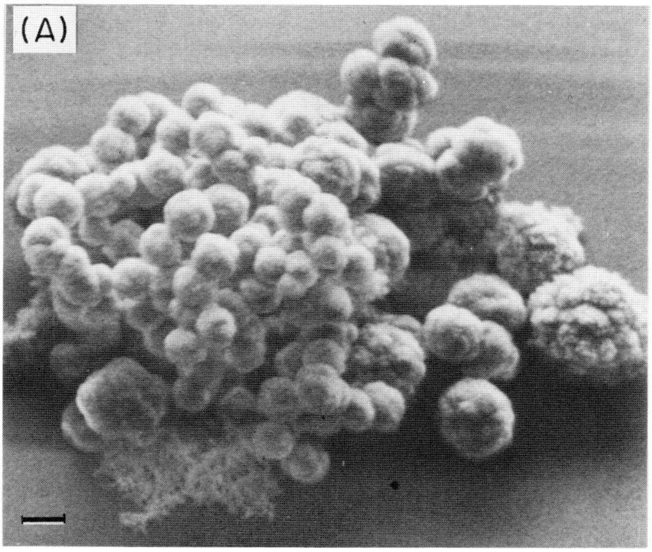

Fig. VI.84 *Scanning Electron Micrographs of Poly(1,4-epoxycyclohexane) Crystallized during Polymerization.*

Monomer concentration 2.05 mole/kg; catalyst concentration 0.007 mole/kg; −15°C; reaction see Eq. (138); solvent, methylene chloride. (A) Spherical particles at conversion 0.7% ($\overline{M}_n = 2000$, $\overline{M}_w/\overline{M}_n = 1.15$) made up of lamellar substructure of 108 Å thick oligomer lamellae. (B) Dendritically developed particles at conversion 4.5%($\overline{M}_n = 25\,000$, $\overline{M}_w/\overline{M}_n = 1.61$), at this time a high molecular weight fraction has developed. Electron micrographs by Stejny (1973). Scale bar 1 μm.

chloride solution, initiated with methyloxonium hexachloroantimonate (Stejny, 1973).

Initiation

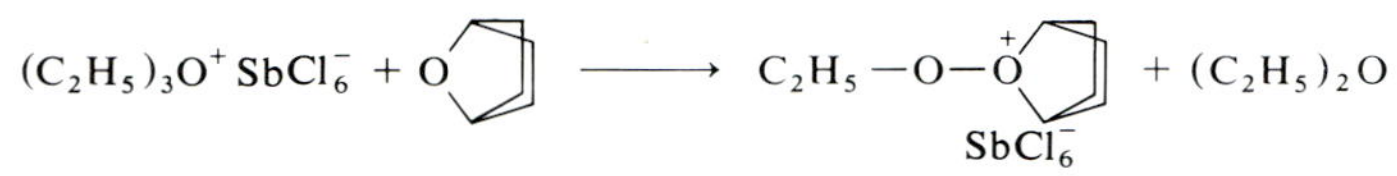

Propagation (138)

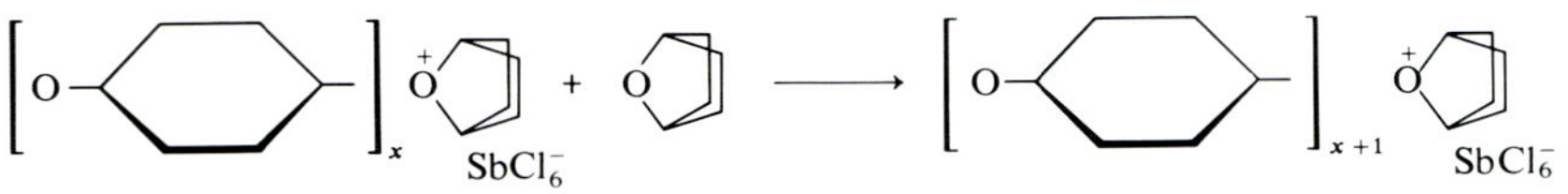

In this reaction, the crystallization starts almost immediately. The initial particles are spherical as shown in Fig. VI.84A. Their number stays almost constant at later stages of the reaction. The initial unimodal molecular weight distribution corresponds approximately to the constituent fully extended chain oligomer lamellae of the spherical particles (detected by low angle X-ray diffraction, ~108 Å). Further oligomer produced homogeneously adds to the lamellae laterally. This further growth changes the spherical superstructure into increasingly more dendritic structures (Fig. VI.84B). In addition to the growth of more oligomer, further polymerization is possible on the oligomer crystal surfaces; this leads to production of higher molecular weights. Some low molecular weight material remains throughout the reaction, most likely due to occlusion. Overall, the crystallization site must be closer to the polymerization site than in the polytetrahydrofuran case.

Errede *et al.* (1960) studied the polymerization of *p*-xyxylene in 0.03 *M* hexane and toluene solutions at −78°C. In contrast to the polymerization from the gas phase, described in Sect. 6.4.1, the reaction is very slow in solution at this temperature. A high percentage of macromolecules (92%) is formed and precipitates in a highly swollen form as soon as the oligomers produced in the initial, homogeneous reaction phase reach a degree of polymerization of about 20. Brief warming of the solution speeds up the reaction by increased initiation, but complete reaction at higher temperature leads only to oligomer. The reaction to polymer is thus governed by a sufficiently slow initiation rate followed by fast reaction in the precipitated state. The morphology of the precipitate was not analyzed, but may well be similar to the poorly crystallized material obtained at −78°C from gas phase polymerization (Sect. 6.4.1). No chain transfer reaction was observed at this low temperature, so that a slow decrease in rate has been linked to occlusion of chain ends. Iguchi (1970) has fitted the kinetic data to a one dimensional Avrami expression (see Table VI.6). He assumed that the growth of polymer chains can be treated similarly to fibrillar crystal growth. A similar treatment

was applied by Iguchi (1970) to the data of the epichlorohydrin polymerization by Dimonie and Gavăt (1968).

A final example of change of homogeneous polymerization to heterogeneous polymerization is the polyfluoral $[(CF_3{-}CHO{-})_x]$ formation described by Busfield and McEwen (1973). Polymerization was carried out at 31°C with boron trifluoride diethyl etherate as catalyst in dichloromethane as solvent. The polymer separated at an average degree of polymerization of about 22 (at 1 *M* remaining monomer; lower monomer concentration reduces the separation length). The reaction rate increased as soon as a separate polymer phase was observed as long as the separated polymer was liquid. On solidification, occlusion seemed to cause a loss of active chain ends.

6.4.2.6 *Conclusions*

The examples of crystallization during polymerization from the liquid state described in Sect. 6.4.2 demonstrate the wide range of detailed mechanisms possible. Each system needs a clear analysis of the chemical mechanism as well as the physical changes occurring on phase transformation. The given examples show that simultaneous polymerization and crystallization is possible and occurs usually when the chemical reaction takes place close to the ceiling temperature (see Sect. 6.1.8). At conditions far from the equilibrium, successive polymerization and crystallization is more likely. In the simultaneous polymerization and crystallization case, the initial oligomer lamellae may grow into extended chain crystals. In the successive polymerization and crystallization case, the morphologies observed range from lamellar to fibrillar and the macroconformation of the chains vary from regularly folded, to irregularly folded, to fringed micellar. The major influence of the crystallization on the chemical reaction is a change in rate, change in mechanism, and the possible termination of the polymerization due to occlusion of the active chain ends. In every case, oligomers of sufficient concentration must be produced in a homogeneous reaction phase (see Sect. 5.1.2). In step reactions, the oligomer concentration decreases after nucleation and the mechanism changes to an addition reaction. Oligomers can be used as molecular nuclei and can also be added to other growing chains so that finally a complete transformation to polymer is possible. In addition reactions, once-made oligomer can only grow to polymer by further monomer addition. Large initial oligomer concentrations may cut the yield of polymer.

Only recently has the crystallization during polymerization found more intensive investigation by an interdisciplinary approach. Many questions of the complicated process involving the detailed nature of the reaction site on

or close to the crystal surface and the effect of energy transfer remain presently unanswered.

6.4.3 Solid Monomer Crystallization

There are many reports in the literature describing polymerization reactions which start with crystalline monomers and lead to high molecular weight macromolecules.† This area of research has become known as solid state polymerization, irrespective of the detailed reaction mechanism. These reactions are topochemical in the sense that often the chemical nature and always the kinetics of the reaction is influenced by the crystalline nature of the monomer. Few of these polymerizations are topotactic in the sense that the macromolecular product is crystalline and the complete reaction path is governed by the monomer crystal structure (Sect. 6.1.8). In the present section some aspects of topochemical polymerization which are of interest to crystallization during polymerization are described. Details are given for the formation of polyoxymethylene, the four-center-type polymerization of monomers such as 2,5-distyrylpyrazine, and the growth of polymer crystals in substituted diynes. The special morphologies of the macromolecular crystals grown during polymerization are discussed in Sects. 3.8.1, 3.8.3, and 3.9; some general aspects are summarized in Sect. 6.1.8.

An early observation of a polymerization which started with a crystalline monomer was made by Letort (1936) who observed the spontaneous polymerization of acetaldehyde. The resulting polymer was amorphous. Even earlier, Fischer (1906) observed that pentaglycylglycine methyl ester changes on prolonged heating at 100°C, below its melting temperature, to an amorphous, water insoluble, polymeric material. The structure of such polypeptides produced by topochemical reactions have been analyzed in more detail by Slyterman and Veenendaal (1952) and Slyterman and Kooistra (1952).

A large interest was attracted since 1960 by the radiations induced polymerization of solid monomers.‡ Again the reactions were topochemical, but it was proven by Adler (1959) that the early suggestion by Mesrobian *et al.*

† Besides the reviews cited in Sect. 6.1.8 see, for example: Collection of papers in *J. Polym. Sci. Part C*, **4** (2) (1964); Hardy (1969). *Plaste Kaut.* **16**, 723; Eastmond (1970). *Progr. Polym. Sci.* **2**, 1; Herz and Stannett (1968). *Macromol. Revs.* **3**, 1; Abkin and Gerasimov (1971). *Vysokomol. Soedin. Ser. A* **13**, 240; Okamura (1972). *Pure Appl. Chem.* **30**, 181; Collection of papers in the *Proc. Tihany Symp. Radiat. Chem.* (*3rd*) (1971) published 1972, Dobos, ed. Acad. Kiado, Budapest; Osugi (1970). *Revs. Phys. Chem. Japan* **40**, 122; Kargin *et al.* (1964). *J. Polym. Sci. Part C* **4**, 767; Hardy (1973). *J. Polym. Sci. Part C* **42**, 663; and Agnew and Brown (1971). *J. Polym. Sci. Part A*1 **9**, 2561.

‡ See, for example, the reviews by Charlesby (1965). *Repts. Progr. Phys.* **28**, 463; Chapiro (1966). In "Actions Chimiques et Biologigues des Radiations", 10e série (Haissinsky, ed.) Masson et Cie., Paris; and Bruk (1966). *Usp. Khim. Polimerov. Khim. Moscow*, **42**.

(1954) that the product polymers are usually amorphous is correct. The reactions were found to be typically two-phase reactions initiated at mosaic boundaries, dislocations, and regions of stress. Acrylamide, which forms a solid solution with propionamide, and methacrylamide, which is soluble in certain concentrations in isobutyramide, can be polymerized in the crystalline solid solution (Adler and Reams, 1964; Faucitano and Adler, 1970). Since in this case the reactive component cannot diffuse through the crystal, the mechanism of polymerization must involve initiation at a position of accidental high concentration of monomer, followed by surface reaction. The polymer occupies a smaller volume, which enhances surface diffusion of monomer. In addition, the monomer–solvent crystals must continuously recrystallize to permit surface diffusion to the polymer interface for polymerization. Sella and Trillat (1961) and Sella and Bensasson (1962) published photographs of acrylamide crystal surfaces at various stages of irradiation induced polymerization. Globules of 20-Å diameter were found a short time after the start of reaction. The globules were aligned along the crystallographic directions of the monomer crystal. The smallest particles represented single growing macromolecules. In time the globules grew to several hundred angstrom in diameter and fused with neighboring globules to give a fibrillar appearance. X-ray results showed that the produced macromolecules were in all cases amorphous.

Chemical initiation of polymerization of monomer crystals is also possible. These reactions involve the diffusion of the initiator along grain boundaries and reaction from the outside of the crystal. A special method (molecular ray method) of simultaneous vacuum deposition of monomer and initiator on a cold substrate was developed by Kargin *et al.* (1959). Reaction, sometimes explosively, was observed on warming as soon as crystallization occurred.† It is not a general fact, however, that the polymerization reaction is enhanced by crystallization or by starting from the crystalline state of the monomer. Tabata *et al.* (1968) showed that acrylonitrile and propylene oxide dispersed in glassy methyltetrahydrofuran and methylene chloride, respectively, polymerized at the same temperature faster than the pure crystalline monomers.

Polymerization in systems of liquid crystalline nature, monolayers, micelles, and matrices is also possible.‡ Under such condition there may be

† For detail see Kabanov *et al.* (1959). *Vysokomol. Soedin.* **1**, 1422; ibid. 859 [English transl. *Polym. Sci. USSR* **2**, 179 (1960)]; Kargin and Kabanov (1961). *J. Polym. Sci.* **52,** 71; Kargin *et al.* (1961). *Vysokomol. Soedin.* **3**, 426 [English transl. *Polym. Sci. USSR* **3**, 136 (1962)]; Kargin *et al.* (1964). *J. Polym. Sci. Part C* **4**, 767; Papissov *et al.* (1967). *Vysokomol. Soedin.* **A9**, 327 [English transl. *Polym. Sci. USSR* **9**, 365 (1967)].

‡ For a review of polymerization in such ordered systems, see the proceedings of the Midland Macromolecular Meeting 1974, to be published in the *Int. J. Polym. Mater.* (1975).

greater monomer mobility than in the solid state under conditions which maintain some of the beneficial orientation. The produced polymers seem, however, again to be usually amorphous (see also Hardy *et al.*, 1970).

All reactions starting from the crystalline monomer are initially topotactic. Topotactic control is then lost somewhere along the reaction path. For example, the primary radicals of irradiated acrylic acid are arranged at $-100°C$ with their $H_3C—\dot{C}H$ bond parallel to the double bonds of the other monomer molecules (Shioji *et al.*, 1963). Even at this low temperature some monomer units can add to the radical, forming a secondary radical. The secondary radicals are still oriented and have added the monomer in the direction of the two-fold screw axis of the monomer crystal. The reaction stops at the dimer or trimer stage since the intermolecular distances are now out of reach for low temperature reaction (Adler and Petropoulos, 1965). On warming of the crystals, further reaction occurs, but the longer oligomers were shown by ESR to have lost their orientation (see, for example, O'Donnell *et al.*, 1964). Also lost is the tacticity of the polymer chain (see Sect. 1.5), so that secondary crystallization of such vinyl polymer is also not possible.

Somewhat longer topotactic control is possible for special monomer crystal structures. Single crystals of vinyl stearate which have their vinyl groups separated by 25.4 Å have been polymerized after initiation by γ radiation by Morosoff *et al.* (1965). The parallel orientation of the stearate chains is preserved during the reaction. Up to 30% conversion, the discrete monomer phase was retained (monoclinic). At higher conversion a change of the side chain packing to hexagonal was observed. The resulting polymer was, however, not syndiotactic, the condition a topotactic reaction would require; it is even probable that the tacticity is less than in liquid polymerization. Thus holding the vinyl groups via their side chains in fixed layers is not sufficient for topotactic polymerization and crystallization. The orientation of the vinyl groups in the monomer crystals did not lead to orientation (crystallization) of the polymer chains of the product.

Stacks of monolayers of vinyl stearate built up by successive transfer from monolayers spread on a water surface were polymerized with γ irradiation by Cemel *et al.* 1972 (see also Puterman *et al.*, 1974). Each layer polymerized independently of the others. As in the polymerization of single crystals, the polymer is soluble in the monomer crystals, maintaining monomer side-chain packing. Above 30% conversion, change to the polymer side chain packing was observed. Again the maintained orientation is solely due to side-chain crystallization.

Some orientation was also retained on thermal polymerization of *p*-benzamidostyrene when polymerization was carried out below the polymer glass transition temperature (Jakabhazy *et al.*, 1964). The monomer crystal has a layer structure which separates reactive radicals. The reaction proceeded in successive layers of the monomer crystal from the outside to the inside without induction period. However, the retained orientation detected by optical microscopy and infrared dichroism disappears above the glass transition temperature of the polymer (120°C). The polymer produced is slightly less dense than the monomer crystal since it is amorphous. Similar orientation in the amorphous state was found on polymerization of terephthalonitril oxide crystals (Fujimoto, 1967). Oriented polymer was also observed by Matsuda *et al.* (1970) on polymerization of acrylamid and methacrylamide by chlorine gas activated with ultraviolet light.

A topotactic polymerization to oligomers of 3–6 monomers was proven by Hardy *et al.* (1972, 1973) for the polymerization of crystalline acenaphthylene initiated by iodine. The conversion was strictly linear with time when corrected for diffusion effects of the iodine. Relatively little motion is necessary for the monomer to change to the *threo*-diisotactic chain molecules. Similarly, the initial stages of vinylene carbonate polymerization yields topotactic initial oligomeric products. In these case again the final polymers are amorphous (Hardy *et al.*, 1974).

A somewhat unique situation has been studied in the free radical polymerization in urea (White, 1960) and thiourea (Brown and White, 1960) clathrates. Several monomers have been found for which conditions for solid state polymerization to a stereo–regular crystal are fulfilled. The radicals are protected from termination by neighboring radicals by the urea or thiourea walls of the canals. The proper approach of the monomer molecules is achieved by stacking the monomer molecules as outlined schematically in Fig. VI.85. Here the polymer crystal structure is not as critical for the reaction as in a monomer crystal.

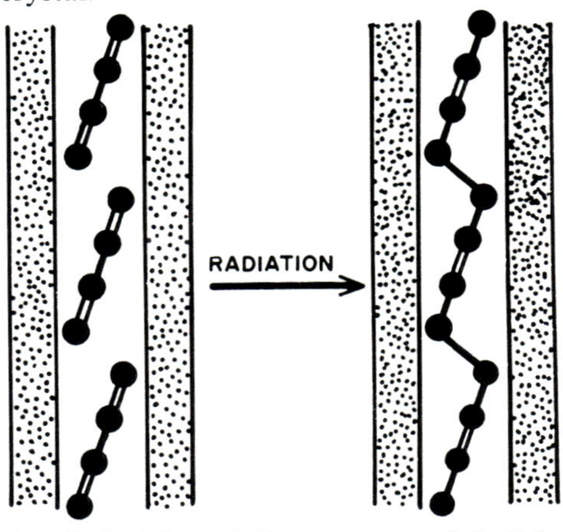

Fig. VI.85 *Schematic Representation of the Polymerization in Clathrates.*
Drawn after Brown and White (1960).

In thiourea the polymerization of 2,3-dimethylbutadiene and 2,3-dichloro-butadiene was investigated by Brown and White (1960) and Chatani *et al.* (1970a) (see also Chatani and Tadokoro, 1973). Both monomers form complexes with thiourea in which they are stacked as shown in Fig. VI.85. The thiourea to monomer ratios are about 3:1. The hexagonal canal in thiourea is 7 Å in diameter while the two monomers are flat molecules of an approximate crossection of 7.6 × 5 Å normal to the canal, causing some deformation.

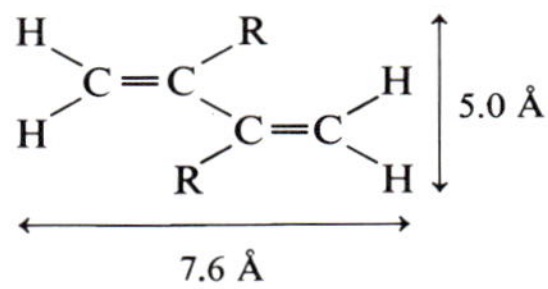

(139)

In the complex crystal, the monomer repeat distance is 6.25 Å. This arrangement permits reaction between the planar free radical and the planar double bond. Initiation of the reaction is achieved by a short burst of high energy radiation. The yield and quality of the polymer produced is independent of temperature and dose rate, which is indicative of isolated sequences achieving polymerization. A great improvement of the yield is possible on annealing of the complex before irradiation. Both polymers have a trans configuration around the remaining double bond. The central single bond shows trans conformation and the two remaining single bonds have clockwise and counter-clockwise rotated positions. As the polymerization progresses, streaks develop in the X-ray diffraction pattern parallel to the layer lines, an indication that the correlation between neighboring macromolecules is not perfect. The periodicity along the molecular chains is 4.80 Å; 1.45 Å shorter than before polymerization. Considerable motion must have thus occurred within the canal. Topotacticity is thus only preserved because of the relatively rigid walls of the clathrate. The polymer complex is stable at room temperature, which requires that there should not be large vacancies in the polymerized canals. The additional monomer needed to fill up the canals may have diffused into the canals during polymerization (Chatani *et al.* 1970a).

Extraction of the thiourea preserves needlelike crystalline aggregates of the polymers with higher melting temperatures than the same polymer produced by conventional means. The high specificity of this polymerization is indicated by the fact that only a small number of molecules enters the thiourea template, and an even smaller number reacts. Reactions which were carried out in molecular sieves which have branching and intersecting canals led only to products without structural regularity (Brown and White, 1960).

In urea, the polymerization of butadiene and vinyl chloride was analyzed in some detail by Sakurada and Nanbu (1959) and White (1960) (see also

Chatani and Tadokoro, 1973). Exact geometries in the about 1 Å smaller canals were not proposed. Again, the polymers were highly stereoregular. The crystalline *trans*-1,4-polybutadiene prepared by removal of the urea with aqueous acetone had melting ranges up to 147–154°C, while the recrystallized polymer melted as conventionally produced polymer at 135–137°C. One must assume that the removal of urea did not allow complete collapse of the extended polymer chains to a random coil before crystallization. The higher melting point would then be due to a thermodynamically more stable crystal morphology with more extended polymer chains, which may in addition show some superheating. An introduction of chain bends in the planar syndiotactic conformation of poly(vinyl chloride) made in urea complexes on hot pressing was observed by infrared techniques (Krimm *et al.*, 1965).

The solid state formation of polyoxymethylene crystals from cyclic monomer crystals has been analyzed in considerable detail. In cases of polymerization from cyclic monomers, topotactic control may be more likely since ring-opening and polymerization breaks and remakes similar bonds so that there is no significant change in the overall bond-length.† Still, the detailed crystal geometry of monomer, intermediate, and polymer must be considered for a possible topotactic reaction. Trioxane was the first solid cyclic monomer whose polymerization to polyoxymethylene was observed. Exposure of the crystals to gaseous formaldehyde leads to polymerization starting at the interface (Kohlschütter, 1930; Kohlschütter and Sprenger, 1932) (see also Sect. 6.4.1). Similarly, cationic initiators will polymerize solid trioxane (Okamura *et al.* 1962b, 1964) and tetroxane (Kobayashi *et al.*, 1971) (see also Sect. 6.4.2.2). Ionizing radiation absorbed at low temperature leads to high molecular weight crystalline polyoxymethylene on postirradiation polymerization at 40–60°C (Okamura *et al.*, 1962a; Hayashi *et al.*, 1964). No such reaction is possible in the molten or dissolved states, even at elevated pressure (Yamazawa *et al.*, 1971). The crystal morphology of the obtained macromolecules is usually fibrillar (fibrillated) as described in Sect. 3.8.3 (see Fig. III.78). The fibers of the polyoxymethylene are oriented

† Other cyclic monomers which have been polymerized from the crystalline monomeric state to macromolecules are, for example: β-propiolactone and diketene (Okamura *et al.*, 1962a); 3,3-bis(chloromethyl)cyclooxabutane (Keggenhoff and Wegner, 1973); trithiane (Lando and Stannett, 1965; Mammi *et al.* 1965; Carazzolo *et al.*, 1967; and Nadkarni and Schultz, 1973); trithiane-trioxane mixed crystals (Lando and Frayer, 1968); 1,3,5-triselenane (Carazzolo and Mammi, 1967); 1,3,5,7-tetraselenocane (Russo *et al.*, 1966); and hexamethyl-cyclotrisiloxane (Lawton *et al.*, 1956). The last monomer reacts to a cross-linked polymer; all others are crystalline. The reactions are not fully analyzed. Similar to the trioxane and related rings described in the text, they show topotactic control only in the early reaction stages.

along the *c* axis (needle-axis) of the trioxane crystal. The crystal structure of the fibers when observed by X-ray diffraction is, however, not single crystalline. Besides the expected macromolecular chain axis orientation along the needle-axis, a certain fraction of the chains (which decreases with increasing polymerization temperature) is oriented 76°7′ to the trioxane axis (Carazzolo *et al.*, 1963). The two orientations of polymer crystals were called *Z* and *W*.

More detailed studies of the mechanism, taking into account crystallographic properties as well as chemical properties, were carried out by van der Heijde *et al.* (1966), Adler (1966), Colson and Reneker (1970, 1973), Muñoz–Escalona and Wegner (1971), Eastmond (1973), and Voigt-Martin (1974). Although not all particulars are solved at present, and differences may exist in the various preparations analyzed, the following picture seem to describe the process. At low temperature (below 25°C), the mobility of the trioxane molecules in the crystal is low and little reaction occurs after irradiation. Above 40°C there is sufficient motion in trioxane crystals to cause the appearance of a narrow line in the NMR spectrum (Sect. 4.1.6). Komaki and Matsumoto (1963) interpret this motion as a rotation of the molecule about the threefold axis of symmetry with a fraction of the molecules adopting the boat conformation.

The low temperature irradiation is assumed to produce activated species which bring about the nucleation step of the macromolecular crystals at high temperature. Nucleation is kinetically detectable through an induction time (about 30 minutes at 48°C). After this induction time a number of small, hexagonal platelets of about 100 Å thickness and up to several 1000 Å in diameter are observed. These initial crystals have been called the precursor crystals. Little is known about the molecular arrangement within these crystals. They seem to appear first at the {001} subgrain boundaries within the trioxane crystals (Bassett, 1967). Out of precursor crystal grow fibers of considerable length in a relatively short time (more than 10 μm in 30 min at 48°C). Eastmond (1973) suggests that the growth occurs in direction of dislocation lines.

Figure VI.86 illustrates the middle portion of a group of such fibers of polyoxymethylene after removal of the trioxane host crystal by sublimation. Many additional precursor platelets are also visible, occurring in some cases as rows of successive platelets (Colson and Reneker, 1973), and on careful nispection reveal an internal microstructure of 400–700 Å diameter (Voigt-Martin, 1974). Any one of the fibers of polyoxymethylene may be entirely of the *Z* or *W* orientation (Colson and Reneker, 1973) or may change in intervals along the *c* axis from one to the other (Voigt-Martin, 1974). Adler (1966) illustrated on polycrystalline samples that the polymer fibrils, once initiated, may continue through several crystals independent of the surrounding monomer crystal. This indicates that topotactic control is lost.

Fig. VI.86 *Polyoxymethylene Crystals Grown inside an Irradiated Trioxane Crystal.*
Long Z and W orientation fibers and isolated as well as repeated (stacked) precursor crystals can be seen. Electron micrograph by Colson and Reneker (1973).

The basic orientations of the Z and W fibers is based on two different growth directions. Both are illustrated in Fig. VI.87. The crystal growth occurs mainly in the direction of the negative end of the trioxane crystal (Colson and Reneker, 1973) in agreement with the assumed cationic mechanism (Jaacks *et al.*, 1973). The geometric restrictions within the monomer crystal complicate the growth stage. Overall, the density increases by 6%, but on polymerization in the c direction (Z orientation, Fig. VI.87A) every fourth trioxane ring has to be removed to match the polyoxymethylene with the trioxane crystal. In return there is a considerable contraction perpendicular to the c direction. In order for the polymerization not to set up stresses in the host crystal, additional molecules of polyoxymethylene have to start growing parallel to the stack of trioxane. Then, growth could occur by sublimation across the void as indicated in the figure. Possibly, the overall energetics are governed largely by the stresses (Voigt-Martin, 1974) and the internal electrical field (Colson and Reneker, 1973).

The W orientation can be achieved by lateral connection of the trioxane rings with initially little movement (Fig. VI.87B). The contraction of the crystal and the resulting voids are in this case above and/or below the drawn-out sheet of polyoxymethylene. With such large voids at the growth front, topotactic control must be lost. Remaining orientation is caused by oriented nucleation rather than by topotaxy.

The conversion of trioxane to polyoxymethylene is schematically shown in Fig. VI.88. Besides the nucleation and longitudinal growth, irregular growth of rather short fibrils is observed along grain boundaries. In addition it was shown by Voigt-Martin (1974) that the later stage of polymerization causes thickening of the original fibers. The molecular weight reaches quickly ($\sim$2 hr) a constant value of usually 10^5, much shorter than the fibrils, which thus contain many chain ends. Once nucleated, the growth does not continue into pure unirradiated trioxane crystals. The fibrils grow only a small distance into areas not irradiated. The initial fibrils seem to be to some degree extended chain crystals (at least in the Z regions), however, the fibers are metastable. Electron microscopy shows a subgrain structure of a few hundred angstroms in diameter and annealing shows fast conversion to regularly folded chain lamellae (see Chapter VII and Fig. III.78).

Of interest in connection with the trioxane polymerization are the results obtained on the cyclic tetramer, pentamer, and hexamer. Chatani *et al.* (1968, 1970b, 1973) and Chatani and Tadokoro (1973) analyzed the crystal interrelations after postirradiation polymerization. Similar to the trioxane case, these cyclic monomers yield also highly crystalline polymer of one major orientation which is crystallographically related to the monomer

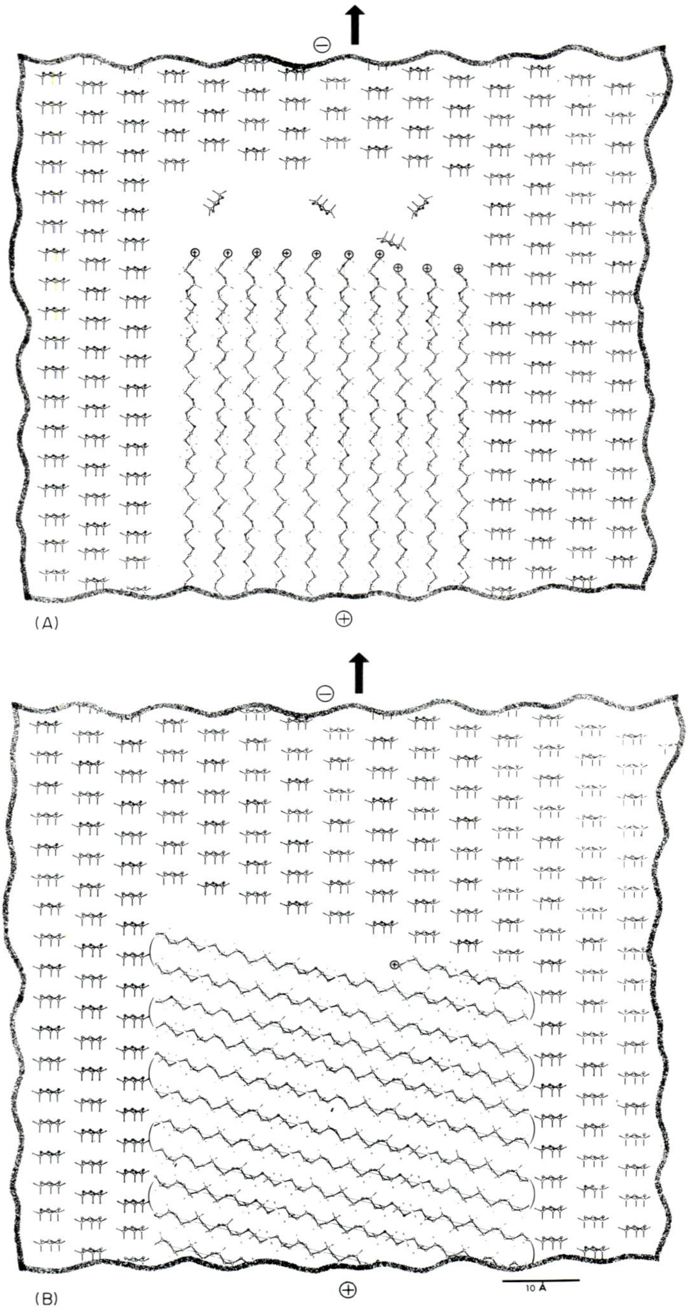
(A)
(B)
10 Å

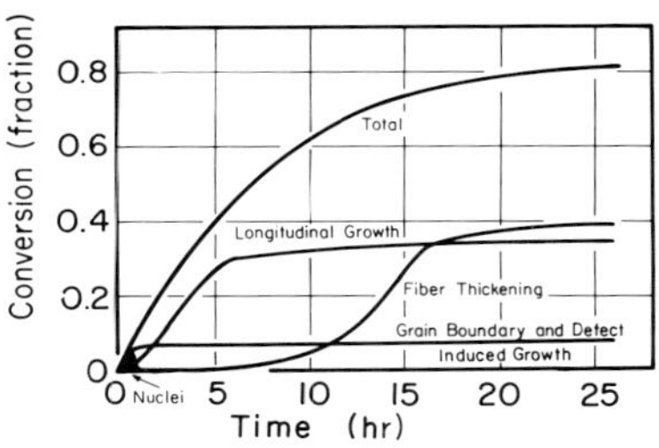

Fig. VI.88 *Conversion of Trixoane into Polyoxymethylene after Irradiation.* Also indicated are the various contributions proposed from microscopic analysis. Drawn after Voigt-Martin (1974) (Approximate conditions: Irradiation for 23 hr with X-rays at a dose of about 10^5 roentgen/min cooled with liquid air, post irradiation polymerization at 50°C).

crystal. In addition, one or more "twin" orientations are found. The ratios of yield between these orientations is variable with temperature. The tetramer has the major fraction of the macromolecular crystal aligned with the chain axis (parallel to the normal of the monomer ring). The major portion of the pentamer polymerizes more closely within the plane of the monomer rings (in the [110] and [$\bar{1}$10] directions of the orthorhombic monomer unit cell). The cyclic hexamer is again trigonal as the trimer crystal, but the major macromolecular component does not polymerize with the polymer crystal *c* axis parallel to the monomer crystal *c* axis, rather it crystallizes parallel to one of the three *a* axis directions ([110], [010], or [$\bar{1}\bar{1}$0]). As a result, the polymer crystals are powderlike rather than fiberlike. For the postirradiation polymerization of the tetramer, temperatures of 60–100°C are most suitable

---

Fig. VI.87 *Polyoxymethylene Crystal Growing within a Trioxane Crystal.*
Schematic views of a (110) plane of the trigonal trioxane ($a = 9.395$ Å, $c = 8.350$ Å, R3c, $\rho = 1.406$ g/cm^3, 6 molecules per unit cell) and a (100) plane of polyoxymethylene I (see Table II.14). (A) Polymerization in the *c* direction (*Z* orientation). (B) Polymerization in the trioxane ring direction (*W* orientation). Since the molecules are longer than the fiber width, folding at the boundary is indicated. The actual width of fibers of the *W*-type is 3–10 times larger than drawn. Figures from Colson and Reneker (1973).

because of the higher monomer melting temperature (114°C). In this case the fibrillar crystals show regularly folded chain macroconformation of 70–240 Å fold period. Muñoz-Escalona and Wegner (1971) suggest that compared with trioxane, the even larger surplus of monomer in the *c* direction and the higher polymerization temperature are the main causes for chain folding.

An example of closer match between monomer and macromolecular crystal is found in the 2,5-distyrylpyrazine and related compounds† and their polymerization products (see tabulation, p. 359, Chapter III). The photopolymerization of this type of crystalline monomer was first reported by Hasegawa and Suzuki (1967) and is described in some detail in Sect. 3.8.4 (Eq. (9), Chapter III) (see also Hasegawa *et al.*, 1973). The isomorphism between monomer and polymer permits a topotactic reaction with only small amounts of loss of crystal perfection through fibrillation. In contrast to the polyoxymethylene formation the maximum polymerization rate is not close to the 2,5-distyrylpyrazine melting temperature (226°C), but far below it. This seems to indicate that for closer topotactic control large monomer mobility is not needed. For polymerization with two neighboring molecules, the center of gravity of the monomer does not have to move.

The progress of the reaction was followed by Nakanishi *et al.* (1972) by X-ray diffraction. A single crystal of 2,5-distyrylpyrazine was irradiated at wavelengths of about 500 μm, in which case the reaction stops at an oligomer stage with an average degree of polymerization of 5. The monomer single crystal turned white after several hours of irradiation, an indication of fibrillation. The oligomer showed a lower density than the monomer (1.177 versus 1.244 g/cm^3). Most of this density change is due to an expansion in the crystallographic *b* direction to about half the difference which is finally observed for the polymer. The oligomer crystals show no contraction in the *a* direction. Further irradiation with higher energy light completes the polymerization, which without prior low energy irradiation would have occurred in one step. This stage of the reaction goes with an increase in density and occurs also without loss of orientation. At least the initial oligomer is dissolved in the monomer crystal as can be seen from the reduction in melting point (Hasegawa *et al.*, 1974). At higher temperatures ($\sim$320°C) the polymer crystals seem to undergo a topotactic depolymerization to the oligomer (Hasegawa *et al.*, 1974).

A series of even more perfect topotactic polymerizations have been observed with crystalline monomers which contain two conjugated triple

† For related compounds, see Iguchi *et al.* (1968). *J. Polym. Sci. Part A1* **6**, 1054; Holm and Zienty (1972). *J. Polym. Sci. Part A1* **10**, 1311; and Iwakura *et al.* (1968). *J. Polym. Sci. Part B* **6**, 463.

bonds as active centers for reaction. The crystal morphologies of two such polymers are described in Sect. 3.9 (Figs. III.142–III.144) and general principles are outlined in Sect. 6.1.8. The conjugated diynes have been known for a long time (Baeyer and Landsberg, 1882). The first suggestion about the structure of the solid state reaction product was that the reaction may lead to a linear polymer (laddertype polymer) as indicated by Eq. (140)

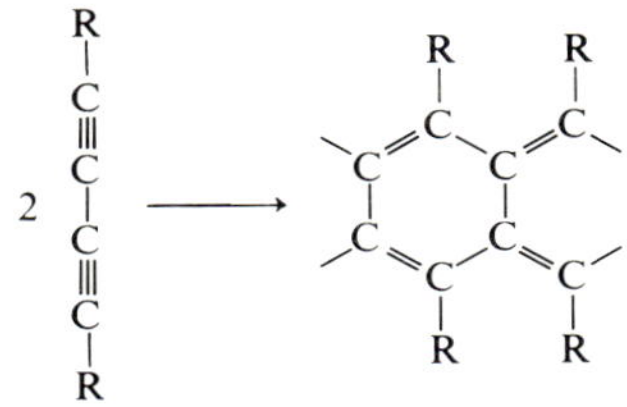

(140)

This reaction was proposed by Seher (1954) and Bohlmann and Inhoffen (1956). For such a reaction considerable contraction on condensation would be expected at right angles to the monomer axis, and the reaction should be of the two-phase type (see Sect. 6.1.8). It was observed, however, that several monomer crystals of this type change without phase separation to the polymer crystals. Making use of crystallographic as well as chemical considerations the following reaction mechanism was proven by Wegner (1969):

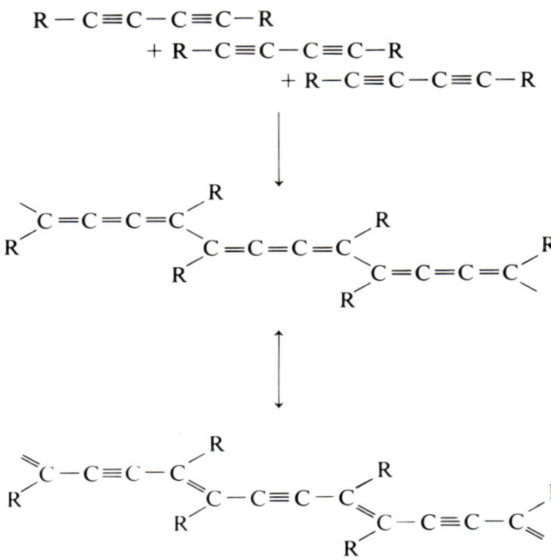

(141)

Thus, the reaction leads to a fully conjugated main chain. The substituents R are in the case of phenyl groups inclined close to 90° to the fiber axis.

A series of monomers with different substituents R and their reactivity are listed in Table VI.15. Inspection of Table VI.15 indicates that the influence of

TABLE VI.15

REACTIVITY OF DIYNES OF STRUCTURE $R_1{-}C{\equiv}C{-}C{\equiv}C{-}R_2$[a]

Substituent[b]	T_m (°C)	Reactivity[c]	T_p (°C)[d]	Color (polymer)
$CH_2{-}C_6H_5$	101	—	—	—
$CH_2{-}OH$	112	+	80	Brown-red
$CH_2{-}O{-}CO{-}C_6H_5$	70	—	—	—
$CH_2{-}O{-}CO{-}O{-}C_6H_5$	109	—	—	—
$CH_2{-}O{-}CO{-}C_6H_4{-}CO{-}OH$ (*o*)	153	+++	95	Blue-violet
$CH_2{-}O{-}CO{-}NH{-}C_6H_5$	172	+++	110	Red-violet
$CH_2{-}O{-}CO{-}NH{-}\alpha{-}$Naphthyl	205	+++	163	Purple red
$CH_2{-}O{-}CO{-}NH{-}C_4H_9$	78	++	60	Purple blue
$CH_2{-}O{-}SO_2{-}C_6H_4{-}CH_3$ (*p*)	96	+++	65	Red
$C(CH_3)_2{-}O{-}CO{-}NH{-}C_6H_5$	160	—	—	—
$(CH_2)_8{-}CO{-}OH$	113	+	90	Blue black
$(CH_2)_8{-}CO{-}O^-\ \frac{1}{2}\ Ba^{2+}$	—	+++	100	Black
$R_1 : CH_3$; $R_2 : CH_2{-}OH$	44	+++	30	Red
$C_6H_4{-}H$	88	—	—	—
$C_6H_4{-}NO_2$ (*o*)	215	++	—	Dark red
$C_6H_4{-}NO_2$ (*m*)	152	++	—	Dark red
$C_6H_4{-}NO_2$ (*p*)	290	—	—	—
$C_6H_4{-}NH{-}CO{-}CH_3$ (*o*)	235	+++	172	Red-violet
$C_6H_4{-}NH{-}CO{-}CH_3$ (*m*)	252	+++	120	Green-blue
$C_6H_4{-}NH{-}CO{-}CH_3$ (*p*)	>345	—	—	—
$C_6H_4{-}NH{-}CO{-}C_6H_5$ (*o*)	216	+	—	Violet
$C_6H_4{-}NH{-}CO{-}C_6H_5$(*m*)	232	++	—	Violet
$C_6H_4{-}NH{-}CO{-}C_6H_5$ (*p*)	308	—	—	—
$C_6H_4{-}NH{-}CO{-}NH{-}C_6H_5$ (*o*)	205	++	150	Light blue
$C_6H_4{-}NH{-}CO{-}NH{-}C_6H_5$ (*m*)	310	+++	140	Green blue
$C_6H_4{-}NH{-}CO{-}NH{-}C_6H_5$ (*p*)	>345	—	—	—

[a] Table by Wegner (1972).
[b] $R_1 = R_2$ unless indicated.
[c] (−) no reactivity; (+) yield <1%; (++) <yield 20%; (+++) yield >20% after 5 hr of uv-irradiation at 25°C with a 200 W high pressure Hg lamp or after 10 hr of annealing below T_m.
[d] For best thermal polymerization.

the substituents on the reactivity is based mainly on their effect on the monomer crystal structure (Wegner, 1972). A favorable arrangement for polymerization of the diyne grouping in neighboring molecules involves stacking in such a manner that tilting about the center of gravity of the molecule without motion of the side groups can accomplish the proper reaction configuration (shearing polymerization). The para-substituted monomers at the bottom of Table VI.15 hinder this motion and are thus unreactive. Depending upon the intramolecular and intermolecular effects after the reaction, phase separation may still occur if the product is metastable (see Figs. VI.40 and VI.41). Typical examples of no phase separation are the poly[2,4-hexadiyne-1,6-bis(*p*-tolylsulfonate)] and the poly[2,4-hexadiyne-1, 6-bis(phenylurethane)] illustrated in Figs. III.142 and III.143, respectively, while less ordered, phase sperated products result from polymerization of 2,4-hexadiyne-1,6-diol (Baughman, 1972). A less reactive polymorph of the 2,4-hexadiyne-1,6-bis(phenylurethane) and 2,4-hexadiyne-1,6-bis(stearylurethane) also lead to less-ordered products (Wegner, 1972).

It is interesting to note that the diyne grouping permits a broad group of molecules to polymerize whenever the proper geometry is achieved. Baughman and Yee (1974) polymerized cyclic diacetylenes of the types which lead to ladder-type polymer (see Fig. I.5). Two examples of such monomers are shown here.

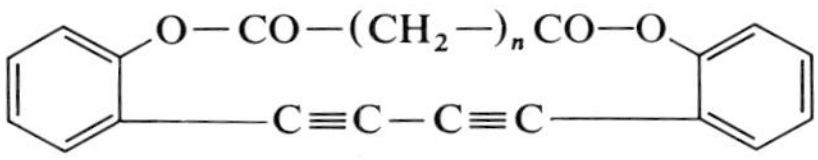

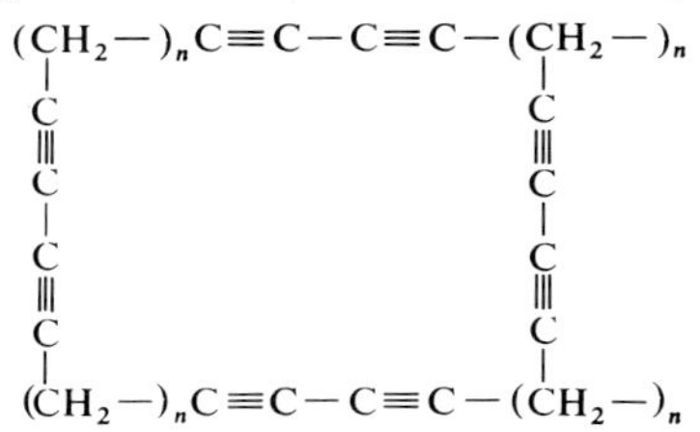

Kiji *et al.* (1973) accomplished triyne polymerization of the type

$$R-O-CH_2-C{\equiv}C-C{\equiv}C-C{\equiv}C-CH_2-O-R$$

The reaction occurs as before in the solid state by 1,4-addition. The residual triple bond is regularly arranged in the all-trans position as part of the substituents. Finally, it was shown by Wegner (1972) that diynes as part of macromolecules can also react in the solid state.

To summarize this section on crystallization during polymerization from solid state monomer, one must observe that all reactions are topochemical, but only in very rare cases is the reaction topotactic. In the large majority of cases topotactic control of the reaction is lost after only few steps of the reaction. This initial influence of the monomer crystal structure on the chemical reaction may be sufficient, however, to accomplish a different reaction path than is observed in the melt or in solution. Also, reaction within the monomer crystal may lead to different polymer macroconformations. Even with total topotactic control it is not guaranteed that the low energy conformation or crystal structure of the resulting macromolecule is produced. Secondary phase separation and polymer recrystallization are possible.

References

Adler, G. (1959). Amorphous and unoriented character of polymers formed by radiation induced solid state polymerization. *J. Chem. Phys.* **31**, 848; see also Adler, G., and Reams, W. (1960). Heterogeneous nature of reaction in radiation-induced solid-state polymerization of acrylamide *ibid.* **32**, 1698.

Adler, G. (1966). Deviations from topotaxy in trioxane polymerization. *J. Polym. Sci. Part A*-1 **4**, 2883.

Adler, G., and Petropoulos, J. M. (1965). Electron spin resonance spectroscopy of irradiated acrylamide. *J. Phys. Chem.* **69**, 3127.

Adler, G., and Reams, W. (1964). Polymerization in solid solutions of acrylamide in propionamide. *J. Polym. Sci. Part A* **2**, 2617.

Andrews, E. H., Owen, P. J., and Singh, A. (1971). Microkinetics of lamellar crystallization in a long chain polymer. *Proc. Roy. London* **A324**, 79.

Arai, H., Wada, T., and Kriyama, I. (1973). New morphology of polyethylene prepared by γ-ray-induced polymerization. *J. Polym. Sci. Polym. Phys. Ed.* **11**, 2297.

Asaubekov, M. A., Krasnikova, N. P., and Kargin, V. A. (1972). Crystallization of polymers in viscous solvents. *Vysokomol. Soedin. Ser. A* **14**, 30 [English transl. *Polym. Sci. USSR* **14**, 33].

Asmussen, F., Schwinon, W., and Ueberreiter, K. (1972). Kristallisationskinetik von Polymeren. XI. Kristallisationsgeschwindigkeit des Hexamethylen-succinat-Polyesters unter erhöhtem Druck. *Kolloid Z.Z. Polym.* **250**, 506.

Avrami, M. (1939). Kinetics of phase change. I. General Theory. *J. Chem. Phys.* **7**, 1103.

Avrami, M. (1940). Kinetics of phase change II. Transformation-time relations for random distribution of nuclei. *J. Chem. Phys.* **8**, 212.

Avrami, M. (1941). Granulation, phase change, and microstructure. Kinetics of Phase Change III. *J. Chem. Phys.* **9**, 177.

Baccaredda, M., Butta, E., and Giusti, P. (1964). Production of polyoxymethylene of high molecular weight and high crystallinity from trioxane. *J. Polym. Sci. Part C* **4**, 953.

Baeyer, A., and Landsberg, L. (1882). Ueber Synthesen mittelst des Phenylacetylens und seiner Derivate. *Ber.* **15**, 57.

Bamford C. H., and Eastmond, G. C. (1969). Solid state addition polymerization. *Quart.* Rev. **23**, 271.

Banks, W., Gordon, M., Roe, R.-J., and Sharples, A. (1963a). The crystallization of polyethylene. I. *Polymer* **4**, 61.

Banks, W., Gordon, M. and Sharples, A. (1963b). The crystallization of polyethylene after partial melting. *Polymer* **4**, 289.

Baranov, V. G., Kenarov, A. V. and Volkov, T. I. (1970). Morphology and kinetics studies of spherulitization of polyethylene terephthalate. *J. Polym. Sci. Part C* **30**, 271.

Barnes, W. J., Luetzel, W. G., and Price, F. P. (1961). Crystallization of poly(ethylene oxide) in bulk. *J. Phys. Chem.* **65**, 1742.

Bassett, D. C. (1967). Solid state polymerization at {00.1} subgrain boundaries of trioxane. *Nature (London)* **215**, 731.

Bassett, D. C., Block, S., and Piermarini, G. J. (1974). A high-pressure phase of polyethylene and chain-extended growth. *J. Appl. Phys.* **45**, 4146.

Bassett, D. C., and Turner, B., (1974). On the phenomenology of chain-extended crystallization in polyethylene. *Phil. Mag.* **29**, 925; see also: On the chain-extended and chain-folded crystallization of polyethylene. *ibid.* **29**, 285.

Baughman, R. H. (1972). Solid-state polymerization of diacetylenes. *J. Appl. Phys.* **43**, 4362.

Baughman, R. H. (1974). Solid-state synthesis of large polymer single crystals, *J. Polym. Sci. Polym. Phys. Ed.* **12**, 1511.

Baughman, R. H., and Yee, K. C. (1974). Solid state polymerization of a cyclic diacetylene. *J. Polym. Sci. Polym. Chem. Ed.* **12**, 2467.

Baur, H. (1965). Zur Theorie der Kristallisation von Copolymeren. *Kolloid Z. Z. Polym.* **203**, 97.

Baur, H. (1966). Zur Frage nach der geordneten Selektion von Sequenzen bei der Kristallisation von Copolymeren. *Kolloid Z. Z. Polym.* **212**, 97.

Baur, H. (1967). Zur Dynamik des Schmelzen und Kristallisierens in Mischungen (Teil I). *Ber. Bunsenges.* **71**, 703.

Baur, H. (1968). Bemerkungen zur Kinetik der Kristallisation von Polymeren. *Kolloid Z. Z. Polym.* **224**, 36.

Bawn, C. E. H., Ledwith, A., and Matthies, P. (1959). The mechanism of the polymerization of diazoalkanes catalyzed by boron compounds. *J. Polym. Sci.* **34**, 93.

Becker, R., and Döring, W. (1935). Kinetische Behandlung der Keimbildung in übersättigten Dämpfen. *Ann. Phys.* **24**, 719.

Beech, D. R., and Booth, C. (1972). Crystallization of high molecular weight fractions of poly(ethylene oxide) from dilute solution in ethanol. *Polymer* **13**, 355.

Beech, D. R., Booth, C., Dodgson, D. V., and Hillier, I. H. (1972a). Crystallization of poly(ethylene oxide) fractions: Crystallization isotherms. *J. Polym. Sci. Part A–2* **10**, 1555.

Beech, D. R., Booth, C., Hillier, I. H., and Pickles, C. J. (1972b). The crystallization of poly(ethylene oxide) fractions: Temperature dependence of growth rate. *Eur. Polym. J.* **8**, 799.

Bekkedahl, N. (1934). Forms of rubber as indicated by temperature-volume relationship. *J. Res. Nat. Bur. St.* **13**, 411. For more detail, see Wood, L. A., and Bekkedahl, N. (1946). Crystallization of unvulcanized rubber at different temperatures. *J. Appl. Phys.* **17**, 362.

Bentley, W. A., and Humphreys, W. J. (1931). "Snow Crystals." McGraw Hill, New York; reprinted (1962), Dover, New York.

Berlin, A. A., Vol'fson, S. A., Oleinik, E. F., and Yenikolopyan, N. S. (1970). Thermodynamics of the formaldehyde-trioxane-polyformaldehyde equilibrium system. *Vysokomol. Soedin.* **A12**, 443 [*English transl. Polym. Sci. USSR* **12**, 506].

Binsbergen, F. L. (1970). Depolarization of linearly polarized light by a polycrystalline specimen. *J. Macromol. Sci.* **B4**, 837.

Binsbergen, F. L., and DeLange, B. G. M. (1970). Heterogeneous nucleation in the crystallization of nucleated polypropylene. *Polymer* **11**, 309.

Blundell, D. J., and Keller, A. (1968). The concentration dependence of the linear growth rate of polyethylene crystals from solution. *J. Polym. Sci. Part B* **6**, 433.

Bohlmann, F., and Inhoffen, E. (1956). Polyacetylenverbindungen, XVI. Mitteil.: Synthese des Anacylins. *Ber.* **89**, 1276.

Boon, J., Challa, G., and van Krevelen, D. W. (1968). Crystallization kinetics of isotactic polystyrene. I. Spherulitic growth rate. *J. Polym. Sci. Part A*-2 **6**, 1791.

Booth, A., and Hay, J. N. (1971). The kinetics of crystallization of polyethylene. *Polymer* **12**, 365.

Booth, A., and Hay, J. N. (1972). An extension of polymer crystallization to a branching mechanism for spherulitic growth. *Brit. Polym. J.* **4**, 9.

Bort, D. N., Minsker, K. S., Okladnov, N. A., Shtarkman, B. P., and Kargin, V. A. (1962). Origin of secondary structures in polyethylene during polymerization. *Dokl. Akad. Nauk SSSR* **145**, 787.

Boyles, J. G., and Toby, S. (1967). Kinetics of polymerization of gaseous formaldehyde. *J. Polym. Sci. Part A*1 **5**, 1705; see also (1966). The mechanism of the polymerization of gaseous formaldehyde. *J. Polym. Sci. Part B* **4**, 411.

Brandrup, J. and Immergut, E. H. (1966). "Polymer Handbook." Wiley (Interscience), New York.

Brenschede, W. (1949). Sphärolitische Struktur synthetischer Hochpolymerer. *Kolloid Z.* **114**, 35.

Briegleb, G. (1929). Die dynamisch-allotropen Zustände des Selens. *Z. Phys. Chem.* **A144**, 321.

Brown, J. F., Jr., and White, D. M. (1960). Stereospecific polymerization in thiourea canal complexes. *J. Amer. Chem. Soc.* **82**, 5671.

Brown, N. (1967). Polymerization of formaldehyde. *In* "Polyaldehydes" (O. Vogl, ed.). Arnold Publ. London and Dekker, New York; see also *J. Macromol. Sci.* **A1**, 209.

Buls, V. W., and Higgins, T. L. (1970). A uniform site theory of Ziegler catalysis. *J. Polym. Sci. Part A*1 **8**, 1025; A particle growth theory for heterogeneous Ziegler polymerization. *ibid.* 1037.

Buls, V. W., and Higgins, T. L. (1973). Kinetics of Ziegler propylene polymerization. *J. Polym. Sci. Polym. Chem. Ed.* **11**, 925.

Burke, J. G. (1966). "Origins of the Science of Crystals." Univ. of California Press, Berkeley, California.

Burton, W. K., and Cabrera, N. (1949). Crystal growth and surface structure. Parts I and II. *Discuss. Farday Soc.* **5**, 33, 40.

Burton, W. K., Cabrera, N., and Frank, F. C. (1951). The growth of crystals and the equilibrium structure of their surfaces. *Phil. Trans. Roy. Soc. London* **A243**, 299.

Burnett, B. B., and McDevit, W. F. (1957). Kinetics of spherulite growth in high polymers. *J. Appl. Phys.* **28**, 1101.

Busfield, W. K., and McEwen, I. J. (1973). The cationic polymerization of fluoral-II. Kinetics and mechanism of the polymerization initiated by boron trifluoride etherate in dichloromethane solution. *Eur. Polym. J.* **9**, 1127.

Calvert, P. D., and Uhlmann, D. R. (1972). Theory of polymer crystallization without chain folding. *J. Appl. Phys.* **43**, 944.

Carazzolo, G., and Mammi, M. (1967). Crystal structure of orthorhombic polyselenomethylene. *Makromol. Chem.* **100**, 28.

Carazzolo, G., Leghissa, S., and Mammi, M. (1963). Polyoxymethylene from trioxane by solid state polymerization: A case of twinning in a synthetic polymer. *Makromol. Chem.* **60**, 171.

Carazzolo, G., Mammi, M., and Valle, G. (1967). Radiation polymerization of crystalline trithiane. *Makromol. Chem.* **100**, 295.

Carslaw, H. S., and Jaeger, J. C. (1959). "Conduction of Heat in Solids." 2nd ed. Oxford Univ. Press, London and New York.

Cemel, A., Fort, T., Jr., and Lando, J. B. (1972). Polymerization of vinyl stearate. *J. Polym. Sci. Part A*-1 **10**, 2061.

Chanzy, H. D., and Marchessault, R. H. (1969). Spontaneous catalyst fracture and its influence on nascent polyethylene morphology. *Macromolecules* **2**, 103.

Chanzy, H. D., Day, A., and Marchessault, R. H. (1967). Polymerization on glass-supported vanadium trichloride; Morphology of nascent polyethylene. *Polymer* **8**, 567.

Chanzy, H. D., Revol, J. F., Marchessault, R. H., and Lamandé, A. (1973). Nascent structures during the polymerization of ethylene. I. Morphology and model of growth. *Kolloid Z. Z. Polym.* **251**, 563.

Chanzy, H. D., Bonjour, E., and Marchessault, R. H. (1974). Nascent structures during the polymerization of ethylene. II. Calorimetric study. *Colloid Polym. Sci.* **252**, 8.
Charlesby, A. (1965). Solid-state polymerization induced by radiation. *Rep. Progr. Phys.* **28**, 463.
Chatani, Y., and Tadokoro, H. (1973). Solid-state polymerization and polymer crystallization. *J. Macromol. Sci.* **B8**, 203.
Chatani, Y., Uchida, T., Tadokoro, H., Hayashi, K., Nishii, M., and Okamura, S. (1968). X-ray crystallographic study of solid state polymerization of trioxane and tetraoxymethylene. *J. Makromol. Sci.* **B2**, 467.
Chatani, Y., Nakatani, S., and Tadokoro, H. (1970a). Structural study of the polymerization of 2,3-dichlorobutadiene in the thiourea canal complex. Structural change during canal polymerization and the crystal structure of the resultant polymer. *Macromolecules* **3**, 481.
Chatani, Y., Kitahama, K., Tadokoro, H., Yamauchi, T., and Miyake, Y. (1970b). Solid-state polymerization of pentoxane $(-CH_2-O-)_5$. *J. Macromol. Sci.* **B4**, 61.
Chatani, Y., Ohno, T., Yamauchi, T., and Miyake, Y. (1973). Crystal structure and radiation-induced solid-state polymerization of 1,3,5,7,9,11-hexoecane. *J. Polym. Sci. Polym. Phys. Ed.* **11**, 369.
Colson, J. P., and Reneker, D. H. (1970). Polyoxymethylene crystals grown within irradiated trioxane crystals. *J. Appl. Phys.* **41**, 4296.
Colson, J. P., and Reneker, D. H. (1973). Growth direction of polyoxymethylene crystals inside irradiated trioxane crystals. *J. Appl. Phys.* **44**, 4243.
Cooper, M., and Manley, R. St. J. (1973). Growth kinetics of polyethylene crystals from dilute xylene solution. *J. Polym. Sci. Polym. Lett. Ed.* **11**, 363.
Coughlin, M. C., and Wunderlich, B. (1972). Extended-chain crystals. IX. The morphology and mechanism of formation of extended chain crystals of selenium. *Kolloid Z. Z. Polym.* **250**, 482.
Coughlin, M. C., and Wunderlich, B. (1973). Polymerization during crystallization of selenium. *J. Polym. Sci. Polym. Phys. Ed.* **11**, 1735.
Crabtree, J. R., Grimsby, F. N., Nummelin, A. J., and Sketchley, J. M. (1973). The role of diffusion in the Ziegler polymerization of ethylene. *J. Appl. Polym. Sci.* **17**, 959.
DeCandia, F., Taglialatela, A., and Vittoria, V. (1973). Influence of inorganic salts on the kinetic and crystallization mechanism of nylon 6. *Kolloid Z. Z. Polym.* **251**, 61.
Dent Glasser, L. S., Glasser, F. P., and Taylor, H. F. W. (1962). Topotactic reactions in inorganic oxy-compounds. *Quart Rev.* **16**, 343.
Devoy, C., and Mandelkern, L. (1969). Effect of molecular weight on the radial growth rates of polymer spherulites. *J. Polym. Sci. Part A*-2 **7**, 1883.
Devoy, C., Mandelkern, L., and Bourland, L. (1970). Crystallization kinetics of dilute polyethylene solutions. *J. Polym. Sci. Part A*–2, **8**, 869.
Dimonie, M., and Gavăt, I. (1968). Polymerization of epichlorohydrin by using the triethylaluminum-water catalyst system. *Eur. Polym. J.* **4**, 541.
Dole, M. (1959). Thermodynamic properties of high polymers as function of their pretreatment as determined by specific heat measurements. *Kolloid Z.* **165**, 40.
Doll, W. W., and Lando, J. B. (1968). The polymorphism of poly (vinylidene fluoride). II. The effect of hydrostatic pressure. *J. Macromol. Sci.* **B2**, 219.
Dörffurt, H. (1956). Paper at the 3rd Aachener Kunststoff-Kolloquium, reported in *Kunststoffe* **46**, 114.
Eastmond, G. C. (1973). A proposed mechanism for the solid-state polymerization of cyclic monomers in lattice defects. *Polym. J.* **4**, 392.
Ehinger, H. (1961). Kristallformen des trigonalen Seleniums. *Z. Krist.* **115**, 235.

Eisenberg, A., and Tobolsky, A. V. (1960). Equilibrium polymerization of selenium. *J. Polym. Sci.* **46**, 19; see also Eisenberg, A. (1963). Equilibrium polymerization of selenium. A re-evaluation of the enthalpy of initiation from viscosity data. *J. Polym.* Sci. *Part B* **1**, 33.

Ergoz, E., Fatou, J. G., and Mandelkern, L. (1972). Molecular weight dependence of the crystallization kinetics of linear polyethylene. I. Experimental results. *Macromolecules* **5**, 147.

Errede, L. A., and Szwarc, M. (1958). Chemistry of p-xylylene. Its analogues and polymers. *Quart. Rev.* **12**, 301.

Errede, L. A., Gregorian, R. S., and Hoyt, J. M. (1960). The chemistry of p-xylylene. VI. The polymerization of p-xylylene. *J. Amer. Chem. Soc.* **82**, 5218.

Evans, U. R. (1945). The laws of expanding circles and spheres in relation to the lateral growth of surface films and the grain size of metals. *Trans. Faraday Soc.* **41**, 365.

Ewers, W. M., Zachmann, H. G., and Peterlin, A. (1972). Berechnung der Schlaufenabstandsverteilung bei der Bildung von Kristallen aus Kettenmolekülen. *Kolloid Z. Z. Polym.* **250**, 1187; see also *J. Macromol. Sci.* **B6**, 695.

Fatou, J. M. G., and Barrales-Rienda, J. M. (1969). Rates of crystallization of polyethylene fractions by calorimetry. *J. Polym. Sci. Part A*–2 **7**, 1755.

Faucitano, A., and Adler, G. (1970). Annealing and recrystallization in the solid state. Polymerization in solid solutions of methacrylamide and isobutyramide. *J. Macromol. Sci.* **A4**, 261.

Fischer, E. (1906). Synthesen von Polypeptiden. *Ber.* **39**, 453.

Fischer, E. W. and Lorenz, R. (1963). Über Fehlordnungen in Polyäthylen-Einkristallen. *Kolloid Z. Z. Polym.* **189**, 97.

Flory, P. J. (1953). "Principles of Polymer Chemistry." Cornell Univ. Press, Ithaca New York.

Flory, P. J. (1955). Theory of crystallization in copolymers. *Trans. Faraday Soc.* **51**, 848.

Flory, P. J. (1962). On the morphology of the crystalline state in polymers. *J. Amer. Chem. Soc.* **84**, 2857.

Frank, F. C. (1958). Discussion to Mandelkern (1958), p. 496.

Frank, F. C., and Tosi, M. (1961). On the theory of polymer crystallization. *Proc. Roy. Soc. (London)* **A263**, 323.

Frenkel, J. (1946). "Kinetic Theory of Liquids." Oxford Univ. Press, London and New York; reprinted (1955) by Dover, New York.

Fujie, A., Komoto, T., Oya, M., and Kawai, T. (1973). Crystallization of polypeptides in the course of polymerization. III. Further studies on the growth mechanism of poly (*L*-alanine) crystals. *Makromol. Chem.* **169**, 301.

Fujimoto, S. (1967). A topochemical study of the polymerization of terephthalonitril oxide. *J. Polym. Sci. Part B* **5**, 301.

Fujisaki, H., Westmore, J. B., and Tickner, A. W. (1966). Mass spectrometric study of subliming selenium. *Can. J. Chem.* **44**, 3063.

Furukawa, J., and Saegusa, T. (1963). "Polymerization of Aldehydes and Oxides." Wiley (Interscience), New York.

Furuta, N. (1964). Dendrite crystals of metallic selenium grown from vapor. *Jap. J. Appl. Phys.* **3**, 237.

Gandica, A., and Magill, J. H. (1972). A universal relation for the crystallization kinetics of polymeric materials. *Polymer* **13**, 595.

Geil, P. H. (1963). "Polymer Single Crystals." Wiley (Interscience), New York.

Gent, A. N. (1954). Crystallization and the relaxation of stress in stretched natural rubber vulcanizates. *Trans. Faraday Soc.* **50**, 521.

Gent, A. N. (1955). Crystallization in natural rubber. IV. Temperature dependence. *J. Polym. Sci.* **18**, 321.

Gent, A. N. (1965). Crystallization in stretched polymer networks. I. trans-polychloroprene. *J. Polym. Sci. Part A* **3**, 3787.

Gent, A. N. (1966). Crystallization in stretched polymer networks. II. trans-Polyisoprene. *J. Polym. Sci. Part A2* **4**, 447.

Georgiadis, T., and Manley, R. St. J. (1972). The morphology of nascent polyethylene prepared with the soluble catalyst $(C_5H_5)_2TiCl_2 - R_2AlCl$. *Kolloid Z. Z. Polym.* **250**, 557.

Gibbs, J. W. (1878). On the equilibrium of heterogeneous substances. *Trans. Conn. Acad.* **III**, 343-524; reprinted in (1961). "The Scientific Papers of J. Willard Gibbs, Ph.D., LL.D," p. 325, Dover, New York, which in turn is a reprint of the first edition published by Longmans, Green, London, 1906.

Godovskii, Yu. K. (1969). The effect of temperature and molecular structure on the rate of crystallization of polymers. *Vysokomol. Soedin. Ser. A* **11**, 2129 [*English Transl.*: *Polym. Sci. USSR* **11**, 2423].

Godovskii, Yu. K., and Slonimskii, G. L. (1966). Calorimetric study of isothermal crystallization kinetics of isotactic polypropylene. *Vysokomol. Soedin.* **8**, 403 [*English transl.*: *Polym. Sci. USSR* **8**, 441]; (1969). Nature of the crystallization nuclei present in a polypropylene (PP) melt. *Vysokomol. Soedin. Ser. A.* **11**, 1285 [*English transl.*: *Polym. Sci. USSR* **11**, 1461].

Godovskii, Yu. K., Slonimskii, G. L., and Garbar, N. M. (1972). Effect of molecular weight on the crystallization and morphology of poly(ethylene oxide) fractions. *J. Polym. Sci. Part C* **38**, 1.

Gogolewski, S., and Pennings, A. J. (1973). Crystallization of polyamides under elevated pressure. Nylon 6. *Polymer* **14**, 463.

Gogolewski, S., and Turska, E. (1972). Isothermal crystallization of copolyamides. I. Kinetics of crystallization of nylon 6-piperazine adipate and nylon 6-piperazine terephthalate random copolyamides. *J. Appl. Polym. Sci.* **16**, 1959.

Göler, V., Sachs, Frh. v., and Sachs, G. (1932). Zur Kinetik von Kristallisationsvorgängen. *Z. Phys.* **77**, 281.

Gordon, M., and Hillier, I. H. (1965). Mechanism of secondary crystallization of polymethylene. *Phil. Mag.* **11**, 31.

Gorham, W. F., and Niegisch, W. D. (1971). Xylylene polymers. *Encycl. Polym. Sci. Technol.* **15**, 98.

Gornick, F., and Mandelkern, L. (1962). Effect of noncrystallizable components on the crystallization kinetics of polymers. *J. Appl. Phys.* **33**, 907.

Graff, R. J. L., Kortleve, G., and Vonk, C. G. (1970). On the size of the primary particles in Ziegler catalysts. *J. Polym. Sci. Part B* **8**, 735.

Griffith, J. H., and Rånby, B.G. (1959). Crystallization kinetics of isotactic polypropylene. *J. Polym. Sci.* **38**, 107.

Guttman, J. Y., and Guillet, J. E. (1970). Mechanism of propylene polymerization on single crystals of α-titanium trichloride. *Macromolecules* **3**, 470.

Haas, T. W., and Maxwell, B. (1969). Effects of shear stress on the crystallization of linear polyethylene and poly-1-butene. *Polym. Eng. Sci.* **9**, 225.

Ham, G. E. (1959). Helical stereospecific polymerization. *J. Polym. Sci.* **40**, 569.

Ham, G. E. (1960). Idealized helical stereospecific polymerization. *J. Polym. Sci.* **46**, 475.

Ham, G. E. (1962). The role of symmetry in crystallization and polymerization phenomena. *J. Polym. Sci.* **61** 293.

Hardin, I. R., and Yeh, G. S. Y. (1973). Strain induced crystallization and reversible phenomena in transpolyisoprene. I. Electron microscopic studies. II. X-ray studies. *J. Macromol. Sci.* **B7**, 375, 393.

Hardy, Gy., Cser, F., Kalló, A., Nyitrai, K., Boder, G., and Lengyel, M. (1970). Investigations in the field of radiation induced solid state polymerization XXVI. Structural studies related to the polymerization of cholesteryl acrylate. *Acta. Chim.* (*Budapest*) **65**, 287.

Hardy, Gy., Kovács, G., Koszterszitz, Gy., and Cser, F. (1972). Investigations in the field of solid state polymerization. XXVIII. Investigation of iodine initiated solid state oligomerization of acenaphthylenene. Kinetics and mechanism of the reaction. *Acta Chim.* (*Budapest*), **75**, 69.

Hardy, Gy., Kovács, G., Cser, F., and Koszterszitz, Gy. (1973). Investigations in the field of solid state polymerization. XXIX. Investigation of solid state oligomerization of acenaphthylene initiated by Iodine. Structure of the monomer and oligomers. *Acta Chim.* (*Budapest*) **75**, 303.

Hardy, Gy., Cser, F., and Nyitray, K. (1974). Investigations in the field of solid state polymerization. XXXII. Topotactic solid state polymerization of vinylene carbonate. *Acta Chim.* (*Budapest*) **80**, 41.

Hartley, F. D., Lord, F. W., and Morgan, L. B. (1954). Crystallization phenomena in polymers. III. Effect of melt condition and the temperature of crystallization on the course of crystallization in polyethylene terephthalate. *Phil. Trans. Roy. Soc. London* **A247**, 23.

Hartman, P. (1963). Crystal form and crystal structure. *In* "Physics and Chemistry of the Organic Solid State" (D. Fox, M. M. Labes, and A. Weissberger, eds.), Vol. 1, p. 369. Wiley (Interscience), New York.

Hartman, P., and Perdok, W. G. (1955). On the relationship between structure and morphology of crystals. I, II, and III. *Acta Cryst.* **8**, 49, 521, 525.

Harvey, E. D., and Hybart, F. J. (1970). Rates of crystallization of copolyamides. II. Random copolymers of nylon 6.6 and 6. *J. Appl. Polym. Sci.* **14**, 2133.

Hasegawa, M., and Suzuki, Y. (1967). Four-center type photopolymerization in the solid state: Poly-2,5-distyrylpyrazine. *J. Polym. Sci. Part B* **5**, 813.

Hasegawa, M., Suzuki, Y., Nakanishi, H., and Nakanishi, F. (1973). Four-center type photopolymerization in the crystalline state. *Progr. Polym. Sci. Japan* **5**, 143.

Hasegawa, M., Nakanishi, H., Yurugi, T., and Ishida, K. (1974). A reversible topochemical process in the crystalline state: photopolymerization of 2,5-distyrylpyrazine (DSP) and thermal depolymerization of poly-DSP. *J. Polym. Sci. Polym. Lett. Ed.* **12**, 57.

Hasegawa, R., Tanabe, Y., Kobayashi, M., and Tadokoro, H. (1970). Structural studies of pressure-crystallized polymers I. Heat treatment of oriented polymers under high pressure. *J. Polym. Sci. Part A*-2 **8**, 1073.

Hatakeyama, T., Kaneda, H., Hashimoto, T. and, Kanetsuna, H. (1974). Effect of pressure on the crystallization kinetics of polyethylene. Unpublished; for earlier work see also: Kanetsuna, H., Mitsuhashi, S., Iguchi, M., Hatakeyama, T., Kyotani M., and Maeda, Y. (1973). Effect of pressure on the crystallization of polyethylene. *J. Polym. Sci. Polym. Symp.* **42**, 783.

Hay, J. N., and Booth, A. (1972). The effect of a secondary process on the course of polymer crystallization. *Brit. Polym. J.* **4**, 19.

Hayashi, K., Ochi, H., and Okamura, S. (1964). Radiation-induced post polymerization of trioxane in the solid state. *J. Polym. Sci. Part A* **2**, 2929.

Helfand, E., and Lauritzen, J. I., Jr. (1973). Theory of copolymer crystallization. *Macromolecules* **6**, 631.

Helms, J. B., and Challa, G. (1972). Infrared spectroscopic study of the crystallization of isotactic polystyrene from solution. *J. Polym. Sci. Part A–2* **10**, 761.

Heyer, H. (1966). Kinetics of crystal growth. *Angew. Chem. Int. Ed.* **5**, 67 (German ed. **78**, 130).

Hillier, I. H. (1965). Modified Avrami equation for the bulk crystallization kinetics of spherulitic polymers. *J. Polym. Sci. Part A* **3**, 3067.

Hillig, W. B. (1966). A derivation of classical two-dimensional nucleation kinetics and the associated crystal growth laws. *Acta Met.* **14**, 1868.

Hine, J. (1966). The principle of least motion. Application to reactions of resonance-stabilized species. *J. Org. Chem.* **31**, 1236.

Hirai, N., and Eyring, H. (1958). Bulk viscosity of liquids *J. Appl. Phys.* **29**, 810.

Hirth, J. P. and Pound, G. M. (1963). Condensation and evaporation; nucleation and growth kinetics. *Progr. Mater. Sci.* **11**, 1.

Hoffman, J. D., and Weeks, J. J. (1962). Rate of spherulitic crystallization with chain folds in polychlorotrifluoroethylene. *J. Chem. Phys.* **37**, 1723.

Hoffman, J. D., Lauritzen, J. I. Jr., Passaglia, E., Ross, G. S., Frolen, L. J., and Weeks, J. J. (1969). Kinetics of polymer crystallization from solution and the melt. *Kolloid Z. Z. Polym.* **231**, 564.

Hoffman, J. D., Lauritzen J. I., Jr., Ross, G. S., and Frolen, L. (1973). Unpublished data, to be published at a later date.

Holland, V. F., and Lindenmeyer, P. H. (1962). Morphology and crystal growth rate of polyethylene crystalline complexes. *J. Polym. Sci.* **57**, 589.

Hoshino, S., Meinecke, E., Powers, J., Stein, R. S., and Newman, S. (1965). Crystallization kinetics of polypropylene fractions. *J. Polym. Sci. Part A* **3**, 3041.

Huber, A. (1935). Zur Kinetik von Kristallisationsvorgängen. *Z. Phys.* **93**, 227.

Hybart, F. J., and Pepper, B. (1969). Rates of crystallization of copolyamides. I. Random copolymers of nylon 6 and 11. *J. Appl. Polym. Sci.* **13**, 2643.

Iguchi, M. (1970). A comment on the kinetics in a heterogeneous polymerization system. *J. Polym. Sci. Part A–1* **8**, 1013.

Iguchi, M. (1972). A branching model for crystallization kinetics. *Brit. Polym. J.* **4**, 401; reprinted with only minor changes (1973). *J. Macromol. Sci.* **B8**, 259.

Iguchi, M. (1973). Growth of needle-like crystals of polyoxymethylene during polymerization. *Brit. Polym. J.* **5**, 195.

Iguchi, M., Kanetsuna, H., Komoto, T., and Kawai, T. (1969a). Formation of nylon 4 spherulites in the course of polymerization. *Kolloid Z. Z. Polym.* **235**, 1246.

Iguchi, M., Kanetsuna, H., and Kawai, T. (1969b). Growth of polyoxymethylene crystals in the course of the polymerization of trioxane in solution. *Makromol. Chem.* **128**, 63.

Iguchi, M., Kanetsuna, H., and Kawai, T. (1970). Formation of polymer crystals during polymerization. *Kobunshi* **19**, 577; translated with additions (1971). *Brit. Polym. J.* **3**, 177.

Ishibashi, T., and Tani, Y. (1969). Crystallization of nylon 6. IV. Dependence of isothermal crystallization rate on the molecular weight and temperature. *Kobunshi Kagaku* **26**, 199.

Iwamoto, R., Bopp, R., and Wunderlich, B. (1974). Crystallization during polymerization of poly-p-xylylene. III. Crystal structures, thermal analysis, and morphology as a function of crystallization conditions. Manuscript, to be published, *J. Polym. Sci. Polym. Phys. Ed.*

Jaacks, V., and Kern, W. (1962). Über die Polymerisation bei der Sublimation des Trioxans. 17. Mitt. über Polyoxymethylene. *Makromol. Chem.* **52**, 37.

Jaacks, V., and Kern, W. (1963). Initiatoren für die Polymerisation des Trioxans. 20. Mitt. über Polyoxymethylen. *Makromol. Chem.* **62**, 1.

Jaacks, V., Boehlke, K., and Kern, W. (1973). On the mechanism of chain propagation in radiation-induced polymerization of trioxane in the solid state. *Makromol. Chem.* **165**, 51.

Jackson, J. B., and Longman, G. W. (1969). The crystallization of poly(ethylene terephthalate) and related copolymers. *Polymer* **10**, 873.

Jackson, F., Hsu, T. S. and, Brasch, J. W. (1972). Direct microscopic observation of the crystallization process of polyethhlene at high pressure: Acicular bladed crystals. *J. Polym. Sci. Part B* **10**, 207; see also *Nature* (*London*) *Phys. Sci.* **239**, 45.

Jaffe, M. and Wunderlich, B. (1969). Superheating of extended-chain polymer crystals. *In* "Thermal Analysis" (R. F. Schwenker and P. D. Garn, eds.) Vol. I, p. 387. Academic Press, New York.

Jain, N. L. (1969). Crystallization of polymer-diluent systems. IV. Polyethylene oxide-diethyl sebacate system. *Makromol. Chem.* **125**, 192.

Jakabhazy, S. Z., Morawetz, H., and Morosoff, N. (1964). Polymerization in the crystalline state. V. Oriented chain growth in the thermally initiated polymerization of p-acetamidostyrene and p-benzamidostyrene. *J. Polym. Sci. Part C* **4**, 805.

Jenckel, E. (1955). Kinetik der Kristallisation. *In* "Die Physik der Hochpolymeren" (H. A. Stuart, ed.), p. 521. Springer Verlag, Berlin.

Johnsen, U., and Lehmann, J. (1969). Die Kinetik der Kristallisation des Poly-3,3-(chloromethyl)-oxacyclobutans (Penton) aus verdünnter m-Xylol-Lösung. Teil I: Anfangsbereich der Kristallisation; Teil II: Endbereich der Kristallisation. *Kolloid Z. Z. Polym.* **230**, 317, 325.

Johnsen, U., Nachtrab, G., and Zachmann, H. G. (1970). Untersuchung der isothermen Kristallisation von Äthylen-Vinylacetat-Copolymeren mit Hilfe der Differential-Scanning-Calorimetrie. *Kolloid Z. Z. Polym.* **240**, 756.

Johnson, W. A., and Mehl, R. T. (1939). Reaction kinetics in processes of nucleation and growth. *Trans. AIME* **135**, 416; see also (August 1939). *Metals Technol. AIME Tech. Publ.* 1089.

Jones, D. H., Latham, A. J., Keller, A., and Girolamo, M. (1973). Fold length of single crystals of polystyrene: A conflict with crystallization theories at high supercoolings. *J. Polym. Sci. Polym. Phys. Ed.* **11**, 1747.

Kaiser, J., Wegner, G., and Fischer, E. W. (1972). Das Wachstum von Polymerketten im Kristallgitter des Monomeren. *Kolloid Z. Z. Polym.* **250**, 1158.

Kardos, J. L., and Christiansen, A. W., and Baer, E. (1966). Structure of pressure-crystallized polypropylene. *J. Polym. Sci. Part A–2* **4**, 777.

Kargin, V. A., Kabanov, V. A., and Zubov, K. P. (1959). Polymerization in systems produced by the molecular ray method. *Vysokomol. Soedin.* **1**, 265.

Kargin, V. A., Bort, D. N., Shtarkman, B. P., and Minsker, K. S. (1964). Supermolecular structures arising directly in the polymerization process. *Vysokomol. Soedin.* **6**, 189 [*English transl.*: *Polym. Sci. USSR* **6**, 218].

Katchalski, E., and Sela, M. (1958). Synthesis and chemical properties of poly-α-amino acids. *Advan. Protein Chem.* **13**, 243.

Kawai, T. (1973). Crystallization of polypeptides during crystallization. *Kobunshi* **22**, 22.

Kawai, T., and Komoto, T. (1973). Crystallization during polymerization with special reference to synthetic polypeptides. *J. Macromol. Sci.* **B8**, 189.

Keggenhoff, B., and Wegner, G. (1973). Solid-state polymerization of oxetanes. I. Lattice parameters and packing properties of the monomers. II. Investigation of the growth of the polymer phase as related to the mechanism of polymerization. *J. Polym. Sci. Polym. Phys. Ed.* **11**, 759, 769.

Keii, T. (1972). "Kinetics of Ziegler Natta Polymerization." Kodansha, Tokyo, and Chapman and Hall, London.

Keith, H. D. (1969). Crystallization of polymers from the melt and the structure of bulk semicrystalline polymers. *Kolloid Z. Z. Polym.* **231**, 421.

Keith, H. D., and Padden, F. J. Jr. (1964). Spherulitic crystallization from the melt. II. Influence of fractionation and impurity segregation on the kinetics of crystallization. *J. Appl. Phys.* **35**, 1286.

Keller, A., and Pedemonte, E. (1973). A Study of growth rates of polyethylene crystals. *J. Crystal Growth* **18**, 111.

Keller, A., and Wilmouth, F. M. (1969). On the morphology and origin of the fibers observed in nascent Ziegler polyethylene. *Makromol. Chem.* **121**, 42.

Keller, A., Lester, G. R., and Morgan, L. B. (1954). Crystallization phenomena in polymers. I. Preliminary investigation of the crystallization characteristics of polyethylene terephthalate. *Phil. Trans Roy. Soc. London* **A247**, 1.

Kern, W., and Jaacks, V. (1960). Some kinetic effects in the polymerization of 1,3,5-trioxane. *J. Polym. Sci.* **48**, 399.

Kiji, J., Kaiser, J., Wegner, G., and Schulz, R. C. (1973). Solid-state polymerization of derivatives of 2,4,6-octatriyne: 9. Topochemical reaction of monomer with conjugated triple bonds. *Polymer* **14**, 433.

Kilian, H. G. (1965). Partiell kristallisierende Copolymere als "pseudoeutektische Systeme." *Kolloid Z. Z. Polym.* **202**, 97.

Kim, H.-G., and Mandelkern, L. (1968). Crystallization kinetics of stretched natural rubber. *J. Polym. Sci. Part A*–2 **6**, 181.

Kitamaru, R., and Hyon, S.-H. (1974). Size and orientation of cristallites in lightly cross-linked polyethylene, crystallized from the melt under uniaxial compression. *Makromol. Chem.* **175**, 255.

Klement, J. J., and Geil, P. H. (1972). Deformation and annealing behavior. III. Thin films of polycarbonate, isotactic poly(methyl methacrylate) and isotactic polystyrene. *J. Macromol. Sci.* **B6**, 31.

Kobayashi, E., Higashimura, T., and Okamura, S. (1971). Cationic polymerization of tetraoxane [tetroxocane] in the solid state. I. Suspension polymerization of tetraoxane in the solid state catalysed by boron trifluoride etherate. II. Polymerization of tetraoxane in the solid state catalysed by boron trifluoride. III. Orientation of poly(oxymethylene) obtained by the cationic polymerization of tetraoxane and trioxane in the solid state. *Kobunshi Kagaku* **28**, 987, 992, 997.

Kobayashi, K., and Nagasawa, T. (1970). Crystallization of sheared polymer melts. *J. Macromol. Sci.* **B4**, 331.

Kohlschütter, H. W. (1927). Ueber topochemische Reaktionen. *Kolloid Z.* **42**, 254.

Kohlschütter, H. W. (1930). Zur Morphologie hochmolekularer Stoffe. I. Faserbildung mit Polyoxymethylen. *Ann.* **482**, 75.

Kohlschütter, H. W., and Sprenger, L. (1932). Über die Umwandlung kristallisierten Trioxymethylens zu hockmolekularem Polyoxymethylen. *Z. Phys. Chem.* **B16**, 284.

Kolmogoroff, A. N. (1937). On the statistics of crystallization processes in metals. *Isvest. Akad. Nauk SSSR Ser. Math.* **1**, 335.

Komaki, A., and Matsumoto, T. (1963). Nuclear magnetic resonance of solid trioxane. *J. Polym. Sci. Part B* **1**, 671. (Full details were apparently not published as promised.)

Komoto, T., and Kawai, T. (1973). Crystallization of polypeptides in the course of polymerization, 4. Studies on the molecular arrangement in oligo(L-alanine) crystals as polymerized. *Makromol. Chem.* **172**, 221.

Komoto, T., Iguchi, M., Kanetsuna, H., and Kawai, T. (1970). Formation of spherulites during polymerization of lactams. *Makromol. Chem.* **135**, 145.

Komoto, T., Akaishi, T., Oya, M., and Kawai, T. (1972). Crystallization of polypeptides in the course of polymerization. I. Growth mechanism of poly-L- and DL-alanine crystals. *Makromol. Chem.* **154**, 151.

Komoto, T., Kim, K. Y., Minoshima, Y., Oya, M., and Kawai, T. (1973). Crystallization of polypeptides in the course of polymerization. 2. Chain folded crystals of polyglycine and poly-β-alanine *Makromol. Chem.* **168**, 261.

Konrad, G., and Zachmann, H. G. (1971). Untersuchungen der Absolutintensität der Röntgenkleinwinkelstreuung an Polyäthylenterephthalat in Abhängigkeit von den Kristallisationsbedingungen. *Kolloid Z. Z. Polym.* **247**, 851,

Kossel, W. (1927). Zur Theorie des Krystallwachstums. *Nachr. Ges. Wis. Göttingen* 135.

Kovacs, A. J. (1955). Cinétique de cristallisation du polyéthylène. *Ric. Sci.* (*Suppl.*) **25**, 668.

Kovacs, A. J., and Gonthier, A. (1972). Crystallization and fusion of self-seeded polymers. II. Growth rate, morphology and isothermal thickening of single crystals of low molecular weight poly(ethylene-oxide) fractions. *Kolloid Z. Z. Polym.* **250**, 530.

Kovacs, A. J., Manson, J. A., and Lévy, D. (1966). Propriétés des copolymères biséquencés polyoxyéthylène-polystyrène. III. Cinétique de croissance des monocristaux en solution. *Kolloid Z. Z. Polym.* **214**, 1.

Kovacs, G., Kovacs, E., and Morawetz, H. (1966). Polymerization in the crystalline state. VIII. Polymerization in N-carboxy anhydrides of γ-benzyl glutamate, γ-methyl glutamate, and ε-carbobenzoxylysine. *J. Polym. Sci. Part A*–1 **4**, 1533.

Kozlowski, W. (1972). Kinetics of crystallization of polyamide 6 from the glassy state. *J. Polym. Sci. Part C* **38**, 47.

Kraus, G., and Gruver, J. T. (1972). Kinetics of strain-induced crystallization of a trans-polypentenamer. *J. Polym. Sci. Polym. Phys. Ed.* **10**, 2009.

Krimm, S., Shipman, J. J., Folt, V. L., and Berens, A. R. (1965). Infrared evidence for solid state constraints on chain conformation in poly(vinyl chloride). *J. Polym. Sci. Part B* **3**, 275.

Kubo, S. (1971). Crystallization during polymerization of polyparaxylylene. Thesis, Dept. of Chemistry, Rensselaer Polytech. Inst., Troy, New York.

Kubo, S., and Wunderlich, B. (1971). Morphology of poly-p-xylylene crystallized during polymerization. *J. Appl. Phys.* **42**, 4558.

Kubo, S., and Wunderlich, B. (1972). Crystallization during polymerization of poly-p-xylylene. *J. Polym. Sci. Polym. Phys. Ed.* **10**, 1949.

Lanceley, H. A., and Sharples, A. (1966). A kinetic study of polymer crystallization from solution. *Makromol. Chem.* **94**, 30.

Lando, J. B., and Frayer, P. (1968). The copolymerization of trioxane and trithiane in the solid state. *J. Polym. Sci. Part B* **6**, 285.

Lando, J. B., and Stannett, V. (1965). Radiation polymerization of crystalline trithiane. *J. Polym. Sci. Part A* **3**, 2369.

Lauritzen, J. I., Jr. (1973). Effect of a finite substrate length upon polymer growth rate. *J. Appl. Phys.* **44**, 4353.

Lauritzen, J. I. Jr., and Hoffman, J. D. (1959). Formation of polymer crystals with folded chains from dilute solution. *J. Chem. Phys.* **31**, 1680.

Lauritzen, J. I. Jr., and Hoffman, J. D. (1960). Theory of formation of polymer crystals with folded chains in dilute solution. *J. Res. Nat. Bur. Std.* **64A**, 73.

Lauritzen, J. I., Jr., and Hoffman, J. D. (1973). Extension of theory of growth of chain-folded polymer crystals to large undercoolings. *J. Appl. Phys.* **44**, 4340.

Lauritzen, J. I., Jr., and Passaglia, E. (1967). Kinetics of crystallization in multicomponent systems: II. Chain-folded polymer crystals. *J. Res. Nat. Bur. Std.* **71A**, 261.

Lauritzen, J. I., Jr., DiMarzio, E. A., and Passaglia, E. (1966). Kinetics of growth of multicomponent chains. *J. Chem. Phy.* **45**, 4444.

Lawton, E. J., Grubb, W. T., and Balwit, J. S. (1956). A solid state polymerization initiated by high-energy electrons. *J. Polym. Sci.* **19**, 455.

Lee, T. D., and Yang, C. N. (1952). Statistical theory of equations of state and phase transitions. II. Lattice gas and Ising model. *Phys. Rev.* **87**, 410.

Leese, L., and Baumber, M. W. (1965). Kinetics and mechanism of trioxanpolymerization. *Polymer* **6**, 269.

Letort, M. (1936). Sur un nouveau polymère de l'acétaldéhyde. *C.R. Acad. Sci. Paris* **202**, 767; see in particular the later publications M. Letort and X. Duval (1943). Sur la formation et la préparation d'un polymère nouveau de l'acétaldéhyde. *ibid.* **216**; and M. Letort, X. Duval, and J. Pollin (1947). Crystallization et polymérisation. Formation du polyacétaldéhyde. *C.R. Acad. Sci. Paris* **224**, 50.

Leugering, H. J., and Kirsch, G. (1973). Beeinflussung der Kristallstruktur von isotaktischem Polypropylene durch Kristallisation aus orientierten Schmelzen. *Angew. Makromol. Chem.* **33**, 17.

Liberti, F., and Wunderlich, B. (1968). Melting of caprolactam. *J. Polym. Sci. Part A*-2 **6**, 833.

Lindenmeyer, P. H., and Holland, V. F. (1964). Relationship between molecular weight, radial growth rate, and the width of the extinction bands in polyethylene spherulites. *J. Appl. Phys.* **35**, 55.

Lovering, E. G. (1970). Effect of molecular weight on the growth rate of low melting spherulites of trans-1,4-polyisoprene. *J. Polymer Sci. Part A*–2 **8**, 747.

Luch, D. and Yeh, G. S. Y. (1972). Morphology of strain-induced crystallization of natural rubber. I. Electron microscopy on uncrosslinked thin film. *J. Appl. Phys.* **43**, 4326.

Macchi, E. M. (1972). Crystallization and polymerization of 6-aminocaproic acid as simultaneous processes. *J. Polym. Sci. Part A*–1 **10**, 45.

Macchi, E. M. (1973). The direct conversion of 6-aminocaproic acid vapor into polymer crystals. *Makromol. Chem.* **165**, 313.

Macchi, E. M., Morosoff, N., and Morawetz, H. (1968). Polymerization in the solid state. X. The solid state conversion of 6-aminocaproic acid into oriented nylon 6. *J. Polym. Sci. Part A*–1 **6**, 2033.

Mackley, M. R., and Keller, A. (1973). Flow induced crystallization of polyethylene melts. *Polymer* **14**, 16.

Magat, M. (1962). Polymerization in the solid state. *Polymer* **3**, 449.

Magill, J. H. (1964, 1967, 1969) Crystallization of poly(tetramethyl-p-silphenylene)siloxane (TMPS). Parts I, II, and II. *J. Appl. Phys.* **35**, 3249; *J. Polym. Sci. Part A*–2 **5**, 89; **7**, 1187.

Magill, J. H., Pollack, S. S., and Wyman, D. P. (1965). Glass temperature and crystal modification of linear polyethylene. *J. Polym. Sci. Part A* **3**, 3781.

Magill, J. H., Li, H.-M., and Gandica, A. (1973). A corresponding states equation for crystallization kinetics. *J. Crystal Growth* **19**, 361.

Magill, J. H., Kojima, M. and Li, H.-M. (1974). Crystallization behavior of polysiloxanes. Presented at the *IUPAC Symp. Macromol. Aberdeen, Sept.* 1973 (Abstract B–33) to be published later.

Majer, J. (1960a). Sekundäre Kristallisation eines Polyäthylens vom Ziegler-Typ. *Collect. Czech. Commun.* **25**, 2454.

Majer, J. M. (1960b). Einfluß der Schmelzbedingungen auf die Kristallisation von Polypropylen. *Kunststoffe* **50**, 565.

Mammi, M., Carazzolo, G., and Valle, G. (1965). Relative orientation of the twinned polythiomethylene crystals from trithiane by γ-irradiation. *J. Polym. Sci. Part B* **3**, 863.

Mandelkern, L. (1955). Crystallization kinetics in high polymers. II. Polymer diluent mixtures. *J. Appl. Phys.* **26**, 443.

Mandelkern, L. (1958). Crystallization kinetics in polymeric systems. *In* "Growth and Perfection of Crystals" (R. H. Doremus, B. W. Roberts, and D. Turnbull, eds.), p. 487. Wiley, New York and Chapman and Hall, London.

Mandelkern, L. (1964a). "Crystallization of Polymers." McGraw Hill, New York.

Mandelkern, L. (1964b). The crystallization kinetics of polymer-diluted mixtures: The temperature coefficient of the process. *Polymer* **5**, 637.

Mandelkern, L., Quinn, F. A., Jr., and Flory, P. J. (1954). Crystallization kinetics in high polymers. I. Bulk Polymers. *J. Appl. Phys.* **25**, 830.

Mandelkern, L., Jain, N. L., and Kim, H. (1968). Temperature dependence of the growth rate of spherulites. *J. Polym. Sci. Part A–2* **6**, 165.

Marchessault, R. H., and Chanzy, H. D. (1970). Polymerisation de polyethylene sur trichlorure de vanadium depose sur support. Morphologie du polymere naissant. *J. Polym. Sci. Part C* **30**, 311.

Marchessault, R. H., Fisa, B., and Chanzy, H. D. (1972). Nascent morphology of polyolefins. *Crit. Rev. Macromol. Sci.* **1**, 315.

Marker, L., Hay, P. M., Tilley, G. P., Early, R. M., and Sweeting, O. J. (1959). *J. Polym. Sci.* **38**, 33.

Martin, G. M., and Mandelkern, L. (1963). Effect of hydrostatic pressure on the crystallization kinetics of natural rubber. *J. Appl. Phys.* **34**, 2312.

Mateva, R., Wegner, G., and Lieser, G. (1973). Growth of polyoxymethylene crystals during cationic polymerization of trioxane in nitrobenzene. *J. Polym. Sci. Polym. Lett. Ed.* **11**, 369.

Matsuda, T., Higashimura, T., and Okamura, S. (1970). Solid state polymerization of acrylamide and methacrylamide by chlorine under ultraviolet irradiation. *J. Macromol. Sci.* **A4**, 1.

McHugh, A. J., and Schultz, J. M. (1973). Extensional flow induced crystallization from solution. *Kolloid Z. Z. Polym.* **251**, 193.

Mehta, A. (1973). Thesis, Dept. of Chemistry, Rensselaer Polytech. Inst., Troy, New York.

Mehta, A., and Wunderlich, B. (1974). Detection of tie-molecules by thermal analysis. *Makromol. Chem.* **175**, 977.

Melillo, L., and Wunderlich, B. (1972). Extended chain crystals. VIII. Morphology of polytetrafluoroethylene. *Kolloid Z. Z. Polym.* **250**, 417.

Mesrobian, R. B., Ander, P., Ballantine, D. S., and Dienes, G. J. (1954). Gamma-ray polymerization of acrylamide in the solid state. *J. Chem. Phys.* **22**, 565.

Miller, E., Fankuchen, I., and Mark, H. (1940). Polymerization in the solid state. *J. Appl. Phys.* **20**, 531.

Miller, P., Jackson, J. F., and Porter, R. S. (1973). Infrared dichroism of polyethylene crystallized by orientation and pressure in a capillary viscometer. *J. Polym. Sci. Polym. Phys. Ed.* **11**, 2001.

Miyamoto, Y., Nakafuku, C., and Takemura, T. (1972). Crystallization of polychlorotrifluoroethylene. *Polym. J.* **3**, 120.

Morawetz, H. (1963). Thermal reactions of organic solids. *Phys. Chem. Org. Solid State* **1**, 287–328; (1965) **2**, 853–872; see also (1966). Polymerization in the solid state. *J. Polym. Sci. Part C* **12**, 79.

Morawetz, H. (1966). Oriented chain growth by polymerization in the crystalline state and by related processes. *Pure Appl. Chem.* **12**, 201.

Morgan, L. B. (1954). Crystallization phenomena in polymers. II. The course of the crystallization. *Phil. Trans. Roy. Soc. (London)* **247**, 13.

Morosoff, N., Lim, D., and Morawetz, H. (1964). Preparation of biaxially oriented polycaproamide by the solid state polycondensation of ε-aminocaproic acid. *J. Amer. Chem. Soc.* **86**, 3167.

Morosoff, N., Morawetz, H., and Post, B. (1965). Polymerization in the solid state. VII. A crystallographic study of the radiation initiated polymerization in single crystals of vinyl stearate. *J. Amer. Chem. Soc.* **87**, 3035.

Mucha, M., and Wunderlich, B. (1974). Crystallization during polymerization of diazomethane. I. The boron trifluoride catalyzed reaction. *J. Polym. Sci. Polym. Phys. Ed.* **12**, 1993.

Mullins, W. W. (1959). Analysis of the linear cooperative problem as a Markoff process. *Phys. Rev.* **114**, 389.

Muñoz-Escalona, A., and Wegner, G. (1971). Kinetische Untersuchungen zur strahlungsinduzierten Festkörperpolymerisation von Trioxan und Tetroxan. IV. Mitt. der Reihe: "Kinetische und morphologische Untersuchungen zur Polymerisation von oxacyclischen Monomeren im festen Zustand." *Makromol. Chem.* **149**, 197; see also **148**, 331.

Nachtrab, G., and Zachmann, H. G. (1970). Zur Kinetik der Kristallisation von Copolymeren. *Ber. Bunsenges. Phys. Chem.* **74**, 837.

Nadkarni, V. M., and Schultz, J. M. (1973). A study of the solid-state polymerization of trithiane. *J. Mater. Sci.* **8**, 525.

Nakamura, K., Katayama, K., and Amano, T. (1973). Some aspects of nonisothermal crystallization of polymers. II. Consideration of the isokinetic condition. *J. Appl. Polym. Sci.* **17**, 1031.

Nakanishi, H., Hasegawa, M., and Sasada, Y. (1972). Four-center type photopolymerization in the crystalline state. V. X-ray crystallographic study of the polymerization of 2,5-distyrylpyrazine. *J. Polym. Sci. Part A–2* **10**, 1537.

Nagasawa, T., and Shimomura, Y. (1974). On the forming mechanism of shishkebab structure. *J. Polym. Sci. Polym. Phys. Ed.* **12**, 2291.

Newton, Sir Isaac (1730). "Opticks." London; Book 3, query 31; reprinted frequently, for example: Hutchins, R. M. (ed.) (1952). "Great Books of the Western World." Vol. 34, p. 536.

O'Donnell, J. H., McGarvey, B., and Morawetz, H. (1964). Polymerization in the solid state. VI. ESR-study of the propagating species in a polymerizing single crystal of barium methacrylate dihydrate. *J. Amer. Chem. Soc.* **86**, 2322.

Okamura, S., Hayashi, K., and Kitanishi, Y. (1962a). Radiation-induced solid-state polymerization of ring compounds. *J.* Polym. Sci. **58**, 925.

Okamura, S., Higashimura, T., and Takeda, K. (1962b). Cationic polymerization of trioxane in the solid state. *Makromol. Chem.* **51**, 217.

Okamura, S., Kobayashi, E., Takeda, M., Tomikawa, K., Higashimura, T. (1964). Cationic polymerization of trioxane in solid phase. *J. Polym. Sci. Part C* **4**, 827.

Onsager, L. (1944). Crystal statistics. I. A two-dimensional model with an order-disorder transition. *Phys. Rev.* **65**, 117.

Oono, R., Miyasaka, K., and Ishikawa, K. (1973). Crystallization kinetics of biaxially stretched natural rubber. *J. Polym. Sci. Polym. Phys. Ed.* **11**, 1477.

Padden, F. J., Jr., and Keith, H. D. (1959). Spherulitic crystallization in polypropylene. *J. Apply. Phys.* **30**, 1479.

Parker, R. L. (1970). Crystal growth mechanisms: energetics, kinetics, and transport. *Solid State Phys.* **25**, 152–299.

Pechhold, W., Liska, E., and Baumgärtner, A. (1972). Defektkristall und Mäandermodell. *Kolloid Z. Z. Polym.* **250**, 1017.

Pennings, A. J., and Pijpers, M. F. J. (1970). On the kinetics of crystallization of polymers from stirred solutions. *Macromolecules* **3**, 261.

Pennings, A. J., van der Mark, J. M. A. A., and Kiel, A. M. (1970). Hydrodynamically. induced crystallization of polymers from solution. III. Morphology. *Kolloid Z. Z. Polym.* **237**, 336.

Pennings, A. J., Zwijnenburg, A., and Lageveen, R. (1973). Longitudinal growth of polymer crystals from solutions subjected to simple shear flow. *Kolloid Z. Z. Polym.* **251**, 500.

Perret, R., and Skoulious, A. (1972). Étude de la cristallisation des copolymerès trisequéncés poly-ε-caprolactone/polyoxyéthylène/poly-ε-caprolactone. *Makromol. Chem.* **162**, 147, 163.

Peterlin, A. (1964). Secondary crystallization and annealing of polyethylene. *J. Appl. Phys.* **35**, 75.

Peterlin, A., and Roeckl, E. (1963). NMR and dilatometric studies of polyethylene recrystallization. *J. Appl. Phys.* **34**, 102.

Phillips, P. J., and Andrews, E. H. (1972). Crystallization of natural rubber at high pressures. *J. Polym. Sci. Polym. Lett.* **10**, 321.

Poisson, S. D. (1837). "Recherches sur la Probabilité des Jugements en matières criminelle et en matière civile." p. 206, Bachelier, Paris.

Poland, D., and Scheraga, H. A. (1970). "Theory of Helix-Coil Transitions in Biopolymers." Academic Press, New York.

Price, F. P. (1959). Growth habit of single polymer crystals. *J. Chem. Phys.* **31**, 1679.

Price, F. P. (1960). The growth habit of single polymer crystals. *J. Polym. Sci.* **42**, 49.

Price, F. P. (1961). Markoff chain model for growth of polymer single crystals. *J. Chem. Phys.* **35**, 1884.

Price, F. P. (1964). Theoretical aspects of polymer morphology-crystallization from solution. *SPE Trans.* **4**, 151.

Price, F, P. (1965a). Some comments on the "Avrami" equation. *J. Appl. Phys.* **36**, 3014.

Price, F. P. (1965b). A phenomenological theory of spherulite crystallization: Primary and secondary crystallization processes. *J. Polym. Sci. Part A* **3**, 3079.

Price, F. P., and Kilb, R. W. (1962). The morphology and internal structure of poly-(ethylene oxide) spherulites. *J. Polym. Sci.* **57**, 395.

Price, F. P., and Thornton, J. M. (1973). The effect of impingement of rods on the Avrami equations. *Bull. Amer. Phys. Soc.* **18**, 295.

Price, M. B., and McAndrew, F. B. (1967). The polymerization and copolymerization of trioxane. *J. Macromol. Sci.* **A1**, 231.

Privalko, V. P., and Lipatov, Yu. S. (1973). Effect of molecular packing density on the crystallization rates of linear polymers from the melt. *Kolloid Z. Z. Polym.* **251**, 583; *Vysokomol. Soedin. Ser. B* **15**, 682.

Puterman, M., Fort, T., Jr., and Lando, J. B., (1974). The polymerization and structure of mixed multilayers of ethyl and vinyl sterate. *J. Colloid Interface Sci.* **47**, 705.

Rabesiaka, J., and Kovacs, A. J. (1961). Isothermal crystallization kinetics of polyethylene. III. Influence of sample preparation. *J. Appl. Phys.* **32**, 2314.

Rahl, F. J., Evanc, M. A., Fredericks, R. J., and Reimschuessel, A. C. (1972). Morphology of emulsion-grade polytetrafluoroethylene. *J. Polym. Sci. Part A-2*, **10** 1337.

Rees, D. V., and Bassett, D. C. (1971). Crystallization of polyethylene at elevated pressures. *J. Polym. Sci. Part A-2* **9**, 385.

Reynolds, G. F. (1963). Crystal growth. *Phys. Chem. Org. Solid State* **1** Chapter 3.

Richards, R. B. (1945). The melting of polythene. *Trans, Faraday Soc.* **41**, 127.

Richardson, M. J., Flory, P. J., and Jackson, J. B. (1963). Crystallization and melting of copolymers of polymethylene. *Polymer* **4**, 221.

Robeson, L. M. (1973). Crystallization kinetics of poly-ε-caprolactone/poly(vinyl chloride) solutions. *J. Appl. Polymer Sci.* **17**, 3607.

Robinet, J. B. (1761). "De la nature," p. 290. Amsterdam.

Rodriguez *et al.* (1966). Studies on Ziegler-Natta catalysts. *J. Polym. Sci. Part A*–1 **4**; Rodriguez, L. A. M., van Looy, H. M., and Gabant, J. A. Part I. Reaction between trimethylaluminum and α-titanium trichloride, p. 1905; Part II. Reactions between α- or β-$TiCl_3$ and $AlMe_3$, $AlMe_3$, $AlMe_2Cl$, or $AlEt_3$ at various temperatures, p. 1917; van Looy, H. M., Rodriguez, L. A. M. and Gabant, J. A. Part III. Composition of the nonvolatile product of the reaction between titanium trichloride and trimethylaluminum or dimethylaluminum chloride, p. 1927; Rodriguez, L. A. M., and van Looy, H. M. Part IV. Chemical nature of the active site, p. 1951; Part V. Stereospecificity of the active center, p. 1971.

Rohleder, J., and Stuart, H. A. (1960). Haupt und Nachkristallisation von 4,6-Polyurethan. *Makromol.* Chem. **41**, 110.

Russo, M., Mortillaro, L., Credali, L., and DeChecchi, C. (1966). New crystalline modification of polyselenomethylene by polymerization of 1,3,5,7-tetraselenocane. *J. Polym. Sci. Part A*–1 **4**, 248.

Rybnikar, F. (1960). Kristallisation von Polyäthylenterephthalat. *Coll. Czech. Chem. Commun.* **25**, 1529.

Rybnikar, F. (1962). Kinetics of the crystallization of poly-6-caproamide. *Collect. Czech. Chem. Commun.* **27**, 106.

Sakurada, I., and Nanbu, K. (1959). Polymerization of vinyl chloride in urea complex by γ-irradiation. *Kogyo Kagaku Zasshi* **80**, 307.

Sanchez, I. C. (1974). Modern theories of polymer crystallization. *Rev. Macromol. Chem.* **C10**, 113.

Sanchez, I. C., and DiMarzio, E. A. (1971a). Dilute solution theory of polymer crystal growth: A kinetic theory of chain folding. *J. Chem. Phys.* **55**, 893.

Sanchez, I. C. and DiMarzio, E. A. (1971b). Dilute-solution theory of polymer crystal growth. Some thermodynamic and predictive aspects for polyethylene. *Macromolecules* **4**, 677.

Sanchez, I. C., and Eby, R. K. (1973). Crystallization of random copolymers. *J. Res. Nat. Bur. Std.* **77A**, 353.

Schultz, J. M., and Scott, R. D. (1969). Temperature dependence of secondary crystallization in linear polythylene. *J. Polym. Sci. Part A*–2 **7**, 659.

Seher, A. (1954). Die Konstitution der Isan- und Isanolsäure. *Ann.* **589**, 222.

Sella, C., and Bensasson, R. (1962). Sur la polymérisation a l'état solide de l'acrylamide. *J. Polym. Sci.* **56**, S1.

Sella, C., and Trillat, J. J. (1961). Étude par microscopie électronique de la polymérisa-, tion radiochemique à l'état solide de l'amide acrylique. *C. R. Acad. Sci. Paris* **253** 1511.

Seto, T., and Mori, N. (1969). The concentration dependence of the growth rate of polyethylene crystals from solution. *Rep. Progr. Polym. Phys. Japan* **12**, 157.

Shah, J. K., and Lahti, L. E. (1973). Crystallization kinetics of polypropylene from dilute solutions. *Ind. Eng. Chem. Prod. Res. Devel.* **12**, 304.

Shioji, Y., Ohnishi, S. I., and Nitta, I. (1963). ESR study of irradiated acrylic acid and reaction of the produced radical. *J. Polym. Sci. Part A* **1**, 3373.

Siegmann, A., and Harget, P. J. (1973). High pressure crystallization of polyethylene terephthalate. *Bull. Amer. Phys. Soc. II* **18**, 346.

Slyterman, L. A. Ae., and Kooistra, M. (1952). Reaction of polypeptide esters in the solid state, II. A few cases of prevailing polycondensation. Remarks concerning the reaction mechanism. *Rec. Trav. Chim.* **71**, 277.

Slyterman, L. A. Ae., and Veenendaal, H. J. (1952). Reaction of polypeptide esters in the solid state. I. Migration of methyl groups. *Rec. Trav. Chim.* **71**, 137.

Smith-Johannsen, R. (1971). Polymer whiskers grown from methyl-2-cyanoacrylate vapor. *Science* **171**, 1246.

Southern, J. H., and Wilkes, G. L. (1973). Transparent high density polyethylene films crystallized under orientation and pressure. *J. Polym. Sci. Polym. Lett. Ed.* **11**, 555.

Southern, J. H., Weeks, N., Porter, R. S., and Crystal, R. G. (1972). Unique polyethylene morphologies produced under extrusion conditions. *Makromol. Chem.* **162**, 19.

Staudinger, H., Johner, H., Signer, R., Mie, G., and Hengstenberg, J. (1927). Der polymere Formaldehyd, ein Modell der Zellulose. *Z. Phys. Chem.* **126**, 425.

Stefan, J. (1891). Ueber die Theorie der Eisbildung, insbesondere über die Eisbildung im Polarmeere. *Ann. Phys. Chem.* **42**, 269 (*Ann. Phys.* **278**).

Stein, R. S., and Powers, J. (1962). Microscopic measurements of spherulite growth rates. *J. Polym. Sci.* **56**, S9.

Steno, N. (1669). "De solido intra solidum naturaliter contento dissertationis prodomus." Florence. Translations: J. G. Winter, Macmillan, New York, 1916, reprinted 1968, Hafner Publ., New York; (1923). K. Mieleitner, "Ostwalds Klassiker der exacten Wissenschaften." Leipzig.

Stejny, J. (1973). Polymerization and crystallization of polytetrahydrofuran and poly-(1,4-epoxycyclohexane). *J. Macromol. Sci.* **A7**, 1435.

Stranski, I. N. (1928). Zur Theorie des Kristallwachstums. *Z. Phys. Chem.* **136**, 259.

Straumanis, M. (1935). Das Wachstum von Metallkristallen im Metalldampf. *IV. Z. Phys. Chem.* **30B**, 132.

Straumanis, M. (1940). Die Gitterkonstanten und Ausdehnungskoeffizienten des Tellurs und Selens. *Z. Krist.* **102**, 432.

Stuart, H. A. (Ed.) (1955). "Die Physik der Hochpolymeren. Dritter Band. Ordnungszustände und Umwandlungserscheinungen in festen Hochpolymeren Stoffen." Springer, Berlin.

Szwarc, M. (1947). Some remarks on the 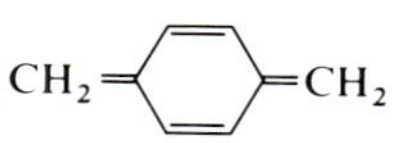 molecule. *Discuss. Faraday Soc.* **2**, 46.

Szwarc, M. (1965). The kinetics and mechanism of N-carboxy-α-amino-acid anhydride (NCA) polymerization to polyamino acids. *Fortschr. Hochpolym.-Forsch.* **4**, 1.

Tabata, Y. (1968). Solid state polymerization. *Advan. Macromol. Chem.* **1**, 283.

Tabata, Y., Ishigure, K., Fujita, Y., and Oshima, K. (1968). Solid-state polymerization in glassy state at low temperature. *J. Polym. Sci. Part C* **23**, 663.

Tait, P. J. T. *et al.* (1972). Ziegler-Natta catalysis. *Polymer* **13**; Burfield, D. R., McKenzie, I.D., and Tait, P. J. T. 1. A general kinetic scheme, p. 302; McKenzie, I.D., Tait, P. J. T., and Burfield, D. R. 2. A kinetic investigation, p. 307; Burfield, D. R., and Tait, P. J. T. 3. Active center determination, p. 315; Burfield, D. R., Tait, P. J. T., and McKenzie, I. D. 4. Quantitative verification of kinetic scheme, p. 321.

Tammann, G. (1903). "Krystallisieren und Schmelzen." Barth, Leipzig; (1923). "Aggregatzustände." Barth, Leipzig, translated by R. F. Mehl (1925). "The States of Aggregation." Van Nostrand Reinhold, Princeton, New Jersey.

Thomas, J. M., and Williams, J. O. (1971). Dislocations and the reactivity of organic solids. *Progr. Solid State Chem.* **6**, 119.

Tobin, M. C. (1974). Theory of phase transition kinetics with growth site impingement. I. Homogeneous nucleation. *J. Polym. Sci. Polym. Phys. Ed.* **12**, 399.

Tomka, J., and Šebenda, J. (1972). On the structure and properties of polyamides. XXXII. Influence of noncrystallizable units on the crystallization of polycaprolactam. *Collect. Czech. Chem. Commun.* **37**, 453

Tomka, J., Šebenda, J., and Wichterle, O. (1967). On the structure and properties of polyamides XXVII. Crystallization of polycaprolactam without a thermal history. *J. Polym. Sci. Part C* **16**, 53.

Toy, M. S. (1971). Surface-polymerization of tetrafluoethylene on fluorine-activated metal substrates. *J. Polym. Sci. Part C* **34**, 273.

Treiber, G., Melillo, L., and Wunderlich, B. (1973a). The effect of solvent concentration on the crystallization of polyethylene under elevated pressure. *J. Polym. Sci. Polym. Lett. Ed.* **11**, 435.

Treiber, G., Boehlke, K., Weitz, A., and Wunderlich, B. (1973b). Crystallization during polymerization of poly-p-xylylene. II. Polymerization at low temperature. *J. Polym. Sci. Polym. Phys. Ed.* **11**, 1111.

Treolar, L. R. G. (1958) "Physics of Rubber Elasticity." 2nd ed. *Oxford Univ. Press, London and New York.*

Turska, E., and Gogolewski, S. (1971). Study on crystallization of nylon 6 (polycapramide): Part 1. Isothermal crystallization. Part 2. Effect of molecular weight on isothermal crystallization kinetics. *Polymer* **12**, 616, 629.

van Antwerpen, F., and van Krevelen, D. W. (1972). Influence of crystallization temperature, molecular weight, and additives on the crystallization kinetics of poly(ethylene terephthalate). *J. Polym. Sci. Polym. Phys. Ed.* **10**, 2423.

van der Heijde, H. B. *et al.* (1966). The polymerization of crystalline trioxane. *Phil. Mag.* **13**; van der Heijde, H. B., and Nauta, H. I. Evidence for a cooperative mechanism p. 1015; Nauta, H. II. Experiments with irradiated trioxane (and tetroxane). Calculation of the crystal field of trioxane, p. 1023; van der Heijde, H. B., and van Kasteren, P. H. G. III. Experiments with irradiated trioxane single crystals, p. 1039; van der Heijde, H. B. IV. Spontaneous polymerization, p. 1055.

Vidotto, G., Lévy, D., and Kovacs, A. J. (1969). Cristallisation et fusion des polymères autoensemencès. I. Polybutène-1, polyéthylène et polyoxyéthylène de haute masse moléculaire. *Kolloid Z. Z. Polym.* **230**, 289

Voigt-Martin, I. (1974). Some kinetic aspects regarding the solid state polymerization of trioxane. *Makromol. Chem.* **175**, 2669.

Volmer, M. (1939). "Kinetik der Phasenbildung." Steinkopff Verlag, Dresden and Leipzig.

von Falkai, B. (1960). Schmelz- und Kristallisationserscheinungen bei makromolekularen Substanzen. I. Kristallisationskinetische Untersuchungen an isotaktischem Polypropylene. *Makromol. Chemie* **41**, 86.

von Falkai, B., and Stuart, H. A. (1959). Zur Kristallisation von isotaktischem Polypropylen. *Kolloid Z.* **162**, 138.

Walker, J. F. (1964) "Formaldehyde," 3rd ed. Van Nostrand Reinhold Princeton, New Jersey.

Wegner, G. (1969). Topochemische Reaktionen von Monomeren mit konjugierten Dreifachbindungen. I. Mitt.: Polymerisation von Derivaten des 2.4-Hexadiin-1.6-diols im kirstallinen Zustand. *Z. Naturforsch.* **246**, 824.

Wegner, G. (1972). Topochemical polymerization of monomers with conjugated triple bonds. *Makromol. Chem.* **154**, 35.

Weir, C. E. (1953). Transitions and phases of polytetrafluoroethylene (Teflon). *J. Res. Nat. Bur. Std.* **50**, 95; see also **53**, 245 (1954).

Wells, A. F. (1946). Crystal growth. *Chem. Soc. (London) Ann. Rep.* **43**, 62.

Wereta, A., Jr., and Gogos, C. C. (1971). Crystallization studies on deformed polybutene-1 melts. *Polym. Eng. Sci.* **11**, 19.

Wessling, R. A., Oswald, J. H., and Harrison, I. R. (1973). Effect of polymerization conditions on the morphology of as-polymerized poly(vinylidene chloride). *J. Polym. Sci. Polym. Phys. Ed.* **11**, 875.

White, D. M. (1960). Stereospecific polymerization in urea canal complexes. *J. Amer. Chem. Soc.* **82**, 5678.

Wichterle, O., Šebenda, J. and Králíček, J. (1961). The anionic polymerization of caprolactam. *Fortschr. Hochpolym.-Forsch.* **2**, 578.

Wolpert, S. M., Weitz, A., and Wunderlich, B. (1971). The time-dependent heat capacity in the glass transition region. *J. Polym. Sci. Part A–2* **9**, 1887.

Wristers, J. (1973a). Nascent polypropylene morphology: Polymer Fiber *J. Polym. Sci. Polym. Phys. Ed.* **11**, 1601.

Wristers, J. (1973b). Direct examination of polymerization catalyst by electron microscopy. *J. Polym. Sci. Polym. Phys. Ed.* **11**, 1619.

Wunderlich, B. (1958). Theory of cold crystallization of high polymers. *J. Chem. Phys.* **29**, 1395.

Wunderlich, B. (1963). The effect of pressure on the crystallization of polyethylene from dilute solution. *J. Polym. Sci. Part A* **1**, 1245.

Wunderlich, B. (1968). Crystallization during polymerization. *Fortschr. Hochpolym.-Forsch.* **5**, 568; See also *Angew. Chem.* **80**, 1009; *Angew. Chem. Int. Ed.* **7**, 912.

Wunderlich, B., and Arakawa, T. (1964). Polyethylene crystallized from the melt under elevated pressure. *J. Polym. Sci. Part A* **2**, 3697.

Wunderlich, B. and Cormier, C. M. (1966). Seeding of supercooled polyethylene with extended chain crystals. *J. Phys. Chem.* **70**, 1844.

Wunderlich, B., and Liberti, F. (1966). Zone polymerization of caprolactam to the crystalline state. *Bull. Amer. Phys. Soc.* **11**, 248.

Wunderlich, B., and Melillo, L. (1968). Morphology and growth of extended chain crystals of polyethylene. *Makromol. Chem.* **118**, 250.

Wunderlich, B., Hellmuth, E., Jaffe, M., Liberti, F., and Rankin, J. (1965). Crystallization of Linear High Polymers from the monomer. *Kolloid Z. Z. Polym.* **204**, 125.

Wunderlich, B., Melillo, L., Cormier, C. M., Davidson, T., and Snyder, G. (1967). Surface melting and crystallization of polyethylene. *J. Macromol. Sci.* **B1**, 485.

Yamaoka, H., Hayashi, K., and Okamura, S. (1964). Gamma-ray induced polymerization of formaldehyde, *Makromol. Chem.* **76**, 196.

Yamazawa, Y., Iwai, T., Ito, A., and Hayashi, K. (1971). Radiation-induced solid-state polymerization of trioxane under higher pressure. *J. Polym. Sci. Part A–1* **9**, 257.

Yasuniva, M., Nakafuku, C., and Takemura, T. (1973). Melting and crystallization of polyethylene under high pressure. *Polym. J.* **4**, 526.

Yeh, G. S. Y. (1972). A structural model for the amorphous state of polymers: Folded-chain fringed micellar grain model. *J. Macromol. Sci.* **B6**, 462.

Yeh, G. S. Y., and Geil, P. H. (1967). Crystallization of polyethylene terephthalate from the glassy amorphous state. *J. Macromol. Sci.* **B1**, 235.

Yeh, G. S. Y., and Lambert, S. L. (1971). Strained-induced crystallization of isotactic polystyrene from the glassy and rubbery states. *J. Appl. Phys.* **42**, 4614.

Yeh, G. S. Y., and Lambert, S. L. (1972). Crystallization kinetics of isotactic polystyrene from isotactic-atactic polystyrene blends. *J. Polym. Sci. Part A–2* **10**, 1183.

Zachmann, H. G. (1964). Das Kristallisations- und Schmelzverhalten hochpolymerer Stoffe. *Fortschr. Hochpolym.-Forsch.* **3**, 581.

Zachmann, H. G. (1969). Statistische Thermodynamik des Kristallisierens und Schmelzens von hochpolymeren Stoffen. *Kolloid Z. Z. Polym.* **231**, 504.

Zachmann, H. G., and Stuart, H. A. (1960). Haupt und Nachkristallisation von Terylen aus dem Glaszustand. *Makromol. Chemie* **41**, 131.

Ziabicki, A. (1974). Theoretical analysis of oriented and non-isothermal crystallization. I. Phenomenological considerations. Isothermal crystallization accompanied by simultaneous orientation or disorientation. II. Extension of the Kolmogoroff-Avrami-Evans theory onto processes with variable rates and mechanisms. *Colloid Polym. Sci.* **252**, 207, 433.

CHAPTER VII

The Annealing of Crystals

7.1 General Description of Annealing

This chapter on annealing of macromolecular crystals presents an effort to summarize the field which forms the logical conclusion to the series of chapters describing crystals of linear macromolecules. Crystal structure, morphology, and defect description (Chapters II–IV) are basic to the understanding of crystals. Nucleation and crystal growth (Chapters V and VI) are the processes involved in the initial crystal formation. Annealing describes the changes possible after initial crystallization.

Section 7.1 deals with the general description, definitions, and some initial theories of annealing of macromolecular crystals. In Sects. 7.2 and 7.3 the observations on annealing of solution grown crystals and melt grown crystals are discussed. Finally, Sect. 7.4, is a first effort to describe the long neglected subject of changing of macromolecular crystals by chemical reactions involving the backbone chain. Overall, annealing is a field of macromolecular physics less well understood since it is concerned with the processes involving an initial state which is frequently not well characterized. Further changes of this state complicate matters even more.

7.1.1 Definitions and Processes of Annealing

The word "annealing" derives from the old English *aelan*, to bake or burn and the prefix *an*, on. Over the centuries this word was applied to the treatment of ceramics, glasses, metals, and now it is also applied to macromolecules. The term annealing implies in a narrow sense the imparting of a certain property by heat treatment without large-scale melting. In a broader sense other treatments, such as chemical reaction and mechanical deformation, can also achieve annealing.

An extensive knowledge has been collected during the last 100 years about the annealing of metals, and in particualr steel. Macroscopically annealing has metallurgically the object to impart ductility, extensibility, and a certain grade of softness. The opposite properties are achieved by "hardening." Hardening in the shop sense signifies, for example, the making of a piece of steel as hard as possible. Intermediate stages between the hardened and annealed state are reached by "tempering." Microscopically annealing, hardening, and tempering can involve changing of crystal perfection, grain sizes and structure, chemical composition, and state of stress.

In analogy to the metallurgical terminology, "annealing" is used in polymer science to describe the improvement of crystallization by heating to temperatures below the melting point which should lead to the growing of crystalline areas, perfection of crystals, and a change to more stable crystal structures (Stuart, 1955).†

In the plastics industry, annealing is used to improve the heat resistance and dimensional stability of amorphous and crystalline polymers when exposed to elevated temperatures. Furthermore, annealing frequently improves the impact strength and prevents crazing and cracking of excessively stressed items. Amorphous, rigid polymers are mainly annealed for stress relief, while crystalline polymers show, in addition, changes in the nature of their crystalline state (Jastrzebski, 1965).

Heat treatment of separate grains (powder) of a material may lead to agglomeration and compaction to a dense body, a process which is called "sintering." This process is basic for ceramic processes and has also found application for metals, glasses, and more recently, for macromolecular materials (Lontz, 1963). Besides the changes found in annealing, sintering is usually found at $\frac{2}{3}$ to $\frac{4}{5}$ of the melting temperature in degrees Kelvin.

A final term used in the overall discussion of annealing is "recrystallization." It describes crystallization of a partially molten or dissolved material using the remaining crystal parts as substrate. Depending on the degree of melting between the initial and final crystalline state, recrystallization is either more closely related to sintering or normal crystallization.

To some degree similar processes to those of annealing, sintering and recrystallization described in this section are also observed on initail crystallization and nucleation. Self-nucleation, described in Sect. 5.1.4, is akin to recrystallization. Secondary crystallization and crystal perfection, treated in

† Note that the German equivalent of annealing is *Tempern*, while the equivalent for the metallurgical term tempering is *Anlassen*.

Sect. 6.1.6, are similar to annealing. Finally, the impingement of crystals on normal crystallization, as discussed in Sect. 6.1.3, may show some of the effects of sintering.

Annealing by heating, chemical reaction, and mechanical treatment must provide a path to an overall system of lower free enthalpy under the given conditions. In the discussion of structures of minimum free enthalpy of Sect. 2.3, it is concluded that for the nearest-neighbor interaction the following hierarchy of effects must be considered for the approach to the lowest free enthalpy: covalent bonding, rotational isomerism, and close packing. On the next larger structure level, crystal defects and high free enthalpy conformations in the amorphous areas described in Chapter IV are to be discussed. Since many crystal defects in macromolecular materials are of the nonequilibrium type, their reduction in number should be favored on annealing. The even larger grain-size level of effects concerns the macroconformations of the molecules (Fig. III.5 and Sect. 3.1.2) and the crystal size and shape (Fig. III.4 and Sects. 3.2 and 5.1). Finally, an overall effect, is the possibility of polymorphism. Any combination of changes of the just listed effects may occur during annealing.

A major conclusion of the discussion of most macromolecular materials is that they are far from the equilibrium structure. Annealing helps these materials approach closer to but leaves them usually far from equilibrium. Thus the study of annealing requires the characterization of an initial, metastable state; the final, less metastable state; the mechanism of the change; and, for reference, the equilibrium state. Rarely is annealing literature found which includes all these studies on any one sample.

7.1.2 Changes in Crystal Shape

The study of macromolecular crystal morphologies led to the general principle that a sufficiently regular, flexible linear macromolecule crystallized from the mobile random state will always crystallize first in a chain-folded macroconformation (Sect. 3.2.2.1). The most observed fold-lengths vary from fifty to a few hundred angstrom (see Table III.3 and Figs. III.42–III.51). Such crystals are metastable despite their possibly large lateral dimensions. The frequent folding gives rise to a large fold surface which increases the free enthalpy and consequently reduces the melting temperature (Fig. IV.13, see also Sect. 3.2.2.5 and Chapter XI). Whenever the molecular mobility is not sufficient for chain folding, a fringed micellar structure of small crystal sizes in all dimensions results (Fig. III.7, see also Sect. 6.1.7). Such crystals are even smaller in size and usually also more defective than folded chain lamellae (Sect. 4.2.1). Finally, crystallization during polymerization leads frequently to fibrillar crystals (Sects. 3.8.1 and 3.8.3). These fibrils may contain more extended chain conformations, but when sufficiently thin (and defect), they are also metastable. The three basic macromolecular crystal habits, lamellar, fringed micellar, and fibrillar, are thus metastable and susceptible to change on annealing. Isometric crystals of macroscopic size are rare (Sect. 3.9, Figs. III.141–143).

The equilibrium shape of a given crystal volume V is determined by the ratio of the surface free energies. The major differences are observed between surfaces which are parallel to the chain direction (surface free energy γ) and those which cut across the chains and contain chain folds, ends, tie molecules, or loops (surface free energy γ_e). Since γ_e is according to nucleation experiments perhaps five to forty times γ (see Sect. 5.1.2.2), the equilibrium crystal habit should be prismatic (see Fig. III.3). The surface free energy of a crystal which for simplicity has been assumed to have a quadratic cross section of dimension a and length l in the chain direction is given by [see also Eq. (11) in Chapter V].

$$G_s = 4al\gamma + 2a^2\gamma_e \quad (1)$$

$$G_s = 4(lV)^{1/2}\gamma + (2V\gamma_e/l) \quad (2)$$

The equilibrium dimensions l_e and a_e and their ratio can be obtained by finding the minimum of G_s

$$l_e = (\gamma_e^2 V/\gamma^2)^{1/3} \quad (3)$$

$$a_e = (\gamma V/\gamma_e)^{1/3} \quad (4)$$

$$l_e/a_e = \gamma_e/\gamma \quad (5)$$

Figure VII.1 illustrates that on annealing a typical extended chain fiber of 20 000 Å length and 150 Å width, as may exist, for example, on the fracture surface of polyethylene (see Fig. III.32), and a typical small folded chain lamella of 130 Å fold-length and 877 Å average diagonal dimension (see Fig. III.51) should reach the common folded-chain habit shown in the center as long as no volume change by material transport is permitted (see Sect. 7.1.7). Indeed, annealing rough fracture surfaces of extended chain crystals of polyethylene and polytetrafluoroethylene or fibrillar crystals of polyoxymethylene leads to less extended chain crystals; however, the process is a recrystallization one (see Sect. 3.4.1, Figs. III.74–78) which produces folded chain epitaxial lamellae. There seems to be no mechanism for a gradual decrease of the crystal dimensions in the macromolecular chain direction by a large amount. (For smaller, reversible changes of fold-length see for example Dawkins *et al.*, 1968). The opposite process of thickening of folded chain lamellae is more easily achieved and is of major concern for the further discussion of annealing of macromolecular crystals (Sects. 7.2–7.4).

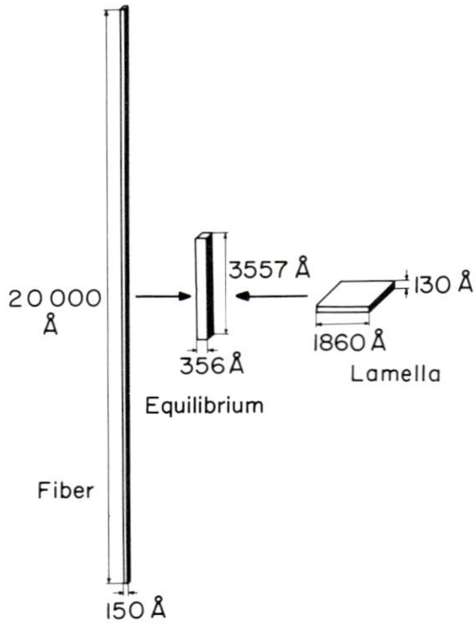

Fig. VII.1 *Schematic Drawing of Annealing of an Extended Chain Fiber and a Folded Chain Lamella of Constant Volume.*
Volume 450×10^6 Å^3. Assumed ratio $\gamma_e/\gamma = 10$, for simplicity drawn cross sections are quadratic. Final dimensions at equilibrium $l_e = 3557$ Å and $a_e = 356$ Å.

The time and temperature dependence of the crystal dimensions during annealing were discussed by Sanchez *et al.* (1973) based on the phenomenological laws of irreversible thermodynamics. By analogy to other irreversible processes, the rate of thickening of a lamellar crystal is taken to be proportional to the surface free energy gradient

$$a^2\, dl/dt = -k\, dG_s/dl \tag{6}$$

By setting

$$x \equiv (l/l_e)^{1/2} \tag{7}$$

and

$$\tau = l^2/k\gamma_e \tag{8}$$

we can integrate Eq. (6) to obtain

$$f(x) - f(x_0) = (t - t_0)/\tau \tag{9}$$

with

$$f(x) = -x + \frac{1}{6}\ln\left(\frac{1 - x^3}{(1 - x)^3}\right) + \frac{1}{\sqrt{3}}\tan^{-1}\left(\frac{2x + 1}{\sqrt{3}}\right)$$

Figure VII.2 illustrates the changes in thickness ratio for the initial to final thickness 0.1 and 0.5. The curves are typically sigmoidal and show in the time interval of strongest increase a long, almost linear portion when plotted versus the logarithm of time. The parameter τ contains the temperature dependence and is linked to the microscopic molecular mechanism involved in the annealing process (see Sect. 7.1.3).

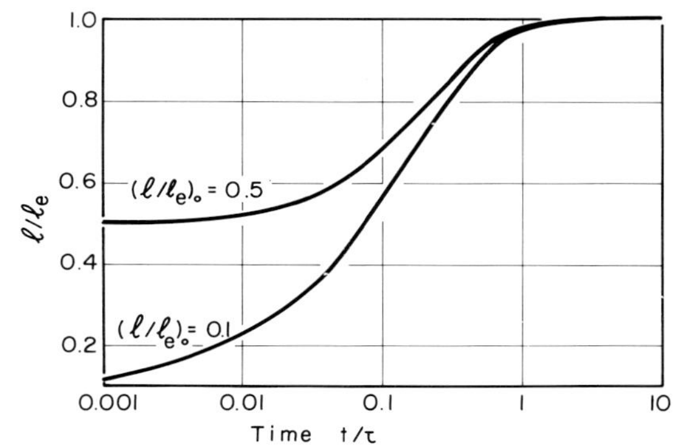

Fig. VII.2 *Change in Lamellar Thickness Ratio as a Function of Reduced Time.*
Calculated from Eq. (9) for initial ratios of thickness to equilibrium thickness of 0.1 and 0.5.

Sanchez *et al.* (1973, 1974) tried to fit Eq. (9) to experimental data on polymer single crystals (see Sect. 7.2). For additional insight, the experimental observation that annealing is strictly a function of ΔT was interpreted as the need to supply the free enthalpy of fusion to a certain number of adjacent chain units v. The rate constant $k\gamma_e$ can then be written [see Eq. (8)]

$$l_e^2/\tau = k_0 \exp(-v\Delta H \Delta T/RTT_m^0) \tag{10}$$

where k_0 may only be little dependent upon temperature. ΔH is the enthalpy of fusion and T_m^0 the equilibrium melting temperature. An estimate for v is given by the number of chain units in one traverse of the crystal (fold-length). This assumption introduces a length dependence into τ which slows the thickening. As a result, the linear portion of Fig. VII.2 is extended, but the overall sigmoidal shape remains (see also Sect. 7.1.3). Another variable is the total volume V which enters into the rate expression through Eq. (3). It is observed that large single lamellae break-up on annealing to smaller domains (see Sect. 7.2, Fig. VII.13). Thermodynamically a crystal broken into smaller domains of increased thickness may still be more stable than the original crystal, further annealing, however, would slow down because of the decreased l_e [see Fig. VII.2 and Eq. (3)]. If sintering (see Sect. 7.1.7) were possible, i.e., the coherently annealing volume increases, annealing would speed up and the final length l_e would increase, an effect, pointed out by Sanchez *et al.* (1973), which may be enhanced by larger hydrostatic pressure (see also Sect. 7.3.3).

The discussion of the changes in crystal shape illustrates the major macroscopic effects on annealing of macromolecular crystals. Fibrillar crystals seem to have no mechanism for continuous reduction of their length; rather, they melt and recrystallize. The shape of the small fringed micelle crystals is so little defined that almost nothing is known at present about their dimensional changes on annealing. It is expected that the major annealing effects are the changes in defect concentration (see Sect. 7.1.4) and the regularization of the surfaces. The thickening of folded chain, lamellar crystals is possible in a more continuous fashion. Irreversible thermodynamics may provide the framework for its macroscopic descriptions.

7.1.3 Changes in Macroconformation

The change of a lamellar habit on annealing to a prismatic habit has the major microscopic problem of fold-length extension for macromolecular crystals. Only for a few crystals has it been possible to increase the fold-length by chemical reaction through breaking and remaking backbone chain bonds on the crystal surface (see Sect. 7.4). In most cases, thickening in the molecular chain direction must involve major mass transport along the molecules through the crystal to adjust to the new dimensions.

An early suggestion of a mechanism for mass transport in polyethylene along the molecular chain was made by Reneker (1962). He proposed that point defects† which contain one extra CH_2 group could be generated at the chain end and moved along the molecule to finally accumulate at folds (Sect. 4.4.1). More quantitative experiments (see Sect. 7.2.1) suggest that this mechanism is not sufficient to explain all observations. Most importantly, it was observed that the chain ends which are in polyethylene and several other macromolecules largely on the crystal surface (Sect. 4.3.4), decrease only slightly in number on annealing (Keller and Priest, 1970).

Observations of preferred doubling of the initial fold-length in nylons by Dreyfuss and Keller (1970) and Burmester *et al.* (1972) led to the proposal of a simpler mechanism which involved rearrangement of portions of the

† For description of point defects see Sect. 4.3.4, for specific kink defects which are able to accomodate added mass see Sect. 4.2.3.

molecule only, without coordinated displacement of the molecule as a whole. Schematically this process is shown in Fig. VII.3. The fold a is consumed by folds b and c in their drive to a larger fold length. This process creates point defects which can be called "fold dislocations" since they correspond to edge dislocations in two dimensions (missing rows of motifs instead of planes, Sect. 4.3.3). Further stages of the refolding process represents climb of the dislocation. Whenever fold a leaves the original lamellar surface, the dislocation energy may be largely regained and a natural explanation for stopping of thickening, at least temporarily, at double the initial fold-length is given. Further thickening requires a new fold dislocation for quadrupling the initial fold-length. This basic mechanism of fold dislocation climb applies for an isolated portion of the lamella, for the complete description of fold extension it must be considered in context of the overall crystal structure. The extension of neighboring folds will be energetically preferred and will influence the geometry; the initial fold surface structure (Sect. 3.2.2.4) will be of importance for the rate of generation of fold dislocations, perhaps through buried folds; for multilayer lamellae (Sect. 3.3.3) the thickening may be enhanced in the case of proper interlamellar contact so that the receding fold may be matched by an advancing fold of the next layer; presence of solvent will change the rate of refolding by increasing the mobility.

The experimentally observed linear logarithmic increase of fold length with annealing time [Sect. 7.2.1, Eq. (16)] was interpreted by Hirai *et al.* (1961a, b) to be indicative of nucleation control. It was assumed that initially an island of increased thickness d_0, one crystallographic repeat distance, and

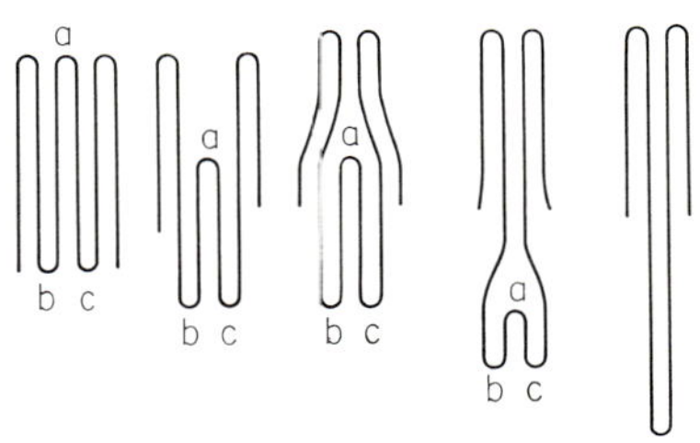

Fig. VII.3 *Schematic Drawing of Successive Stages of the Fold Increase on Annealing.*
The initial fold a becomes a "fold dislocation" on growth of folds b and c. On completion, the dislocation energy is regained and an energetically favored state of double fold-length is reached. Symmetric growth of b and c is assumed for simplicity. Drawing according to Dreyfuss and Keller (1970).

crossection a^2 must be formed on the lamellar surface. Subsequent further growth of the thickened lamella of fold-length $(l + d_0)$ is fast. The free enthalpy of nucleation of the thickened portion from the lamella of thickness l can be written in analogy to Eq. (11) in Chapter V as

$$\Delta G = 4ad_0\gamma - 2a^2 d_0 \gamma_e / l \tag{11}$$

The first part of the right-hand side of Eq. (11) describes the increase in side surface free energy due to formation of the thicker nucleus, and the second part, the decrease in fold surface due to lateral contraction of the thickened lamella. The critical nucleus width a^* is obtained by finding the maximum of Eq. (11)

$$a^* = \gamma l / \gamma_e \tag{12}$$

Insertion of Eq. (11) for the critical dimension a^* into the rate expression for nucleation [Eq. (5), Chapter V] leads in connection with reasonable assumptions of the growth geometry to the rate of thickening

$$dl/dt = A \exp(-\Delta G^*/kT) = A \exp(-Bl) \tag{13}$$

where A and B can be evaluated from Eqs. (5), Sect. 5.1.1 and (11) and the assumed geometry. The experimental increase in fold-length at constant temperature can be made to match Eq. (13) with reasonable values for the adjustable parameters, the strong temperature dependence on approach to the melting temperature, however, is not explained. Also, one should note that a log t dependence over a long time span is already required by the change in shape expressed by Eq. (9). Hirai and Yamashita (1964) tried to rectify the shortcoming of the nucleation concept by taking into account partial melting. They introduced a supercooling dependent crystallinity term in ΔG^* which would decrease the free enthalpy barrier close to the melting temperature linearly with ΔT [compare also with Eq. (10)]. Peterlin (1963) modified Eq. (13) by introducing an explicit expression for the mass transport mechanism to the fold planes. He assumed that the sliding diffusion suggested by Hirai *et al.* (1961a) occurs in collective jumps as an entity in steps of d_0, the repeat distance of the crystal structure. The activation energy of such a jump would be lE_m/d_0, with E_m representing the activation energy for a single unit, single repeat jump. E_m was given the proper temperature dependence to account for the experimental values. It is reasonable to assume, at least qualitatively, that E_m decreases rapidly as the melting temperature is approached.

The problem remains to identify and combine all major effects occurring on fold-length increase for any particular polymer. The diffusion from the chain ends is certainly not the only process since it would lead to a too great molecular weight dependence (proportional to l/M^3 according to Peterlin, 1965). The fold dislocation process, on the other hand, may show a too small molecular weight dependence. The influence of lateral transport necessary to close up created vacancies is not understood at present. It must lead to the frequently observed break-up of lamellae on annealing (see Fig. VII.13) and is considered in the macroscopic discussion of the changes of crystal shape (Sect. 7.1.2). Other problems involve the degree of recrystallization (Sect. 7.1.8) and the transport of chains between grains (Sect. 7.1.7). Knowledge of the detailed defect structure, as discussed in Chapter IV, is also a necessary prerequisite for detailed analysis of the nucleation and growth of the thickened lamella out of its thinner starting material. It is of interest to note that immediately after crystallization, the thickening process is enhanced. Thickening is particularly fast on crystallization of polyethylene and polychlorotrifluoroethylene under elevated pressure and temperature and, in case of polytetrafluoroethylene, on slow cooling at high temperature (see Sect. 6.3.3). One may draw the conclusion that right after crystallization, thickening is easier because of an initially less perfect crystal (and possibly initial formation of a metastable crystal form) (see Sect. 6.1.6).

7.1.4 Changes in Defect Concentration

The description of the defect solid state of macromolecular crystals given in Chapter IV shows that many defects are nonequilibrium defects. Annealing should have the result of a decrease in concentration of such defects. The annealing mechanisms discussed in Sect. 7.1.3, on the other hand, would need an increased number of defects for their operation. Furthermore, the point defects classified as kinks are expected to exist at sufficiently high temperatures in equilibrium concentration (Sect. 4.2.3, p. 451) and thus show changes as long as equilibrium can be attained. Quantitative information on all these defects is limited.

Čačković *et al.* (1969) illustrated that one can estimate from the paracrystalline defect parameter g (see Sect. 4.2.2) with simple geometry assumptions the lower limit of kink concentrations γ

$$g = \frac{\alpha[\gamma(1-\gamma)]^{1/2}}{1+\alpha\gamma} \tag{14}$$

The parameter α is determined by the influence of a defect on the separation of the given crystallographic planes [for a $2g1$ kink α is about 0.5 for the (110) planes, see Table IV.12]. The data on solution crystallized polyethylene before and after annealing and on fast and slow crystallization given in Sect. 4.2.2 (Table IV.11 and p. 445) show little change in g. Čačković *et al.* (1969) estimate γ to be about 0.002, similar in order of magnitude to the estimate of kink concentration given in Sect. 4.2.3 (p. 451). In the event that kinks do reach equilibrium concentrations, raising the temperature should increase their number, but subsequent lowering would decrease the number of defects or freeze a certain number at a given temperature. Quantitative verification of this effect is still outstanding. A heat capacity increase above the value expected from the vibrational spectrum (see Chapter XII) which could perhaps be caused by the reversible heat absorption for defect formation has been observed below the melting point (Bares and Wunderlich, 1973).

Quantitative assessment of dislocations is possible for edge dislocations within single crystal lamellae and screw dislocations between lamellae. The method developed involves electron microscopic observation of moiré patterns of double layer crystals and is described in Sect. 4.3.3. Holland (1964a) measured an increase in intralamellar edge dislocation density from 10^7 lines/cm^2 lamellar surface to 10^{10} lines/cm^2 on annealing of solution-grown single crystals of polyethylene at 95°C for 1 hr. In addition, he observed that the moiré pattern was highly distorted after annealing and that at the crystal edges more dislocations are found; also, evidence was presented for the monoclinic phase of polyethylene (see Table II.9). Niinomi and Takayanagi (1971) analyzed these annealing changes in more detail and linked them with local shearing caused by misfit between adjacent lamellae. They observed that the distortion of the moiré pattern tends to start at the crystal edges where deformation during annealing is easy. Later on, the distortion spreads over the whole crystal. The increase in dislocations occurs at annealing temperatures much below those needed for increase in crystal thickness, so that earlier assumptions that the increase in lamellar thickness is coupled with this increase in intralamellar edge dislocations are not justified. Figure VII.4 shows

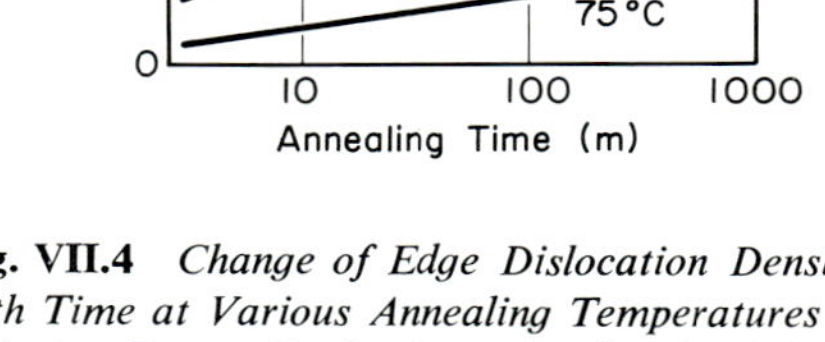

Fig. VII.4 *Change of Edge Dislocation Density with Time at Various Annealing Temperatures in Solution-Grown Single Crystals of Polyethylene.* Note that all temperatures are above the original crystallization temperature (70°C) and below the temperature of initial crystal thickening (100°C). Data by Abe *et al.* (1970). *J. Macromol. Sci.* **B4**, 87.

the time-dependent increase of dislocation density for different temperatures. No change was observed for annealing at 70°C, the original crystallization temperature, unless annealing was carried out in the presence of a solvent. The moiré lines were observed to change their spacing and orientation in limited regions of several thousand angstrom diameter. Around these regions coupled edge dislocations were observed with opposite sign. In the region between the coupled edge dislocations, newly generated by annealing, the moiré pattern changes frequently into the profile of an interlamellar screw dislocation. Figure VII.5 illustrates the possible changes in one of the crystals

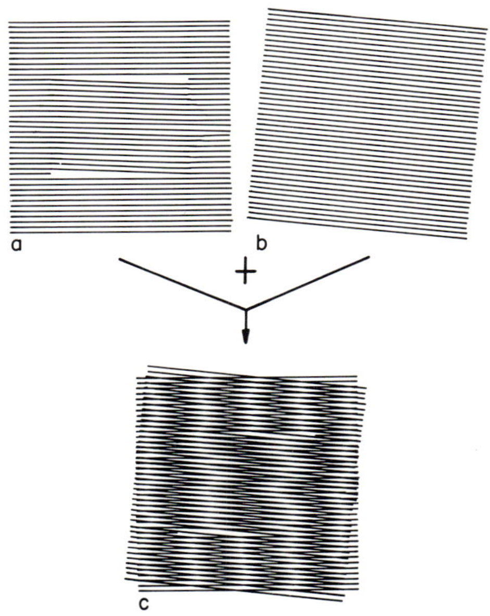

Fig. VII.5 *Schematic Drawing of Crystal Planes with a Pair Intralamellar Edge Dislocations (a) Which Reduces the Misfit Angle with a Second, Perfect Lamella (b).*
Drawing (c) illustrates the resulting moiré pattern. The structure can be further altered if the changed misfit angle is below the critical angle for screw dislocation network formation (see Fig. I. V.53)

(the other is assumed perfect) and the resulting moiré pattern. Typical changes in the sheared region between the coupled dislocations are a few tenths of one percent in lattice spacings, but a 40–50% decrease in the misfit angle between orientation of the two superimposed crystals. Such a large decrease in angle can reduce the misfit angle below the critical angle θ^* and lead to interlamellar screw dislocation networks (see Sect. 4.3.3, p. 482, Fig. IV.53). The conclusion from the observations is that low temperature annealing brings about local shearing between superimposed crystals along a direction which decreases the misfit angle θ. Niiomi and Takayanagi (1971) could justify the observed changes energetically and show, although the involved changes are small, irreversible effects on the dynamic tensile loss modulus. The intensity of the γ absorption at $-130°C$ increased by about 40%, while that of the α_c absorption decreased slightly after annealing at 85°C and 92°C, temperatures at which no thickening can occur.

7.1.5 Polymorphic Changes

A number of macromolecular crystals can exist in different polymorphs (see Sect. 2.4). At any one temperature only one of the polymorphs is stable. The metastable polymorphs can be changed frequently by proper annealing to the stable polymorph through a solid–solid transition. Reversible transitions between polymorphs occurring at different temperatures may contribute to overall annealing through the period of greater mobility that occurs during the transition. Although this increased mobility can lead to a reduction in defects, it is frequently observed that the geometrical constraints of the new phase break up the crystal grain or form multiple twins because of the stresses during the transformation. The nucleation and growth mechanisms of a new phase within a crystal have been discussed by Dehlinger (1958). The major driving force for the formation of the new phase derives from the lower free enthalpy of the more stable phase. Formation of the nucleus of the new phase and possibly also further growth causes, however, considerable positive strain energy (free enthalpy) which is divided among the host crystal, the new phase, and the interface. In case there exists a second, metastable crystal structure of intermediate free enthalpy which has a closer geometric relationship to the initial metastable crystal, i.e., it shows a lower strain energy on initial formation within the host crystal, then this intermediate crystal is formed first, and the final, stable crystal derives from it in a subsequent, additional phase change. This sequence of phase changes was already expressed by Ostwald's law of successive states which says that a phase will occur step-by-step through successively more stable polymorphs.

An example of such behavior is found in the crystallization and annealing of poly-1-butene. Holland and Miller (1964) have shown that poly-1-butene crystallizes at 58°C from amyl acetate after dissolution at 90–120°C to a 0.01–0.1 wt. % solution in form of a metastable orthorhombic crystal structure.† On heating the crystals for one hour in air at 90°C, the fold period

† The crystals are lath-shaped, folded chain single crystals; other crystallization conditions may give the other polymorphs directly.

increases as usual for folded chain lamellae, but at the same time the crystal structure changes to the tetragonal polymorph which is also metastable (see Table II.11, form II). The annealed crystals still retain single crystal character. On crystallization from the melt the tetragonal polymorph is obtained directly (Boor and Mitchell, 1963). On storing the tetragonal crystals for long periods of time (days) at room temperature, they change to the stable, trigonal crystal structure described in Table II.10 (form I). This change in crystal structure destroys the single crystal nature and leads to multiple twinning.

Another example of phase transition on annealing is observed on poly-*p*-xylylene. The stable crystal has the monoclinic α form (Iwamoto and Wunderlich, 1973). On heating above 220°C these α crystals change into the trigonal high-temperature β form (Niegisch, 1971) which is difficult to revert back to the α form by mere cooling. Lieser (1971) grew folded chain single crystals from α-chloronaphthalene of the α as well as β forms (0.05–0.1 wt. % polymer, 190–210°C; see also Fig. III.50). β-crystals grow first, particularly at the higher temperatures, but change, later into the α form without losing the β crystal habit (Fig. III.50). The density difference between the two crystal forms causes cracks in the crystal. Without presence of solvent this β–α transition has only been reported for poorly crystallized low temperature material crystallized and polymerized from the gas phase (Niegisch, 1971).†

A more drastic annealing effect is the topochemical or topotactic reaction of a monomer crystal to the macromolecular state. This change can be initiated by chemical reaction, radiation, or heating. A description of the process is given in Sect. 6.4.3. In most cases the changes during polymerization are so large that the initial crystal structure is lost. In some cases the strain induced during the polymorphic transition leads only to fibrillation along the molecular chain axis, and in very rare cases is there a conversion to the lower free enthalpy macromolecular crystal without serious effect on crystal perfection and size (see Sect. 3.9, Figs. III.142 and III.143). The role of defects in the crystals of reactive small molecules as initial reaction sites has been discussed by Thomas and Williams (1971).

The observation of the metastable monoclinic polymorph on annealing of polyethylene single crystals described in Sect. 7.1.4, which on first inspection violates thermodynamic principles, found its explanation in the fact that the polymorph was caused by deformation. Annealing caused interlamellar stresses to be relieved by deformation. On deformation, however, phase transformations are frequently observed as one of the mechanisms to reduce stress on single crystals (Sect. 4.4.3).

† The β crystals have been described by Niegisch (1971) as γ crystals, but could be shown later to be a mixture of β crystals and trimer. [Kubo and Wunderlich (1972). *Makromol. Chemie* **157**, 299].

7.1.6 Relief of Internal Stresses

The stress relief in other than macromolecular crystals, has been connected with the motion of grain boundaries and dislocations and recrystallization (Friedel, 1964). Macromolecular crystals should show similar effects. An indication is given by the study of interlamellar stress relief by dislocation formation using moiré patterns observed by electron microscopy (Sect. 7.1.4). A much more prominent effect in macromolecular crystals is caused, however, by internal stresses connected with changes in the macroconformation of amorphous macromolecules or the amorphous portion of a partially crystalline macromolecule. Such changes in macroconformation may be produced during solidification while the molecules are deformed by external forces, by nonuniform cooling, or through partial crystallization. Internal stresses of this type can also be introduced subsequent to solidification by macroscopic deformation (drawing, rolling, etc., see Sects. 4.4.3 and 4.4.4). Annealing is expected in such samples by heating to a temperature where sufficient molecular motion is possible. This motion is on a several chain atom scale and involves fully amorphous molecules or the amorphous parts of partially crystalline molecules which are involved in amorphous defects (Sect. 4.3.1) or in disordered interfaces (Sect. 4.3.2). The onset of sufficient molecular motion is usually indicated by the glass transition temperature. Often the glass transition temperature is observed between one-half and two-thirds the melting temperature in degrees Kelvin (Boyer, 1963). Annealing is thus possible much below the melting temperature.

Macroscopic stress relaxation in fully amorphous materials is usually described using the viscoelastic approximation (see, for example, Ferry, 1970). Partially crystalline materials behave in a more complicated manner. Relaxation processes are in this case affected by crystallinity, crystal size distribution, phase boundary structure, crystal orientation, crystallite correlation by connecting molecules, type and concentration of crystal defects and conformation of the amorphous portions of molecules (Illers, 1969). The simplest model of viscoelastic relaxation in amorphous materials is that suggested by Maxwell.† It consists of a spring and a dashpot coupled in series. The relaxation of a stress s at constant deformation is expressed by the exponential equation

$$s = s_0 \exp(-t/\tau) \tag{15}$$

where τ is the relaxation time determined by the ratio of the viscosity of the dashpot and the elastic modulus of the spring. Real amorphous materials show a more complicated relaxation behavior which can only be approximated by means of a series of exponential functions which need then a relaxation time spectrum for their description (Ferry, 1970).

† Maxwell (1866, 1867). *Phil. Trans. Roy. Soc. London* **156**, 249; **157**, 52.

Figure VII.6 illustrates schematically the change of the thermodynamic functions enthalpy H, entropy S, and volume V and the derivative functions heat capacity C_p, expansion coefficient α, and compressibility β of a glass on cooling and heating at different rates through the glass transition region. On

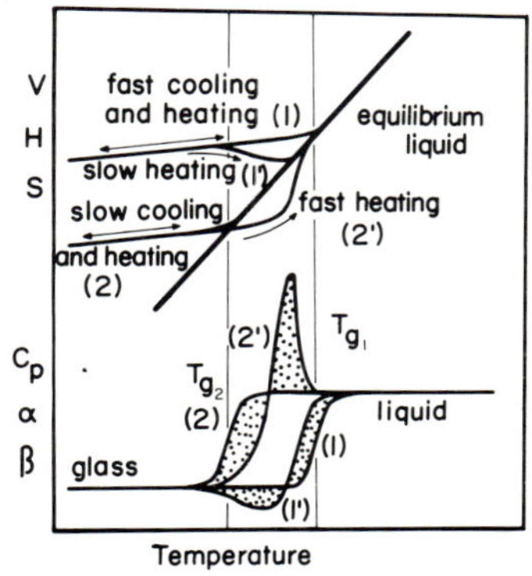

Fig. VII.6 *Schematic Changes in Enthalpy H, Entropy S, Volume V, and the Derivative Properties Heat Capacity* C_p, *Expansion Coefficient* α, *and Compressibility* β *on Heating and Cooling.*

Paths 1: fast cooling and heating; paths 2: slow cooling and heating; path 1′: fast cooling followed by slow cooling; path 2′: slow cooling followed by fast heating.

fast cooling, higher values of H, S, and V are frozen-in (path 1) which on slow heating or isothermal treatment show an annealing effect (path 1′) which moves the sample to the more stable state, otherwise reached on slow cooling (path 2). Analysis of the glassy state is possible by heating quickly and following one of the derivative functions, most conveniently, the heat capacity. A more stable, slowly cooled or annealed glass follows on fast heating the path 2′, while a fast cooled, quenched glass follows on fast heating the path 1. The apparent heat capacities are shown at the bottom of Fig. VII.6. Such analysis curves measured by differential thermal analysis are shown in Fig. VII.7 for polystyrene glasses which were produced with increasing internal stresses by cooling under increasing hydrostatic pressures (Weitz and Wunderlich, 1974). Similar behavior is observed on heating of drawn glassy polymeric samples (Tautz *et al.*, 1964; Grewer and Wilski, 1968; Park and Uhlmann, 1973). Figure VII.8 shows that on annealing at 70°C, the lower limit of the glass

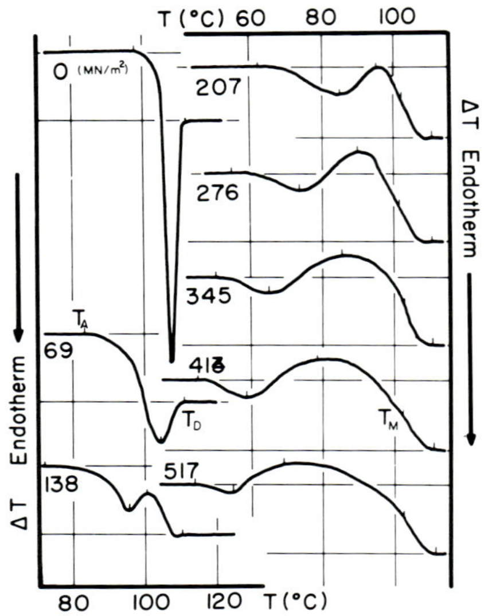

Fig. VII.7 *Differential Thermal Analysis Curves of Glasses of Polystyrene Produced by Slow Cooling (5°C/hr) at Various Pressures.* Pressures given in MN/m² (1 MN/m² = 0.01 kb). Analysis after release of pressure, indicating increasing amounts of internal stress. Heating rate 5°C/min. ΔT is proportional to heat capacity (Weitz and Wunderlich, 1974).

transition region, the more stable state achieved by slow cooling at atmospheric pressure is reached finally. The enthalpy and volume relaxations were exponential, and could not be described by the same relaxation times.

The relaxation behavior of semicrystalline samples is mainly complicated by the effect of the crystalline parts of a molecule on its noncrystalline parts (see Sect. 4.2). Surface defects (Sect. 4.3.2) are frequently the location of internal stresses caused by the continuation of macromolecules from crystalline to amorphous regions. Depending upon the detailed structure, the mobility of sections of molecules adjacent to crystals is more or less restricted (see Sect. 4.1.6). A gross assessment can be obtained through the measurement of the glass transition temperature as a function of crystallinity (Illers, 1969). Partial crystallization increases the glass transition temperature of polystyrene. Poly(ethylene terephthalate) and polypropylene show a maximum in glass transition temperature at intermediate crystallinity. Practically no change in glass transition temperature with crystallinity was observed for poly(propylene oxide).

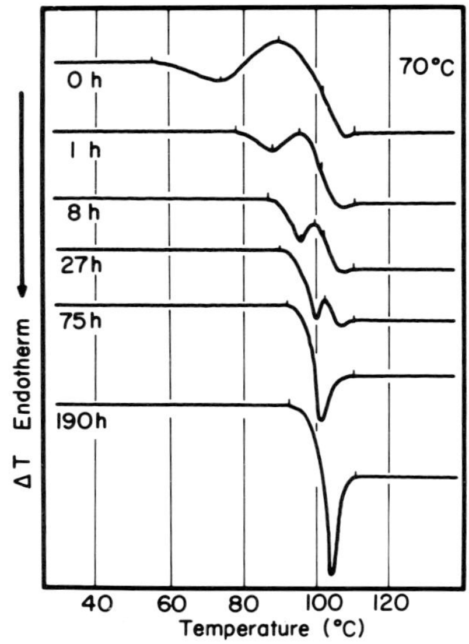

Fig. VII.8 *Differential Thermal Analysis of Polystyrene Glass Produced by Cooling at 417 MN/m² Pressure Annealed for Different Times at 70°C.*
The annealing changes the sample to the more stable state otherwise produced on slow cooling at atmospheric pressure (see Fig. VII.7) (Weitz and Wunderlich, 1974).

Crystallinity alone is not sufficient to characterize the interaction between crystals and amorphous area. Poly(ethylene terephthalate–sebacate) copolymers of 80/20 and 60/40 mole ratio show, for example, glass transitions at 22.8 and 5°C when crystallized at low temperatures to about 0.15 and 0.08 crystallinity, respectively (Wunderlich, 1957). On slow cooling from the melt, the glass transition temperatures are -0.2 and -11°C at crystallinities of about 0.23 and 0.13, respectively. The decrease in glass transition temparature observed is more than one would expect due to the increased removal of crystallizable (high glass transition) ethylene terephthalate sequences from the left-over amorphous phase. One may conclude that due to the slower crystallization, sufficient time was available to produce interfaces between crystals and amorphous areas relatively free of internal stresses.

7.1.7 Transport of Matter

The discussion of annealing in Sects. 7.1.2–7.1.6 is based on the assumption that there is no transport of matter across subgrain or grain boundaries.

There is ample evidence, however, that such transport is possible and must be taken into account in the description of annealing.

The change in size of mosaic blocks in mats of solution grown polyethylene crystals was followed by Čačković *et al.* (1969) as a function of annealing time. On this subgrain level an increase of volume is observed. Figure VII.9 illustrates the changes in the mosaic block dimensions in the [001], [100], and[110]

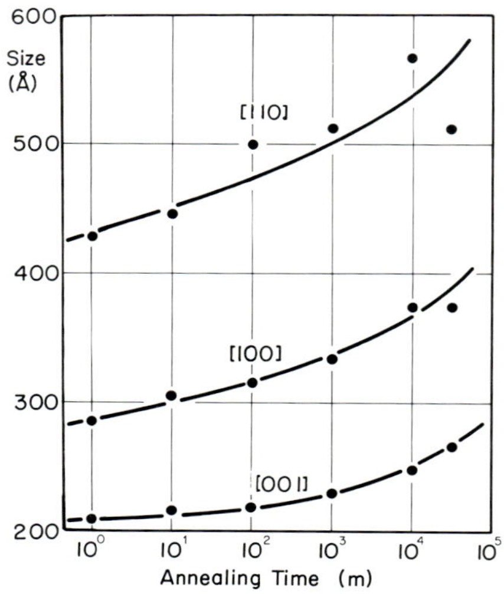

Fig. VII.9 *Increase in Mosaic Block Dimensions as a Function of Annealing Time.*
Linear polyethylene of $M_w \sim 10^5$ crystallized at 85°C from 0.01 wt. % solution in xylene, annealed at 125°C for the given time. Initial mosaic block dimension in direction [001]: 125Å; [110]: 440Å; [100]: 310Å. Data by Čačković *et al.* (1969).

directions. The size in the [001] direction represents the fold-length measured by low-angle X-ray diffraction and corresponds to the lamellar thickness consisting of crystal and surface layer (Sect. 3.2.2.4). The other two dimensions were derived from the analysis of integral breadth of successive orders of the appropriate wide-angle X-ray reflections (see Sect. 4.2.2). A new shape of much larger fold-length and slightly smaller lateral dimensions is quickly established (~1 min). The subsequent mosaic block growth is slower and occurs with about equal speed in all dimensions, so that the shape does not change much further. Čačković *et al.* (1969) suggest that these mosaic dimensions, rather than the grain (lamellar) sizes, may actually reach the equilibrium shape. To justify the experimental shape, a high defect (kink) concentration is proposed to exist in the subgrain boundaries to account for the large lateral dimension (relative to the fold-length, compare with Sect. 7.1.2, Fig. VII.1). These experimental data indicate clearly a mosaic block growth to about 3.5 times the initial volume, matter transport must have occurred.

Across the fold surfaces such matter transport is connected with the fold extension mechanism (Sect. 7.1.3). The hole created in a crystal on fold extension can enable a fold or a cilium of neighboring lamellae to move across the fold surface and thus join separate grains. Such motion was proven by Bank and Krimm (1969) by annealing physical mixtures of separately crystallized polyethylene and perdeuteropolyethylene. The interpenetration was quantitatively followed by measurement of the splitting of the bending and rocking modes of the CH_2 and CD_2 by infrared absorption (see Sect. 3.2.2.3). Initial small scale interpenetration was already possible during the initial filtering of the crystals. 6–12.5% of foreign components could be detected by the decrease in splitting of the major component frequency. Solvent still present during filtering was thought to enhance this initial interpenetration which is most likely caused by cilia movement (see Sect. 3.2.2.4). Annealing at 110°C reverses this process, but as soon as fold extension is observed (on annealing at 115–120°C), the splitting is again decreased, to reach finally (on annealing at 120–125°C) the value of cocrystallized polyethylene and perdeuteropolyethylene.

Matter transport across subgrain boundaries seems to be possible to about the same extent as transport across fold surfaces (see Fig. VII.9). It may involve only the diffusion of the defects making up the subgrain boundary (see Sect. 4.3.2), a process which should involve less actual matter transport than the just discussed interpenetration across the fold surfaces.

The growth of grains can occur either by joining of grains or by growth at the expense of other grains which must shrink and finally disappear in order to show the observed increase in average size. The grain growth is one of the key processes of sintering (Kuczynski, 1973). The other process of sintering is pore volume shrinkage, a process which is intimately connected with the grain growth. The basic theory of particle coarsening and increasing of pore sizes, accompanied by pore number reduction and final pore volume decrease, is based on Ostwald ripening (Fischmeister and Grimvall, 1973).

A different type of grain growth in macromolecular crystals involves chemical reaction of the backbone chain. In crystals of chemically reactive chain molecules like selenium, polyester, or polyamids, folds may open up or chain ends may react with small molecules and bridge between neighboring grains by chemical reaction (see Sect. 7.4).

7.1.8 Recrystallization

Not all annealing processes involve only motion of defects or surface diffusion. Partial melting followed by recrystallization is also frequently observed as can be seen from the reviews of Stuart (1955), Geil (1963), Zachmann (1964), Mandelkern (1964) and Fischer (1969). Figure VII.10 illustrates the

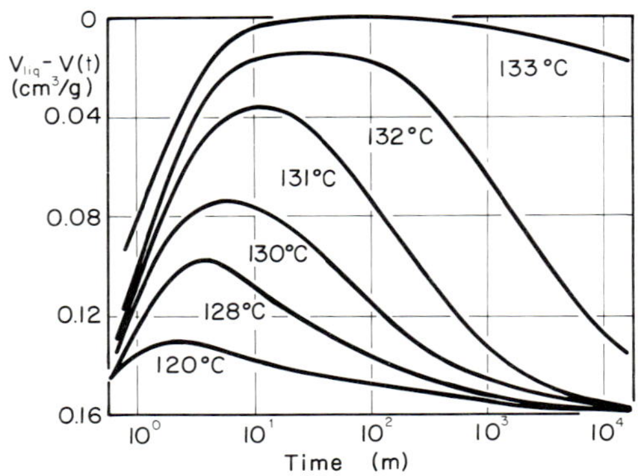

Fig. VII.10 *Change in Specific Volume of Melt Grown Crystals of Linear Polyethylene on Annealing at Various Temperatures.*
Initial crystallization by quenching from 170°C to 25°C. Drawn after unpublished data by M. Gubler, J. Rabesiaka, and A. Kovacs (quoted by Geil, 1963, Fig. V-48, p. 363).

changes in density of a melt crystallized linear polyethylene quenched from 170°C to 25°C on annealing at the given temperatures. Figure VII.11 shows similar data of the annealing of solution-grown crystals of linear polyethylene. In both cases increasing melting occurs above 120°C (see also Chapter IX) before a decrease in specific volume indicates recrystallization. Depending on crystal perfection and initial fold-length, the samples annealed at 132–135°C reach almost amorphous density, leaving only self-nuclei (Sect. 5.1.4.2). Depending upon the molecular weight distribution, the upper limit of remaining self-nuclei may be close to the equilibrium melting temperature (141.4°C). Figure V.27 illustrates a sample of polyethylene which retained self-nuclei at temperatures up to 138.5°C.

Recrystallization on crystal surfaces is also illustrated by epitaxy of the molten surface material on fracture surfaces of polyethylene and polytetrafluoroethylene described in Sect. 3.4.1 (Figs. III.74 to III.77). The observed fold-length of the recrystallized lamellae correspond to the melt crystallized

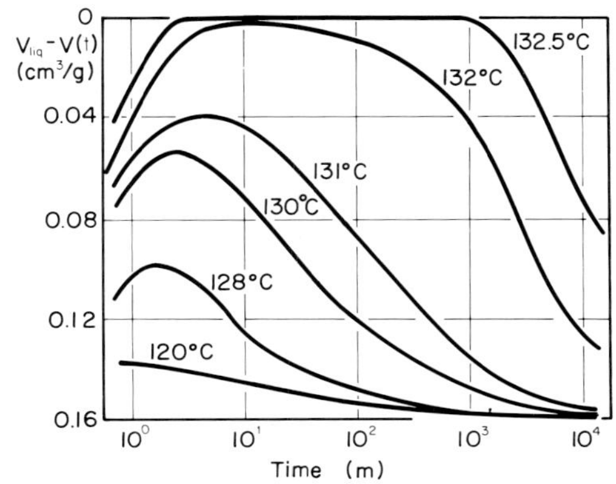

Fig. VII.11 *Change in Specific Volume of Solution Grown Crystals of Linear Polyethylene on Annealing at Various Temperatures.*
Initial crystallization at 80°C from xylene from dilute solution. Drawn after unpublished data by E. W. Fischer and A. Kajiura (quoted by Fischer, 1969, Abb. 12, p. 473).

samples, so that recrystallization can be looked upon as normal crystallization, modified by the ready availability of crystal surfaces as primary nuclei. Recrystallization is thus expected to be governed entirely by the remaining active crystal surfaces and the usual molecular nucleation (Sect. 5.3.4).

Recrystallization follows thus the same sequence of initial crystal growth (Sect. 6.1.4) followed by secondary crystallization and crystal perfection described in Sect. 6.1.6. On heating above the recrystallization temperature, further annealing of the newly grown crystals has been observed by Wunderlich and Melillo (1968). Figure VII.12A shows the polyethylene fracture surface after recrystallization at 130°C. Initial perfection can be observed on annealing at 132°C. Figure III.75 illustrates such a sample after 16 hr of annealing. An increase in fold-length on annealing at 134.5°C coupled with a break-up of the lamellae can be seen in Fig. VII.12B. Final smoothing of the lamellae on the underlying extended chain crystals occurs on further annealing at 135°C (Fig. VII.12C). Figure VII.12D shows finally that, although complete smoothing seems to be achieved on annealing at 135°C, the new surface layer melts at a lower temperature than the extended chain crystals. Annealing did not reach full chain extension.

7.1.9 General Summary of Annealing Processes

The general descriptions of the various processes observed on annealing in Sects. 7.1.2 to 7.1.8 represent an effort at systematizing the many effects which can occur simultaneously on annealing macromolecular samples. Typically a portion of a sample may melt on annealing, while other portions change crystal habit or increase their grain size mainly through changes in fold-length. The latter process starts during heating to the annealing temperature. The defect concentration changes simultaneously with the changes in shape and macroconformation and internal stresses are removed. In time, recrystallization of initially molten material takes place, frequently complicated by molecular weight segregation and nucleation effects. Changes of crystal structure and chemical reactions may also occur under proper conditions.

As long as the various processes cannot be followed simultaneously, interpretations of annealing remain uncertain. The major observations are that there seems to be no continuous path to a decrease of the fibrillar crystals of small diameter to the equilibrium shape as suggested by Fig. VII.1. Melting, followed by folded chain recrystallization is observed, which again is a metastable structure susceptable to annealing (Sect. 7.1.2, and see Fig. VII.12). The lamellar folded chain crystals, in contrast, are observed to increase their fold-length either continuously or in steps. The kinetics observed can most likely be brought into agreement with simple irrervesible thermodynamics of approach to equilibrium size (Sect. 7.1.2) or various nucleation controlled mechanisms (Sect. 7.1.3). The thermodynamic approach suffers from the difficulty of definition of the coherent crystal volume involved in the change of shape. Breaking up of crystals and fusion of grains or subgrains through sintering are difficult to assess (Sect. 7.1.7) and should affect the description drastically [Eqs. (3) to (5), (7) and (9)]. The proposed miscroscopic mechanisms of defect or fold dislocation climb (Sect. 7.1.3) governed by nucleation [Eq. (13)] are still not fully clarified. Ultimately both approaches should be able to describe the major annealing effect of folded chain macromolecular crystals.

The study of changes in concentration of certain defects on annealing and the internal stress relief are closely coupled, but only a beginning has been made to sort out the various effects (Sects. 7.1.4 and 7.1.6). Overall, changes in defect concentration and internal stress level are usually of lower importance than crystal shape and fold-length increases as long as normal solution or melt-grown crystals are concerned. Special effects on annealing of deformed crystals are discussed in Sects. 7.2.3 and 7.3.2.

The question of transport of matter (Sect. 7.1.7) is not only of importance for the increase in ultimate fold-length obtainable for a given mass of polymer, but it also leads to the possible understanding of the process of sintering. As crystals of macromolecules improve or higher melting polymers are developed which cannot be molten without decomposition or irreversible loss of crystal structure, sintering should become more important as a technological tool to improved polymeric materials. Little detailed knowledge of this topic has reached the scientific literature.

Finally, phase transitions (Sect. 7.1.5) and chemical reactions (Sect. 7.4) can further complicate the annealing behavior. In the following sections the experimental evidence on annealing will be reviewed in the light of this general description of annealing effects.

7.2 Crystals from Solution

Most detailed information on annealing could be gained from folded chain single crystals grown from solution. In this case the crystals can be well defined (Sect. 3.3.2) and have been analyzed in detail by electron diffraction and microscopy and X-ray diffraction (see Sect. 4.1.2). Of all polymers analyzed, polyethylene received as usual first and most detailed attention. Keller and O'Connor (1958) observed first that on annealing near the melting point the usual 120 Å spacing due to the original chain folding on crystallization nearly disappeared in the low angle X-ray diffraction diagram and that at the same time another spacing of 200–300 Å appeared. The wide angle diffraction diagram changed also, indicating a randomization about the crystallographic *b*-axis. This first annealing experiment may have been an example of considerable recrystallization (see Sect. 7.1.8). Next, Rånby and Brumberger (1960) measured the increase in X-ray low-angle spacing over a wider temperature range. The first more detailed investigations were by Statton and Geil (1960), Statton (1961), and Fischer and Schmidt (1962) who discovered the breaking up of single crystals on annealing (as illustrated in Fig. VII.13) and

Fig. VII.12 *Annealing of Polyethylene Recrystallized on the Surface of Extended Chain Lamellae.*

Electron micrographs of annealed fracture surfaces (see also Sect. 3.4.1). A. Fracture surface as in Fig. III.32 annealed at 130°C for 5 min which leads to recrystallization to folded chain lamellae of 180–225 Å fold-length (see Fig. III.13). Figure III.75 shows a sample as in A after annealing for 16 hr at 132°C; one can see little change in fold-length, but some increase in crystal perfection. B: Sample as shown in Fig. III.75 after a second annealing for 18 hr at 134.5°C. The lamellae have thickened substantially and show signs of breaking up. The irregular structures have a thickness between 270 and 450Å. C: Sample as shown in C, annealed for a third time for 20 hr at 135°C. Further thickening has lead to an almost smooth surface. D: Sample as shown in C, heated for 1 hr to 138°C and quickly cooled, reveals that the annealed material still melts below the extended chain lamellae. The new folded chain lamellae recrystallized on fast cooling. The base extended chain crystals are not visibly affected by the heat treatment. Electron micrographs by Wunderlich and Melillo (1968).

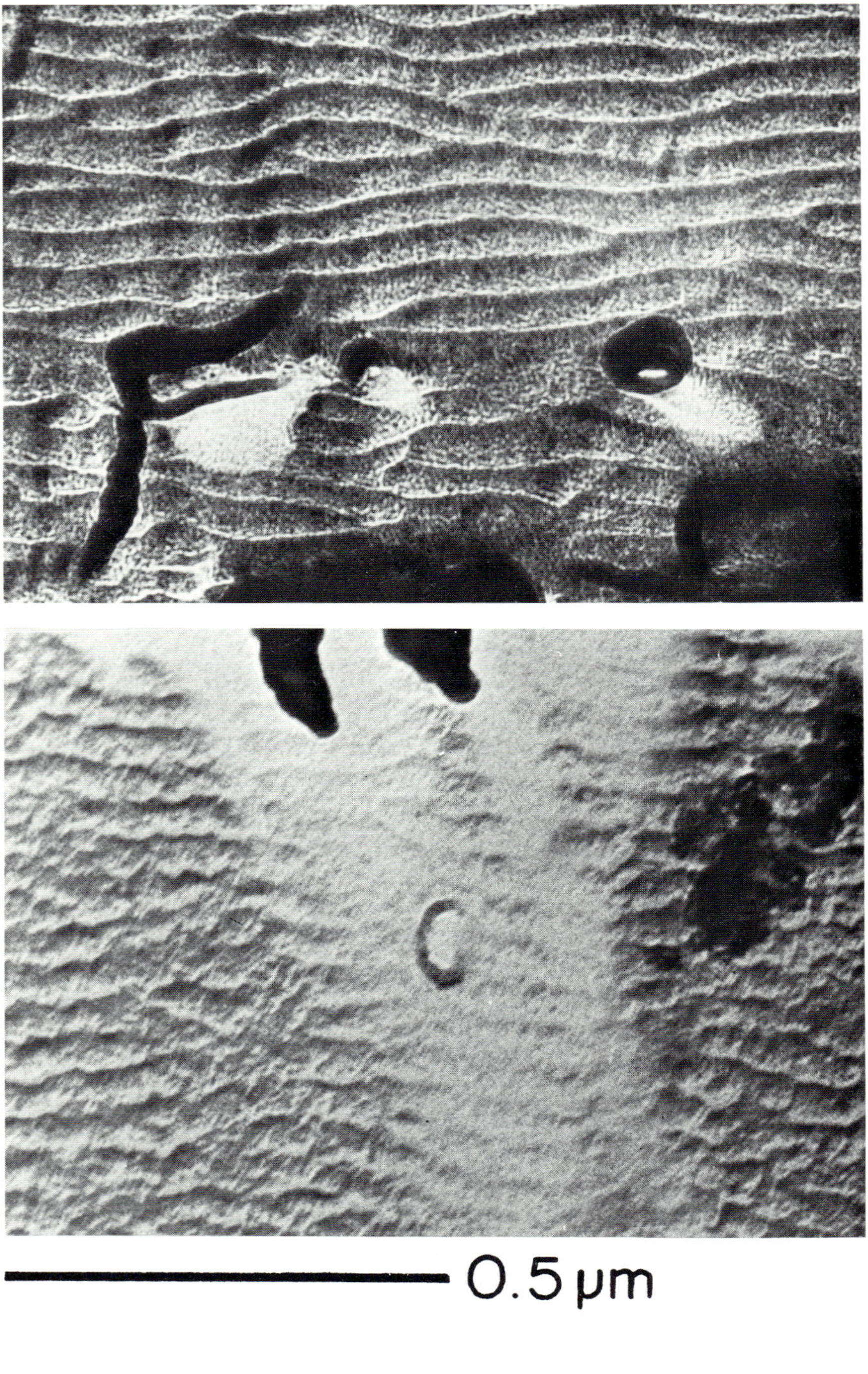
0.5 μm

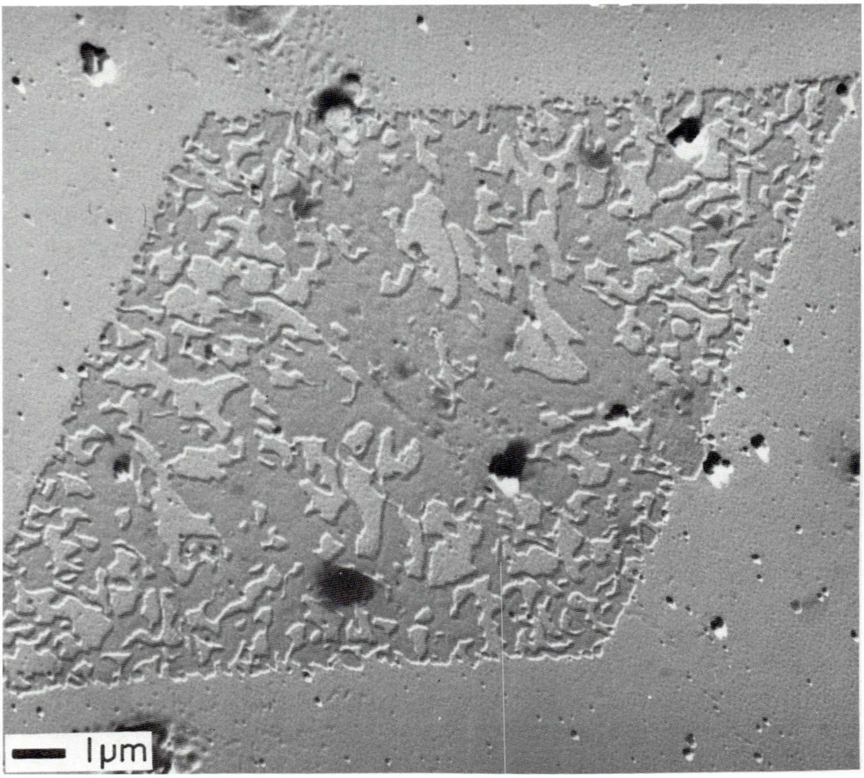

Fig. VII.13 *Electron Micrograph of an Annealed Single Crystal of Polyethylene.* Annealing time 35 min at 125.7°C. Original crystal grown from *p*-xylene at 85°C. Scale bar 1 μm. Figure by Statton and Geil (1960).

studied the kinetics of the annealing. A comprehensive review of the early investigations has been published by Geil (1963, Chapter V).

In this section, the annealing of polyethylene and other single crystals is described. Experimentally the annealing can be separated into at least three different temperature regions:†

1. Changes in defect concentration and stress relief without chain fold increase (low temperature region of annealing, see also Sects. 7.1.4 and 7.1.6);

† Takayanagi and Nagatoshi (1965) propose a region I and II corresponding to the recrystallization and the chain fold increase without melting respectively; Nagai and Kajikawa (1968) distinguish five stages of annealing. Stage 1 and stage 5 correspond to the regions 1 and 3 described below, while stages 2–4 illustrate various effects seen in the intermediate annealing temperature region.

2. Chain fold increase and changes in crystal size without melting (intermediate temperature region of annealing, see also Sects. 7.1.2 and 7.1.3);

3. Recrystallization (high temperature region of annealing, see also Sect. 7.1.8).

This discussion is followed in Sect. 7.2.2 by the description of the effects of liquid on the annealing. Finally, in Sect. 7.2.3 it is shown that solution grown shish-kebobs (Sect. 3.8.2) follow a different annealing path than solution-grown single crystals.

7.2.1 Dry Crystal Annealing

The annealing of dried single crystals grown from solution is discussed first on the example of polyethylene which has been investigated most extensively. The knowledge about other crystals is summarized at the end of this Section (Table VII.2).

Polyethylene single crystals heated to the low temperature annealing region show only little outward changes (Nagai and Kajikawa, 1968). The unit cell dimensions were shown by Davis *et al.* (1968) to change slightly (linearly) with annealing temperature. The *a*-axis length increases, depending upon initial crystallization conditions (up to 0.5%), the *b*-axis length decreases by about the same percentage, while the *c*-axis length remains constant. Iohara *et al.* (1973) confirmed these measurements and interpreted them as being caused by an increasing amount of chain disorder. Another effect observed on some samples by Baltá Calleja *et al.* (1963) is a small decrease (up to 15 Å) in X-ray long period of polyethylene single crystals grown at 90°C from 0.1 wt. % solution in xylene after annealing at temperatures of 95–122°C. These crystals grew as hollow pyramids (see Sect. 3.3.2) and when sedimented, settled with various degrees of tilt of the molecular chain axis relative to the substrate. On annealing, it was observed that the tilt of the molecules is increased for samples initially settled with little tilt, thus causing the decrease in X-ray long period. Samples settled originally with more tilt, decreased less in long period. Besides such changes, which are caused by stress relief within the lamellae, effects due to stress relief between lamellae have been observed. Holland (1964a) discovered an increase in moiré pattern distortion and an increase in terminating moiré fringes (Sect. 4.3.3) between pairs of crystal lamellae. Niinomi and Takayanagi (1971) linked these changes with the stress relief between lamellae through formation of coupled pairs of edge dislocations (Fig. VII.5 and Sect. 7.1.4).

The region of annealing at intermediate temperatures involves chain fold increases without melting. Best understood is the annealing of isolated single crystal lamellae floating in a nonsolvent so that substrate and multilayer effects (see below) are eliminated. The formation of thicker portions of the crystal should cause a shrinkage of the overall dimensions, instead, the laterally large lamellae develop holes (see Fig. VII.13). The formation of holes on isothermal annealing without support was studied by Roe *et al* (1973) (see also Geil, 1963, Fig. V.11). Initially small holes of about 300 Å diameter with raised edges develop. The raised edge areas are about double as thick as the initial lamella. Most of the holes are aligned in groups along {310} planes, the same planes which are involved in the collapse of the pyramidal structures (see Figs. III.56 and III.57) and twinning (see Sect. 3.5). As annealing time increases, the initial holes elongate normal to the {110} fold planes (see Fig. III.52) as can be seen from Fig. VII.14 and to some degree also in Fig. VII.13.

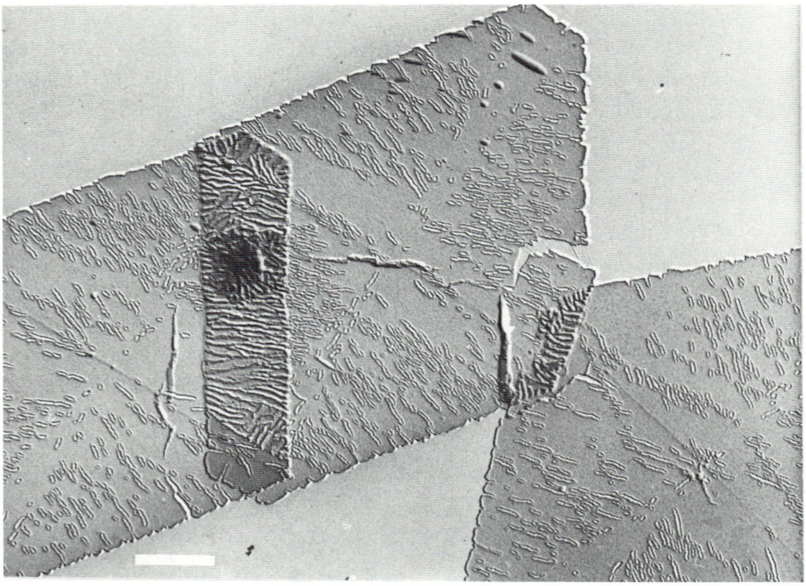

Fig. VII.14 *Electron Micrograph of an Annealed Single Crystal of Polyethylene.* Annealing time 4 days at 118.7°C (suspended in silicone oil). Original crystal grown from xylene at 85°C. Lamellar thickness about 120 Å, thickened areas 240–280 Å, width of holes about 300 Å, width of thickened area about 90 Å. Scale bar 1 μm. Figure by Roe *et al.* (1973).

Their numbers remain largely constant with time at the value established during the first few hours of annealing. The number of holes as a function of annealing temperature is shown in Fig. VII.15 for the chosen crystal preparation. The curve in Fig. VII.15 permits a delineation of the three temperature regions of annealing. Below 118°C is the low temperature annealing region with no fold length change; between 118 and 123°C is the intermediate temperature region of annealing, showing the development of holes as presently

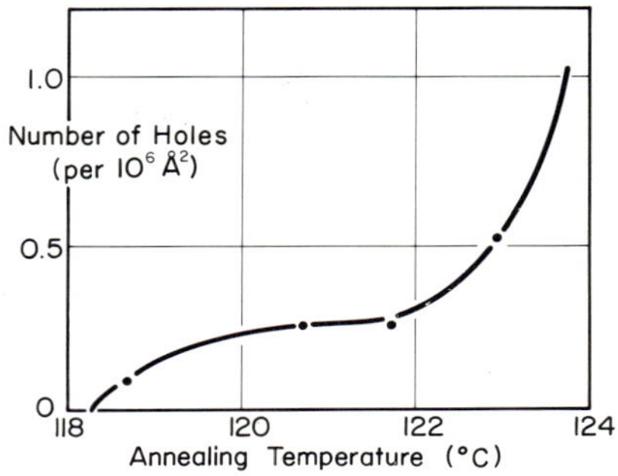

Fig. VII.15 *Number of Holes Developed on Annealing of Polyethylene Single Crystal as a Function of Temperature.*
For crystals see Fig. VII.14; broad molecular weight sample of M_n about 10^4 crystallized from 0.01 to 0.05 wt. % solution xylene at 85°C. Holes counted on samples annealed for 22 hr, but little change in number occurs after the first few hours. (Crystals grown at higher temperature show fewer holes on annealing). Roe *et al.* (1973).

discussed; and, finally, above 123°C the number of holes increases to very large values, fusion of crystals is observed along with severe distortions, an indication of the high temperature annealing region which is characterized by increasing recrystallization.

For samples of different fold-length and crystal perfection the different regions of annealing vary in temperature. Some changes, particularly in number of holes were also observed in seemingly identical samples. The trends are, however, always similar as shown in Fig. VII.15. Table VII.1 contains a collection of experimental data on the temperature of initial thickening for different crystals.† All experimental data on crystals annealed above this temperature are dependent on the time in which the annealing temperature is reached since on slow heating some annealing to larger fold-length may occur during the heating.

The end of the intermediate temperature region for annealing, which is indicated by the large increase in the number of holes in Fig. VII.15, and also by an initial density decreases on annealing (see Fig. VII.11), is given approximately by the melting temperature of the initial lamellae. It was shown by

† Note that Mandelkern *et al.* (1969) identified this temperature with the melting point of the lamellae, which actually occurs, however, several degrees higher as discussed next (see also Fig. VII. 15).

TABLE VII.1

TEMPERATURE OF INITIAL THICKENING ON ANNEALING OF SINGLE CRYSTALS[a,b]

Crystallization temperature (°C)	Initial lamellar thickness (Å)	Temperature of thickening (°C)
75	115 ± 1	114.0 ± 2.0
85[c]	128 ± 2	119.5 ± 1.0
88	145 ± 1	122.5 ± 0.5
90	154 ± 1	124.2 ± 0.5
92.5	172 ± 1	126.8 ± 0.5

[a] Data by Mandelkern *et al.* (1969) on a fraction of 365 000 molecular weight crystallized from 0.08 wt. % solution in xylene.

[b] Annealing time 48 hr.

[c] A lower molecular weight fraction (46 500) showed a thickness of 129.2 ± 2Å and an initial temperature of thickening of 119.5 ± 1°C.

Wunderlich *et al.* (1963) through optical microscopy (see Fig. IV.13) that this melting point can be determined by fast heating which avoids prior annealing (see also Chapter IX). Typical crystal lamellae (growth spirals) of 130 to 150 Å fold-length melt at a temperature of 117.4–122.8°C when heated faster than about 25°C/min. On slower heating, the melting temperature increased in these experiments to about 128°C because of annealing during heating (Fig. IV.13). For the crystals shown in Fig VII.14 a melting temperature of about 125°C was estimated by Roe *et al.* (1973) through comparison with melting temperatures of radiation cross-linked lamellae whose annealing is inhibited.

The upper region of annealing starts thus close to the melting point of the original crystals on fast heating and represents a region of increasingly predominant recrystallization (Sect. 7.1.8). Figure VII.11 shows the degree of melting before recrystallization. On sufficiently slow heating, or multiple step annealing at increasing temperatures, this upper region of annealing can be extended for polyethylene to temperatures above 135°C (as is illustrated by Fig. VII.17). Experimental data in this temperature region are extremely dependent upon the rate of initial heating. On very fast heating, no annealing, but only melting may be observed; maximum annealing is reached by slow heating. Note that a recrystallized portion of the sample may anneal faster through crystal perfection immediately after crystallization (see Sect. 6.1.6), and various heating rates may cause recrystallization at different temperatures and initial degrees of perfection. All these effects may easily offer explanations for often apparently contradictory experimental data.

The mechanisms of annealing in the intermediate temperature region where hole formation predominates can be described in more detail. Geil (1963, p. 325) suggested that the holes are started at defect sites. It has been shown previously that these defects are concentrated in certain {310} planes which are also involved in the collaps of pyramidal crystals (see also Nagai and Kajikawa, 1968). Removal of the pyramidal shape by application of high hydrostatic pressure (see Sect. 7.3.3) also removes the alignment of initial holes (Roe *et al.*, 1973). Most likely certain {310} planes are already strained in pyramidal crystal suspensions and serve on sedimentation as initial slip planes (see Fig. III.57) and on annealing as preferred sites for initiation of thickening. Once a hole is created, molecules can thicken along the original fold planes {110} by one of the mechanisms discussed in Sect. 7.1.3 (Fig. VII.3). Figure VII.16 shows schematically the thickening by pulling away from the

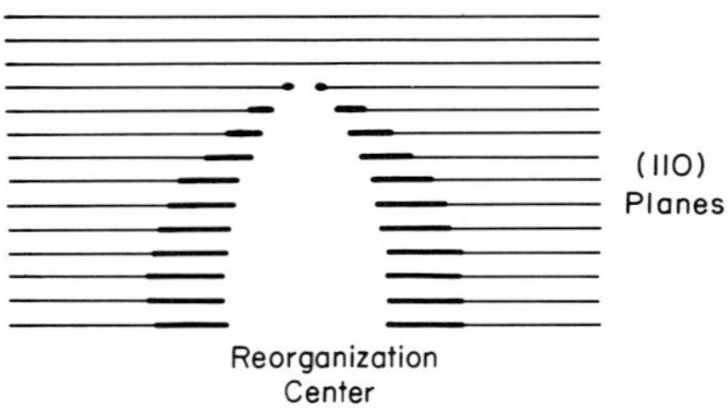

Fig. VII.16 *Schematic Drawing of the Thickening by Transport along the (110) Fold Planes.* Projection along *c*. Thick lines indicate approximately double the fold-length. Width of thickened region about 90 Å which corresponds to a region invaded by only one molecules. Drawn after Roe *et al.* (1973).

initial hole along fold planes which leads on prolonged annealing to formation of elongated holes (Fig. VII.14). A similar mechanism of formation of thickened fronts which move towards the undisturbed thinner interior from a surface was earlier suggested by Schultz (1970) as an explanation of "picture frame" morphology. The stopping at the given thickened border width may find its explanation in the length of the molecule. It is shown in the discussion of self-nucleation in Sect. 5.1.4.1 that single molecules seems to anneal to higher perfection than aggregates of multiple molecules. The new fold-length of double the initial fold-length may find its explanation in the mechanism favoring doubling (Fig. VII.3). Throughout this process of hole formation and thickening, complete crystallographic register is kept by all material of the crystal (Statton and Geil, 1960).

Changing the physical conditions of annealing alters the observed features and may cause variations to the above mechanism. In Fig. VII.13, for example, which shows a crystal which was annealed flattened on a substrate, one can see none of the thickened rims around the holes, the remaining crystal portions seem to have thickened uniformly. Also, some of the regularity and orientation of the holes is lost. It was suggested that thickening is enhanced by contact with some substrates (see Garber and Geil, 1968). A slight acceleration of the annealing processes may permit rearrangement beyond one molecule and remove the inhibition to thickening beyond fold period doubling and account thus for the difference between Figs. VII.13 and VII.14.

On annealing, multilayer crystals (Sect. 3.3.3) behave somewhat differently from monolayers as long as intimate contact is kept between the lamellae. Figure VII.14 shows that in the area of overlap of the lamellae, thickening progressed further. Raised ridges of directions dictated by the fold planes can be seen. In multilayer crystals intermeshing of molecules from different layers can avoid the formation of holes and gain the surface free energy for faster annealing (sintering, see Sect. 7.1.7). Statton (1967) produced coherent macroscopic films of carefully stacked aggregates of single crystal lamellae sedimented on filtering by annealing between 120 and 137°C. After annealing these films were ductile and malleable and represented one single mosaic crystal of unique *c*-axis orientation normal to the film surface.†

Measurement of the Raman-active longitudinal acoustical vibrational mode of polyethylene revealed that on annealing there is not only a fold-length increase measured by a frequency shift, but also a decrease in the uniformity of fold-length, indicated by a 2.5-fold broadening of the line width (Koenig and Tabb, 1974). Melt crystallized polyethylene has initially already a broader distribution which narrows somewhat on annealing.

Unannealed single crystal mats show brittle fracture at low strain due to interlamellar slip. There is little resemblance of such mats to melt crystallized polyethylene films. Blackadder and Lewell (1970) found on crystals mats grown at 75°C from 1 wt % solution in *p*-xylene that density and tensile modulus increased on annealing up to 117–120°C. The swelling in *p*-xylene at 70°C decreased after annealing up to an annealing temperature of 117°C. At higher temperatures, these trends reversed because of increasing recrystallization instead of interpenetration of adjacent lamellae. The analyzed samples melted completely at 129°C under the chosen conditions.

Changes in the unit cell dimensions after annealing in the intermediate temperature range are much smaller than the observed density changes.

† Note that drawn or rolled films (Sect. 4.4.3 and 4.4.4) as well as films crystallized by epitaxy or transcrystallization (Sect. 3.4) have the opposite orientation (*c*-axis in the plane of the film).

Iohara *et al.* (1973) observed a maximum in unit cell volume after heating for 24 hr close to the beginning of the intermediate annealing region. Beyond this temperature (~110°C for 80°C crystallized polyethylene from 0.05 wt % solution in xylene), the unit cell volume decreased because of a decrease in *a*-axis dimension. The *b*-axis length was observed to keep on decreasing continuously on annealing at higher temperature without change in slope as it did in the low temperature annealing region (see previous text).

Analysis of the fold-lengths of single crystal mats as a function of annealing temperature after a long time yields data as illustrated in Fig. VII.17.

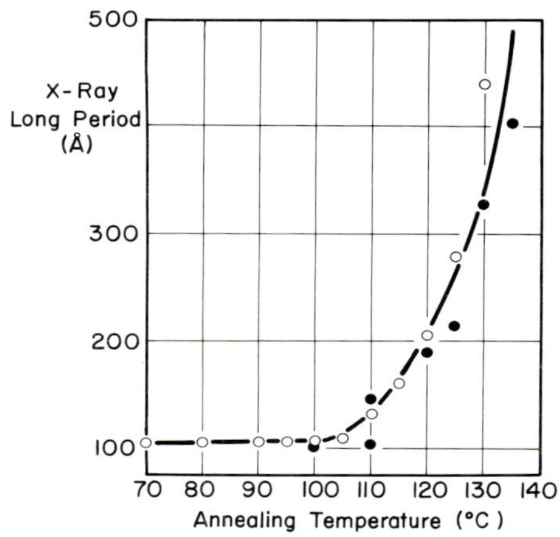

Fig. VII.17 *Dependence of the X-ray Long Period of Mats of Polyethylene Single Crystals after Long-Time Annealing at the Given Temperature*
Open circles: single crystals grown at 70°C from tetrachloroethylene, annealed for 24 hr (Fischer and Schmidt, 1962). Filled circles: single crystals grown on cooling tetrachloroethylene solution, annealed for 15 hr (Statton and Geil, 1960).

The begin of the intermediate annealing temperature range is clearly marked (see also Table VII.1). The increase in long period after sufficient annealing time is approximately continuous with temperature and reaches a (metastable?) limit at any one temperature (see also Fig. III.13). Only data by Statton and Geil (1960) indicate at low annealing temperatures evidence of two coexisting thicknesses, perhaps indicative of the partially thickened isolated lamellae (see Fig. VII.14). There is no break in the curve at the beginning of the upper temperature region of annealing. The time dependence of the increase in fold length was expressed by Fischer and Schmidt (1962) through the equation (see also Fig. III.18)

$$l = l_0 + B(T)\log[(t/t_0) + 1] \tag{16}$$

Newer observations (Fischer, 1969) extended these data to shorter times (0.1 min) and showed that at long times a limiting thickness is reached (see also Sects. 7.1.2 and 7.1.3). Crystals of sharper fractions of polyethylene reach this limiting thickness faster (Takayanagi and Nagatoshi, 1965; Fischer, 1969). A similar limiting thickness is reached on isothermal crystallization from the melt at the same temperature (see Fig. III.13). Annealing and crystal perfection on crystallization (Sect. 6.1.6) are thus seemingly related processes. Nakajima *et al.* (1969) suggested that polyethylene fractions crystallized between 77 and 107°C from different solvents increase in fold-length only, if the annealing is done at a supercooling ΔT less than that experienced by the crystals during crystallization. The connection between annealing and supercooling becomes particularly obvious on annealing at elevated pressure described in Sect. 7.3.3.

As mentioned already above, the beginning of the high temperature annealing region starts at a temperature were complete melting becomes possible on fast heating (zero entropy production melting, see Chapter IX). It has been shown above that density, modulus, and swelling reverse their trends at this temperature to indicate less overall perfection of the recrystallized material under the given annealing conditions. Fold-length and unit cell volume do not seem to show any noticable discontinuity in their trends with increasing recrystallization (see Fig. VII.17). The observation on single crystal lamellae by electron microscopy show increaingly severe breaking-up of the lamellae and recrystallization to new lamellae (see for example, Fischer, 1969, Abb. 4, and Geil, 1963, Fig. V.18). The (100) fold sector, when present, shows the earliest signs of recrystallization. Also observed in the high temperature region of annealing is the possibility of molecular reorientation (Baltá Calleja *et al.* 1963; Bassett and Keller, 1959). Crystal mats with X-ray long spacings above 120 Å (crystallized above 80°C from xylene) showed in these investigations a rotation of the unit cell about the *b*-axis, crystals with long spacings below 120 Å showed a rotation of the unit cell about the *a*-axis. Such reorientations bring the molecular chain axes into the film plane and are most likely caused by recrystallization on the substrate rather than on remaining crystal portions. The orientation of the molecular chain in the substrate plane is expected for such epitaxy (see Sect. 3.4).

Annealing experiments on polyethylene and fifteen other macromolecular folded chain single crystals are summarized with pertinent references in Table VII.2. These data show that there exists a large degree of similarity in the

TABLE VII.2

ANNEALING BEHAVIOR OF SOME FOLDED CHAIN SINGLE CRYSTALS

Macromolecule	Conditions and effects	References
Polyethylene CH_2—	Stress relief, dislocation increase, *a*-dimension increase without thickening at low temperature annealing (<115 to 125°C depending on initial thickness, see Table VII.1); at higher temperature hole formation (Figs. VII.13 and VII.14), thickening and interpenetration in case of multilayer samples (for fold-length see Fig. VII.17); above the melting temperature of the initial lamella (zero entropy production melting) increasing recrystallization with further fold-length extension is found; on slow heating, annealing without complete melting can be extended to temperatures above 135°C; for details see Sects. 7.1.2–7.1.8 and 7.2.1.	*See text*
Polypropylene CH_2—$CHCH_3$—	Crystals of isotactic polymer of 600 000 molecular weight were crystallized at 120°C from 0.005 wt % solution in α-chloronaphthalene to lath-shaped crystals and annealed after drying on the substrate; initial thickness 100Å; 150°C annealing, little change, drawn fibrils from the preparation recrystallized; 160°C, hole formation, no change in orientation, 350 Å fold-length, holes of angular shape, elongated along the lath length.	*Kojkima (1968). J. Polym. Sci. Part A2* ***6****, 1938.* *For syndiotactic polypropylene see Marchetti and Martuscelli (1974). J. Polym. Sci. Polym. Phys. Ed.* ***12****, 1649.*
Poly-1-butene CH_2—$CH(C_2H_5)$—	Various crystal forms (fold-length 70–110 Å) were produced from dilute solutions in phenyl acetate and amyl acetate by crystallization after heating to different temperatures; crystallization temperatures 45°C, 50°C, and 57°C; the hexagonal	*Woodward (1968). J. Polym. Sci. Part A2,* ***6****, 1987; Bair, Salovey (1967). J. Polym. Sci. Part B* ***5****, 429.*

(Continued)

TABLE VII.2 *(continued)*

Macromolecule	Conditions and effects	References
	lamellae (trigonal) and lath-shaped (orthorhombic) morphologies thickened on annealing above 70°C and developed holes; square lamellae (tetragonal) showed little changes up to 79°C at the chosen annealing times of 21.5 hr; the annealed thicknesses were up to 250 Å; polymorphic changes occurring during annealing are described in Sect. 7.1.5.	
Poly(4-methyl-1-pentene) $-CH_2-CH(CH_2-CH(CH_3)_2)-$	Square lamellae of 90–150 Å fold-length from dilute solution in xylene, toluene, or pentyl acetate were studied; annealed on a substrate by electron microscopy; holes developed with a preference of elongation normal to the fold-plane, as in polyethylene, leading to concentration of polymer along the diagonals of the lamellae and picture frame appearance; start of major thickening at about 200°C; at 250°C fold-length of 250–380 Å.	*Morrow et al. (1967). J. Polym. Sci. Part A2* **5**, *493; Frank et al. (1959) Phil. Mag.* **4**, *200.*
Poly(vinyl alcohol) $CH_2-CHOH-$	Parallelogram-shaped lamellae from dilute solution in triethylene glycol showed on annealing for 30 sec at 230°C oval to parallelogram-shaped holes aligned along corrugations caused by collapse of lamellae on the substrate.	*Tsuboi (1968). J. Macromol. Sci.* ***B2*** *603.*
trans-1,4-Polybutadiene $CH_2-CH=CH-CH_2-$	Rounded hexagonal lamellae of 72 Å crystal thickness and 100 Å X-ray low-angle spacing (74% crystallinity) were grown at 18°C from 0.05 wt % solution and annealed as single crystals and sedimented mats for 1 hr at various temperatures; doubling of the fold-length occurs at the polymorph I to II transition which for solution-grown single crystals is lowered to 57°C (bulk samples at 76°C); holes develop during the transition which on multilayer crystals lead to interpenetration; no change in orientation is observed; after annealing, the polymorph transition, which is reversible,	*Tatsumi et al. (1967). J. Macromol. Sci.* ***B1****, 459.*

	is shifted to higher temperature; a second increase in fold-length occurs on annealing close to the melting temperature (136°C); NMR and mechanical loss measurements indicate a tighening of the loose chains on annealing.	
Polyoxymethylene $CH_2{-}O{-}$	Single crystal lamellae from 0.01 wt % or 0.05 wt % solution in bromobenzene were annealed on various substrates [mica, Al, NaCl, and poly(ethylene terephthalate) and glass]; annealing through hole formation and thickening (see also Geil, 1963, Fig. V-20); thickened rims about the holes and elongation normal to the fold plane can be detected; the first three substrates show faster annealing; multilayers also anneal usually faster.	*Garber and Geil (1968). Makromol. Chem.* ***113***, *246.*
Poly(ethylene oxide) $CH_2{-}CH_2{-}O{-}$	Crystals from dilute solution in xylene or amyl acetate of various molecular weights were annealed; high molecular weights (32 000) showed between 49°C and 58°C an approximately doubled X-ray long spacing (80–170 Å); at 58°C a break in the long spacing–annealing temperature curve occurred, jumping the long spacing to 250 Å and increasing then at higher temperatures further to 280 Å at 60°C (Fischer); for 4000 and 6000 molecular weight no change in long period was observed on annealing as close as 1°C to the melting temperature (see also the stepwise changes in long period on crystallization and annealing of bulk samples, Figs. III.17, VI.23, and VI.27, Sects. 6.1.6, 6.3.1.3 and Table VII.3).	*Baltá Calleja et al. (1969). J. Polym. Sci. Part C,* ***16***, *4329; Baltá·Calleja and Keller (1964). J. Polym. Sci. Part A* **2**, *2171; see also Fritzsche and Fischer (1973). Kolloid Z. Z. Polym.* ***251***, *721, and Fischer (1972) Pure Appl. Chem.* ***31***, *113.*
Poly[3,3-bis(chloromethyl)oxacyclobutane] $C(CH_2Cl)_2{-}CH_2{-}CH_2{-}O{-}$	Single crystals from 0.05 wt % solution in *n*-butyl acetate or xylene solution were crystallized at 104°C and 95°C, respectively; the orthorhombic crystals were annealed at temperatures differing by 5°C; electron microscopy of the *n*-butyl acetate crystallized samples (115 Å fold-length) showed no changes	*Garber and Geil (1969). Kolloid Z. Z. Polym.* ***229***, *140; Witenhafer and Koenig (1968). J. Macromol. Sci.* ***B2***, *235.*

(Continued)

TABLE VII.2 *(continued)*

Macromolecule	Conditions and effects	References
	up to 160°C, followed at 160–163°C by a slight thickening and roughening of the growth faces, at 164–166°C the edges were rounded, at 168°C all crystals were molten without change in fold-length; infrared studies on the xylene crystallized samples showed an increase in crystallinity on annealing (5%) and an increase in melting point; all annealing must thus be only crystal perfection without change of shape or fold-length.	
Polyesters $(CH_2-)_6O-CO-(CH_2-)_8CO-O-$ $(CH_2-)_{20}O-CO-(CH_2-)_{12}CO-O-$	Single crystals from ethanol developed on annealing above 58°C wrinkles along the crystallographic *b*-axis and showed holes and areas thickened to 160 Å (from the original 100 Å) at 61°C; crystals from amyl acetate increased on annealing from 60 to 110 Å in thickness, developed however, no holes; a 4% increase in density was observed on annealing.	*Gumargaliyeva and Belavtseva (1966). Vysokomol. Soedin.* ***8****, 1604; English transl. Polym. Sci. USSR* ***8****, 1769.*
Nylon 6 $(CH_2)_5CO-NH-$	Single crystals annealed at 150°C developed large holes, leaving a thickened picture-frame-like morphology; at lower annealing temperatures crystals from 0.04 wt % in 1,4-butane diol of 70 Å X-ray long spacing decrease due to better packing on 30 min annealing at 197–198°C to 66 Å; annealing at 216–217, 221–222, and 227–228°C for 30 min increases the fold-length in steps of approximately 1, 2, and 4 unit cell length (see Table II.16).	*Burmester et al. (1972). J. Polym. Sci. Polym. Lett. Ed.* ***10****, 769 (see also Geil, 1963 Fig. V.-21).*
Nylon 6.6, 6.10, and 6.12 $(CH_2-)_6NH-CO-(CH_2-)_4CO-NH-$ $(CH_2-)_6NH-CO-(CH_2-)_8CO-NH-$ $(CH_2-)_6NH-CO-(CH_2-)_{10}CO-NH-$	Crystals from dilute solution in 1,4-butanediol (0.04 wt. %) were grown at 120 to 130°C; the X-ray long spacings of nylon 6.6, 6.10, and 6.12 were 58–59, 76, and 78 Å, respectively; annealing at low temperature decreased the long spacings slightly due to better packing; nylon 6.6 showed doubling and quadrupling of the long spacing on annealing at 254°C (30–60 min); nylon 6.10 showed no change in long	*Dreyfuss and Keller (1970). J. Macromol. Sci.* ***B4****, 811; J. Polym. Sci. Part B,* ***8****, 253.*

	period up to 218.5°C; nylon 6.12 doubles its long spacing on annealing for 30 min at 214°C; for exact condition of doubling the long period see also Burmester *et al.* (1972).	
Poly(*trans*-cyclododecene) CH=CH—$(CH_2—)_{10}$	Single Crystals were grown of 96% trans configuration polymer from 0.10 wt. % solution in *n*-octane at 23°C to 110 Å fold-length; after a small increase in fold-length with annealing temperature at lower temperature, a steep rise in fold-length occurs on annealing above 67°C (measured up to 240 Å).	*Martuscelli and Vittoria (1972). Polymer* ***13****, 360.*

annealing of the various materials, but special effects arise for different molecules. Fold length-extension is the most general feature, although some macromolecules are more difficult to extend than others. Poly[3,3-bis(chloromethyl)oxacyclobutane] seems to be unable to thicken; nylon 6.10, tetragonal poly-1-butene, and some low molecular weight poly(ethylene oxide)s showed little inclination to thicken. Unique is the low temperature doubling in fold-length of *trans*-1,4-polybutadiene during I $\rightarrow$ II polymorph transition, which is brought to further extension only at an about 80°C higher temperature. Frequently the fold-length extension can be carried out in distinct steps which may be related to the repeating unit length (nylon 6), the initial fold-length [nylon 6.6, nylon 6.12, poly(ethylene oxide)], and possibly the length of the molecular fractions [see melt-crystallized poly(ethylene oxide), Sect. 7.3.1]. All changes in fold-length could be accomplished without loss in orientation. In most cases holes developed in single-layer crystals and interpenetration is likely in multilayer crystals. The shapes of the holes is characteristic for the polymer in question [angular for polypropylene, oval or parallelogram shaped for poly(vinyl alcohol)] and the elongation of initial holes normal to the fold plane observed in polyethylene is also found in polyoxymethylene and poly(4-methyl-1-pentene). Thickening seems also to be possible without hole formation (polyester from amyl acetate solution) or by leaving a picture frame thickened outline of the initial crystal [nylon 6 and poly(4-methyl-1-pentene)]. Evidence for the easier initiation of holes along crystal planes involved in collaps of nonplanar single crystals is, besides for polyethylene, seen in poly(vinyl alcohol) single crystals. Thickened rims around initial holes rather than complete thickening was also seen in polyoxymethylene. The differences of annealing rates for different substrates were explored on polyoxymethylene.

Although the annealing of other than polyethylene single crystals has not nearly been studied to the degree that polyethylene crystals have, it seems possible to generalize the annealing process to some degree. For each specific material special effects must, however, be considered. The subdivision of the experimental data into a low temperature region of annealing which causes inter and intralamellar stress relief without much change in fold-length (Sects. 7.1.4 and 7.1.6), an intermediate temperature region of annealing with chain fold increase and crystal grain or subgrain shape changes without recrystallization (Sects. 7.1.2 and 7.1.3), and a high temperature region of annealing with chain fold increase and increasing recrystallization (Sect. 7.1.8) seems to be a useful one. Polymorphic changes (Sect. 7.1.5) need to be considered separately; they may also enhance other annealing processes during the period of higher mobility connected with the transition. Finally, transport of matter (Sect. 7.1.7) which leads to sintering is possible in both the intermediate and the high temperature region of annealing.

7.2.2 Annealing of Crystals in Contact with the Solvent

Crystallization of folded chain single crystals from solution is shown in Sect. 6.2.1 to occur with little or no crystal perfection (see also Sects. 6.1.6 and 6.2.3). It is thus of interest to find out possible changes which may occur when annealing is carried out in the presence of a solvent. The two basic additional effects caused by the presence of the solvents are a lowering of the melting temperature (see Chapter IX) and a swelling of amorphous layers (see Sect. 4.3.2). The swelling of the surface will remove interlamellar stresses, and stresses in the surface layer itself may also be removed. As long as the solvent cannot enter into the crystal, the chain diffusion process necessary for thickening should be affected only through greater mobility in the folds. The lowering of the melting temperature will shift the complete removal of chains from the crystal to a lower temperature. Overall, one would expect thus an enhancement of annealing by an increase in rate and a shift to lower temperature of the low and high temperature regions of annealing (Sect. 7.2.1). The intermediate region of annealing, which has as its major effect lamellar thickening caused by diffusion of chains within the crystal, should be less affected and can possibly be eliminated due to the shift of the high temperature annealing region to lower temperature; this will be shown below for polyethylene.

Early observations of faster annealing of polyethylene single crystals immersed in relatively poor solvents (ethylene glycol and glycerine) were proven by Statton (1961) not to be plastization effects since liquids like Wood's metal have the same accelerating effect. Most likely, the better contact of the heating liquid caused a larger portion of the sample to melt and recrystallize. The enhancement of annealing comes about because crystal perfection immediately after crystallization is often faster than later annealing (see Sect. 6.1.6). A lower temperature of recrystallization in the presence of even small amounts of tetrachloroethylene (a good solvent at the annealing temperature) was proven by Geil (1963, Chapter V). More subtle effects may occur through changes of interlamellar contact by the adsorbed solvent (Burmester *et al.*, 1972). Close contact between lamellae, reached by complete removal of solvent, may allow, for example, easier interpenetration of chains between adjacent lamellae, but may prevent doubling by the mechanism illustrated in Fig. VII.3.

More details of the annealing in contact with solvents were generated by studying the dissolution process. Bassett and Keller (1962) and Holland (1964b) showed by electron microscopy of solution-grown polyethylene single crystals (from 0.01 wt % in xylene) dispersed in fresh solvent, that on slow heating, partial dissolution occurs somewhat above the crystallization temperature; this is followed by recrystallization in the form of a border around the original crystal, and in the formation of new crystals of the same thickness as the border. Recrystallization at lower temperature, reached by cooling after partial dissolution, led to thinner borders; at higher temperatures, the same process led to thicker borders. One expects this behavior from the data of Fig. III.12. Prolonged heating after the formation of a thicker border dissolves finally the thinner interior of the crystal and leaves "picture frame" crystals. No thickening after initial recrystallization or multiple hole formation as in dry annealing was observed. Dendrites showed similar partial dissolution above their crystallization temperature, which was followed by recrystallization (Wunderlich *et al.*, 1963). Final dissolution of the more defect and thinner dendrites occurred on continued temperature increase at 95–97°C (in toluene), a temperature close to the dissolution temperature of the annealed single crystal lamellae.

This early work was extended to different solvents and supplemented by differential thermal analysis of the dissolution and recrystallization process by Blackadder and Schleinitz (1966) (see also Koenig and Carrano, 1968 and Schreiber, 1972). Light scattering observations during dissolution by Peterlin and Meinel (1964) also confirmed partial dissolution and recrystallization at a higher degree of perfection. In all cases dissolution preceded the formation of recrystallized borders or new crystals. Figure VII.18 illustrates the changes

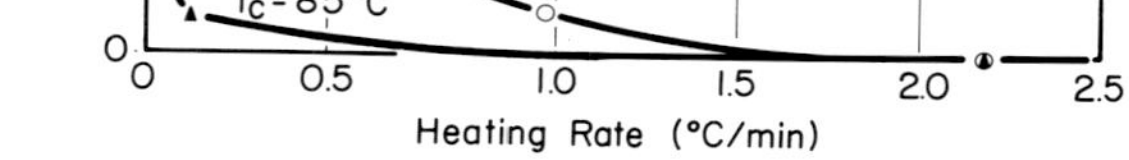

Fig. VII.18 *Fraction of Polyethylene Single Crystals Recrystallized before Final Dissolution as a Function of Heating Rate.* Crystals of various perfection and fold-length were obtained by crystallization at different temperatures (from 0.1 wt. % in *p*-xylene for 1000 min), separated at the crystallization temperature and resuspended in fresh *p*-xylene for dissolution (1.5 wt. % crystals). Data by differential thermal analysis (Blackadder and Schleinitz, 1966).

in fraction of recrystallized material with changes in heating rate of crystals suspended in fresh solvent. Heating rates of 1–2°C/min are sufficient to suppress annealing of crystals grown at 80–85°C, this compares with 25°C/min heating rates needed to suppress annealing without presence of solvent (Sect. 7.2.1 and Fig. IV.13) to a degree of constant melting temperature (approximation of zero entropy production melting or dissolution, see Chapter IX). Figure VII.19 shows the changes in initial crystals, recrystallized polymer,

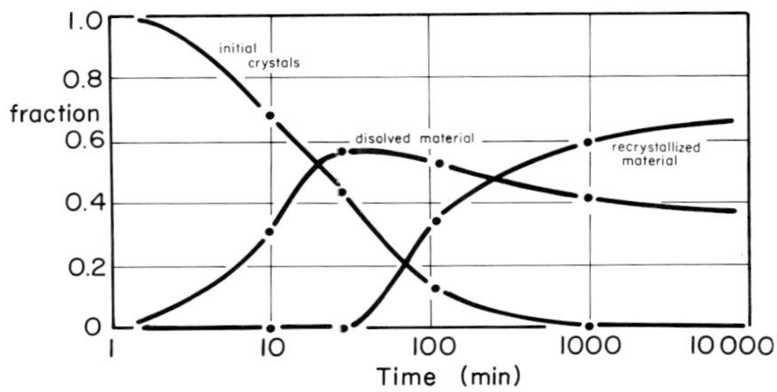

Fig. VII.19 *Distribution of Polyethylene between Initial Crystals, Recrystallized, and Dissolved Material as a Function of Annealing Time at 93°C.*
Crystals grown at 85°C from 0.1 wt. % in *p*-xylene for 1000 min, separated from the mother liquor at 85°C and resuspended in pure *p*-xylene to 1.5 wt % for annealing studies. Data by differential thermal analysis and mother liquor analysis (Blackadder and Schleinitz, 1966).

and dissolved molecules as a function of time on isothermal annealing. The annealing process in the presence of a good solvent is thus best described by the term Ostwald ripening (Sect. 7.1.7). Apart from polyethylene no crystals have been analyzed in detail.

7.2.3 Shish-Kebob Annealing

Shish-kebob-type fibrillar crystals grown from stirred solution (Sect. 3.8.2, Figs. III.79 and III.136–138, and Sect. 6.2.2, Fig. VI.64) or on crystallization during polymerization (Sect. 3.8.1, Fig. III.134) have a compound macroconformation. The inner core has a more extended structure with various poorly characterized defects. The outermost layers are epitaxial, solution-grown folded chain lamellae (Sect. 3.8.1). Annealing should lead to complicated effects. The degree of chain extension of the thin inner core should only be changed by recrystallization (see Sects. 7.1.2 and 7.1.8, Fig. VII.12). The folded chain portions should anneal as described in Sect. 7.2.1.

Annealing of polyethylene shish-kebobs was studied through thermal analysis by Wunderlich *et al.* (1967), and that of polystyrene shish-kebobs by Pelzbauer and Manley (1973). Typically the melting of shish-kebob samples grown from solution begins at a temperature where normal solution grown crystals melt. The main melting occurs in between the melting peak temperatures of solution and melt crystallized samples. In addition, a small portion (typically 5–20%) of the shish-kebob materials shows superheating (see Chapter IX). The superheatable portion has been identified with the fibrous core of the shish-kebobs. On annealing of polyethylene shish-kebobs grown at 96°C from 1 wt % solution in xylene (washed at 99°C) for 2880 min at 129.3°C, the melting peak (measured at 5°C/min heating rate) is shifted from 129.5°C to 138.1°C, a much higher temperature than solution-grown, folded chain crystals reach when similarly annealed. The zero entropy production maximum melting temperature, which is a measure of the most perfect crystal in the analyzed crystals, increases much less. It changed from 134.1°C to 136°C. Most of the annealing effect seems to be thus a recrystallization or crystal perfection which keeps sufficient tie molecules and chain extension to show superheating (see Chapter IX). To some degree the annealing of shish-kebobs is thus related to the annealing of deformed materials (Sect. 7.3.2).

The analyzed polystyrene shish-kebobs (Pelzbauer and Manley, 1973) were grown at 25°C and 100°C from 0.5 wt % solution in mesitylene. The samples showed little melting below 210°C, increasing melting as the temperature increases, and complete melting on quick heating to 241°C when enough time was given at this temperature. Up to 20°C superheating was possible when heating at 40°C/min. Annealing for about 1 hr at 150, 170, 190, and 210°C created a small new low temperature melting peak of increasing melting temperature when analyzed after cooling. This effect is also observed on melt crystallized polystyrene and attributed to melting of increasing amounts of poorly stereospecific polymer in defect crystals. Annealing at 230°C shows initially a major drop in crystallinity, followed by an increase. The recrystallized polymer shows increasing perfection with time as judged by thermal analysis, similar to the observations in polyethylene. Annealing above the melting temperature for 6 min followed by cooling (20°C/min) to 120°C and normal thermal analysis revealed less and less crystallinity of lower perfection grown at the low temperature (melting temperature 221°C). The remaining shish-kebob material can still be made out as a small, high melting peak (melting temperature 243°C at 5°C/min). As soon as the last shish-kebob material is molten, no recrystallization occurred at the lower temperature and the analyzed sample remained completely amorphous as expected for unseeded polystyrene melts under the given conditions.

The special morphology of shish-kebob polymers has thus been shown to lead to different annealing behavior. Specific details however, are, not available at present.

7.3 Crystals from the Melt

Crystallization from the polymer melt is the most common mode of macromolecular crystallization (Sect. 6.3.1). The morphology is usually folded chain lamellar (Sect. 3.3.2) and has frequently a spherulitic superstructure (Sect. 3.7, Table III.5). In contrast to crystallization from solution, lamellae grown from the melt may show considerable crystal perfection after initial growth (see Sect. 6.1.6). It was shown that quick chain fold extension can occur during the perfection step, particularly on melt crystallization of polyethylene at elevated pressure (see Fig. VI.60). In this section the annealing at a higher temperature than the crystallization temperature is treated. Annealing of normal folded-chain, melt grown crystals is discussed in Sect. 7.3.1 and the annealing of deformed crystals in Sect. 7.3.2. A comparison of extended chain crystal growth of polyethylene at elevated pressure and annealing of folded chain crystals at elevated pressure is the main subject of Sect. 7.3.3. Finally, in Sect. 7.3.4 some evidence for annealing by chemical reaction is reviewed.

7.3.1 Annealing of Normal Crystals

As in the discussion of annealing of solution-grown crystals (Sect. 7.2.1), the annealing of folded chain, melt-grown crystals has been studied by far the most intensely for polyethylene. The early work on polyethylene has been reviewed by Geil (1963). It was concerned mainly with the increase in X-ray long period and density as a function of annealing temperature and time (see Fig. III.18). The same logarithmic time dependence of X-ray long period as for solution-grown crystals was observed [Eq. (16)] by Fischer and Schmidt (1962). Figure VII.10 shows that on annealing at temperatures of 120°C and higher, increasing amounts of melting occurs, followed by recrystallization; which again, is similar to solution-grown crystals of polyethylene (Fig. VII.11).

The three temperature regions of annealing which could be delineated for solution grown crystals of polyethylene (Sect. 7.2.1) have not been defined as accurately for melt crystallized samples. One of the reasons is the present inability to study defects within and between isolated melt crystallized lamellae which could reveal changes in the state of stress at low temperatures. Heat capacity and density analysis of polyethylene crystals of different crystallinity by Wunderlich *et al.* (1963) showed considerable deviations from additivity at temperatures as low as room temperature. These effects were interpreted as increases in disorder and/or sizes of defects which are initially reversible, but should lead to stress relief as found in solution-grown crystals.

The intermediate and high temperature region of annealing are discovered easily through the fold-length increase. The temperature of separation of the regions in which fold-length extension occurs with or without recrystallization is, however, more difficult to find. This difficulty has brought several attempts to explain all annealing effects by recrystallization only (see for example Mandelkern and Allou, 1966 and Kawai, 1965, 1969). From graphs like Fig. VII.10 (see also Peterlin, 1964 and Fischer and Schmidt, 1962), one would place the point of some initial melting on quick heating somewhere in the region of 115–125°C, depending upon the initial sample perfection. Since melt crystallized materials are usually less uniform than solution grown-crystals, there is a broader melting region. Also, melt crystallized macromolecules may show with increasing temperature increasing amounts of surface melting which increases the amorphous surface layer of crystals, but contributes very little to recrystallization at the temperature of annealing (Fischer, 1969). In view of the results on solution-grown crystals, it is perhaps useful to take the temperature between the intermediate and high temperature region of annealing at the melting temperature for the average fold-length as defined by Fig. V.43.

The similarity of annealing and crystal perfection on isothermal crystallization was illustrated by Blackadder *et al.* (1972). Annealing for periods of up to two years enabled solution-grown crystals of polyethylene to approach the density, melting temperature and mechanical properties (brittleness) of pressure crystallized extended chain lemellae. The temperature was slowly raised during the annealing time, to reach, finally, 135.8°C. Similar crystal perfection was reported by Mandelkern *et al.* (1966) on isothermal crystallization at 130°C for 20–40 days. On a shorter time scale experiment, the change in density in g/cm^3 as a function of time is shown in Fig. VII.20 for annealing and melt crystallization at the same temperatures. Such experiments lead also to similar final lamellar thicknesses as illustrated in Fig. III.13. Peterlin (1964) attributes the remaining differences to variations in the initial crystals grown. One effect may be the production of largely different initial grain or subgrain sizes which should lead to changed annealing rates as long as transport of matter is limited (see Sects. 7.1.2 and 7.1.7). Another effect may be connected with molecular weight fractionation which is most complete on slow, isothermal crystallization (see Sect. 5.3.3). Annealing a quenched sample produces little fractionation on initial crystallization and leaves in the high temperature annealing region a portion of the crystals unmolten; these can then not participate in fractionation (see also Harland *et al.*, 1972). Fischer (1969) suggests that the final density increases in Figs. VII.10 and VII.20, which are proportional to the increase in fold-length, can be understood by assuming that the interlamellar amorphous layers keep an approximately constant thickness. The final density increase should thus be a result of the decrease in lamellar surface area on thickening.

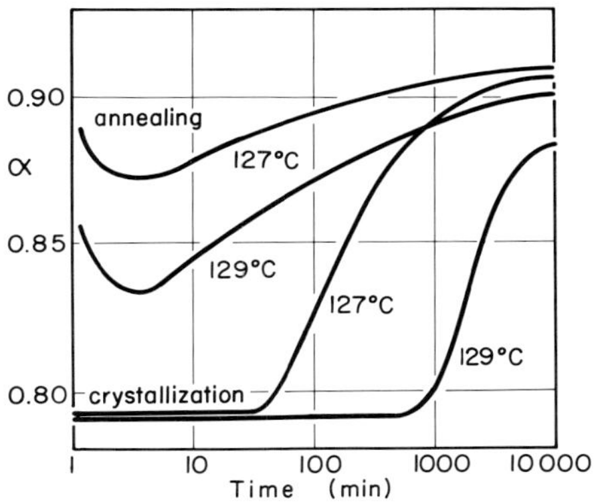

Fig. VII.20 *Density of Annealed and Isothermally Crystallized Polyethylene as a Function of Time.* Identical polymers, the annealed samples were initially quenched in liquid air. Drawn after data presented by Peterlin (1964).

Electron microscopic studies of the annealing of bulk crystallized polyethylene by Geil (1963) have shown that during thickening no gross change in structure and orientation of the crystals occurs. Quenched, nonspherulitic samples were still nonspherulitic after annealing (100 hr at 125°C and 131°C) and possessed random lamellar orientation. Spherulitic samples remained spherulitic with the same diameter and ring spacing on annealing (see Sect. 3.7.2). Even at these high annealing temperatures where considerable recrystallization occurs, large scale formation of new crystal aggregates was thus impossible. Only a small amount of separation of shorter length molecules into new crystals was found. McHugh and Schultz (1971) annealed isothermally grown polyethylene (110°C) at 118°C, 123°C, and 126°C for 3 hr and 72 hr. Electron microscopy of fracture surfaces and etched fragments seemed to indicate hole formation and thickened crystal edges as found on annealing of solution grown crystals (Sect. 7.2.1). Finally, inspection of Fig. VII.12B shows that thickening occurs by breaking up the continuity of the melt crystallized lamellae (hole formation) rather than by wholesale melting and recrystallization as is observed on the initial annealing of the extended fibers in the fracture surface (Figs. III.74 and VII.12A).

In general, there are good reasons to assume that annealing of melt-grown lamellar crystals is similar to annealing of solution-grown crystals. The annealing description is aided as before by separation of three temperature

TABLE VII.3

ANNEALING BEHAVIOR OF SOME MELT CRYSTALLIZED SAMPLES

Macromolecule	Conditions and effects	References
Polyethylene CH_2-	At low annealing temperatures small density and heat capacity deviations from additivity of amorphous and crystalline areas indicate increasing mobility which should lead to stress relief; at higher temperatures fold-length increase is observed along with breaking up of initially lamellar crystals; the initial orientation and gross morphology is not changed; crystallization is spread over a broad temperature range because of nonuniform crystals; similarities exist with crystal perfection on initial crystallization (Sect. 6.1.6) and annealing of solution-grown crystal mats (Sect. 7.2.1); the final crystal thicknesses and densities are not much different from long time isothermal crystallization (Figs. III.13 and VII.20).	*See text and Geil (1963, Chapter V).*
Polypropylene CH_2-CHCH_3-	Oriented (drawn 400–700% at room temperature) and unoriented specimens of 250 000 molecular weight were annealed between 120–170°C; as the annealing temperature increases, the chains which are disordered with respect to isoclined and anticlined helices (see Sect. 2.4.2, Table II.10) transform irreversibly to a stable, ordered crystal structure which has isoclined helices on individual (040) planes and anticlined helices in adjacent (200) planes (Fig. II.42B); the domain sizes increased on annealing;	*Hikosaka and Seto (1973). Polym. J.* ***15****, 111.*
	on annealing after low temperature quenching, fold-length increases leading to 120–500 Å X-ray long periods were observed; isothermal crystallization at the annealing temperature led to approximately the same fold-length; in the final annealing stages the long period increased linearly with the logarithm of time; low temperature annealing was shown to be the cause of aging (see text).	*Blais and Manley (1967). J. Macromol. Sci.* ***B1****, 525.*

Polystyrene $CH_2-CH(C_6H_5)-$	Crystals of a low angle X-ray spacing of 120–310 Å were produced by annealing glassy or lower temperature crystallized samples; on annealing at higher temperature a slow increase to the new value was observed; isothermal crystallization from the melt led to the same thickness.	*Blais and Manley (1967). J. Macromol. Sci.* ***B1****, 525.*
Copolymer of tetrafluoroethylene and hexafluoropropylene $(CF_2-CF_2-)_x$ $(CF_2-CFCF_3-)_y$	Samples of 5–7 CF_3/100C were annealed between 235 and 270°C; the X-ray long period increased continuously with increasing temperature (at a fixed time) from 230–293 to 290–365 Å; at constant temperature the long period increased linearly with the logarithm of time; the unit cell dimensions decreased on annealing similar to the homopolymer.	*Sanches et al. (1973). J. Appl. Phys.* ***44****, 4332.*
Polyoxymethylene CH_2-O-	Films of 45 000 molecular weight were quenched to −78°C and annealed up to 176°C; the observed X-ray long period changed from about 150 Å at low temperature to 480 Å (see also Geil, 1963); similar spacings were found on isothermal crystallization from the melt; on 1°C stepwise annealing, the new thickness was attained without recrystallization; above 171°C extensive recrystallization after fast heating (birefringence decreases first practically to zero); the time dependence is linear with the logarithm of time over a long interval (for data see Sanchez *et al.*, 1973).	*Blais and Manley (1967). J. Macromol. Sci.* ***B1****, 525.*
Poly(ethylene oxide) CH_2-CH_2-O-	A 10 000 and a 20 000 molecular weight fraction sample were annealed after crystallization to different X-ray long periods; as shown in Fig. III.17, the crystals grow in thicknesses which are fixed multiples of the molecular length (see also Sect. 6.3.1.3); annealing at 62°C of the 10 000 molecular weight sample crystallized to lamellae with $\frac{1}{2}$, $\frac{1}{3}$, and $\frac{1}{4}$ the molecular length, the thickness shows little change for the thickest lamellae, the $\frac{1}{3}$ thickness crystal changes abruptly to $\frac{1}{2}$ thickness after 10 hr; while the $\frac{1}{4}$ thickness crystal is already changed to $\frac{1}{3}$ after a few minutes, then also changes to $\frac{1}{2}$ thickness after 2.5 hr, which is the thickness obtained directly on isothermal crystallization; these changes	*Spegt (1970). Makromol. Chemie* ***139****, 139.*

(Continued)

TABLE VII.3 *(continued)*

Macromolecule	Conditions and effects	References
	on annealing are accompanied by complex changes in specific volume, an initial increase due to partial melting occurs before rapid change in fold-length (coupled with a partial fold-length increase) followed by a decrease; the same initial crystals show the abrupt increase in fold-length by one multiple at increasing times for decreasing annealing temperatures (see also Sect. 6.1.6 and Fig. VI.27 for the discussion of crystal perfection in steps on initial crystallization).	
Poly[3,3-bis(chloromethyl)-oxacylcobutane] $C(CH_2Cl)_2-CH_2-CH_2-O-$	This macromolecular material seems to show no increase in lamellar thickness on annealing (see Table VII.2); melt crystallized samples grown quickly (or under elevated pressure) develop the metastable β structure; on annealing above 130°C, there is a partial irreversible phase change to the α structure, the degree of conversion depending upon temperature and crystallization history.	*Witenhafer and Koenig (1968). J. Macromol. Sci.* ***B2****, 247.*
Poly(ethylene terephthalate) $CH_2-CH_2-O-CO-C_6H_4-CO-O-$	Annealing after crystallization at 140°C reveals increasing melting followed by recrystallization (200–247°C); annealing at 220°C for 24 hr after prior crystallization at 110°C (30 min) shows an increase in lamellar thickness from about 130 Å to 210 Å and possible fold regularization (concluded from etching experiments); annealing of a well-crystallized sample at 260°C (48 hr) increases the X-ray long period from 160 Å to 172 Å while doubling molecular weight (annealing under vacuum) (see also Sect. 7.3.4).	*Zachmann and Stuart (1960). Macromol. Chem.* ***41****, 148; Mehta and Bell (1973). J. Polym. Sci. Polym. Phys. Ed.* ***11****, 1793; Miyagi and Wunderlich (1972). J. Polym. Sci. Polym. Phys. Ed.* ***11****, 2085.*
Nylon 6 $(CH_2-)_5CO-NH-$	Quenched films were annealed for 1 hr by immersion in silicon oil or Wood's metal between 100°C and 215°C; the X-ray long period increased slightly from the initial 65 Å up to annealing at 180°C, at which point a strong increase was observed reaching finally 180 Å.	*Hirami (1967). Makromol. Chem.* ***105****, 296.*

Nylon 6.6 $(CH_2-)_6NH-CO-(CH_2-)_4CO-NH-$	Annealing of unoriented nylon 6.6 monofilaments (0.5–1 mm diameter and of 0.8854 specific volume) were carried out up to 260°C; from the separation of the 100 and 010 X-ray diffraction angles as function of annealing temperatures, it is found that the (010) separation decreases continuously from about 3.97 Å to 3.71 Å on annealing between 80°C and 260°C, the 010 X-ray diffraction peak width decreases over the same temperature interval of annealing by 50% and the density increases from 1.13 to 1.16 g/cm^3; low-angle X-ray data are to be published later.	*Hinrichsen (1972). Kolloid Z. Z. Polym.* ***250****, 1162.*
Poly(tetramethyl-*p*-silphenylene siloxane) $Si(CH_3)_2-C_6H_4-Si(CH_3)_2-O-$	A series of fraction of 27 000 to 431 000 molecular weight was annealed between 0 and 130°C; the X-ray low-angle spacing increased from 80 Å to 130 Å in this temperature range; prolonged annealing did not show any further effect.	*Pollack and Magill (1969). J. Polym. Sci. Part A2* ***7****, 55.*

regions of annealing, but at present very little is known about the low temperature region of annealing and the temperature separating the two upper regions. The expected differences between annealing of mats of solution-grown single crystals and melt grown, stacked lamellae arise from differences in crystal perfection, increased number of tie molecules, and more intimate contact on crystallization from the melt. Raman spectroscopy of the longitudinal acoustical vibrational mode during annealing of melt crystallized polyethylene revealed besides the increase in fold-length (frequency shift) a narrowing of the line width which is indicative of increasing uniformity of the fold-length (Koenig and Tabb, 1974). The solution-grown crystals were initially more uniform and broadened their line width on annealing (see Sect. 7.2.1).

Major difficulties in the interpretation of annealing data reported in the literature result from variations in crystal perfection of the starting materials which is usually not analyzed in detail. Larger changes in the annealing behavior occur as soon as the melt-grown crystals approach the fringed micellar or partially extended chain macronformations (see Sect. 7.3.2).

Annealing of polyethylene crystals grown from the melt through contact with solvent (*o*-xylene) was observed by Fuhrmann *et al.* (1970) through a study of the refractive index in the vicinity of the interface. At 84°C, below the dissolution temperature of the lamellar crystals, the refractive index of the solid was shown to increase, indicating a crystal perfection. The solvent refractive index, in turn, decreased very little, showing only a small amount of dissolution. At 96°C a small amount of temporary annealing was observed before complete dissolution.

The study of annealing of other melt-grown crystals of linear macromolecules has frequently been limited to the observation of increased melting temperatures and development of multiple melting peaks on thermal analysis which is discussed in more detail in Chapter IX. The annealing behavior of some ten macromolecular crystals for which more structural detail is available is summarized in Table VII.3. As for melt-grown polyethylene crystals and solution grown single crystals (Table VII.2), the major effects on annealing at increasing temperatures are usually an increase in lamellar thickness, density, melting temperature, and a decrease in unit cell volume. The changes are irreversible and frequently, at least over a reasonable time span, linear with the logarithm of time. Solid–solid phase transitions occur whenever metastable crystals are initially grown (see also Sect. 7.1.5) or changes to high temperature crystal forms are possible, The frequent similarity of the ultimate fold-length obtained on annealing with the fold-length resulting from isothermal crystallization at the same temperature is linked to the similarity of annealing and crystal perfection processes. Whenever differences in X-ray long-periods have been reported (Blais and Manley, 1967), the annealed crystals seem to be slightly thinner, an effect which points to more efficient thickening immediately after crystallization (see also Sect. 7.3.3). Several polymers show special features. For example, poly[3,3-bis(chloromethyl)-oxacyclobutane] shows no thickening on annealing (as in the solution case, Sect. 7.2.1), while poly(ethylene oxide) fractions anneal largely in distinct steps (Spegt, 1970).

The different stages of annealing illustrated on solution-grown crystals are most likely also present for the melt-grown crystals. Little is known, however, about the low temperature annealing. Some information has been provided by Kapur and Rogers (1972) who studied the low temperature annealing (aging at 25°C) of polypropylene crystallized by quenching to 0°C. No changes in crystallinity, crystal structure, or crystal size were observed, but the yield stress increased by 16%, the drawing stress by 150% and the initial modulus by 185% on annealing for 100 hr. Even more changes could be observed in the inert gas transport through the sample. Attempts have been made to explain these measurements on the basis of changes in the intercrystalline regions. Relaxation of residual stresses caused by the initial quenching, decrease in chain mobility due to secondary crystallization, and collapse of microvoids have been considered. Partial melting followed by recrystallization has been observed for several of the macromolecules listed in Table VII.3, indicative of the high temperature annealing region.

7.3.2 Annealing of Deformed Materials

The study of annealing of deformed materials, particularly of drawn fibers, is made easier because of the decreased stability of crystals after deformation which leads to more pronounced annealing effects. The orientation in drawn fibers brings about clear and easy to recognize low-angle X-ray diffraction patterns; similarly, the interpretation of electron micrographs is helped through the unique fiber axis (see Fig. IV.9).

Early work by Hess and Kiessig (1943, 1944) gave already the proper explanation for the X-ray low-angle diffraction (see also Sect. 4.1.2), namely, that it is a measure of the separation of the centers of gravity of successive crystals which with their molecular chain axes are aligned along the draw direction (see Sect. 3.8.2). In the same investigations it was shown that drawn nylon 6 increased its X-ray low-angle spacing from 74 Å to 123 Å when annealed at 214°C and drawn poly(ω-hydroxyundecanoic acid), from 139 Å to 210 Å on annealing at 80°C. These measurements indicate clearly an increase in crystal size in the molecular chain direction (see also Hess and Kiessig, 1953). After these initial investigations, much effort was spent deriving details for the annealing of drawn polyethylene which is discussed next. Data on other macromolecules will be given at the end of this section (Table VII.4).

The first model for the macroconformation of the molecules in drawn fibers was proposed by Hess and Kiessig (1944). It was based on the fringed micelle model (Fig. III.5). Sufficient orientation was assumed to align all crystals so that the molecular axes are parallel to the fiber direction, and the molecular chains were taken to traverse crystalline and amorphous regions alternatingly without changing their overall direction. After the discovery of prominent chain folding in melt and solution crystallization of macromolecules (Sect. 3.2.1), this model was modified by the inclusion of some chain folding at the crystal–amorphous interfaces (Keller and O'Connor, 1957; Bonart and Hosemann, 1959). Fischer *et al.* (1968) could show by low angle X-ray diffraction of samples with enhanced contrast, achieved by absorption of iodine vapor into the amorphous areas, that folding at the interface is likely.† Without folding, the density difference between amorphous and crystalline areas would cause an expansion at the interface (Statton, 1959) or would force the creation of oblique surfaces relative to the fiber axis (Bonart, 1964; see also Sect. 5.1.2.1).

The present model of the structure of drawn material is based on proposals by Bonart and Hosemann (1962), Beresford and Bevan (1964), Fischer and Goddar (1965), Peterlin (1965), and Baltá Calleja *et al.* (1966) [see also Bonart (1969) and Sects. 3.8.2 and 4.4.3]. Figure VII.21 shows a schematic representation of the model (see also Figs. IV.26 and IV.70). The fiber is represented by an almost homogeneous crystalline matrix with a high defect concentration. Fracture surfaces, when observed by electron microscopy are mainly characterized by fibrillation down to diameters of about 100 Å without much additional internal details. The defects are comprised of chain folds, loops, tie molecules, and chain ends. They are not randomly distributed, but cause the low angle X-ray diffraction (Fischer, 1969; Fischer *et al.*, 1971). A typical diffraction pattern is shown in Fig. VI.22A. The diffraction along the meridian indicates the crystal dimension along the chain axis, the diffraction along the equator is indicative of the fibrillation. The volume of the crystalline area can be measured by a combination of wide- and low-angle X-ray diffraction (Sect. 4.1.2). Table III.6 lists typical data, showing that on drawing the volume of crystal grains or subgrains is reduced to 10% or less of their original value (to about 5×10^{-18} cm^3). The final shape is 100–200 Å in all dimensions. Such a shape does not correspond to equilibrium (Sect. 7.1.2, Fig. VII.1).

† The X-ray small-angle intensity at the meridian decreases on addition of small amounts of iodine before increasing on addition of larger amounts. This intensity change is proportional to the change in electron density projection onto the fiber axis on going from the crystalline to the amorphous area. The electron density is higher for the amorphous area than the crystalline area if all chains continue into the next crystalline area, and lower if many chains fold. Iodine raises the electron density and thus proves by causing an initial intensity decrease that folding is likely (Fisher *et al.*, 1968).

Fig. VII.21 *Schematic Drawing of the Macroconformation of Drawn Material.* Drawn after Fischer and Goddar (1965).

The major annealing effects expected should thus be a change in shape to a more extended chain conformation (Sect. 7.1.3) and an increase in volume by alignment of neighboring grains or mass transport across the grain boundaries (Sect. 7.1.7).

The structure of fibrillar extended chain crystals shows superficial resemblance to the drawn fiber structure, it represents a defect-free fiber. The major difference between these two structures is the presence of frequent folds in drawn fibers. The folds permit annealing of the crystal grains by chain fold extension which, however, never reaches the level of extended chain crystals (Sect. 3.2). The extended chain fiber of small diameter has, in contrast, no mechanism of decrease in extension to reach an equilibrium shape (Fig. VII.1), it must melt first, and then recrystallize (see Sect. 7.2.1). The recrystallization leads, as discussed in Chapter VI, first to chain folding (see also Figs. III.74 and III.78).

As in the annealing of solution and melt crystallized crystals (Sects. 7.2.1 and 7.3.1), it is possible to subdivide the annealing temperatures into three regions. Loboda-Čačković *et al.* (1969, 1971) could show that a low temperature region of annealing of their polyethylene sample (drawn 10× at 0.6 mm/min at 80°C) reached up to about 110°C. The next change in annealing behavior was found at about 130°C, with a final limit of the annealing region at 135°C, above this temperature recrystallization with complete loss of orientation was observed. Although these three temperature regions may correspond formally to those of solution and melt crystallized samples, differences arise because of the largely different structure of the starting material. A clear separation of the three regions is presently not possible because little attention has been directed in the past to the discussion of the effects of annealing in the low and high temperature regions and differently drawn materials should show different limits of the regions of annealing.

As in the undeformed crystals, one would expect that the low temperature region of annealing to be connected mainly with stress relief. This stress relief should occur in the amorphous area and should be coupled to minor changes within the crystals. The amorphous areas are held in the drawn material in a state of much higher stress than in undeformed materials. Following a moderate deformation process of low density polyethylene (50% elongation), Oda and Stein (1972) could show by X-ray and birefringence measurements that orientation of the amorphous areas precedes the crystal deformation. The metastable balance between crystal deformation and amorphous orientation is reached only after 0.1 sec and 100 sec, depending upon temperature. Heating the samples above the draw temperature must cause, thus, some further stress relief. The nonequilibrium state of the amorphous areas was shown by thermal analysis (Peterlin and Meinel, 1965; Fischer and Hinrichsen, 1966). The enthalpy of drawn samples (10–20 ×) is of the order of magnitude of 10–20 cal/g less than a supercooled melt at the same temperature. The mobility of the amorphous portions of the molecules, when measured by NMR (see Sect. 4.1.6), and their ability to take up solvent is also much less after drawing (Peterlin and Olf, 1966; see also Fischer *et al.*, 1968). These measurements and the observation of shrinkage of drawn fibers on annealing are the basic evidence for the large amount of stress in the amorphous areas.† The observation of small-angle X-ray diffraction and density by Loboda-Čačković *et al.* (1969) on a polyethylene drawn 10× at 85°C in the low temperature annealing region indicate a continuous, slow logarithmic increase in density of the overall sample with time. The X-ray low-angle spacing increased by about 10%, to reach a limit after 10^3–10^4 min. From the change in the equatorial X-ray diffraction, it was concluded that neighboring crystalline areas align laterally in groups of 4 and 9 without changing the ultimate fibrillar diameter. Much of the increase in density may also be caused by the slow loss of voids (Fischer and Hinrichsen, 1966).

† On annealing at 120°C and 125°C sorption and diffusion of C_2Cl_4 vapors increase despite an increase in crystallinity. The values for completely relaxed amorphous components are reached [Peterlin and Williams (1972), *Brit. Polymer J.* **4**, 271].

The lower limit of the intermediate temperature range of annealing was set by Loboda-Čačković *et al.* (1971) at the temperature of first absorption of heat of fusion on thermal analysis. This temperature can not be related to the melting temperature of the metastable crystals in the actual sample.†

The major experimental observations on annealing in the intermediate temperature range are distinct changes in the low-angle X-ray diffraction pattern, development of a transverse lamellar structure in electron micrographs, increase in density, increasing changes of the mechanical relaxation‡ behavior towards the undrawn sample, changes in the NMR spectrum, and shrinkage of the drawn sample, reversing the deformation process.

Figure VII.22B shows that on annealing, the diffraction along the meridian becomes more intense and moves to lower diffraction angles. Typically, two

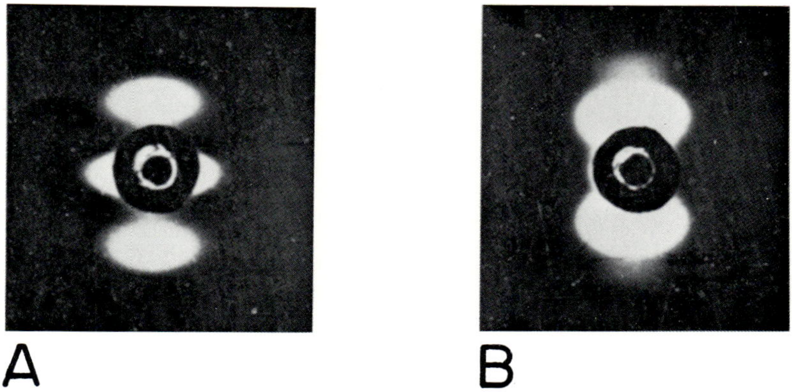

Fig. VII.22 *X-ray Small Angle Diffraction Patterns of Drawn Polyethylene.* A: Drawn 15× at 70°C, unannealed; B: Annealed for 1 hr at 25°C. (Fischer *et al.*, 1968).

† Loboda-Čačković *et al.* (1971) set the lower limit of the intermediate temperature range of annealing at 110°C for a polyethylene sample drawn 10× at 80°C at 0.6 mm/min. This sample showed an X-ray long spacing of 235Å. A large folded chain lamella of this thickness would melt at about 132°C (zero entropy production melting temperature). A large, fully extended chain crystal of this thickness would melt at 127°C (see Fig. V.43). The additional large side surfaces and added defects in the crystals of the drawn material could perhaps reduce the melting temperature somewhat from that of solution grown crystals, but the remaining stress in the amorphous areas should keep the melting temperature higher until a large fraction of the crystals is molten. Further complications which are not fully resolved are the possibility of prior annealing during thermal analysis and partial melting of the surface layers of the crystals (see Chapter IX).

‡ For a general discussion of the changes of mechanical properties as a function of annealing of drawn materials see for example Fischer *et al.* (1971), Illers (1969), and G. Meinel and A. Peterlin (1967), *J. Polymer Sci. Part B* **5**, 613.

orders of diffraction can be seen after annealing in the intermediate temperature range (Loboda-Čačković *et al.*, 1971). The change in the X-ray long period derived from such data is shown in Fig. VII.23. The increase depends

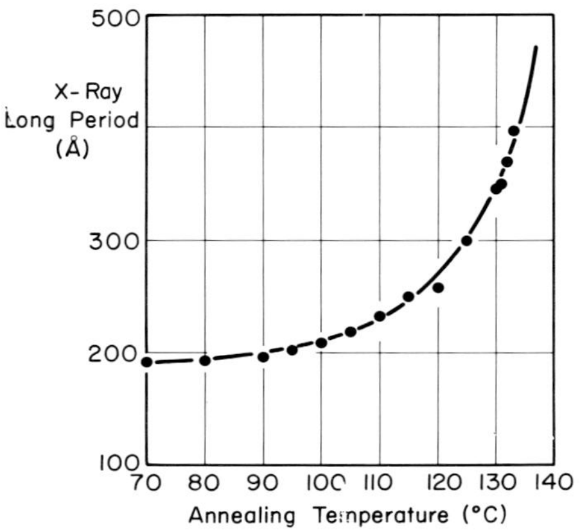

Fig. VII.23 *Temperature Dependence of the X-ray Long Period of a Linear Polyethylene.*
The samples were drawn 17× at 70°C and annealed at the given temperature for 24 hr. (Fischer and Schmidt, 1962). Sakurai *et al.* (1974) have shown a similar long period change on continuous fast heating (18°C/min) of a 6× drawn polyethylene.

approximately linearly upon the logarithm of time. For higher degrees of deformation and on keeping the sample under external stress, a slower increase of the long period is observed (Fischer, 1969). The increase in intensity was used by Fischer *et al.* (1968) to calculate the electron density fluctuations $\langle \Delta\eta^2 \rangle$ between the crystalline and amorphous regions along the draw direction. The relationships between the electron density fluctuation and the crystal structure is simple if one assumes a two-phase structure. The correlation between the electron density fluctuation and mass density fluctuation $\Delta\rho$ is given by (see also Eq. III.3)

$$\langle \Delta\eta^2 \rangle = (\Delta\rho)^2 v^c (1 - v^c)(\Sigma\, Z_i/M_0)^2 \tag{17}$$

where v^c represents the volume fraction crystallinity [Eq. (8), Chapter IV] and $(\Sigma\, Z_i/M_0)$ is the conversion factor from mass to electron density. Figure VII.24 shows some typical data. The density difference between crystalline and amorphous area in drawn polyethylene is much smaller than usually

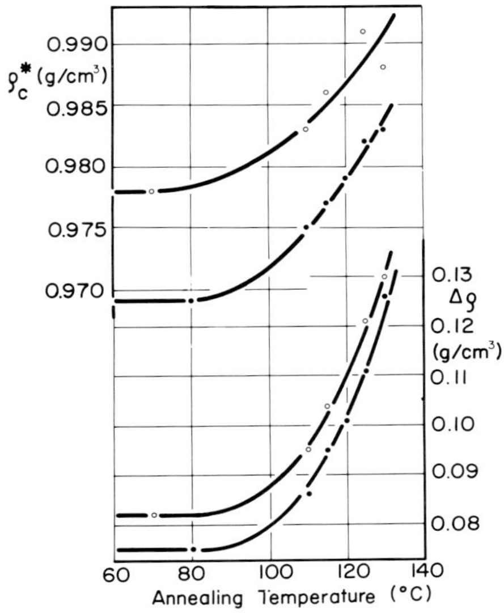

Fig. VII.24 *Density Difference between Amorphous and Crystalline Areas and Effective Density of the Crystalline Areas of Drawn Polyethylene.* Samples drawn 15× at 70°C (open circles) and 8× at 80°C (filled circles), annealed for 60 min at the given temperatures (Data by Fischer *et al.*, 1968). Note that the ρ_c* scale is expanded by a factor of 2.5.

assumed for the normal two-phase model (see Table IV.1). Since only little change is noted in the crystalline unit cell dimensions (<0.5%), most of the effect is due to an increased amorphous density, an observation which agrees with the observed lower enthalpy† (see previous discussion). On annealing, the density difference increases towards the value found for melt grown crystals (0.143 g/cm^3). Combination of density measurements and X-ray low-angle and wide-angle data permits also to calculate the effective crystalline area density ρ_c* (Fischer *et al.*, 1968). Figure VII.24 shows its increase as a function of annealing temperature. The effective crystalline density after drawing is less than calculated from the unit cell dimensions which suggests considerable defects in the crystals which heal out on annealing. Note that the

† These observations agree also with the reduced mobility in the amorphous areas measured by NMR and the reduced solubility of small molecules in drawn polyethylene just discussed and also with a decrease in meridional low-angle X-ray intensity on increasing draw ratio observed by Peterlin and Corneliussen (1968).

density decrease of the crystalline area on drawing is less than the density increase of the amorphous. On annealing, one finds that the amorphous density decreases and the crystalline density increases. The crystallinity calculated with the true average densities of the amorphous and crystalline areas may, because of these opposing trends, increase on annealing, stay the same, or possibly even decrease.

Electron microscopy of fracture surfaces or etched surfaces of annealed drawn fibers or films reveals strongly lamellar appearance as shown in Fig. IV.9 (see also Sakaoku and Peterlin, 1967). The lamellar structure agrees in orientation and spacing with the small-angle X-ray results. The fibrillar structure of the unannealed drawn samples disappears, and with it disappears also the low-angle X-ray scattering along the equator (see Fig. VII.22B). Lateral aggregation was also derived from low-angle X-ray data on the basis of a two-dimensional paracrystalline structure of the amorphous and crystalline areas by Loboda-Čačković *et al.* (1969, 1971) (see Sect. 4.2.2). The major portion of lateral alignment seems to occur during a well-defined time interval during which little increase in density is observed and the X-ray long period increases 10–20 Å. The overall fibrillar diameter of usually about 100 Å does not change significantly during this alignment.

Direct information on the motion in drawn and annealed polyethylene has been gained through NMR by Fischer *et al.* (1971). The total NMR signal was fitted to four component signals (see Sect. 4.1.6, compare to Eq. (42), Chapter IV).

$$Y_{\mathrm{total}} = [a_{\mathrm{b}}\, y_{\mathrm{b}}(X, M) + a_{\mathrm{mo}}\, y_{\mathrm{mo}}(X, \beta_{\mathrm{m}}) + a_{\mathrm{mu}}\, y_{\mathrm{mu}}(X, \beta_{\mathrm{m}}) + a_{\mathrm{n}}\, y_{\mathrm{n}}(X, \beta_{\mathrm{n}})]N \tag{18}$$

The subscripts *b*, *m*, and *n* refer to the broad, intermediate, and narrow components, *X* represents the dependence on the external field, *M* allows for the small reduction in line width arising from thermal expansion, *N* is the normalization factor, and *a* and *β* are the mass fraction and the effect of the properly averaged local field. Relative to undrawn polyethylene, the intermediate component is separated into an oriented (subscript o) and an unoriented (subscript u) part. The broad component is a direct measure of the crystallinity (see Table IV.9). The intermediate component was calculated assuming rotational motion of the CH_2 groups about the molecular chain axis. The contribution Y_{mo} arises from chain segments aligned parallel to the fiber axis during drawning, and Y_{mu} from randomly distributed chain segments. The CH_2 groups contributing to the narrow component are able to carry out relatively unhindered micro-Brownian motion and are considered truly amorphous. Figure VII.25 shows the changes in the various components as a function of annealing temperature. The broad component CH_2 groups do not change in amount up to about 125°C. Since the overall second moment

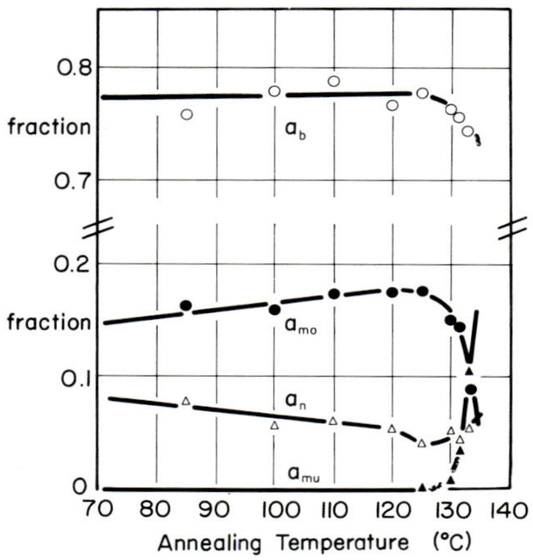

Fig. VII.25 *Changes of Mass Fractions of Drawn Polyethylene on Annealing as Measured by NMR.* Linear polyethylene drawn at 70°C 10× at 10 mm/min. a_b: broad component mass fraction ($\sim w^c$); a_{mo}: oriented intermediate component (stressed, oriented amorphous); a_{mu}: unoriented intermediate component (hindered amorphous); a_n: narrow component (mobile, amorphous). For detailed definition see Eq. (18) and Sect. 4.1.6. (Data by Fischer *et al.*, 1971).

of the sample increases, however, starting at about 100–110°C, increasing perfection of the crystalline material must occur without large change in amount. The narrow component CH_2 groups decrease slowly to a shallow minimum at 125°C. Their mobility increases, however, constantly as measured by a decreasing β_n. Unoriented intermediate component CH_2 groups appear only at about 125°C at the expense of CH_2 groups from the broad and oriented intermediate components.

All these measurements point to the fact that in the intermediate region of annealing temperature no large scale melting and recrystallization occurs, this occurs only in the little investigated high temperature region of annealing above 125–130°C. In this temperature region the low-angle X-ray diffraction pattern becomes even sharper. Loboda-Čačković *et al.* (1971) report at least three orders of diffraction. The lower enthalpy of the amorphous portion of the drawn material increases in this temperature region rapidly to the melt-extrapolated value (Fischer and Hinrichsen, 1966). The oriented intermediate component of the NMR signal drops to zero (see Fig. VII.25) and the unoriented intermediate component and the narrow component increase. The crystallinity starts decreasing. The upper limit of this high temperature region of annealing is reached when orientation of the sample after cooling is lost and the drawn sample is completely relaxed.

The structure model which accounts for the processes in the important intermediate region of annealing is schematically shown in Fig. VII.26. The

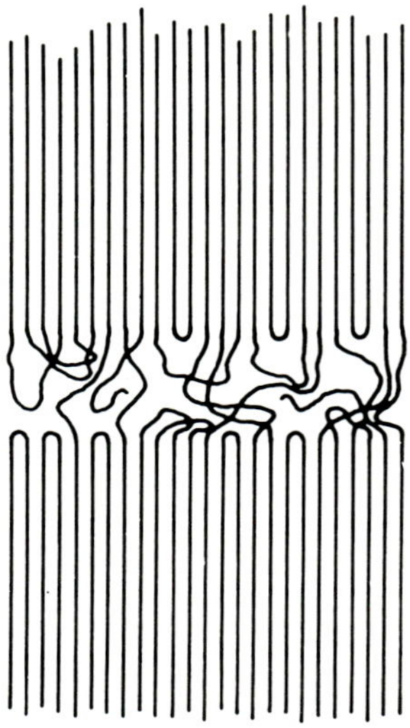

Fig. VII.26 *Schematic Drawing of the Macroconformation of Annealed Drawn Polyethylene.*
Compare to Fig. VII.21 which shows the unannealed sample. (Drawn after Fischer and Goddar, 1965).

major processes in the crystal are thus a diffusion of defects out of the crystalline region (density increase at constant crystallinity), an increase in the fold length, perhaps by the fold diffusion process described in Fig. VII.3 (increase in X-ray low-angle spacing), and alignment of neighboring crystalline regions (change in equatorial low-angle scattering and electron microscopy, Fig. IV.9). In the amorphous areas, the density decreases (Fig. VII.24), causing an increase in the mobility of the CH_2 segments contributing to the narrow NMR component. Their total fraction does not increase, however, until the limit of the intermediate annealing temperature region is reached; rather, there seems to be a small decrease which may result from a compromise in the production of more extended crystalline areas (see Fig. VII.25). The final macroconformation after annealing proposed by Fischer and Goddar (1965) is shown in Fig. VII.26. Annealing transforms the structure shown in Fig. VII.21 to that of Fig. VII.26.

Data on annealing of ten drawn macromolecules are summarized in Table VII.4. Comparison of the different materials reveals great similarities. In all described cases there is an increase in density, X-ray long spacing, and in low-angle diffraction intensity. The changes are frequently proportional to the logarithm of time, with increasing slopes at higher annealing temperatures. Depending upon the samples, special effects such as phase transformation and new crystallization may complicate the process. In general, however, the overall mechanism described for polyethylene, with small variation to accommodate the different macromolecules, seems to be applicable to annealing of many drawn materials (see Fischer, 1969 and Baltá-Calleja and Peterlin, 1971). The structure of the drawn material consists of a less dense than normal (defect) crystalline portion and a more dense and less mobile than normal amorphous portion. The crystallites are small and oriented, with a fold-length set by the drawing conditions (mainly temperature), and the amorphous part is highly ordered. The overall structure is fibrillar with tie molecules between crystalline and amorphous areas and between fibrils (Figs. VII.21 and IV.70). Relaxation of the tie molecules is connected with shrinkage of the fiber and decrease in modulus. The alignment of crystal defects increases the density of the crystalline area and the mobility of the connected amorphous portions of the molecules; it increases the density difference between crystalline and amorphous areas and with it the X-ray low angle scattering intensity. By elimination of microcracks (or voids) along the fiber direction and alignment of crystalline areas between fibrils, the equatorial low-angle X-ray scattering and the horizontal width of the meridional diffraction is decreased, and finally a lamellar superstructure is produced. The shape of the crystalline areas may approach equilibrium dimensions for the given mass so that thickening by fold extension may be limited by the increase in volume of the crystals. Annealing under condition of fixed length retards many of the effects, but hinders the disorientation of the crystals relative to the fiber axis. Little is known about the first annealing process at lower temperature and the high temperature annealing which must lead to increasing recrystallization and partial melting.

Observations which have not found explanations include a reversible change of X-ray long period and intensity which was found, by O'Leary and Geil (1967) in addition to the irreversible portion just described. On nylon 6, poly(ethylene terephthalate), and polyoxymethylene, it was shown that the long period increases with temperature, but on cooling, only a portion is irreversible due to annealing. On a second cycle, the increase is close to fully

TABLE VII.4

ANNEALING BEHAVIOR OF SOME DRAWN MACROMOLECULAR SAMPLES

Macromolecule	Conditions and effects	References
Polyethylene CH_2—	The annealing has been studied by many techniques; most information has been derived for the intermediate temperature range of annealing; the X-ray low-angle spacing and the density increases logarithmically with time and the diffraction intensity is enhanced (see Figs. VII.22 and 23); a transverse alignment of crystals of the crystalline areas is visible by electron microscopy (Fig. IV.9); the density of the crystalline areas increases and that of the amorphous decreases (Fig. VII.24); overall, no large-scale recrystallization occurs in this temperature region, rather, the increase in fold-length and the loss of defects occurs by diffusion (Fig. VII.25); neighboring crystalline areas align without loss of the elemental fibrillar diameter (about 100 Å) as is shown schematically in Fig. VII.26; on reaching the limit of the high temperature annealing region, the stressed amorphous portion is lost (enthalpy and NMR measurement) and further fold-length increase and perfection in lamellar morphology occurs; at the highest temperatures, increasing recrystallization and loss of orientation is expected.	*See text; for branched polyethylene see Loboda-Čačković et al. (1973). J. Polym. Sci. Symposium No.* ***42****, 577.*
Polypropylene, isotatic CH_2—$CHCH_3$—	Samples drawn at 20°C and 100°C to a ratio 6 and 10 were annealed in fixed and free conditions between 40°C and 155°C for times between 1 and 1000 min; X-ray long period, intensity, and sample density increase logarithmically with time; the t intercepts A of the logarithmic expressions $X = A + B \log(t + C)$ increase with annealing temperature and are somewhat higher for freely contracting samples; the difference from the fixed length annealed samples increases with annealing temperature (particularly	*Baltá-Calleja and Peterlin (1971). Makromol. Chem.* ***141****, 91; (1969). Macromol. Sci.* ***B4****, 519; Morosoff et al. (1972). J. Polymer Sci. Part A2,* ***10****, 1221, see also Table VII.3.*

	for the 100°C drawn sample); the slopes *B* increase also with annealing temperature, but only for X-ray long period and intensity; the slope of the density curve is constant or decreases somewhat with increasing temperature, and only the low-angle X-ray intensity shows a larger *B* for the freely contracting samples; the equatorial small-angle X-ray scattering disappears on annealing for freely contracting samples; the wide-angle X-ray difraction shows at higher annealing temperatures a more oriented fiber structure for the fixed annealed samples, but the sizes of the coherently scattering crystalline areas are not affected; the shrinkage is almost instantaneous (slightly time dependent at lower annealing temperatures) and strongly annealing temperature dependent, the percentage shrinkage is practically independent on draw ratio and shows about 10% at 80°C and 72% at 155°C annealing temperature; measurement of the mosiac block dimensions (see Sects. 4.1.2 and 4.2.2) L_{110}, L_{130}, L_{040} reached quickly constant and annealing temperature independent ratios relative to the low-angle X-ray spacing on annealing at 120, 140, and 155°C (6× sample, drawn at 21°C), which may be an indication of achievement of close to equilibrium shape (see Sect. 7.1.2); all these effects seem to agree well with the mechanism discussed for polyethylene.	
Poly(vinylidene fluoride) CH_2—CF_2—	Samples drawn 300% at various temperatures were annealed in fixed and free conditions and analyzed by low and wide X-ray diffraction; as usual, the long period increases with increasing draw temperature to about 190 Å at 160°C; in addition, below about 90°C only the β crystal polymorph was formed (neck formation), while at higher temperature increasing amounts of α are retained (continuous deformation without neck); *fixed length annealing* increases the long period logarithmically with time, the crystal orientation improves, remaining α form decreases in	*Kosmyinin et al. (1972). Vysokomol. Soedin.* ***14A****, 1365; English transl. Polym. Sci. USSR* ***14****, 1530.*

(*Continued*)

TABLE VII.4 *(continued)*

Macromolecule	Conditions and effects	References
	amount, the density increases, the low-angle intensity increases, and the diffraction due to longitudinal microcracks disappears; annealing (at 160°C) samples with primarily α crystals, a 30–35 Å greater long period is reached than with samples of β (which melts 10–15°C higher); in mixed α–β specimens, annealing gives only one long period, increasing with the amount of α; the β crystals of different draw temperature yield about the same size and crystallinity after annealing (at 160°C); the increasing amount of α crystals at higher draw temperature anneal at the constant temperature to larger sizes and higher crystallinity; it is concluded that the vicinity of (more highly stressed) β crystals retards annealing of α crystals; increasingly *free contraction* annealing (at 160°C?) shows decreasing perfection of orientation, indicating some α formation in β samples and changes from a fibrillar to a layer structure.	
Polyacrylonitrile $CH_2{-}CHCN{-}$	Commercially drawn fibers and solution cast films drawn between 90°C and 140°C from 3 to 10× were investigated by low- and wide-angle X-ray diffraction; contrast of the otherwise only poorly analyzable substances could be enhanced by thermal oxidation; the drawn materials have a two-phase structure, characterized by a 100–150 Å X-ray long spacing; the ordered phase is more disturbed than in other macromolecules since it is set up only by the syndiotactic sequences of the molecule; annealing shows a strong increase in low-angle X-ray intensity and an increase in long spacing, both of which are temperature and time dependent.	Hinrichsen *(1972)*. *J. Polym. Sci. Part C*, **38**, *303;* *(1973)*. *J. Appl. Polym. Sci.* **17**, *3305*.

Polyoxymethylene $CH_2{-}O{-}$	Strips of melt crystallized spherulitic samples were drawn at 128°C at a rate of 0.13 mm/min up to 200% deformation and annealed at 145–158°C for 5 min; up to about 100% deformation, many of the effects of draw are reversible on annealing; above this temperature, less reversibility is observed. The low angle X-ray diffraction shows an increase in intensity and long period; wide angle X-ray diffraction shows a loss of orientation and electron microscopy, the development of lamellar structures normal to the fiber directions; shrinkage is almost instantaneous.	*Siegmann and Geil (1969). J. Macromol. Sci.* ***B4****, 557.*
Poly(ethylene terephthalate) $(CH_2{-})_2O{-}CO{-}C_6H_4{-}CO{-}O{-}$	Drawn samples have been annealed free or fixed at temperatures from 100°C to 206°C; an increase in X-ray long spacing (increasing from 100 to 195 Å) and intensity has been observed with increasing annealing temperature (samples of 175 Å initial fold-length showed only little increase in long spacing on annealing for 1 min); at low annealing temperatures (<110°C), the crystallinity decreases first before increasing and the amorphous density seems to remain higher than extrapolated from the melt; at higher annealing temperature the crystallinity increases rapidly, followed by a logarithmically linear increase over several decades similar to crystallization from the melt; the initial crystallization is faster than from the unstrained melt (at about 200°C the crystallinity of the analyzed sample increases from about 20% to 60%); samples free to shrink reach somewhat larger long periods; quick heating by immersion in liquids yields higher long periods, larger shrinkage, but lower crystallinity; measurement of the infrared absorption at 988 cm^{-1} indicates regular chain folded material on annealing (more for freely contracting fibers then for constant length annealing); NMR shows similarly a substantial increase in groups with liquid-like motion (again less for fixed length annealing).	*Dumbleton (1969). Polymer* ***10****, 539; see also Spruiell et al. (1972). Trans. Soc. Rheol.* ***16****, 535; Statton (1959). J. Polym. Sci.* ***41****, 143; J. Appl. Phys.* ***41****, 4290.*

(Continued)

TABLE VII.4 *(continued)*

Macromolecule	Conditions and effects	References
Nylon 6 $(CH_2-)_5$ CO—NH—	Fast heating (18°C/min) of a 3× drawn fiber pre-annealed at 200°C shows a slow increase in X-ray long period from 105 Å to 115 Å up to 211°C, then a quick rise to 150 Å after melting begins; melt quenched film and cold drawn fibers have been compared in their annealing behavior by heating for 1 hr in oil at 100–215°C; the X-ray long period increases for both samples from 65 Å to about 120 Å for a freely relaxing fiber and to about 100 Å for a fixed fiber (all at 215°C).	*Sakurai et al. (1974). J. Polym. Sci. Polym. Phys. Ed.* ***12****, 1587; Hirami (1967). Makromol. Chem.* ***105****, 296.*
Nylon 6.6 $(CH_2-)_6NH-$ CO—$(CH_2-)_4$ CO—NH—	Generally poorly characterized fibers, filaments, or strips were drawn at 50–180°C 2–5× or higher and annealed up to about 260°C; the low angle X-ray intensity increased strongly with annealing temperature, the lateral width of the diffraction spots decreased, and the long period changed typically from about 80 Å at low temperature to about 150 Å after annealing at 250°C; only little disorientation has been observed on annealing of freely contracting samples; the density increases and after annealing, an earlier onset of fluid-like motion of part of the sample was observed by NMR; the sonic modulus decreased drastically; infrared analysis shows that at low annealing temperatures (~180–210°C) the regular fold content increases parallel with crystallinity (new crystallization); at intermediate annealing temperatures there is a faster increase in fold content than in crystallinity (regularization of folds), at higher temperatures (±250°C) a saturation or decrease of fold content occurs, most likely because the increase in fold period exceeds the regularization process; samples annealed under free shrinkage always have a higher regular fold content; the degree of crystallinity is changed very little by drawing and annealing as judged by calorimetry;	*Dismore and Statton (1966). J. Polym. Sci. Part C* ***13****, 133; Dumbleton and Buchanan (1968). J. Polym. Sci. Part A2* ***6****, 1527; Koenig and Itoga (1972). J. Macromol. Sci.* ***B6*** *309, 327; Hinrichsen (1972). Kolloid Z. Z. Polym.* **250,** *1162.*

	the largest portion of the density change has been attributed to changes in the packing within the crystalline areas.	
Polyurethane $(CH_2-)_6NH-CO-O-(CH_2-)_4O-CO-NH-$	On annealing of a sample of 7–12 000 molecular weight between 100–170°C an increase in the intensity of the X-ray low-angle scattering and long period was observed; the X-ray long period increased from 75 to 120 Å; on annealing at fixed length the (002) and (200) plane separation decrease by up to 4%; on annealing under free retraction, the (002) separations stay constant (3.76 Å) while the (200) separations decrease as before; annealing in the presence of water leads to somewhat larger X-ray long periods.	*Zahn and Winter (1952). Kolloid Z.* ***128****, 142.*
Poly(tetramethyl-*p*-silphenylene siloxane) $Si(CH_3)_2-C_6H_4-Si(CH_3)_2-O-$	Oriented films were annealed up to 130°C; the low angle X-ray diffraction intensity increased more than for bulk annealed material (see Table VII.3); also the X-ray long period increased perhaps slightly more (10 Å) than identically annealed bulk crystallized polymer.	*Pollock and Magill (1969). J. Polym. Sci. Part A2* **7**, 551.

reversible. The magnitude seems to be beyond that expected from thermal expansion and partial melting. For example, in the case of poly(ethylene terephthalate) a drawn sample of an initially 130 Å X-ray long period on heating to 250°C increases to 214 Å, on cooling, it reduces its long period to 155 Å, showing a 60 Å reversible change.

7.3.3 Annealing under Elevated Pressure

Annealing under elevated pressure permits the application of higher temperature to a sample without melting.† For many low and high molecular weight organic molecules the rise in melting temperature is about 20°C/kb. Most affected by the high pressure is the amorphous phase. Data on compressibility of the amorphous and crystalline phases of polyethylene up to 5 kb have been reported for example by Hatakeyama *et al.* (1974).

The study of annealing under elevated pressure became of particular interest when it was found that polyethylene and polychlorotrifluoroethylene can crystallize under elevated pressure in the extended chain macroconformation (see Sects. 3.3.1 and 6.3.3). In this section, annealing of solution and melt-grown crystals of polyethylene will be treated first, followed by a discussion of the annealing of oriented polyethylene and polypropylene. Effects of pressure on the annealing of other macromolecules will be mentioned last.

The first annealings of polyethylene crystals under pressure were carried out by Osugi *et al.* (1964) (up to 2 kb pressure) and Brown and Eby (1964) (up to 0.6 kb pressure). Both groups of authors analyzed the effect of pressure on the annealing of solution-grown single crystals of polyethylene of initially about 100 Å thickness. At constant temperature, increasing pressure decreases the amount of thickening obtained. For a similar fold-length increase on annealing as obtained at atmospheric pressure, a shift to higher temperatures, parallel to the increase in melting temperature with pressure, was necessary. The ultimate fold-lengths obtained in these experiments were about 400 Å, a value which could also be obtained (and somewhat exceeded) on annealing at atmospheric pressure at 124°C (see also Sect. 7.2.1).

† The melting temperature in degrees Celsius of polyethylene as a function of pressure up to about 5 kb is in this discussion taken to be (pressure p in bar) $143.7 + 2.673 \times 10^{-2}p - 1.475 \times 10^{-6}\ p^2$ for extended chain crystals and $134.9 + 1.963 \times 10^{-2}p - 0.958 \times 10^{-6}\,p^2$ for slowly cooled folded chain crystals (Davidson and Wunderlich, 1969). These values were obtained by differential thermal analysis at 4°C/min (peak temperatures) and may be subject to superheating, reorganization, sample composition and crystal perfection (see Chapters VIII and IX). For the frequently discussed pressures here, the following extended and folded chain crystal melting peak temperatures can be calculated: 3.0 kb, 211°C, 185°C; 4.2 kb, 230°C, 200°C; 5.0 kb, 240°C; 209°C, 5.1 kb, 242°C, 210°C; 5.34 kb, 244°C, 212°C. Besides the uncertainties introduced by the variation of samples, temperature and pressure measurements in the usual apparatus needed for such high compression is not always reliable.

Higher pressures were applied by Fischer and Puderbach (1969) (up to 4.2 kb pressure). The X-ray long spacing increased in these experiments from 110 Å to 350 Å at the highest pressure. This increase is of the same order of magnitude as has been observed for atmospheric pressure annealing. It was suggested, in addition, that the rate of long period increase was even somewhat slower than under similar supercooling at atmospheric pressure.

More detail was provided by electron microscopy of single lamellae after annealing at elevated pressure by Roe *et al.* (1973). Suspensions of single crystal lamellae in silicone oil were pressurized to 6 kb and annealed at 255°C for 22 hr. The features which developed in the lamellae were similar to those found on annealing at 120.7°C at atmospheric pressure, but seemed to occur at smaller degree of supercooling (see Sect. 7.2.1 and Fig. VII.14). The raised edges around the holes developed on annealing were practically identical to those at atmospheric pressure. A major difference to atmospheric pressure annealing is an increase in the number of developed holes by a factor of 3–4 (compared to Fig. VII.15) and a change in the alignment of the holes along the *b*-axis rather than the {310} planes (Fig. VII.14). On further annealing, elongation of the holes is still normal to the {110} fold planes as before. The change in the alignment of the holes was linked to a loss of the pyramidal shape of the single crystal lamellae on annealing at elevated pressure. The increase in pressure on annealing of solution grown-single crystals raises thus the temperature of annealing to a level of similar supercooling as needed for annealing at atmospheric pressure, increases the number of centers were annealing can start (holes), flattens the originally tent-like crystals, and causes a change in the alignment of the holes.

A comparison of the pressure annealing results with the atmospheric pressure data discussed in Sect. 7.2.1 suggests that the analyzed temperature region corresponds only to the intermediate temperature region of annealing. At somewhat higher temperature one might still expect annealing without loss of orientation through increasing amounts of recrystallization which could lead to the formation of extended chain crystals as suggested originally by Rees and Bassett (1968, 1969). The low temperature region of annealing has also not been studied at elevated pressures.

Melt crystallized polyethylene was shown by Rees and Bassett (1968, 1969) to increase in melting temperature on annealing at 300–310°C and 9 kb pressure without losing the original ringed spherulitic organization (see Fig. III.116). More details were provided by analysis of several differently crystallized polyethylenes annealed at 5.1 kb by Gruner *et al.* (1969). Table VII.5 lists the densities and melting peak temperatures after annealing for a quickly cooled, folded chain crystallized polyethylene and a sample crystallized in the extended chain form. Up to about 205°C no changes could be observed on annealing these samples. At higher annealing temperatures the melting peak

TABLE VII.5

ANNEALING OF POLYETHYLENE AT 5.1 kb PRESSURE[a]

Annealing temperature (°C)	Folded chain crystals[b] Density (g/cm³)	Folded chain crystals[b] Melting peak[d] (°C)	Extended chain crystals[c] Density (g/cm³)	Extended chain crystals[c] Melting peak[d] (°C)
Reference	0.965	131.4	0.991	141.9
185	0.965	131.7	—	—
190	0.965	131.5	0.992	142.2
195	0.965	131.6	—	—
200	0.965	131.6	—	—
205	0.965	132.4	—	—
210	0.965	132.1	0.992	142.2
213	—	—	0.992	143.2
215	0.967	133.1	0.993	143.3
217	—	—	0.993	143.2
220	0.975	134.6	—	—
221	—	—	0.992	142.7
225	0.980	135.7	—	—
226	—	—	0.991	142.7
230	0.984	137.8	0.990	142.9
235	0.989	139.3	—	—
238	0.990	143.0	—	—
240	0.991	143.2	—	—
242	0.991	143.0	—	—
245	0.986	142.9	—	—
249	—	—	0.981	142.5

[a] Linear polyethylene of M_W 153 000 and polydispersity 18, annealed at the given temperature for 20 hr, followed by cooling to room temperature within 3 hr and release of pressure for analysis. Data by Gruner *et al.* (1969).

[b] Fast melt quenched crystals; spherulitic samples as shown in Fig. III.116.

[c] Pressure crystallized sample as illustrated in Fig. III.29.

[d] Measured by differential thermal analysis, 10°C/min heating rate. The melting peak shows at the higher densities (chain extension) considerable degree of superheating (see Chapter IX). In case of double peaks (see text) only the high temperature melting peak is listed.

of the folded-chain crystals becomes narrower and moves to higher temperature. At an annealing temperature of 235°C a sharp melting peak at 139.3°C is observed as shown in Fig. VII.27. The start of significant melting occurs at 135°C, a temperature higher than even the equilibrium melting temperature of the low molecular weight portion of the sample (see Fig. V.43). On initial crystallization by quenching, this sample showed little segregation (see Sect.

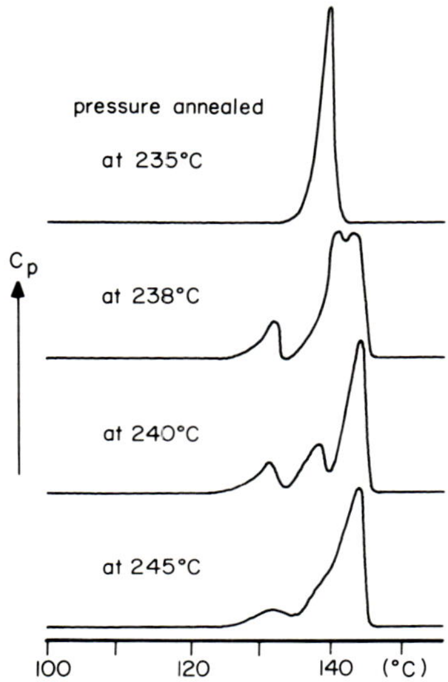

Fig. VII.27 *Differential Scanning Calorimetry Curves of Chain Folded Melt Crystallized Polyethylene after Annealing for* 20 *hr at* 5.1 *kb at the Indicated Temperatures.* For other details see also Table VII.5. Data by Gruner *et al.* (1969).

5.3.3) and the annealing must have occured by reorganization without rejection of the shorter molecules which remained part of the mixed crystal and have thus a higher melting temperature (see Chapter VIII). In agreement with these observations, there were no macroscopic signs of melting or flow on these samples after annealing. This picture changed drastically on annealing at 238°C (see Fig. VII.27). At this temperature a second melting peak at distinctly higher temperature developed, assigned to extended chain crystals. In addition, a third melting peak occurred at lower temperature, assigned to the now-segregated low molecular weight fraction. Such changes could only be caused by partial melting and recrystallization. Macroscopically the sample, which was originally in the form of small pellets, was fused to one solid sample after annealing at 238°C. After annealing at 240°C almost complete recrystallization to extended chain crystals was observed, and after annealing at 245°C the crystals became poorer again when judged by density (Table VII.5) and the amount of segregation of the low molecular weight fraction (Fig.

VII.27). The latter observations are an indication of incomplete recrystallization at the annealing temperature, the poorer crystals were formed by quick crystallization at a lower temperature on cooling the sample. Electron microscopy supports this interpretation. Fracture surfaces show the typical irregular surface caused by microdrawing of folded chain lamellae (see Fig. III.39) on annealing up to 230°C. Occasional lamellae of 200–700 Å fold-length could be seen on annealing at 230°C. After annealing at 235°C, relatively narrow 600–1500 Å striated lamellae were observed. Only after annealing at 238°C were there any laterally large lamellae with a chain extension of about 5000 Å. The crystals observed on annealing at 240°C and 242°C are indistinguishable from normal, pressure crystallized polyethylene with extended chain lamellae of a chain extension of up to 20 000 Å (Fig. III.32) and lateral widths of hundreds of micrometer. At higher temperature, the crystals became poorer again. Annealing of the same initial crystals at atmospheric pressure (under vacuum) did not produce the perfection of the 235°C, 5.1-kb annealed samples. Even on multiple annealings at increasing temperatures, the samples remained at lower density and melting peak temperature. The best samples had a density of 0.977 g/cm^3 and a melting peak at 137.2°C (compare with Table VII.5). Much longer annealing times or crystal perfection on isothermal crystallization may, however, yield similar fold lengths (see Fig. III.13). Annealing of the initially extended chain crystals under elevated pressure shows little change in crystal perfection. In fact, above 221°C a slight decrease in density is seen in Table VII.5. This is due to the melting of the low molecular weight fraction, which is well segregated in these samples, and recrystallizes to poorer crystals on cooling.

Comparing the annealing of melt-grown crystals under elevated pressure to the annealing at atmospheric pressure, described in Sect. 7.3.1, one recognizes that the uppermost annealing temperature (243°C at 5.1 kb) is close to the melting temperature of the extended chain lamellae (see footnote on page 418). The melting temperature of the folded chain lamellae (without reorganization) occurs at about 210°C, somewhat below the temperature where initial perfection of the crystals occurs in density and melting point (see Table VII.5). Identifying this temperature as before (Sects. 7.2.1 and 7.3.1) as the upper limit of the intermediate temperature region of annealing suggests that pressure retards annealing below this temperature. Substantial annealing is only observed well into the upper annealing temperature region. Recrystallization, which can lead to laterally large lamellae through matter transport, occurs only at a smaller degree of supercooling ($\sim$236°C, $\Delta T \sim$ 6°C) relative to atmospheric pressure annealing ($\sim$125°C, $\Delta T \sim$ 13°C, see Fig. VII.10). The magnitude of reorganization and the rate, however, are much enhanced at these relatively high temperatures. The close connection of the temperatures of initial perfection with the melting temperatures explains why the shift of annealing effects with pressure goes approximately parallel to the shift in melting temperatures. The formation of extended chain crystals of polyethylene on annealing under pressure is thus most likely based on a mechanism similar to the one for crystallization from the melt under elevated pressure. Only the annealing through recrystallization leads to extended chain crystals. Annealing without matter transport at higher degrees of supercooling are retarded by pressure, but may be enhanced at lower degrees of supercooling.

The annealing of drawn polyethylene crystals at elevated pressure was studied in more detail by Zubov *et al.* (1972) and Bassett and Carder (1973a, b). Similarities to the annealing of drawn crystals at atmospheric pressure (Sect. 7.3.2) as well as to the annealing of normal crystals under elevated pressure have been discovered. Fischer and Puderbach (1969) found little difference in annealing at atmospheric pressure in a temperature range up to 220°C and a pressure up to 4.2 kb (the sample was drawn to 15 × at 70°C). Similarly, Zubov *et al.* (1972) saw little increase in crystal dimensions ($L_{002} = 300$ Å, $L_{110} = 130$ Å, $\rho = 0.968$, X-ray long period 350 Å, drawn 16 × at 95°C at 10 mm/min) on annealing for 2 hr at up to 250°C at 7 kb. The latter authors did notice, however, an increasing change of the orthorhombic polyethylene to the monoclinic crystal form II (see Table II.9). Starting at 260°C, the crystal dimensions increased quickly to reach optimum annealing at 270°C (2 hr) without loss of orientation. Above 270°C, increased loss was observed. The maximum crystal parameters were $L_{002} > 1000$ Å (limit of X-ray camera resolution), $L_{110} = 700$ Å, $\rho = 0.996$ g/cm^3. The initially fibrillar structure observed by electron microscopy changed to the typical extended chain crystal structure, oriented with the crystallographic c-axis parallel to the original drawing direction. The monoclinic crystal structure did not develop to any degree at these higher annealing temperatures. Above 280°C the perfection decreased because of complete melting of the sample under annealing conditions. The optimum temperatures on annealing at 5 kb and 3 kb were reported to be 260°C and 215°C, respectively. Bassett and Carder (1973b) observed similar morphologies on annealing at 5.34 kb pressure (samples cold drawn 8.1 ×, annealed for 15 min after slow heating). The low-angle X-ray spacing and density started to increase at about 235°C and values above 1500 Å (electron microscopy) and 0.993 g/cm^3 were reached above 245°C. The melting endotherms of the annealed, drawn samples were comparable to those of the undrawn samples after annealing (see Fig. VII.27). On annealing at 238°C the first segregation of low molecular weight occurred parallel with the increase in the major melting peak to extended chain level. The density and lamellar thickness increased at 245°C logarithmically with time (Bassett and Carder, 1973a). A comparison with identical material annealed at 130°C at atmospheric pressure showed a much slower linear increase in lamellar thickness (78 Å/decade versus 185 Å/decade).

As in undrawn material, the appearance of extended chain crystals seems to be connected with recrystallization, signaled by the rejection of low molecular weight polymer from the initial crystals. In contrast to the research on the annealing of deformed crystals at atmospheric pressure (Sect. 7.3.2) which was carried out mostly in the intermediate annealing temperature region, the pressure annealing has been studied most in the high annealing temperature region. The absence of observed effects on annealing besides the phase transition to the monoclinic form suggests that perhaps the processes otherwise observed in the intermediate region of annealing are retarded under elevated pressure.

Although still somewhat speculative, the data on pressure annealing of polyethylene seem to indicate retardation of annealing effects in the crystalline and amorphous regions when compared at equal distance from the equilibrium melting temperature. At smaller supercooling, and in particular under conditions of recrystallization, extended chain crystals can be produced similarly to the observed formation of extended chain crystals on crystallization from the melt at elevated pressure (Sect. 6.3.3). The role which is played by a possible transition to a high pressure phase is not fully understood at present (Bassett and Carder, 1973b).

The annealing of oriented polypropylene was studied at 8 kb by Zubov *et al.* (1971) on a monoclinic sample drawn 6× at 20°C (10 mm/min) ($L_{110} = 250$ Å, $L_{002} = 160$ Å, low-angle spacing 440 Å, $\rho = 0.907$ g/cm^3). Annealing at 250–285°C for 2 hr at fixed length caused some additional interlamellar crystallization. At 300°C a triclinic crystal modification was observed to form parallel with considerable degradation in the intercrystalline zone. The crystal dimensions increased somewhat ($L_{110} = 355$ Å, $L_{002} =$ 185 Å, low-angle X-ray spacing = 460 Å), but the molecular weight changed from 300 000 to 12 000. After annealing at 325°C the crystals were mainly of the triclinic form which is also observed on crystallization at elevated pressure (see Sect. 6.3.3), but the molecular weight was reduced to 1200 at a density of 0.893 g/m^3.

Another macromolecule which shows decomposition under pressure annealing conditions is polyoxymethylene (Wunderlich and Bopp, 1971). Annealing at 5 kb could only be carried out at temperatures to 160–180°C before large scale decomposition occurred. The upper annealing temperature limit was determined by the type of stabilization used for the polymer. These decomposition temperatures are about 100°C below the expected melting temperature at the chosen pressure and prevent any annealing.

Hasegawa *et al.* (1970) reported on some experiments of annealing of oriented syndiotatic polypropylene, isotactic poly(4-methyl-1-pentene) and poly(vinylidene fluoride) at 4.5 kb. In all cases, phase transitions were observed as a result of the elevated pressure.

7.3.4 Annealing with Changes in the Backbone

The major hindrance to annealing of metastable macromolecular crystals is shown in Sects. 7.2.1–7.3.3 to be the existence of folds and tie molecules. Their removal takes large-scale reorganization of the macroconformation of the molecule. A more efficient way seems to be the reorganization of stressed portions of molecules by chemical reaction at the backbone chain (Wunderlich, 1973). Since stressed molecules in the amorphous region of a semicrystalline sample show an enhanced chemical reactivity, they can be selectivly cut and then permit easier reorganization or further crystallization of the sample. The accompanying loss in molecular weight may be of less significance to the properties of the sample than the possible increase in crystal perfection. Reversible reactions may restore or increase the original molecular length after improvement of the crystals. In this section evidence for annealing by chemical reaction will be discussed. Up to the present, this field of research has been relatively neglected although it holds considerable promise for the improvement of macromolecular crystals.

Early observations made during the study of oxidative degradation of polyethylene led Winslow *et al.* (1963) to conclude that most of the initial oxygen uptake was used for chain cleavage in the amorphous areas, which was then followed by additional crystallization. Typical density changes of a high molecular weight film ($M_w > 10^6$) after prolonged exposure to oxygen at 100°C are tabulated below (Winslow and Matreyek, 1966). On oxidized

Oxygen content (%)	Density at 22°C (g/cm^3)
0.2	0.940
0.9	0.976
1.2	0.983
3.0	1.003

samples of density 0.99 g/cm^3 practically no amorphous areas could be detected by X-ray diffraction, and electron microscopy showed greatly increased lamellar thickness. Parallel to the density increase went an increased heat of fusion. Winslow and Martreyek (1966) called this effect "chemicrystallization." Gruner *et al.* (1969) could show that by isothermal annealing of polyethylene in air, densities as high as 0.991 g/cm^3 and melting peak temperatures by differential thermal analysis (at 10°C/min heating rate) as high as 139.7°C could be obtained. Vacuum annealing of the same material under similar optimum conditions could achieve at best a density of 0.974 g/cm^3 and a melting peak temperature of 135.9°C (20 hr annealing time). The oxygen uptake of the sample annealed in air was so little that on fast cooling after melting its initial density (0.965 g/cm^3) could be recovered. Electron microscopy of the air annealed sample showed a morphology similar to that of the best pressure annealed polyethylene at a temperature low enough not to permit recrystallization to the extended chain form (see Fig. VII.27 and Table VII.5, sample annealed at 235°C).

Other examples of crystal annealing by chemical reaction can be derived from the etching experiments described in Sect. 4.1.4. Most detailed was the information on change of perfection of various cellulose crystals during etching (Table IV.5). The initially poorer crystals of regenerated cellulose increased their perfection when portions of the stressed amorphous sections of the molecules were removed.

Of particular interest for crystal annealing are the chemical reactions which can remake bonds on crystal surfaces after the removal of stressed portions of molecules. Such reactions are the prerequisite for the production of larger aggregates of improved crystals. The model macromolecule showing such a reaction is selenium. The initial crystallization of selenium from the melt occurs by chain-folded growth followed by thickening (see Sect. 6.4.2.4). Adjacent lamellae were shown in later stages to fuse across the fold surfaces. This process was suggested by Crystal (1970) to occur by fold opening and repolymerization between crystals. Figure IV.39 shows a fracture surface of an aggregate of selenium lamellae connected across the fold surfaces.

Other macromolecules which should be suitable for fold opening and stress removal in tie molecules and repolymerization to a higher overall crystal perfection are the polyesters, -amids, -urethanes, and -ureas. At sufficiently high temperatures under appropriate conditions these macromolecules show transesterification and transamidation reactions below the crystalline melting point. In the course of these reactions, which occur in the amorphous areas, stress caused by initial crystallization or on subsequent deformation can be removed and chain folds can be opened.

More details were derived for the example of poly(ethylene terephthalate) by Miyagi and Wunderlich (1972). To have a reproducible starting material, folded chain lamellae of melt crystallized poly(ethylene terephthalate) were first etched to oligomer lamellae by removal of folds through hydrolysis with water (see Sect. 4.1.4 and Figs. III.25 and III.62). This was followed by annealing below the melting temperature of oligomer lamellae in vacuum. Reaction through ester interchange or anhydride formation could occur. Excess of terephthalic acid was noted to sublime away. The increase in molecular weight on annealing at 250°C as a function of time is tabulated below.

Annealing time (hr)	Molecular weight
0	2 300
3	9 700
6	14 200
12	19 600
24	25 200

During the polymerization, there is no reason to reintroduce metastable folds or stressed molecules, so that one expects to produce tie molecules or longer loops. Figure VII.28 illustrates the crystal perfection with annealing. Four differently produced oligomer lamellae (Q, Qe, R, Re) change from a behavior which shows a decreasing melting temperature with increasing heating rate to one with increasing melting temperature. The decreasing melting temperatures are indicative of beginning annealing at the lower heating rates, while the increasing melting temperatures are indicative of superheating caused by tie molecules (Chapter IX). The density decreased during annealing and the X-ray low-angle diffraction did not show any large increase (120–170 Å is the range of low-angle spacings). This can only be interpreted as a formation of rather imperfect boundary layers between the lamellae. Conditions for a more perfect joining of the crystals have not been found as yet. Comparing the annealed oligomer with the directly annealed sample S_3 (Fig. VII.28), shows that even during crystallization or thermal annealing ester interchange is likely, and from all starting materials similar intercrystalline regions are produced.

A direct influence of ester interchange reactions on annealing or later stages of crystallization was observed by Lenz and Go (1973). Random copolymers changed to block copolymers with increased crystallinity. For example, a 70/30 mixture of *cis*- and *trans*-1,4-cyclohexylenedimethylene glycols with terephthalic acid changed on heating for 2 hr at 285°C in the presence of an ester-interchange catalyst (titanium tetrabutoxide, 1%) to a more blocky copolymer which had almost doubled crystallinity and an increased melting temperature. After initial crystallization of the sufficiently long trans sequences (see Sect. 6.3.4) there are additional longer trans sequences formed by the random ester-interchange reaction continuing in the amorphous portion of the polymer. Formed in the proper location, these longer trans sequences are continuously removed from the isomer mixture through crystallization. This produces cis-enriched random amorphous copolymer molecule portions and an increased number of longer trans sequences. A more detailed study of the sequence length distribution could be carried out in the poly(ethylene terephthalate-*co*-2-methylsuccinate). Heating for 30 hr in the presence of 1% titanium tetrabutoxide at 220°C, the average

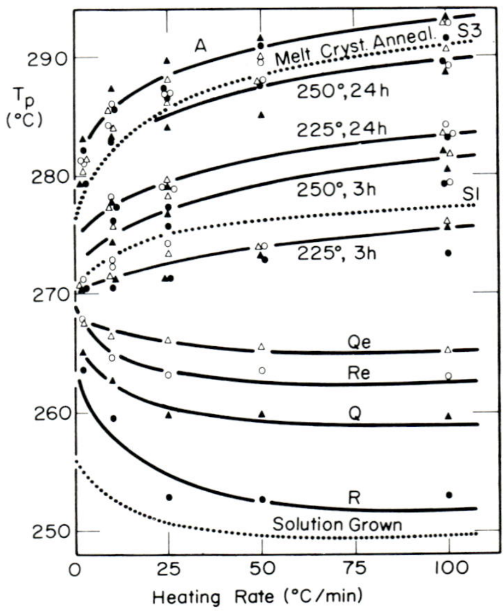

Fig. VII.28 *Melting Peak Temperatures by Differential Thermal Analysis of Differently Annealed Poly(ethylene terephthalate)s.* Samples R, Q, Re, and Qe are oligomer lamellae produced by etching of the melt crystallized poly(ethylene terephthalate) S_1 at 180°C for 320 min and 180 min (R and Q, respectively); samples Re and Qe were etched in the same way and extracted to remove monomers produced during etching. All samples annealed to similar perfection. Dotted curves are reference materials: S_1, starting material crystallized for 48 hr at 250°C; solution-grown crystals from 0.075 wt.% solution in nitrobenzene at 110°C. Sample S_3 was annealed directly from sample S_1 at 260°C for 48 hr. Samples A were stepwise annealed at 240°C, 245°C, and 255°C for 24 hr at each temperature. Data by Miyagi and Wunderlich (1970).

sequence length (78.5% terephphalate) could be increased from 5.3 to 6.8, which led to a 64% increase in crystallinity and a 9°C increase in melting temperature of the copolymer (Lenz and Go, 1974).

In a similar way, isomerization reactions can enhance annealing and further crystallization. The copolymer of the different isomers of the ω-hydroxyester of methylazelaaldehyde glycerol acetal was shown to undergo acetal interchange towards an equilibrium mixture within the molecule in the amorphous state (Lenz *et al.*, 1973a).

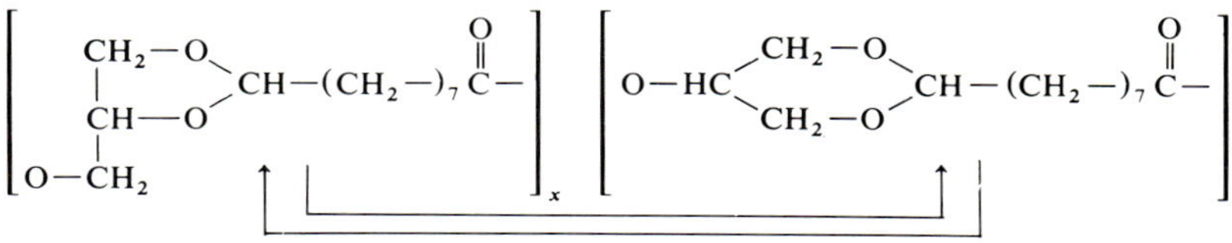

(19)

The five-membered rings are contained in the molecule as the larger fraction. On keeping the copolymer over long time periods (18 months), longer five-membered ring sequences of the molecules crystallize, however, and remove these molecular sections from the equilibrium mixture. Chemical reaction in the amorphous portion reestablishes equilibrium by reducing the remaining six-membered ring fraction. The resulting polymer contains thus more than the equilibrium fraction of the five-membered rings.

Similarly, it was observed that 1,4-polybutadiene which exists as a cis and a trans isomer (which crystallize differently, see Sect. 2.4.3) can be forced from the overall equilibrium ratio by crystallization. The isomerization is in this case achieved by UV-irradiation. The trans-rich polymer was shown initially to form cis repeating units according to the expected approach to equilibrium, but on further crystallization, the reaction direction reversed and the cis content decreased (Lenz *et al.*, 1973b).

A final, special type of annealing by chemical reaction is the expansion of polyoxymethylene folds in contact with a solution of trioxane in the presence of a catalyst as described in Sect. 6.4.2.2. In this case, the whole range from chain folded lamellar crystals to extended chain crystals could be covered (see Fig. VI.76 and Eq. VI.131).

References

Baltá-Calleja, F. J., and Peterlin, A. (1971). Plastic deformation of polypropylene. V. Annealing of drawn polypropylene. *Makromol. Chem.* **141**, 91.

Baltá-Calleja, F. J., Bassett, D. C., and Keller, A. (1963). A study of X-ray long periods produced by annealing polyethylene crystals. *Polymer* **4**, 269.

Baltá-Calleja, F. J., Hosemann, R., and Wilke, W. (1966). Die Struktur von rekristallisiertem und verstrecktem Polyäthylen als Modell für lineare Hochpolymere. *Makromol. Chem.* **92**, 25.

Bank, M. I. and Krimm, S. (1969). Infrared study of lamellar linking by cilia in polyethylene single-crystal mats. *J. Appl. Phys.* **10**, 4248.

Bares, V, and Wunderlich, B. (1973). The question of $C_p - C_v$ for crystalline polyethylene. *J. Polym. Sci. Polym. Phys. Ed.* **11**, 397.

Bassett, D. C. and Carder, D. R. (1973a). Lamellar thickening and chain-extended growth of polyethylene. *Polymer* **14**, 387.

Bassett, D. C., and Carder, D. R. (1973b). Oriented chain-extended polyethylene. I. Formation and characterization. II. Simple mechanical behavior. *Phil. Mag.* **28**, 513, 535.

Bassett, D. C., and Keller, A. (1959). Similarities between the behavior of single crystals and that of bulk material on annealing in polyethylene. *J. Polym. Sci.* **40**, 565.

Bassett, D. C., and Keller, A. (1962). On the habits of polyethylene crystals. *Phil. Mag.* **7**, 1553; see also (1960). *J. Roy. Micros. Soc.* **79**, 243.

Beresford, D. R., and Bevan, H. (1964). The effect of tension and annealing on the X-ray diffraction pattern of drawn 6.6 nylon. *Polymer* **5**, 247.

Blackadder, D. A., and Lewell, P. A. (1970). Properties of polymer crystal aggregates: (1) Comparison of polyethylene crystal aggregates with bulk crystallized polyethylene. (2) Annealing of polyethylene aggregates. Part 3. Comparison of the annealing behavior of bulk-crystallized polyethylene with that of aggregates of polyethylene crystals. *Polymer* **11**, 125, 147, 659.

Blackadder, D. A., and Schleinitz, H. M. (1966). The dissolution and recrystallization of polyethylene crystals suspended in various solvents. *Polymer* **7**, 603.

Blackadder, D. A., Keniry, J. S., and Richardson, M. J. (1972). Annealing of high density polyethylene for very long times at low supercooling under vacuum. *Polymer* **13**, 584.

Blais, J. J. B. P., and Manley, R. St.J. (1967). X-ray long periods in bulk crystalline polymers. *J. Macromol. Sci.* **B1**, 131.

Bonart, R. (1964). Zur Deutung der anomalen Kristallorientierung in linar vertrecktem Polyäthylentherepthalat. *Kolloid Z. Z. Polym.* **199**, 136.

Bonart, R. (1969). Kristall-und Kolloidstrukturen beim Dehnen and Verstrecken. *Kolloid Z. Z. Polym.* **231**, 438.

Bonart, R., and Hosemann, R. (1959). Modellversuche zur Deutung der Röntgen-Langperiodeninterferenzen. IUPAC, Wiesbaden, Abstract I 739; printed in full (1960). *Makromol. Chem.* **39**, 105; see also (1960). *Z. Electrochem.* **64**, 314.

Bonart, R., and Hosemann, R. (1962). Fibrillarstrukturen in kaltverstrecktem linearen Polyäthylen. *Kolloid Z.* **186**, 16.

Boor, Jr., J., and Mitchell, J. C. (1963). Kinetics of crystallization and a crystal-crystal transition in poly-1-butene. *J. Polym. Sci. Part A* **1**, 59.

Boyer, R. F. (1963). The relation of transition temperatures to chemical structure in high polymers. *Rubber Chem. Tech.* **36**, 1303.

Brown, R. G., and Eby, R. K. (1964). Effect of crystallization conditions and heat treatment on polyethylene: Lamellar thickness, melting temperature, and density. *J. Appl. Phys.* **35**, 1156.

Burmester, A. F., Dreyfuss, P., Geil, P., and Keller, A. (1972). On the annealing of polyamide mats. *J. Polym. Sci. Polym. Lett. Ed.* **10**, 769.

Čačković, H., Hosemann, R., and Wilke, W. (1969). Mosaikstruktur und Korngrenzen in Polyäthylen-Einkristallen. *Kolloid Z. Z. Polym.* **234**, 1000.

Crystal, R. G. (1970). Polymeric nature of selenium crystallization. I. Morphology and thermodynamic considerations. II. Crystallization kinetics and secondary crystallization. *J. Polym. Sci. Part A*2 **8**, 1755, 2154.

Davidson, T., and Wunderlich, B. (1969). Differential thermal analysis of polyethylene under high pressure. *J. Polym. Sci. Part A*2 **7**, 377.

Davis, G. T., Eby, R. K., and Martin, G. M. (1968). Variations of the unit-cell dimensions of polyethylene: Effect of crystallization conditions, annealing, and deformation. *J. Appl. Phys.* **39**, 4973.

Dawkins, J. V., Holdsworth, P. J., and Keller, A. (1968). Reversible long period variations with temperature in single crystals of branch-containing polyethylenes. *Makromol. Chem.* **118**, 361.

Dehlinger, U. (1958). Umwandlungen und Ausscheidungen im kristallinen Zustand. *In* "Handbuch der Physik" (S. Flügge, ed.) Vol. VII/2, Kristallphysik II. Springer Verlag, Berlin, Göttingen, Heidelberg.

Dismore, P. F., and Statton, W. O. (1966). Chain folding in oriented nylon 66 fibers. *J. Polym. Sci. Part C* **13**, 133.

Dreyfuss, P. and Keller, A. (1970). A simple chain refolding scheme for the annealing behavior of polymer crystals. *J. Polym. Sci. Part B* **8**, 253.

Ferry, J. D. (1970). "Viscoelastic Properties of Polymers," 2nd ed. Wiley, New York.

Fischer, E. W. (1969). Zusammenhänge zwischen der Kolloidstruktur kristalliner Hochpolymerer und ihrem Schmelz- und Rekristallisationsverhalten. *Kolloid Z. Z. Polym.* **231**, 458.

Fischer, E. W., and Goddar, H. (1965). Elektronenmikroskopische Untersuchungen an verstrecktem Polyäthylen. IUPAC, Prague, Abstract p. 558; printed in full (1969). *J. Polym. Sci. Part C* **16**, 4405.

Fischer, E. W., and Hinrichsen, G. (1966). Untersuchungen zum Schmelzverhalten des verstreckten Polyäthylens mit Hilfe der Differential-Thermo-Analyse. *Kolloid Z. Z. Polym.* **213**, 28.

Fischer, E. W., and Puderbach, H. (1969). Untersuchungen zur Ursache der Entstehung von "Extended Chain"-Kristallen des Polyäthylens. *Kolloid Z. Z. Polym.* **235**, 1260.

Fischer, E. W., and Schmidt, G. F. (1962). Über Langperioden von verstrecktem Polyäthylen. *Angew. Chem.* **74**, 551; *Int. Ed.* **1**, 448.

Fischer, E. W., Goddar, H., and Schmidt, G. F. (1968). Röntgenkleinwinkeluntersuchungen zur Struktur der fehlgeordneten Bereiche in verstrecktem Polyäthylen. Teil I. Absolutintensität der Röntgenkleinwinkelstreuung. Teil II. Messungen an kontrastierten Proben. *Makromol. Chem.* **119**, 170.

Fischer, E. W., Goddar, H. and Piesczek, W. (1971). Relationship between structure of drawn polymers and molecular motion. *J. Polym. Sci. Part C* **32**, 149.

Fischmeister, H., and Grimvall, G. (1973). Ostwald ripening—a survey. *In* "Sintering and Related Phenomena" (G. C. Kusczynski, ed.).Plenum Press, New York.

Friedel, J. (1964). "Dislocations." Pergamon, Oxford and Addison-Wesley, Reading, Massachusetts.

Fuhrmann, J., Driemeyer, M., and Rehage, G. (1970). Nachkristallisation und Aufschmelzen, ausgelöst durch Diffusion von Lösungsmitteln in Hochpolymeren. *Ber. Bunsenges.* **74**, 842.

Garber, C. A., and Geil, P. H. (1968). Polyoxymethylene single crystals. II. The effect of substrate on annealing behavior. *Makromol. Chem.* **113**, 246.

Geil, P. H. (1963). "Polymer Single Crystals." Chapter V. Wiley (Interscience), New York.

Grewer, Th., and Wilski, H. (1968). Die spezifische Wärme des Polyvinylchlorids. *Kolloid Z. Z. Polym.* **226**, 46.

Gruner, C. L., Wunderlich, B., and Bopp, R. C. (1969). Extended Chain Crystals. VI. Annealing of polyethylene under elevated pressure. *J. Polym. Sci. Part A2* **7**, 2099.

Harland, W. G., Khadr, M. M., and Peters, R. H. (1972). High-density polyethylene: Thermal history and melting characteristics. *Polymer* **13**, 13.

Hasegawa, R., Tanabe, Y., Kobayashi, M., Tadokoro, H., Sawaoka, A., and Kawai, N. (1970). Structural studies of pressure-crystallized polymers. I. Heat treatment of oriented polymer under high pressure. *J. Polym. Sci. Part A2* **8**, 1073.

Hatakeyama, T., Hashimoto, T., Kanetsuna, H. (1974). Compressibility of polyethylene under pressure up to 5000 kg/cm^2. *Colloid Polym. Sci.* **252**, 15.

Hess, K., and Kiessig, H. (1943). Zur Kenntnis der Feinstructur der Polyamidfasern. *Naturwissenschaften* **31**, 171.

Hess, K., and Kiessig, H. (1944). Über Langperiodeninterferenzen and micellaren Faserfeinbau bei vollsynthetischen Fasern. *Z. Phys. Chem.* **A193**, 196.

Hess, K., and Kiessig, H. (1953). Über Langperioden-Interferenzen bei synthetischen Fasern. *Kolloid Z.* **130**, 10.

Hirai, N., and Yamashita, Y. (1964). Partial melting and thickening of single crystals of high polymers during heat treatment. *Kobunshi Kagaku* **21**, 173.

Hirai, N., Mitsuhata, T., and Yamashita, Y. (1961a) Thickening process of polyethylene single crystals by heat treatment. I. *Kobunshi Kagaku* **18**, 33.

Hirai, N. Yamashita, Y., Mitsuhata, T., and Tamura, Y. (1961b). Thickening of lamellar crystals of polyethylene by heat treatment. *Rep. Res. Lab. Surface Chem. Okayama Univ.* **2**, 1.

Holland, V. F. (1964a). Dislocations in polyethylene single crystals. *J. Appl. Phys.* **35**, 3235.

Holland, V. F. (1946b). Investigation of the morphology associated with the dissolution process of polymer single crystals. *J. Appl. Phys.* **35**, 59.

Holland, V. F., and Miller, R. L. (1964). Isotatic polybutene-1 single crystals: Morphology. *J. Appl. Phys.* **35**, 3241; see also *J. Polym. Sci. Part B* **2**, 519.

Illers, K.-H. (1969) Mechanisches Relaxationsverhalten und Struktur teilkristalliner Polymerer. *Kolloid Z. Z. Polym.* **231**, 622.

Iohara, K., Imada, K., and Takayanagi, M. (1973). X-ray studies on effect of annealing on crystal structure of polyethylene single crystals. *Polym. J. Japan* **4**, 239.

Iwamoto, R., and Wunderlich, B. (1973). Crystal structure of poly-p-xylylene. I. the α-form. *J. Polym. Sci. Polym. Phys. Ed.* **11**, 2403.

Jastrzebski, Z. D. (1965). Annealing. *Encycl. Polym. Sci. Technol.* **2** 138.

Kapur, S. and Rogers, C. E. (1972). Aging of quenched polypropylene. *J. Poly. Sci. Polym. Phys. Ed.* **10**, 2107.

Kawai, T. (1965). On the thickening mechanism of polymer single crystals on annealing. *Kolloid Z. Z. Polym.* **201**, 104.

Kawai, T. (1969). On the thickening of lamellar crystals during isothermal crystallization. *Kolloid Z. Z. Polym.* **229**, 116.

Keller, A., and O'Connor, A. (1957). Large periods in polyethylene: the origin of low-angle X-ray scattering. *Nature* (London) **180**, 1289.

Keller, A., and O'Connor, A. (1958). Study of single crystals and their associations in polymers. *Disuss Faraday Soc.* **25**, 114.

Keller, A., and Priest, D. J. (1970). On the effect of annealing on the chain ends in single crystals of polyethylene. *J. Polym. Sci. Part B* **8**, 13.

Koenig, J. L., and Carrano, A. (1968). Differential thermal analysis of polyethylene in tetrachlorethylene. I. Morphological effects on solution temperatures. II. Structural effects on solution and crystallization temperatures. *Polymer* **9**, 359, 401.

Koenig, J. L., and Tabb, D. L. (1974). Annealing studies of solution and bulk crystallized polyethylene using the Raman-active longitudinal acoustical vibrational mode. *J. Macromol. Sci.* **B9**, 141.

Kuczynski, G. C. (ed.) (1973). "Sintering and Related Phenomena." Plenum Press, New York.

Lenz, R. W., and Go, S. (1973). Crystallization-induced reactions of copolymers. III. Ester-interchange reorganization of poly (*cis/trans*-1,4-cyclohexylene-dimethylene terephthalate). *J. Polym. Sci. Polym. Chem. Ed.* **11**, 2927.

Lenz, R. W., and Go, S. (1974). Crystallization-induced reactions of copolymers. IV. Ester-interchange reorganization of poly(ethylene terephthalate-*co*-2-methylsuccinate). *J Polym. Sci. Polym. Chem. Ed.* **12**, 1.

Lenz, R. W., Martin, E., and Schuler, A. N. (1973a). Crystallization-induced reaction of copolymers. I. Ring-size isomerization of poly(ester acetals). *J. Polym. Sci. Polym. Chem. Ed.* **11**, 2265.

Lenz, R. W., Ohata, K., and Funt, J. (1973b). Crystallization-induced reaction of copolymers. II. *Cis-trans* isomerization of 1,4-poly-1,3-butadiene. *J. Polym. Sci. Polym. Chem. Ed.* **11**, 2273.

Lieser, G. (1971). Gitterumwandlungen in Einkristallen hochpolymerer Stoffe und ihre Beziehung zur Kolloidstruktur. Dissertation. Naturw. Fakultät, Joh. Gutenberg Univ. Mainz.

Loboda-Čačković, J., Hosemann, R., and Wilke, W. (1969). Röntgenkleinwinkel- und Dichtemessungen an verstrecktem und getempertem linearem Polyäthylen. *Kolloid Z. Z. Polym.* **235**, 1162.

Loboda-Čačković, J., Hosemann, R., and Čačković, H. (1971). Superstrukturänderungen beim Tempern von linearem Polyäthylen. *Kolloid Z. Z. Polym.* **247**, 824.

Lontz, J. F. (1963). Sintering of polymeric materials. *In* "Fundamental Phenomena in the Material Sciences." L. J. Bonis and H. H. Hauser, eds. Vol. 1, Sintering and plastic deformation. Plenum Press, New York.

Mandelkern, L. (1964). "Crystallization of Polymers." McGraw Hill, New York.

Mandelkern, L., and Allou, A. L. Jr. (1966). The fusion of polyethylene single crystals. *J. Polym. Sci. Part B* **4**, 447.

Mandelkern, L., Price, J. M., Gopalan, M., and Fatou, J. G. (1966). Sizes and interfacial free energies of crystallites formed from fractionated linear polyethylene. *J. Polym. Sci. Part A*2 **4**, 385.

Mandelkern, L., Sharma, R. K., and Jackson, J. F. (1969). On the annealing of polyethylene crystals formed from dilute solution. *Macromolecules* **2**, 644.

McHugh, A. J., and Schultz, J. M. (1971). Morphological effects of annealing in linear polyethylene. *Phil. Mag.* **24**, 155.

Miyagi, A., and Wunderlich, B. (1972). Etching of crystalline poly(ethylene terephthalate) by hydrolysis. Annealing during polymerization of crystalline poly(ethylene terephthalate). *J. Polym. Sci. Polym. Phys. Ed.* **10**, 2073, 2092.

Nagai, H., and Kajikawa, N. (1968). Annealing of polyethylene single crystals. *Polymer* **9**, 177.

Nakajima, A., Hayashi, S., and Nishimura, H. (1969). Change in lamellar thickness of polyethylene single crystals during isothermal annealing in bulk. *Kolloid Z. Z. Polym.* **229**, 107.

Niegisch, W. D. (1971). Xylylene polymers, structure. *Enc. Polym. Sci. Technol.* **15**, 113.

Niinomi, M., and Takayanagi, M. (1971). Electron microscopic studies on dislocations in polyethylene single crystals. *Progr. Polym. Sci. Japan* **2**, 199.

Oda, T., and Stein, R. S. (1972). X-ray diffraction relaxation of polyethylene. *J. Polym. Sci. Part A*2 **10**, 685.

O'Leary K., and Geil, P. H. (1967). Reversible long-period change during the annealing of crystalline polymers. *J. Macromol. Sci.* **B1**, 147.

Osugi, J., Hara, K., Hirai, N., and Hikasa, J. (1964). Crystallization of polyethylene under high pressure. *Rev. Phys. Chem. Japan* **34**, 59.

Park, J. B., and Uhlmann, D. R. (1973). Recovery of deformed polymers. III. Thermal properties. *J. Appl. Phys.* **44**, 201.

Pelzbauer, Z., and Manley, R. St. J. (1973). Annealing of polystyrene fibrillar crystals. *J. Macromol. Sci.* **B7**, 345.

Peterlin, A. (1963). Thickening of polymer single crystals during annealing. *J. Polym. Sci. Part B* **1**, 279.

Peterlin, A. (1964). Secondary crystallization and annealing of polyethylene. *J. Appl. Phys.* **35**, 75.

Peterlin, A. (1965). Crystalline character in polymers. *J. Polym. Sci. Part C* **9**, 61.

Peterlin, A., and Corneliussen, R. (1968). Small-angle X-ray diffraction studies of plastically deformed polyethylene. II. Influence of draw temperature, draw ratio, annealing temperature and time. *J. Polym. Sci. Part A2* **6**, 1273.

Peterlin, A., and Meinel, G. (1964). Light scattering by polyethylene single crystal suspensions. *J. Polym. Sci. Part B* **2**, 751.

Peterlin, A. and Meinel, G. (1965). Thermodynamic properties of drawn linear polyethylene. *J. Appl. Phys.* **36**, 3028, see also Heat content of amorphous regions of drawn linear polyethylene. *J. Polym. Sci. Part B* **3**, 783.

Peterlin, A., and Olf, H. G. (1966). NMR observations of drawn polymers. V. Sorption into drawn and undrawn polyethylene. *J. Polym. Sci. Part A2* **4**, 587.

Rånby, B. G., and Brumberger, H. (1960). Observations on chain folding in crystalline polyethylene. *Polymer* **1**, 399.

Rees, D. V., and Bassett, D. C. (1968). Origin of extended-chain lamellae in polyethylene. *Nature* (London) **219**, 368.

Rees, D. V., and Bassett, D. C. (1969). On the formation of extended-chain lamellae in polyethylene. *J. Polym. Sci. Part B* **7**, 273.

Reneker, D. H. (1962). Point dislocations in crystals of high polymer molecules. *J. Polym. Sci.* **59**, S39.

Roe, R.-J., Gieniewski, C., and Vadimsky, R. G. (1973). Lamellar thickening in polyethylene single crystals annealed under low and high pressure. *J. Polym. Sci. Polym. Phys. Ed.* **11**, 1653.

Sakaoku, K., and Peterlin, A. (1967). Surface replicas of drawn polyethylene. II. Annealing of cold-drawn samples with different draw ratios. *J. Macromol. Sci.* **B1**, 103.

Sakurai, K., Miyasaka, K., and Ishikawa, K. (1974). Thickening of crystals during rapid heating. I. Linear polyethylene and nylon 6. *J. Polym. Sci. Polym. Phys. Ed.* **12**, 1587.

Sanchez, I. C., Colson, J. P., and Eby, R. K. (1973). Theory and observation of polymer thickening. *J. Appl. Phys.* **44**, 4332.

Sanchez, I. C., Peterlin, A., McCrackin, F. L., and Eby, R. K. (1974). Theory of crystal thickening. *Bull. Amer. Phys. Soc.* **19**, 217.

Schreiber, H. P. (1972a). Temperature scanning of polyolefin solutions. *Polymer* **13**, 78.

Schreiber, H. P. (1972b). Characterization of polyolefins by solution DTA. *J. Appl. Polym. Sci.* **16**, 539.

Schultz, J. M. (1970). Qualitative theory for the picture frame morphology of annealed lamellar crystals. *J. Macromol. Sci.* **B4**, 775.

Spegt, P. (1970). Etude de l'évolution structurale de polyoxyéthylène pendant leur recui. *Makromol. Chem.* **139**, 139.

Statton, W. O. (1959). Polymer texture: The arrangement of Crystallites. *J. Polym. Sci.* **41**, 143.

Statton, W. O. (1961). Rate of recrystallization of polyethylene single crystals. *J. Appl. Phys.* **32**, 2332.

Statton, W. O. (1967). Coherence and deformation of lamellar crystals after annealing. *J. Appl. Phys.* **38**, 4149; see also (1968). *Polym. Preprint Amer. Chem. Soc. Div. Polym. Chem.* **9**(2), 1200.

Statton, W. O. and Geil, P. H. (1960). Recrystallization of polyethylene during annealing. *J. Appl. Polym. Sci.* **3**, 357.

Stuart, H. A. (1955). Beeinflussung der Kristallisationen durch Abschrecken, Tempern und Quellen. *In* "Die Physik der Hochpolymeren" (H. A. Stuart, ed.), Vol. 3, Ordnungszustände und Umwandlungserscheinungen in festen hochpolymeren Stoffen. Springer Verlag, Berlin; §44.

Takayanagi, M., and Nagatoshi, F. (1965). Effect of heat-treatment upon final thickness of single crystals of polyethylene. *Mem. Faculty Eng. Kyushu Univ.* **24**, 33.

Tautz, H., Glück, M., Hartmann, G., and Leuteritz, R. (1964). Die spezifische Wärme von Hochpolymeren in Abhängigkeit von der Vorgeschichte des Materials. *Plaste und Kautschuk* **11**, 657.

Thomas, J. M., and Williams, J. O. (1971). Dislocations and the reactivity of organic solids. *Progr. Solid State Chem.* **6**, 119.

Weitz, A. and Wunderlich, B. (1974). Thermal analysis and dilatometry of glasses formed under elevated pressure. *J. Polym. Sci. Polym. Phys. Ed.* **12**, 2473.

Winslow, F. H., and Matreyek, W. (1966). Chemicrystallization of polyethylene. *Polym. Preprints Amer. Chem. Soc. Div. Polym. Chem.* **7**, 540.

Winslow, F. H., Aloiso, C. J., Hawkins, W. L., Matreyek, W., and Matsuoka, S. (1963). Dependence of polyolefin oxidation on morphology. Oxidative crystallization of polyethylene. *Chem. Ind.* **533**, 1465.

Wunderlich, B. (1957). Thermodynamics of the Copolymer System Poly(ethylene terephthalate-sebacate). Thesis, Northwestern Univ. Dept. of Chem., Evanston, Illinois.

Wunderlich, B. (1973). Chemical reactions involving the backbone chain of macromolecular crystals. *In* "Reactions of Polymers" (J. A. Moore, ed.). Reidel Publ. Co., Boston, Massachusetts.

Wunderlich, B., and Bopp, R. (1971). Decomposition of stabilized polyoxymethylene. *Makromol. Chem.* **147**, 79.

Wunderlich, B., and Melillo, L. (1968). Morphology and growth of extended chain crystals of polyethylene. *Makromol. Chem.* **118**, 250.

Wunderlich, B., Sullivan, P., Arakawa, T., DiCyan, A. B., and Flood J. Thermodynamics of crystalline linear high polymers. III. Thermal breakdown of the crystalline lattice of polyethylene. *J. Polym. Sci. Part A* **1**, 3581.

Wunderlich, B., Cormier, C. M., Keller, A., and Machin, M. J. (1967). Annealing of stirrer-crystallized polyethylene. *J. Macromol. Sci.* **B1**, 93.

Zachmann, H. G. (1964). Das Kristallisations- und Schmelzverhalten hochpolymerer Stoffe. *Fortschr. Hochpolym.-Forsch.* **3**, 581.

Zubov, Yu. A., Konstantinopol'skaja, M. B., Selikhova, V. I., Korobko, A. P., Sokol'ski, V. A., Terifonov, A. S., Kryukov, A. V., and Kargin, V. A. (1971). The effect of heat treatment under pressure on the structure of oriented polypropylene. *Vysokomol. Soedin. Ser. A* **13**, 1078 [*English transl. Polym. Sci. USSR* **13**, 1212].

Zubov, Yu. A., Selikhova, V. I., Konstantinopol'skaja, M.B., Sukhov, F. F., Slovokhotova, N. A., Bakeyev, N. F., Kryukov, A. V., Sokolskii, V. A., and Belov, G. P. (1972). Effect of annealing under high pressure on the structure of oriented polyethylene. *Vysokomol. Soedin. Ser. A* **14**, 2090 [*English transl. Polym. Sci. USSR* **14**, 2345.]

Author Index

Numbers in italics refer to the pages on which the complete references are listed.

C

I

J

K

N

T

Subject Index

The index is organized according to macromolecules *and* subject matter. For monomer information check the corresponding macromolecule. Subjects with several entries show boldface page numbers to indicate more extensive discussion. Frequently used terms are listed only under the page numbers where definitions can be found.

C

R

V

W

Z

A 6
B 7
C 8
D 9
E 0
F 1
G 2
H 3
I 4
J 5